Insulin: Molecular Biology to Pathology

EDITED BY

Frances M. Ashcroft
*University Laboratory of Physiology,
Parks Road, Oxford OX1 3PT, UK*

AND

Stephen J. H. Ashcroft
*Nuffield Department of Clinical Biochemistry,
John Radcliffe Hospital, Headington,
Oxford OX3 9DU, UK*

IRL PRESS
at
OXFORD UNIVERSITY PRESS
Oxford New York Tokyo

Oxford University Press, Walton Street, Oxford OX2 6DP

Oxford New York Toronto
Delhi Bombay Calcutta Madras Karachi
Petaling Jaya Singapore Hong Kong Tokyo
Nairobi Dar es Salaam Cape Town
Melbourne Auckland

and associated companies in
Berlin Ibadan

Oxford is a trade mark of Oxford University Press

Published in the United States
by Oxford University Press, New York

A catalogue record for this book is available from the British Library

Library of Congress Cataloging in Publication Data
Insulin : molecular biology to pathology/edited by Frances M. Ashcroft and Stephen J.H. Ashcroft.
Includes bibliographical references and index.
1. Diabetes — Molecular aspects. 2. Diabetes — Pathophysiology.
3. Insulin — Physiological effect. I. Ashcroft, Frances M.
II. Ashcroft, Stephen J. H.
[DNLM: 1. Diabetes Mellitus — physiopathology. 2. Insulin — physiology. WK 820 159]
RC660.1473 1992 612.3'4 — dc20 91–46299

ISBN 0–19–963229–4 (h/b)
ISBN 0–19–963228–6 (p/b)

Typeset by
Footnote Graphics, Warminster, Wilts.
Printed and bound in Great Britain by
the Bath Press

Preface

This book is intended to provide an up-to-date account of what is known about insulin, from the synthesis of the hormone through the mechanism of its action to the pathophysiology of diabetes mellitus. It therefore covers recent developments in fields as diverse as those of molecular biology and epidemiology. The book is meant to be accessible to graduate students, research workers, and clinicians, who are interested in insulin and diabetes. One of the problems facing readers seeking to discover information in a different field from their own is that of unfamiliar language. For this reason we have provided a detailed glossary. We have also tried to provide a systematic introduction to each of the major areas that can be understood by readers who have no prior knowledge of the field and we have encouraged authors to describe the experimental procedures used in studies in their field, where possible. Readers who are unfamiliar with an area are therefore advised to read the relevant introduction first.

We have requested authors to confine themselves to major classical references and to review articles. Hence many relevant papers are not quoted directly and we apologize in advance to those authors whose work is omitted.

This book owes much to many individuals. We owe a particular debt to Dr Illani Atwater who introduced us to each other eight years ago, so starting a scientific collaboration that has been both enjoyable and profitable. Finally, we thank our families, friends, and colleagues for their patience and understanding during the preparation of the manuscript.

Oxford
February 1992

F.M.A.
S.J.H.A.

Introduction

The story of insulin has its beginnings in antiquity. There is a reference in the Ebers Papyrus to a condition resembling diabetes mellitus, suggesting that the written record goes back to 1500–3000 BC. The first known clinical description of diabetes appears to have been made by Aulus Cornelius Celsus (*c.* 30 BC–AD 50) but Aretaeus of Cappadocia (*c.* AD 30–90) provided a detailed and accurate account and introduced the name 'diabetes', from the Greek word for 'siphon'.

> Diabetes is a strange disease, which fortunately is not very frequent. It consists in the flesh and bones running together into urine. It is like dropsy in that the cause of both is moisture and coldness, but in diabetes the moisture escapes through the kidneys and bladder. The patients urinate increasingly; the urine keeps running like a rivulet. The illness develops very slowly. Its final outcome is death . . . The patients are tortured by an unquenchable thirst; they never cease drinking and urinating . . . As the illness progresses . . . the whole body wastes away . . . The disease was called diabetes, as though it were a siphon, because it converts the human body into a pipe for the transflux of liquid humors. Now, since the patient goes on drinking and urinating, while only the smallest portion of what he drinks is assimilated by the body, life naturally cannot be preserved very long, for a portion of the flesh also is excreted through the urine. The cause of the disease may be that some malignity has been left in the system by some acute malady, which afterwards is developed into this disease.

There is no mention in European medical literature of the existence of sugar in the urine of diabetics until the seventeenth century, when Thomas Willis, in Oxford, noted the sweet taste of diabetic urine, which he suggested to come from the blood. However, Arab, Chinese, Japanese, and Hindu writings indicate that a disorder in which a sweet urine occurs was known before AD 1000.

During the eighteenth and early nineteenth century it became accepted that glycosuria was a diagnostic feature of diabetes and the disease was regarded as a metabolic derangement. However it was not until 1869 that the islets that now bear his name were discovered in pancreatic tissue by Paul Langerhans. He was never to know the significance of his discovery, as he died in 1888, one year before the key observations of von Mering and Minkowski that removal of the pancreas led to the development of diabetes in dogs. Von Mering had not been interested in diabetes and indeed utilized Minkowski's surgical skill to depancreatize dogs in order to study the role of the pancreas in fat digestion. Minkowski, however, noted the polyuria of the first depancreatized dog, tested for sugar in the urine, and rightly associated the diabetes with removal of the pancreas.

At the beginning of the twentieth century, Opie and others were convinced that the islets of Langerhans had the function of an endocrine gland that produced a secretion, lack of which led to diabetes. However, for twenty years all attempts to isolate the active principle were unsuccessful. Many attempts to administer pancreatic extracts to diabetic patients had to be halted because of the coma that developed: it is probable that this was, in fact, hypoglycaemia due to insulin in the extracts.

The story of the ultimate success of Banting and Best in preparing from pancreas an extract capable of reversing diabetes has been told many times, most recently and dramatically by Michael Bliss in his book *The discovery of insulin*, in which he describes the isolation of insulin as 'one of the greatest achievements of modern medicine'. A key to Banting and Best's success was the use of the duct-ligated pancreas. This caused the acinar pancreas to atrophy, whilst leaving the islets of Langerhans intact, thus facilitating preparation of the active islet principle. The idea was based on a report by Barron of a patient whose pancreatic acini, but not islets of Langerhans, had atrophied following obstruction of the pancreatic duct by a stone. Frederick Banting arrived in Toronto in May 1921, where Macleod made facilities and the assistance of Charles Best available to him. In February 1922 Banting and Best were able to publish the first paper describing the successful lowering of the blood glucose in depancreatectomized dogs injected with pancreatic extracts. Macleod's suggestion of the use of alcohol to destroy pancreatic enzymes in the extract allowed them to make effective extracts from non-duct-ligated pancreases and the chemist J. B. Collip joined the group and made important contributions to the purification procedure.

The first clinical test was performed in January 1922. A 14-year-old diabetic, Leonard Thompson, received injections of the Toronto workers' extracts, which elicited a marked and dramatic improvement in his diabetes. Presentation of the clinical findings received a standing ovation at the meeting of the Association of American Physicians in Washington, in May 1922. Commercial development of insulin preparation with the facilities of the Eli Lilley Company was rapid and effective, and soon the 'miracle' drug was being made widely available to diabetics throughout the world. Banting and Macleod received the Nobel prize on October 25th, 1923. Banting immediately announced that he wished to share his prize with Best; Macleod did likewise with Collip. Although arguments over where the credit for the epoch-making discovery should go led to disagreements and antagonism among the Toronto workers, there is no doubt from our historical perspective that all four workers played key roles, and countless diabetics everywhere are indebted to all the members of the team.

The pace of discovery quickened when it was established that insulin was a protein. The first crystalline insulin was obtained by Abel in 1926 and Scott showed that the crystals were the zinc salt of the protein. The work of Auguste Loubatières between 1942 and 1945 demonstrated the insulin-releasing ability of certain sulphonamides and led to the development of the clinically useful sulphonylurea drugs which are still used today to treat non-insulin-dependent diabetes. Frederick Sanger, in Cambridge, received the Nobel prize in 1959 for his elucidation of the sequence of amino acids in insulin, the first time that a protein had been completely sequenced. Insulin was also the first protein to be chemically synthesized, a notable feat of organic chemistry by Katsoyannis and colleagues in 1963. The pioneering work of Dorothy Hodgkin and her group led to insulin becoming, in 1972, one of the first proteins whose three-dimensional structure was known from X-ray crystallographic analysis. A further Nobel prize went to Rosalyn Yalow who, with Solomon Berson, used insulin to develop the radioimmunoassay technique, which revolutionized endocrinology by providing a sensitive and specific assay for hormones. Biochemical studies of the islets of Langerhans were initiated in 1963 when Jarrett and Keen in England and Claes Hellerström in Sweden showed that it was possible to obtain viable islets of Langerhans free from surrounding exocrine tissue: this enabled detailed knowledge of the biochemical events involved in the synthesis and secretion of the hormone to be obtained. Don Steiner in Chicago showed that insulin was synthesized as a single-chain precursor, proinsulin, later cleaved to insulin during its passage through the cell. Recent years have seen definition of the way in which glucose and sulphonylureas trigger the secretion of insulin.

The revolution in techniques afforded by the growth of molecular biology has been of enormous importance in the study of insulin. The insulin gene was one of the first human genes to be cloned by Bell and colleagues in 1980. The first primary sequence of a peptide hormone receptor was that of the insulin receptor, cloned separately by the groups of Ullrich and Ebina in 1985. The way in which insulin affects intracellular events via binding to its receptor remains an area of intense study. In 1979 human insulin also became the first protein to be produced commercially by recombinant DNA techniques.

Dramatic advances have also occurred in our understanding of the disease diabetes mellitus. It was recognized as early as the end of the last century that the disease had a hereditary tendency, and Naunyn noted the distinction between juvenile and aged diabetes. It is now clear that there are indeed two distinct clinical entities. Definition of the genes involved in pre-disposition to these diseases is making remarkable progress, and the growing knowledge of the nature of the immune system-mediated destruction of the β-cells that occurs in type I diabetes has been made possible by the explosive growth in immunology. It is now feasible to envisage new forms of treatment of type I diabetes involving specific intervention in the disease process itself. Transplantation of islets of Langerhans is also poised to become a clinical reality.

The study of insulin, its synthesis, secretion, and actions, and the consequence to the body of a deficiency in its action or production, encompasses a wide variety of scientific and clinical disciplines. Our primary purpose in producing this book has been to attempt to paint this broad canvas in a way that will allow those working on or interested in one aspect of insulin to obtain a detailed knowledge of other aspects of this fascinating molecule. We hope that both the general reader and the specialist will gain from these pages the most up-to-date and comprehensive knowledge of the field. We believe that the major advances of the last few years make such a volume timely, and we hope that the book will convey at least some of the excitement felt by those working on insulin as each new discovery brings us to a better understanding but also poses new and deeper questions.

Contents

Contributors

FRANCES M. ASHCROFT
University Laboratory of Physiology, Parks Road, Oxford OX1 3PT, UK

STEPHEN J. H. ASHCROFT
Nuffield Department of Clinical Biochemistry, John Radcliffe Hospital, Headington, Oxford OX3 9DU, UK

ELAINE M. BAILYES
Department of Clinical Biochemistry, University of Cambridge, Addenbrookes Hospital, Hills Road, Cambridge CB2 2QR, UK

HEIKE BÄRMEIER
RH Williams Laboratory, Department of Medicine, University of Washington, Seattle, Washington, USA

S. J. BEVAN
Cellular Nutrition Research Group, Department of Biochemistry, University of Oxford, South Parks Road, Oxford, UK

EROL CERASI
Department of Endocrinology and Metabolism, Hebrew University Hadassah Medical Center, Jerusalem, Israel

MICHAEL R. CHRISTIE
Nuffield Department of Clinical Biochemistry, John Radcliffe Hospital, Headington, Oxford OX3 9DU, UK

ANDREW R. CLARK
Department of Medicine, University of Birmingham, Queen Elizabeth Hospital, Birmingham B15 2TH, UK

TERRY L. DELOVITCH
Banting and Best Department of Medical Research and Departments of Pharmacology and Immunology, University of Toronto, Toronto, Ontario, Canada M5G 1L6

RICHARD M. DENTON
Department of Biochemistry, School of Medical Sciences, University of Bristol, University Walk, Bristol BS8 1TD, UK

G. D. DIMITRIADIS
Cellular Nutrition Research Group, Department of Biochemistry, University of Oxford, South Parks Road, Oxford, UK

KEVIN DOCHERTY
Department of Medicine, University of Birmingham, Queen Elizabeth Hospital, Birmingham B15 2TH, UK

SYAMALIMA DUBE
RH Williams Laboratory, Department of Medicine, University of Washington, Seattle, Washington, USA

PAUL C. GUEST
Department of Clinical Biochemistry, University of Cambridge, Addenbrookes Hospital, Hills Road, Cambridge CB2 2QR, UK

WILLIAM HAGOPIAN
RH Williams Laboratory, Department of Medicine, University of Washington, Seattle, Washington, USA

JOHN C. HUTTON
Department of Clinical Biochemistry, University of Cambridge, Addenbrookes Hospital, Hills Road, Cambridge CB2 2QR, UK

PETER IN'T VELD
Department of Pathology, Vrije Universiteit Brussel, Brussels, Belgium

ALLAN KARLSEN
RH Williams Laboratory, Department of Medicine, University of Washington, Seattle, Washington, USA

RONAN P. KELLY
Cellular Nutrition Research Group, Department of Biochemistry, University of Oxford, South Parks Road, Oxford, UK

RITA KIEKENS
Department of Metabolism and Endocrinology, Vrije Universiteit Brussel, Brussels, Belgium

INGRID KOCKUM
RH Williams Laboratory, Department of Medicine, University of Washington, Seattle, Washington, USA

ÅKE LERNMARK
RH Williams Laboratory, Department of Medicine, University of Washington, Seattle, Washington, USA

ANDREW NEIL
Nuffield Department of Clinical Medicine, Oxford University; Diabetes Research Laboratories, Radcliffe Infirmary; and Department of Public Health and Primary Care, Oxford University, Radcliffe Infirmary, Oxford, UK

E. A. NEWSHOLME
Cellular Nutrition Research Group, Department of Biochemistry, University of Oxford, South Parks Road, Oxford, UK

DANIEL PIPELEERS
Department of Metabolism and Endocrinology, Vrije Universiteit Brussel, Brussels, Belgium

MICHA J. RAPOPORT
Banting and Best Department of Medical Research and Departments of Pharmacology and Immunology, University of Toronto, Toronto, Ontario, Canada M5G 1L6

JOHN W. SEMPLE
Department of Immunohematology, St. Michael's Hospital, Toronto, Ontario, Canada M6B 1W8

KENNETH SIDDLE
Department of Clinical Biochemistry, University of Cambridge, Addenbrookes Hospital, Hills Road, Cambridge CB2 2QR, UK

JEREMY M. TAVARÉ
Department of Biochemistry, School of Medical Sciences, University of Bristol, University Walk, Bristol BS8 1TD, UK

ROBERT TURNER
Nuffield Department of Clinical Medicine, Oxford University; Diabetes Research Laboratories, Radcliffe Infirmary; and Department of Public Health and Primary Care, Oxford University, Radcliffe Infirmary, Oxford, UK

RALF WASSMUTH
RH Williams Laboratory, Department of Medicine, University of Washington, Seattle, Washington, USA

PART I THE PANCREATIC β-CELL

Editors' introduction

Insulin is secreted from the β-cells of the islets of Langerhans. The islets were described by Paul Langerhans, their discoverer, as 'small cells, most of them with entirely homogeneous contents and a polygonal shape, having a round nucleus without nucleoli and usually lying side by side in pairs or small groups'. The islets, which range from about 75 to 225 μm in diameter, lie dispersed throughout the exocrine pancreas. The human pancreas has roughly one million islets which make up only about 1–2 per cent of the total pancreatic mass.

The islet comprises a central core of insulin-producing β-cells surrounded by the glucagon-secreting α-cells. Somatostatin (D-cells) and pancreatic polypeptide (PP) secreting cells are also present. The β-cells predominate and in the rat comprise about 70 per cent of islet cells. Neighbouring β-cells influence each other through gap junctions. These dynamic structures permit the passage of ions and small molecules and enable the electrical activity of adjacent β-cells to be synchronized.

The pancreatic blood supply is provided by the coeliac and anterior mesenteric arteries and drains into tributaries of the hepatic portal vein. The islets are richly vascularized. One to three afferent arterioles enter the islet and branch to form an extensive capillary network. Each islet cell appears to be bordered on at least two sides by a capillary. The capillary endothelium is fenestrated and the fenestrae are covered by a thin diaphragm formed from the fused plasma membranes of the endothelial cells. In most cases a double basement lamina is present, one adjacent to the vascular endothelium and the other to the islet parenchyma. Nutrient secretagogues and insulin must cross these basement membranes in order to enter or leave the β-cell. It is therefore tempting to speculate that basement membrane thickening and deposition of amyloid, both characteristics of type II diabetes, could have functional consequences. The flow of blood within the islet is from core to periphery; consequently, the α-cells and D-cells in the periphery are perfused by blood with a high concentration of insulin. On the other hand, only those β-cells near the periphery are subjected to paracrine actions of glucagon and somatostatin. There is also evidence that the α-cells are supplied with blood before the D-cells. These considerations are relevant to the functioning of the islet as a unit.

The islet receives a rich autonomic innervation, the fibres following the arterioles, penetrating deep within the islet and terminating close to the islet cells. Neural influences are thus likely to play a physiological role in regulating insulin secretion. Sympathetic catecholaminergic fibres from the splanchnic nerve enter the islet via the coeliac ganglion and exert an inhibitory effect on insulin release and a stimulatory effect on glucagon release. Sympathetic innervation primarily mediates the islet response to stress, by favouring hyperglycaemia and thus ensuring an adequate supply of glucose to the brain. Stimulation of the parasympathetic system elicits insulin secretion. The pre-ganglionic cholinergic fibres running within the vagus terminate at ganglia within the pancreas. The post-ganglionic fibres innervating the islet cells are mostly cholinergic but some are peptidergic. Cholinergic innervation controls the cephalic phase of insulin release, i.e. the anticipatory response to a meal. A similar early response to feeding has also been demonstrated for pancreatic polypeptide and glucagon secretion. The neuropeptides so far identified in intrapancreatic nerves are calcitonin gene-related peptide, cholecystokinin, galanin, gastrin-releasing polypeptide, leu-enkephalin, substance P, and vasoactive intestinal peptide. All of these peptides have been shown, at least under certain experimental conditions, to be able to modulate the secretion of islet hormones. However, their functional roles remain to be defined.

1 Morphology of the pancreatic β-cell

DANIEL PIPELEERS, RITA KIEKENS, and PETER IN'T VELD

1.1 Localization

1.1.1 In the pancreas

The islet β-cell population comprises less than 2 per cent of the volume of an adult mammalian pancreas (McEvoy 1981; Stefan *et al.* 1982; Rahier *et al.* 1983; Chen *et al.* 1989). Most insulin-containing cells occur in the islets of Langerhans, cell groups that are easily distinguished during microscopical inspection of the gland (Fig. 1). The pancreatic islets are scattered throughout the organ, with often a higher density in the tail region (Hellman 1970). Their size and number varies among species, and seems to be regulated by developmental, nutrient, and hormonal factors (Hellman 1959; Bonnevie-Nielsen *et al.* 1983). In the adult rat pancreas, islet sizes range from less than 50 μm to 500 μm, while estimates of their total numbers have exceeded 10^5 per organ (Lacy and Greider 1972). The small-sized islets are most numerous, but the islets larger than 150 μm represent more than 50 per cent of the total islet volume (Hellman *et al.* 1964). Virtually all *in vitro* studies have been conducted on large-sized islets as larger aggregates are preferentially selected by the isolation procedures developed by Hellerström (1964) and by Lacy and Kostianovsky (1967). It should thus be kept in mind that most *in vitro* data presently available for rodents are representative of the β-cells that occur in the 200–500 largest islets of the pancreas (Fig. 2). Pancreatic β-cells can also be encountered as single units outside the islets. Their occasional identification in the tubular wall of the pancreatic duct has been associated with their ontogenetic origin (Like and Orci 1972; Pictet and Rutter 1972; Falkmer *et al.* 1985) (Fig. 3).

The dispersion of the insulin-containing β-cells within the pancreas can be considered as essential for the regulation of exocrine functions (Henderson and Daniel 1979). It is also conceivable that the overall function of the pancreatic β-cell population is served by this anatomical distribution throughout the gland. There exists, however, no direct evidence in support of this view, and it is still unclear how this can be tested in an intact organ. One can only speculate about the theoretical advantages of distributing the endocrine pancreatic function into multiple units which differ in composition and in internal milieu. A diversity in the function of individual islets could certainly contribute to a tight and multifactorial

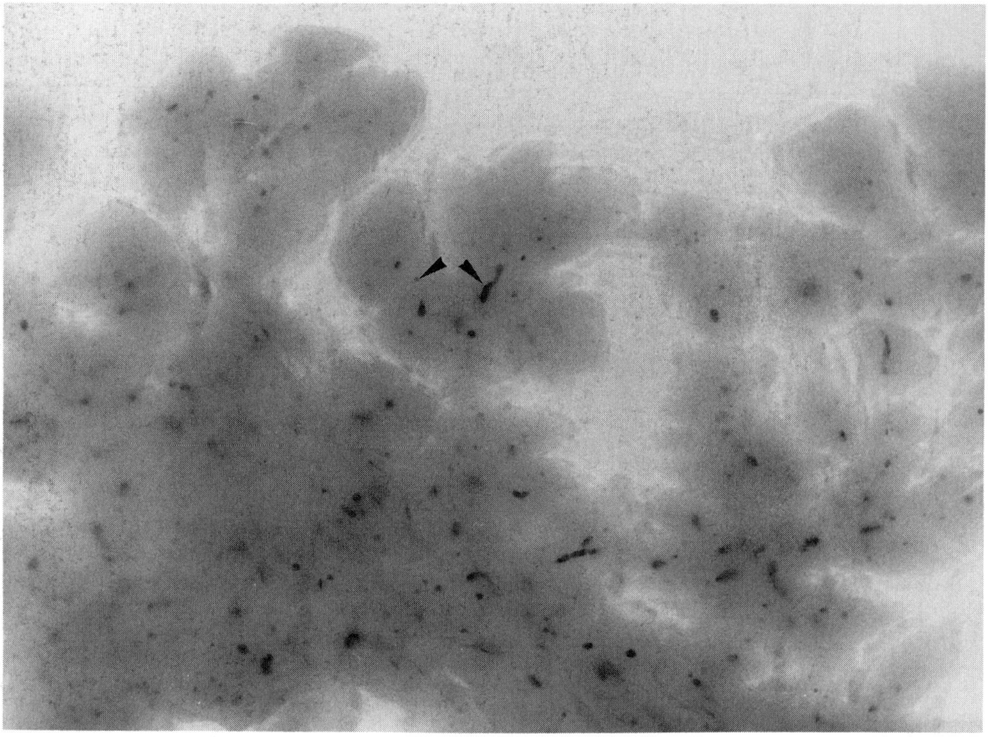

Fig. 1 Part of an adult rat pancreas stained with dithizone. The islets of Langerhans are recognizable as dark spots of variable size and format (×7.5)

control of glucose homeostasis. Its existence is not yet certain but is at least compatible with the observation of functional differences between small and large islets (Kitahara and Adelman 1979; Reaven *et al.* 1979), between islets from ventral and dorsal regions (Trimble *et al.* 1982), and between single and aggregated β-cells (Pipeleers *et al.* 1982).

1.1.2 In the islets of Langerhans

Pancreatic β-cells compose more than 60 per cent of the adult islet volume (Rahier *et al.* 1983). In sections from rodent pancreata, it is readily evident that islet β-cells comprise the central core of the islet structure and that other endocrine cells form the mantle (Ferner 1952; Hellman 1967; Orci *et al.* 1975). An ordered topography was also recognized in other species (Hellman 1967), sometimes only after an elaborate morphological analysis (Grube *et al.* 1983). A common feature is that the insulin-containing β-cells form a homologous area within each islet; only a small proportion of β-cells is juxtaposed to other endocrine cells, such as the glucagon-containing A-cells and the somatostatin-containing D-cells (Orci and Unger 1975; Orci *et al.* 1975). This particular intra-islet organization probably results from the

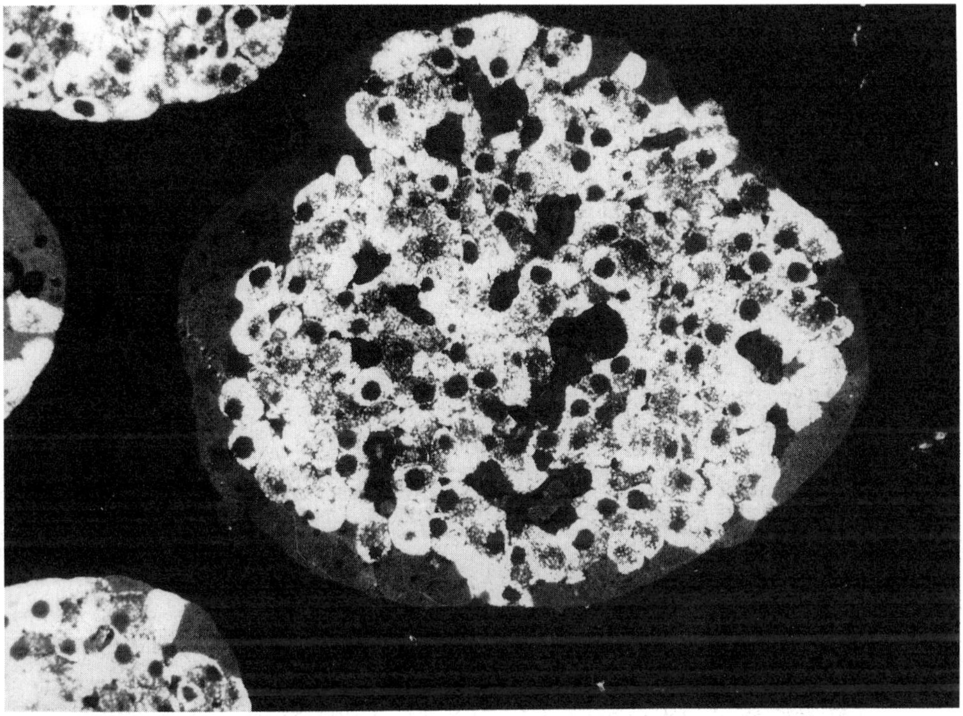

Fig. 2 Semi-thin section of an isolated rat islet of Langerhans after immunofluorescent staining for insulin. The insulin-containing cells occupy the larger part of the tissue and are, over an extensive area, surrounded by one or two layers of insulin-negative cells (×400)

sequence of events in the development of the islet organ. It is also suggestive of the existence of adhesion molecules favouring homologous contacts between pancreatic β-cells. To what extent this organization influences the function of pancreatic β-cells is still unknown (Pipeleers 1984). The obcervation that isolated single pancreatic β-cells release little insulin in response to a glucose stimulus (Pipeleers *et al.* 1982) supports the view that the organization of β-cells within the pancreatic islet may determine the amplitude of their secretory response. From a theoretical viewpoint, it is conceivable that the cells are exposed to signals from nearby endocrine and non-endocrine cells as well as to messages from distal tissues (Pipeleers 1984; Pipeleers 1987). The microanatomy of each islet will, however, determine to what extent each of these possible influences will be implicated in the regulation of the cells *in vivo*.

1.2 Local environment

1.2.1 Endocrine cells and pancreatic hormones

The vast majority of pancreatic β-cells are juxtaposed to other β-cells (Grube *et al.* 1983; Jörns *et al.* 1988). The homologous β-cell groups are ordered as cord-like

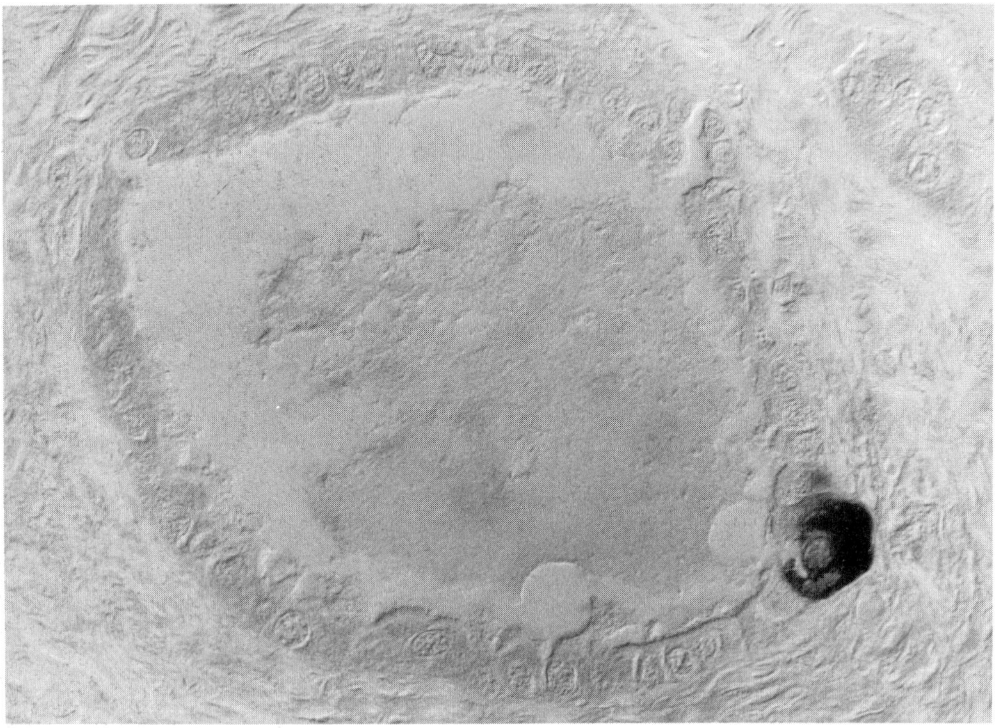

Fig. 3 Section through a human pancreatic duct after immunofluorescent staining for insulin. A strongly positive cell is present within the ductal wall (differential interference contrast, ×750)

structures or appear as compact masses (Lacy and Greider 1972; Jörns *et al.* 1988). The cells exhibit signs of polarity in their intracellular organization (Bonner-Weir 1988); groups of 8–10 β-cells were found to arrange their apical poles around a central capillary while keeping their basal poles in close contact with other capillaries (Fig. 4) (Bonner-Weir 1988). At the sites where several β-cells make contact, microvillous-rich cell surfaces have been recognized (Fujita *et al.* 1981) and found to be associated with a glucose transporter immunoreactivity (Orci *et al.* 1989). Neighbouring β-cells can exchange ions and small molecules via gap junctions which are heterogeneously distributed throughout the total β-cell population (Meda *et al.* 1980, 1982). These junctions do not correspond to fixed anatomical structures but represent dynamic formations which rapidly assemble from, or disassemble into, gap junctional particles (In't Veld *et al.* 1986). Glucose, calcium, and cyclic AMP influence the configuration, the number, and the permeability of these gap junctions between β-cells (Fig. 5) (Meda *et al.* 1979, 1983; Kohen *et al.* 1983; In't Veld *et al.* 1985, 1986). Intercellular communication seems thus a physiologically regulated event in aggregated groups of islet β-cells. The recognition of several signs of inter- and intracellular organization is, of course, not sufficient to attribute a major functional significance to the anatomical arrangement of islet β-cells. However, the

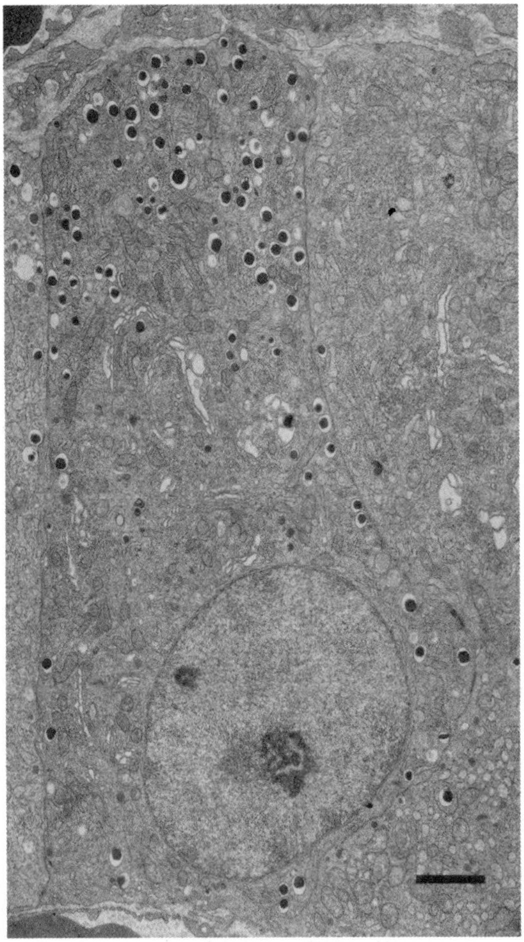

Fig. 4 *In situ* electron micrograph of a rat islet β-cell after sulphonylurea treatment. The degranulated cells exhibit a polarity of their secretory vesicles at the apical pole (×5000) (reprinted from Bonner-Weir 1988)

observation that aggregated β-cells exhibit a markedly higher secretory response than single β-cells (Pipeleers *et al.* 1982) provides one argument in favour of the idea that the formation of homologous β-cell groups may offer a number of functional advantages.

Only a small fraction of islet β-cells is in direct contact with islet endocrine non-B-cells. In the adult human pancreas, it was found that β-cells make anatomical contacts with the majority of glucagon-containing A-cells and of somatostatin-containing D-cells; this contrasts with a relatively low frequency of heterologous contacts between A- and D-cells (Grube *et al.* 1983). The contacts between islet B- and D-cells appear of firm nature, as many resist the procedures of islet isolation and dissociation (Fig. 6) (Pipeleers and Pipeleers-Marichal 1981; Pipeleers 1984). It has been demonstrated that the structural coupling between B- and D-cells (as well as between B- and A-cells) can take the form of gap junctional complexes which allow intercellular exchange of small molecules (Orci *et al.* 1975; Michaels and

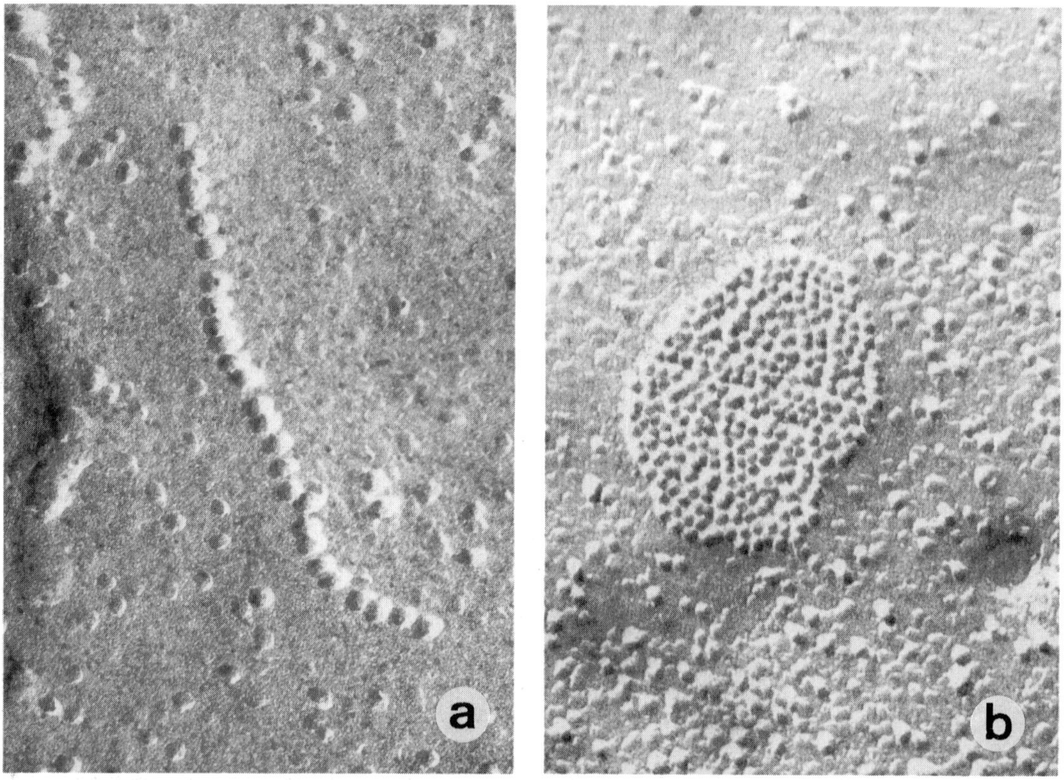

Fig. 5 Linear (a) and polygonal (b) configuration of gap junctional units between centrally located rat islet β-cells. In pancreata perfused at 2.8 mM glucose (a), the majority of gap junctional units appear as linear structures (× 320 000); at 20 mM glucose (b), most units rearrange into a polygonal pattern (×200 000)

Sheridan 1981; Meda *et al.* 1982). The islet β-cell masses may thus establish functional interactions with their neighbouring endocrine non-B-cells via their peripherally located units.

It is also conceivable that interactions between islet cells may also occur via locally released secretory products (paracrine regulation) (Lernmark and Hellman 1970; Orci and Unger 1975). Indeed, both *in vivo* and *in vitro* studies have demonstrated that insulin release can be modulated by pancreatic hormones (review Pipeleers 1984; Samols *et al.* 1986). Glucagon is known for its stimulatory effects (Samols *et al.* 1965; Grodsky *et al.* 1967; Malaisse *et al.* 1967), whereas somatostatin-14, insulin, and pancreatic polypeptide have been described as inhibitory (Koerken *et al.* 1974; Liljenquist *et al.* 1978; Murphy *et al.* 1981; Waldhäusl *et al.* 1982). β-Cells purified from rat pancreas have been used to assess whether these effects result from direct interactions with the insulin-releasing cells and whether they require higher concentrations of the hormone than those found in the periphery. Glucagon turned out to be a powerful stimulus for islet β-cells, exerting its activity via high-affinity receptors at concentrations greater than 10^{-10} M (Schuit and Pipeleers 1985; Van Schravendijk *et al.* 1985; Pipeleers 1987). No effects were noticed with pancreatic

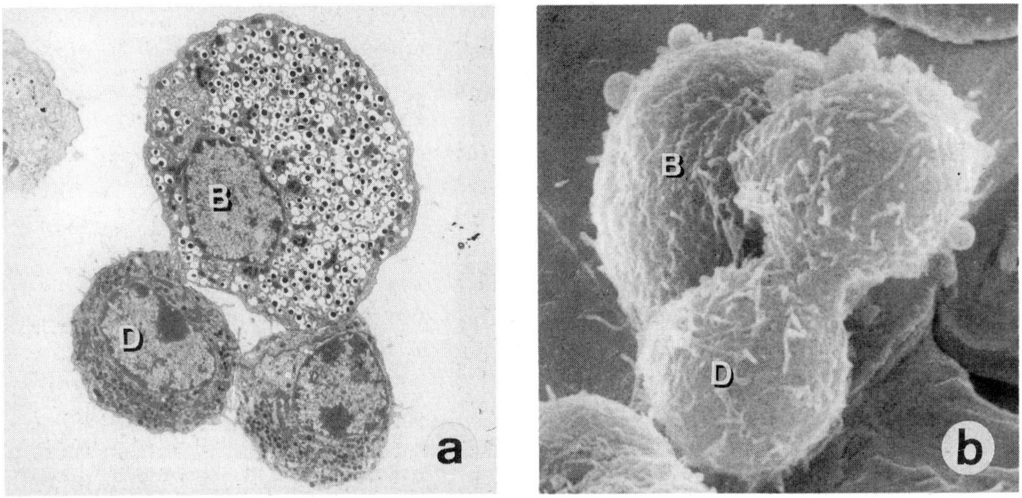

Fig. 6 Electron micrograph of coupled islet cells that have been purified from a dissociated islet cell preparation. The fraction is primarily composed of islet β-cells. Most islet non-β-cells in this fraction correspond to smaller somatostatin-containing D-cells which are structurally coupled to insulin-containing β-cells. The D-cells present characteristic long microvilli. (a) Transmission EM, (b) scanning EM

polypeptide. Inhibitory effects of somatostatin-14 and insulin were detectable on pure β-cells, but appeared not to be mediated by high-affinity interactions: minimal effective concentrations were as high as 10^{-9} and 10^{-7} M, respectively, and tenfold higher levels did not markedly increase the suppressive actions of these peptides (Schuit *et al.* 1989; Van Schravendijk *et al.* 1990). The marginal inhibitory effect of insulin is probably mediated via a low-affinity binding to insulin-like growth factor I (IGF-I) receptors (Van Schravendijk *et al.* 1987). On the basis of these data, we have concluded that rat islet β-cells are sensitive to the glucagon levels which are present in the afferent islet arterioles (10^{-10} M) (Pipeleers 1984). They can, of course, also respond to the locally produced hormone, but it is still unknown whether the released glucagon reaches the receptive pole of the islet β-cells before escaping from the islet interstitial and vascular spaces. The same question relates to insulin and pancreatic somatostatin, whose inhibitory effects in pure β-cell preparations required concentrations that are probably only reached in the proximity of the B- or D-cell secretory poles. More information is thus needed on the composition of the interstitial fluid at the receptor pole of the islet β-cells *in situ*. The data do not exclude the possibility that only a fraction of the pancreatic β-cells is under the influence of locally released glucagon or somatostatin. Such diversity of regulation would certainly contribute to homeostatic control, but has not yet been accessible to experimental testing. Recent studies have, on the other hand, examined whether locally released glucagon and somatostatin can influence the islet β-cells via the vascular route (Samols and Stagner 1988; Samols *et al.* 1988). These experiments consisted of comparing effluent insulin levels from pancreata that were either retrogradely or anterogradely perfused with glucagon or somatostatin, or with their respective antibodies. It was thus shown that insulin release is not influenced by the glucagon and somatostatin levels in the islet capillary network (Stagner *et al.* 1988). The outcome of these functional studies is consistent with the morphological observation that the islet blood flow first irrigates the insulin-releasing β-cells before it reaches the other endocrine islet cells (Bonner-Weir and Orci 1982).

1.2.2 Blood compartment

The vascular structure of the pancreas exhibits multiple capillary glomeruli at the sites of the islets (Lacy and Greider 1972). Ten to twenty per cent of the incoming blood is immediately directed to these glomerular agglomerates, which together irrigate 1 per cent of the pancreatic volume (Lifson *et al.* 1980; Jansson 1984). Each glomerulus is composed of numerous tortuous capillaries which have branched from one or more afferent arterioles (Fujita and Murakami 1973; Bonner-Weir and Orci 1982; Ohtani 1983). In the rat, most of these glomeruli are located within the central β-cell mass and are drained by efferent capillaries that run through the islet mantle of non-B-cells (Bonner-Weir and Orci 1982). The islet capillary network establishes close contacts with the majority of islet β-cells (Grube *et al.* 1983; Bonner-Weir 1988) (Fig. 7). It has been conceived as a framework around which rows or columns of pancreatic β-cells are tightly arranged and oriented (Lacy and Greider 1972; Bonner-Weir 1988). The endocrine cells are separated from the blood compartment by an interstitial space of varying width and by endothelial cells. As in other endocrine tissues, the islet endothelial cells contain numerous circular pores, the fenestrations, which allow rapid exchange of molecules up to M_r 40 000 between the vascular and the interstitial space (Bearer and Orci 1985) (Fig. 8). The microanatomy of the islet tissue seems thus to favour communication between the pancreatic β-cells and the blood compartment, contributing to the rapid recognition of circulating signals and the fast procurement of secretory products. The rapidity of this communication is subject to regulatory mechanisms which are not yet completely elucidated. It has been demonstrated that the flow through the capillaries can vary in intensity as well as in direction (McCuskey and Chapman 1969; Jansson and Hellerström 1983). Of particular interest is the observation of a glucose-induced increase in islet blood flow which appears to be generated by glucose-sensitive sensors in the brain and mediated by the vagus (Jansson and Hellerström 1983, 1986). Cholinergic fibres in the vascular wall may thus induce a selective vasodilation of afferent islet arterioles and thus lead to an increased insulin release (Miller 1981). It is conceivable that such a mechanism is involved in the cephalic phase of insulin release.

1.2.3 Interstitial space

The interstitial space of the islet tissue can only be vaguely described. Its absolute or relative volume has not yet been determined and its composition is largely

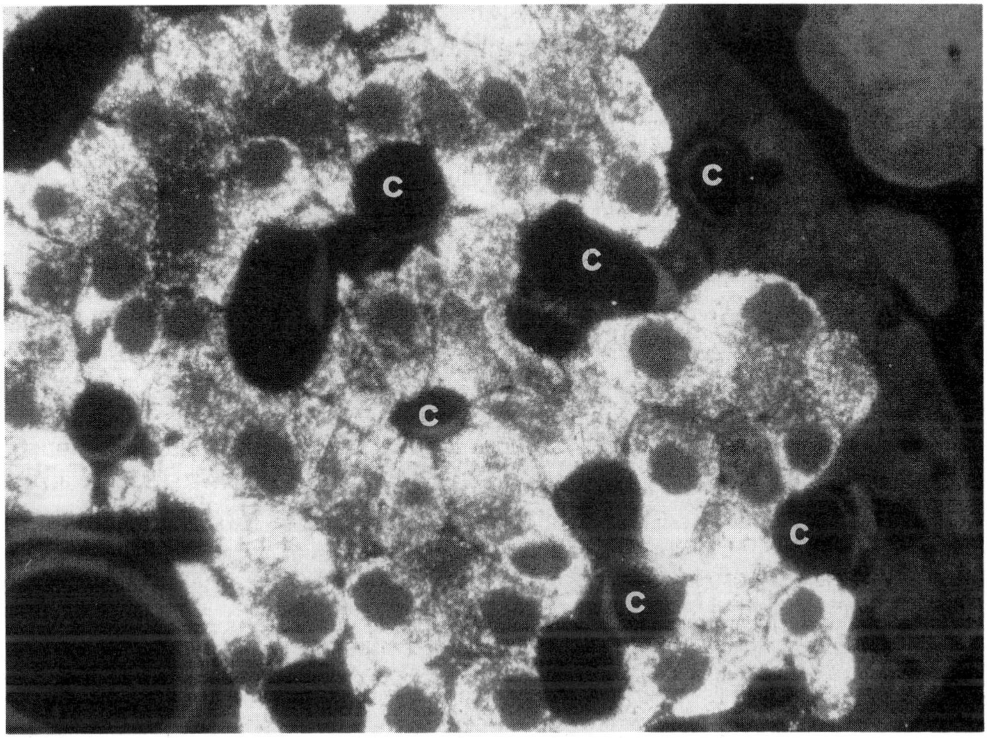

Fig. 7 Semi-thin section of an adult rat islet in intact pancreatic tissue. Insulin-containing cells are stained by an immunofluorescent reaction. Most of them are in direct contact with one or two capillaries (c) (×900)

undefined. Virtually no information exists on the chemical nature of the extracellular matrix: the ground substance is still to be analysed and no fibrous proteins other than collagen have been identified (Fig. 8). Some of the matrix substances may be secreted by the fibroblasts which are located close to the islet tissue, but it is unknown which islet cells contribute to the formation of matrix proteins and polysaccharides. Basal lamina are formed at the interstitial's interface with the endocrine and with the endothelial cells, but it is unclear whether and how these specializations interact with these cell groups.

Several non-endocrine cell types have been identified in the islet interstitial space. Interactions with endothelial and/or endocrine cells have only been documented for some of them. The best evidence has been collected for a role of the interstitial's neural endings. The presence of autonomic extensions is characteristic for all examined mammalian species but their content in cholinergic, adrenergic, and peptidergic neurotransmitters has been found to vary considerably (Miller 1981) (Fig. 9). The islet nerve fibres originate from either intrapancreatic, para-aortic, or central ganglion cells, and reach the islets via the afferent arterioles. Synapses have not been encountered within the islet tissue, but close associations

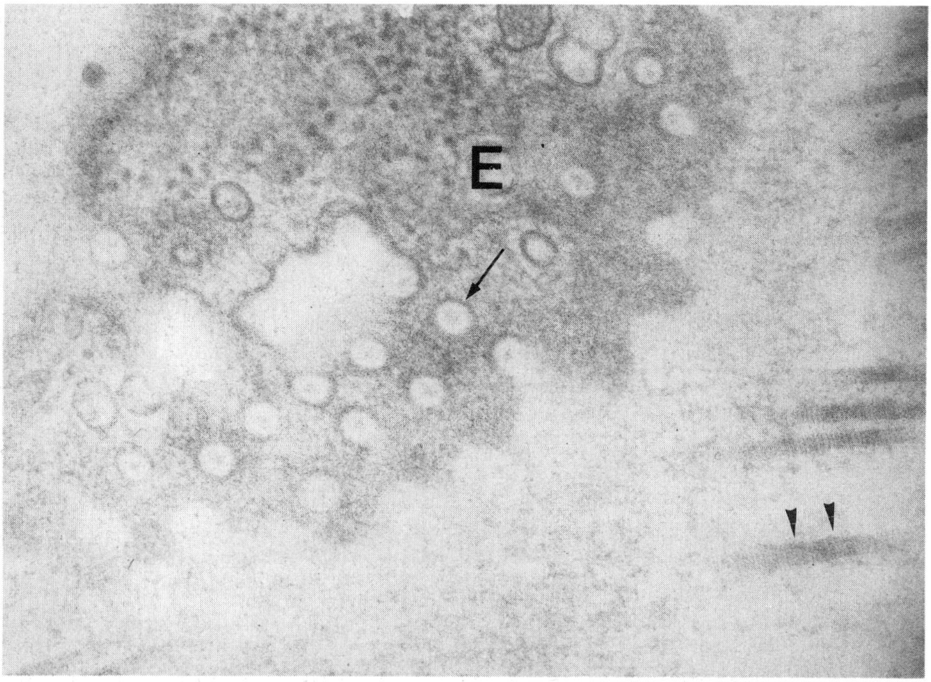

Fig. 8 Electron micrograph of an endothelial cell (E) and surrounding interstitial space with collagen fibres (arrowheads). The endothelial cell presents several fenestrations (arrow) in which the central portion of the diaphragm can be recognized (×50 000)

were noticed between nerve endings and endothelial or endocrine cells (Orci *et al.* 1973). Several of the recognized neurotransmitters were shown to influence insulin release, although often at supraphysiological concentrations. Cholinergic agents exert a stimulatory effect via muscarinergic receptors (Ahrén and Taborsky 1986), whereas adrenergic compounds suppress β-cell secretory activity via α_2-adrenergic recognition sites (Porte 1967; Schuit and Pipeleers 1986). Among the tested neuropeptides, vasoactive intestinal polypeptide, cholecystokinin, and gastrin-releasing polypeptide enhance hormone release, whereas galanin, neuropeptide Y, and calcitonin gene-related peptide were rather inhibitory (reviewed by Ahrén *et al.* 1986).

The interstitial space around the islet capillaries also contains small numbers of mononuclear phagocytes and dendritic cells (Faustman *et al.* 1981; Farr and Anderson 1985). Several of these cells exhibit a strong surface expression of major histocompatibility complex (MHC) class II antigens (Farr and Anderson 1985; Pipeleers *et al.* 1987). Their function may consist of the phagocytosis of damaged cells (Pipeleers *et al.* 1987) or the induction of a local immune reaction (In't Veld and Pipeleers 1988; Kolb-Bachofen *et al.* 1988). It has also been postulated that interleukin-1 secretion from islet macrophages can lead to the selective destruction of pancreatic β-cells (Nerup *et al.* 1988).

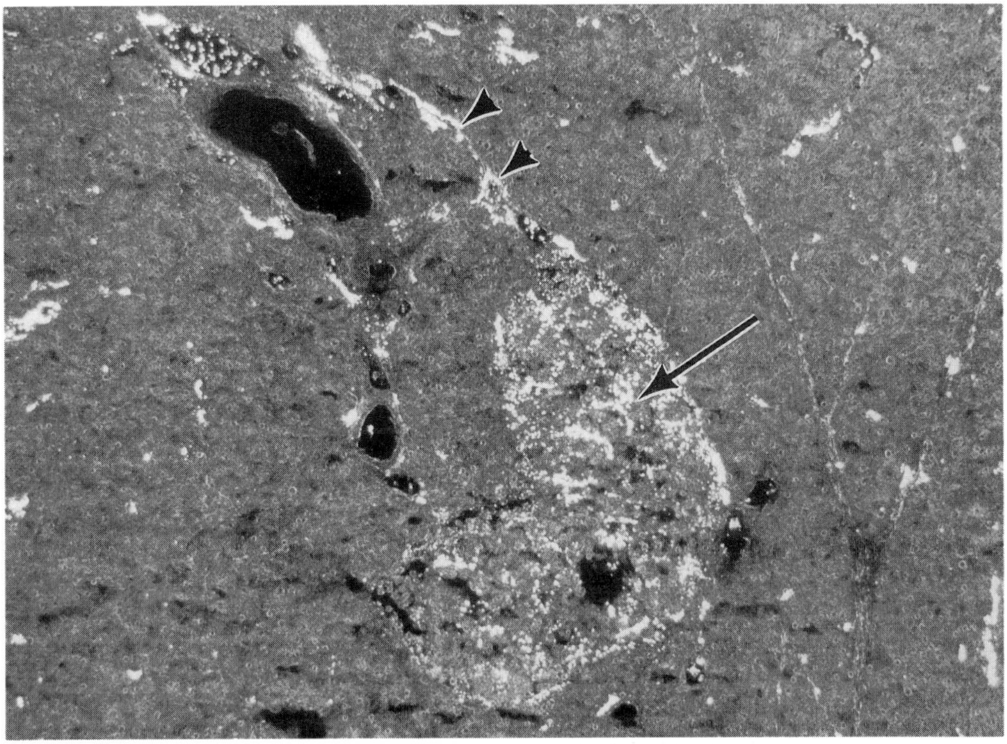

Fig. 9 Histochemical staining for cholinesterase in a cryosection of rat pancreas. A strong reaction is seen in the islet tissue (arrow) and along the blood vessels and septa of the exocrine part (arrowheads). Thus, cholinergic fibres appear to be present primarily in the endocrine tissue (×115)

1.2.4 Exocrine tissue

Insulin-containing β-cells can establish close contacts with pancreatic exocrine cells in ductal epithelia as well as in acini (Bendayan 1987). In both locations, desmosomes were identified between closely apposed endocrine and exocrine cells. There are, however, no data available on the percentage of pancreatic β-cells with such exocrine cell contacts. An abnormally high association has been noticed in spiny mice (*Acomys cahirinus*), where the endocrine islet cell masses appear to be entirely integrated in the exocrine tissue (Malaisse-Lagae *et al.* 1975). The functional significance of these associations is unknown. It is not certain that the single β-cells that occur in ductal epithelia are potent secretory units (Pipeleers *et al.* 1982). Their secretory activity may be less dependent on circulating glucose levels than on the conditions within the epithelium, which may be affected by locally released insulin. Effects of insulin on acinar cell function have been demonstrated (Hellman *et al.* 1964; Saito *et al.* 1980; Korc *et al.* 1981) but it is not evident that they (only) occur between anatomically associated cells. A fraction of the islet hormones could diffuse into the peri-insular areas and hence explain the regional differences in exocrine cell function (Hellman *et al.* 1964; Malaisse-Lagae *et al.* 1975). The released islet hormones can also reach the exocrine tissue via the blood vessels that leave the pancreatic islets (Fujita and Murakami 1973; Bergeron *et al.* 1980).

1.3 Morphological identification

Different techniques can be used to recognize the endocrine β-cells in sections of pancreatic tissue. Staining with aldehyde-fuchsin has been used for more than 30 years to identify β-cells on the basis of the high number of cysteinic S–S bridges in their secretory vesicles (Gomori 1950; Mowry 1983). A more sensitive and specific method consists of the fluorescent or peroxidase labelling of stored insulin with insulin antibodies (Lacy 1959). At the electron microscopical level, the cells can be recognized by the typical features of their secretory vesicles, namely a crystalline, electron-dense core surrounded by a wide halo. The recent development of *in situ* hybridization histochemistry made it also possible to identify β-cells from the presence of mRNA for proinsulin (Chen *et al.* 1989).

The ultrastructure of the pancreatic β-cells has been described in previous reviews (Orci 1974, 1982; Orci *et al.* 1988). Particular attention has been given to the organelles and structures that are involved in the synthesis, processing, and release of the hormonal products (Steiner *et al.* 1972; Orci 1985; Orci *et al.* 1988). Using high-resolution autoradiography, Orci followed the path of newly formed proteins from the rough endoplasmic reticulum to the storage granules in the secretory vesicles (Orci 1985). Fifteen minutes after their synthesis, most of the proinsulin molecules were recovered in the Golgi complex. They rapidly move from the *cis* to the *trans* Golgi cisternae where they are packed into clathrin-coated vesicles. The sorting of proinsulin towards these particular vesicles may involve an interaction with a receptor that is attached to the inner side of the *trans* Golgi membranes (Orci *et al.* 1988). During and shortly after their separation from the Golgi complex, these vesicles appear to be uniformly filled with a proteinous material of moderate electron density. Simultaneous peptide analysis indicated that proinsulin conversion is initiated at this stage (Orci *et al.* 1984, 1988; Orci 1985). This initiation probably depends on the presence of, first, membrane constituents that create the intravesicular conditions for proteolytic conversion and, secondly, of enzymes that cleave the synthesized precursor proteins; the conversion products include β-cell-specific peptides as well as peptides which have also been found in other endocrine cells. The biologically most important specific protein, insulin, crystallizes as a zinc complex which will occupy the central core of the secretory vesicle (Howell 1984) (Fig. 10). One hour after synthesis of its precursor, insulin is almost completely converted and has become the main product of the storage vesicles, which no longer exhibit a clathrin coating. These events are discussed in detail in Chapter 3.

Adult pancreatic β-cells have a relatively large hormone store, providing a mean insulin reserve of 50 pg per cell (Pipeleers 1987). A rise in glucose concentration is thought to induce the cytoplasmic ionic conditions that lead to the fusion of the

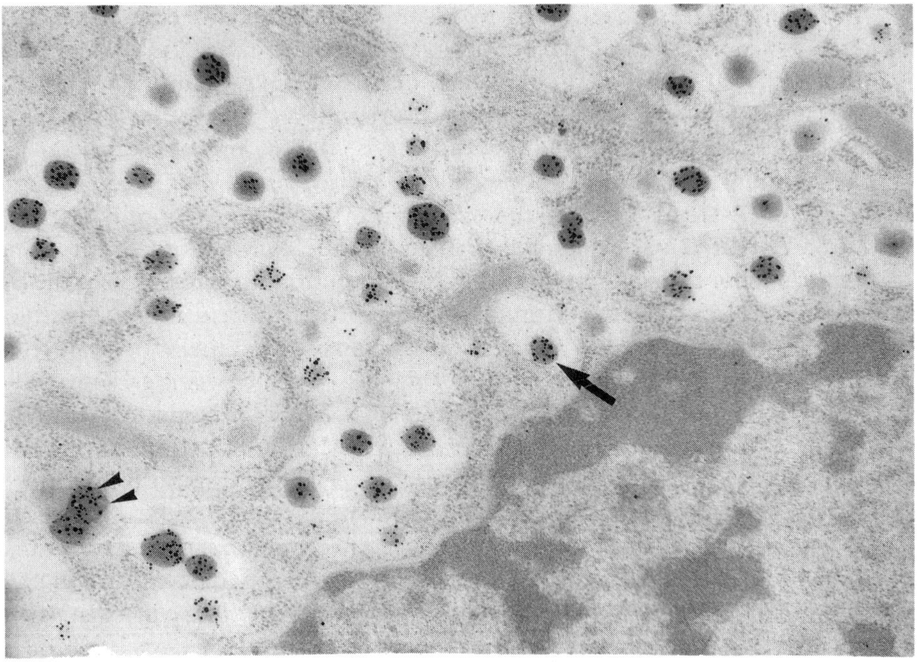

Fig. 10 Ultrathin section of a rat islet β-cell after immunogold labelling for insulin. Most of the insulin immunoreactivity is localized in the central core of the secretory granule (arrow). A positivity is also identified in phagosomes (arrowhead) (×35 000)

secretory vesicles with the plasma membrane, as described in Chapter 4. This fusion leads to an exocytosis of the stored peptides (Lacy 1961; Orci *et al.* 1973) while the vesicle membranes are incorporated in the plasma membrane. Parallel endocytosis is expected to maintain the cellular surface area and to channel the internalized membrane components to an early stage in secretory vesicle formation (Orci *et al.* 1973, 1988).

1.4 Physicochemical characteristics of β-cells

The availability of purified pancreatic β-cells made it possible to measure a number of physical parameters of this cell type (Fig. 11). For adult rats, cellular volume ranged from 700 to 1300 μm^3 (Pipeleers and Pipeleers-Marichal 1981), with an intracellular water space of 0.8 picolitre (8×10^{-13} litre) (Gorus *et al.* 1984). The size of the other endocrine islet cells is markedly smaller, so that differences in sedimentation velocity can be used to purify islet β-cells from a rat islet cell preparation (Pipeleers and Pipeleers-Marichal 1981). The density of adult rat islet β-cells (d = 1.065) is smaller than that of glucagon-containing A-cells and that of somatostatin-containing D-cells (Pipeleers and Pipeleers-Marichal 1981). The differences in cell density are not sufficient to purify islet cell types by density gradient centrifugation, but allow further enrichment of partially purified preparations (Pipeleers and Pipeleers-Marichal 1981). Single islet β-cells rapidly and spontaneously reaggregate into spherical particles which can establish a mantle with the islet endocrine non-B-cells that are present in the preparation (Hopcroft *et al.* 1985; Pipeleers *et al.* 1985; Montesano 1986; Halban *et al.* 1987). Reaggregation of islet β-cells is calcium-dependent and stimulated by glucose and cyclic AMP (Maes and Pipeleers 1984).

Most of the presently known chemical characteristics of islet β-cells have been determined in tissue extracts. Their identification helped clarify a number of processes in the production and release of insulin and in its regulation by glucose (see following chapters). Several of the described molecules and reactions have been associated with morphologically recognizable structures. The glucose transporter, GLUT-2, has been located in microvillous membranes (Orci *et al.* 1989). The processing of newly formed hormone was followed within the different subcellular organelles (Orci 1982; Orci *et al.* 1988) and the discharge of insulin into the medium was localized to specialized areas of the cell membrane (Orci *et al.* 1979). Cytochemistry at the ultrastructural level has thus become a powerful instrument in the unravelling of islet β-cell functions. For many cytochemically defined molecules or reactive determinants, the role or significance is still unknown. In the following section, we will list those that confer some degree of specificity to the adult insulin-containing β-cells.

1.5 β-Cell markers

1.5.1 β-Cell-specific secretory proteins

Insulin

The secretory vesicles are 80 per cent composed of cell specific proteins with insulin immunoreactivity (Hutton 1989). These are principally insulin and C-peptide—both present in equimolar amounts—and a much smaller amount of proinsulin and its intermediate conversion products. Insulin molecules are stored in crystalline form, occupying the electron-dense portion of the secretory granule (Howell 1984). In a minority of islet β-cells, the secretory vesicles contain mostly proinsulin, as if no conversion occurs in these cells before storage and/or release (Orci *et al.* 1985).

Islet amyloid polypeptide (IAPP)

In addition to the derivatives of preproinsulin, the secretory vesicles also contain islet amyloid polypeptide (IAPP) (Westermark *et al.* 1986) or amylin (Cooper *et al.* 1987), which is derived from preproamylin (Sanke *et al.* 1988). This protein was first isolated from the amyloid deposits in insulin-producing tumours: it consists of 37 amino acids and has a molecular weight of 3900. It was also recovered from pancreata of non-insulin-dependent diabetics who are known to exhibit amyloid deposits in the islet tissue (Westermark *et al.* 1987). Normal islet β-cells contain amylin (Lukinius *et al.* 1989) stored in the secretory vesicles at a ratio of 1 to 100

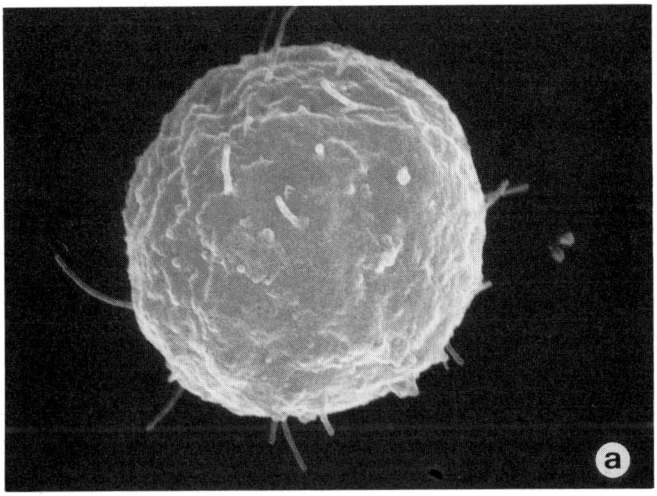

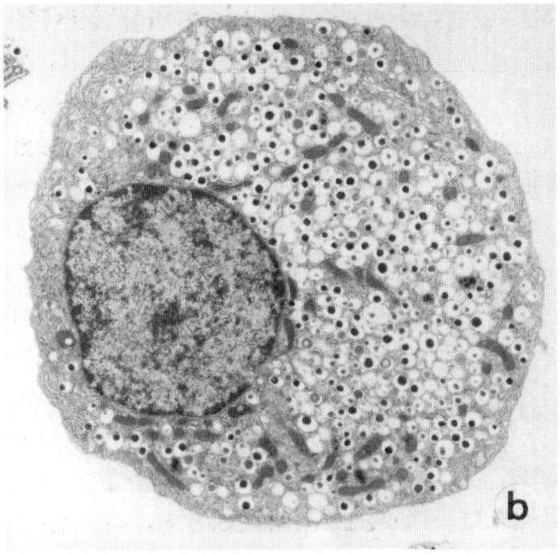

Fig. 11 Scanning (a) and transmission (b) electron micrographs of a single purified β-cell, isolated by autofluorescence-activated cell sorting. The cells appear structurally intact and well granulated (b). Their surface membrane exhibits a few long microvilli (a) (×3500)

with respect to the insulin content. The peptide is co-secreted with insulin (Fehmann *et al.* 1990) and is detectable in the circulation (Nakazato *et al.* 1989). IAPP-immunoreactivity has been noticed in as yet unidentified cells of the gut (Nakazato *et al.* 1989), but has not yet been associated with the presence of amylin.

1.5.2 Other secretory vesicle components

The secretory vesicles also contain a number of components that have analogues in other secretory cells (see review by Hutton 1989). Their concentration and molecular form may differ from one cell type to another, suggesting differences in their function or in their substrates. In the pancreatic β-cells, these vesicle constituents have been located peripherally to the electron-dense core of crystallized insulin, i.e. in the electron-lucent mantle or in the vesicle membrane. They can be classified in different categories.

Ion pumps

Membrane components such as proton-translocating ATPases and calcium-dependent ATPases are expected to generate the ionic conditions that activate proteases for precursor conversion and that induce configurational changes in the produced peptides so that their storage is facilitated (Hutton 1989).

Trapped molecules

The intravesicular milieu can trap small molecules from the cytoplasm and concentrate them prior to their release with stored peptides. The acidic pH may, for example, lead to uptake of bases such as serotonin and dopamine, two substances that have been identified in the secretory vesicles of pancreatic β-cells (Cegrell 1968; Jaim-Etcheverry and Zieher 1968; Lundquist *et al.* 1975). In adult islet β-cells, the accumulation of these two biogenic amines is preceded by cellular uptake of the precursors 5-hydroxytryptophan and L-dopa and by their subsequent decarboxylation. This property of amine precursor uptake and decarboxylation (APUD) is shared with peptidergic cells of the central and peripheral nervous systems (Pearse 1982). As it was also identified in other extraneural peptide-releasing cells, Pearse proposed the existence of a neuroendocrine system that is composed of APUD-cells of neuroectodermal origin (Pearse 1982). This hypothesis has been challenged by Pictet and by Le Douarin, who presented evidence in favour of an endodermal origin of the islet β-cells (Pictet and Rutter 1972; Le Douarin 1988). The increasing number of neural markers that are recognized in islet β-cells keeps open the possibility that the endodermal precursors of pancreatic islet cells have an ectodermal ancestry (Teitelman *et al.* 1987; Alpert *et al.* 1988). Little is known of the role of biogenic amines in islet β-cells. The substances are released into the medium together with the other secretory products, but it is unclear whether they exert paracrine effects as they do in neurotransmission.

Proteolytic enzymes

Like other peptide-releasing cells, islet β-cells contain proteolytic enzymes which cleave the precursor forms of their secretory products (Steiner *et al.* 1972; Docherty and Steiner 1982). The enzymes are probably sorted within the Golgi cisternae into the secretory vesicles. They have been identified as endopeptidases 1 and 2 and as carboxypeptidase H (Davidson and Hutton 1987; Davidson *et al.* 1988). After endoproteolytic cleavage of proinsulin, the carboxypeptide removes the C-terminal basic amino acids as discussed in detail in Chapter 3. The carboxypeptidase is relatively abundant in the insulin secretory vesicles, representing 2–5 per cent of their protein content (Hutton 1984). It is also recovered, at least in part, among the released products (Guest *et al.* 1989).

1.5.3 Non-β-cell-specific secretory proteins

Islet β-cells release several other proteins in addition to the β-cell-specific peptides which were discussed in Section 1.5.1. Only a few have been characterized (Hutton 1989). One of them is chromogranin-A, a 50 kDa protein which is widely distributed in endocrine and neural tissues (reviewed by Simon and Aunis 1989). This protein is converted in parallel to proinsulin, through interaction with the endopeptidase II and carboxypeptidase H. Two cleavage products have been identified in islet β-cells: the 20 kDa betagranin and the 5 kDa pancreastatin (Tatemoto *et al.* 1986; Hutton *et al.* 1988). The role of these peptides is still unknown. The inhibition of glucose-induced insulin release by pancreastatin suggests that this product may exert a biological effect, but does not indicate the molecule's physiological site of action. In other tissues, cleavage may be absent or may occur at a different rate or at different sites, leading to differences in the form of the stored and released chromogranin products (Simon and Aunis 1989). These differences may thus express a cell specificity and could confer cell-specific biological effects.

1.5.4 Neural markers

In addition to possessing secretory vesicle components which are also present in peptidergic neurons, islet β-cells contain a number of molecules in other compartments which are identical or analogous to characteristic constituents of neural tissue. Thus, their cytoplasm was found to generate a glutamate decarboxylase activity which converts glutamate into γ-aminobutyric acid (GABA) (Okada *et al.* 1976). This enzymatic activity has not been detected in other islet endocrine cells (Garry *et al.* 1986), explaining the β-cell-specific location of GABA in islet tissue (Gerber and Hare 1979). It was suggested that GABA exerts intranuclear effects within the pancreatic β-cells (Garry *et al.* 1986); it may also interact with membrane receptors on islet A-cells (Rorsman *et al.* 1989). The enzyme seems to be involved in the immunological events that are associated with insulin-dependent diabetes: studies by Baekkeskov *et al.* (1990) have demonstrated that the β-cell-specific antibodies that occur before or at onset of the disease specifically recognize this particular protein, which has been first denoted as the 64K islet antigen and is discussed further in Chapter 10. Another similarity with neurons is the presence of a synaptophysin-like immunoreactivity in small cytoplasmic vesicles (Wiedenmann *et al.* 1986). Synaptophysin is a 38 kDa glycoprotein, which was originally described in presynaptic vesicles of neurons (Wiedenmann and Franke 1985). Other neuronal markers have also been found to be positive in islet β-cells such as neuron-specific enolase (Schmechel *et al.* 1978).

It is conceivable that these chemical similarities with neural cells indicate a common ancestral origin (Teitelman *et al.* 1981; Pearse 1982; Alpert *et al.* 1988). To what extent they imply functional similarities with neurons is debatable, although this possibility should be seriously considered in view of other indications, such as the electrically excitable nature of the β-cell (Dean and Matthews 1970), the presence of glial cells around the islets (Donev 1984), and the description of neuro-insular complexes (Fujita 1959).

1.6 Functional heterogeneity

Most studies on islet β-cells implicitly assume that all the insulin-containing cells in a preparation are functionally identical. We have recently demonstrated that this is not the case. Experiments on purified islet β-cells indicated that the cells can exhibit marked differences in their individual properties (Pipeleers 1987; Schuit *et al.* 1988). In particular, individual cell responses to glucose varied widely (Van De Winkel and Pipeleers 1983; Schuit *et al.* 1988). This variability turned out to be a major determinant for the sigmoidal shape of the dose–response curves (Schuit *et al.* 1988).

Strong evidence for the concept of functional heterogeneity came from autoradiographs of islet β-cells wherein newly synthesized proteins had been labelled at different glucose concentrations (Schuit *et al.* 1988). Under none of the conditions were all the cells functionally identical. At glucose concentrations as low as 1 mM, some 25 per cent of the cells were in a state of active protein synthesis. An increase in glucose concentration dose-dependently increased the number of biosynthetically active cells (Fig. 12). This phenomenon of glucose-induced recruitment finally resulted in the participation of more than 80 per cent of the β-cells in the protein synthetic response (Schuit *et al.* 1988). Similar observations were made in isolated islet preparations (Schuit *et al.* 1988).

The secretory activity of an islet β-cell preparation may also be composed of

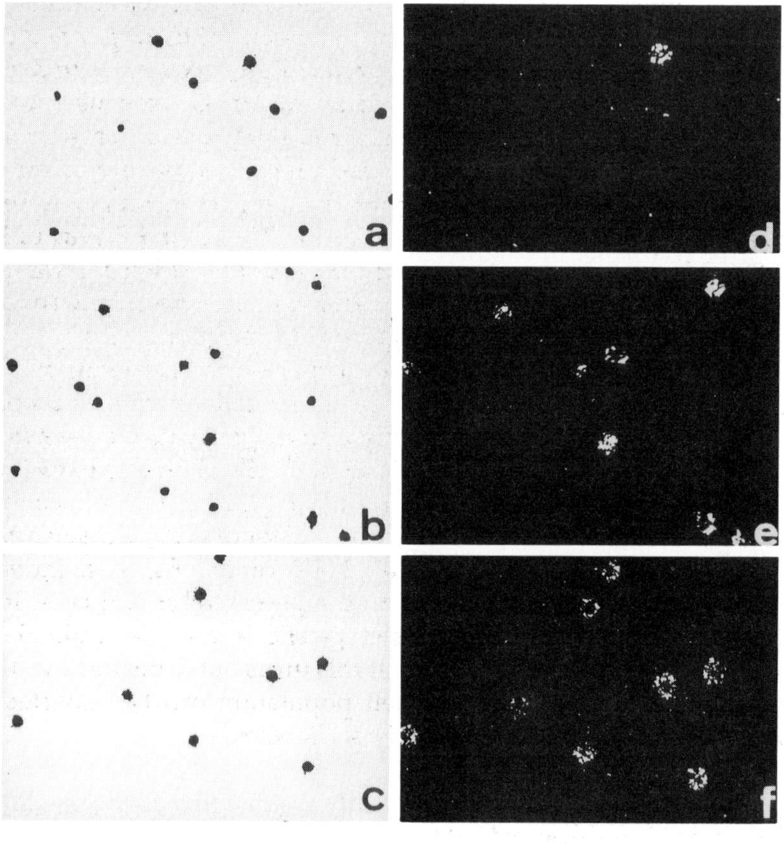

Fig. 12 Glucose-induced recruitment of pancreatic β-cells into biosynthetic activity. Single purified β-cells are shown in transillumination, visualizing all cells (a–c) and in epipolarization, visualizing only ^{3}H-labelled cells (d–f). The cells were incubated with [^{3}H]leucine at 1 mM (a,d), 3 mM (b,e), and 10 mM (c,f) glucose. The percentage of ^{3}H-labelled cells increases with the glucose concentration (d–f) (×1750) (reprinted from Schuit *et al.* 1988)

individually different cell responses. This functional heterogeneity is, however, more difficult to assess in unpurified islet cell preparations, where individual islet β-cells can differ in their state of aggregation or intercellular coupling and in their exposure to pancreatic hormones, conditions that are known to exert profound influences on the rate of hormone release (Pipeleers 1987). Salomon and Meda have nevertheless produced evidence that, in these preparations also, a phenomenon of glucose-induced recruitment of islet β-cells occurs, resulting in a dose-dependent increase in the number of cells with an activated secretory activity (Salomon and Meda 1986). Their conclusions were based on the use of a haemolytic plaque assay which allows detection of hormone release from one cell, as described in Section 4.1.2.

The mechanism underlying the differences in individual β-cell responses are closely related to the cellular handling of glucose (Kiekens *et al.* 1989). Early work on the electrophysiological properties of islet cells had already noticed the existence of intercellular differences in the threshold for glucose-induced electrical activity (Dean and Matthews 1970). The availability of purified β-cells and of techniques to monitor their individual redox state, allowed us to demonstrate that islet β-cells also differ in their individual metabolic responses to glucose (Van De Winkel and Pipeleers 1983; Kiekens *et al.* 1989). Certain cells undergo a shift in their metabolic redox state when glucose concentrations are raised from 1 to 3 mM, others require an increase to 10 mM, and still others remain unresponsive irrespective of the glucose concentration (Van De Winkel and Pipeleers 1983; Kiekens *et al.* 1989). Moreover, when the islet β-cells are subdivided according to this difference in metabolic behaviour, a close correlation is noticed between the metabolic and biosynthetic responsiveness to glucose (Kiekens *et al.* 1989).

It is, of course, essential to document also the existence of a functional heterogeneity *in situ*. Supportive evidence can be derived from a number of observations. It is, for example, known that only a small percentage of islet β-cells can enter the mitotic cycle (Swenne 1982). Individual β-cells differ also in their degree of granulation and proinsulin content (Orci *et al.* 1985). Differences in cellular insulin content become even more pronounced after a prolonged glucose stimulus (Stefan *et al.* 1987). We have also observed the phenomenon of glucose-induced recruitment of β-cells into biosynthetic activity in the perfused rat pancreas. It thus seems likely that the *in vivo* response of islet β-cells is also determined by the functional heterogeneity in the population. If this turns out to be the case, the perspective of a functionally heterogeneous β-cell population can be expected to influence our speculations on the pathogenesis of diabetes.

Acknowledgements

This work was supported by grants from the Belgian Fund for Medical Research (3.0075.88 and 3.0093.90) and from the Belgian Ministry of Science Policy (Gekoncerteerde Aktie 86/91-102). Rita Kiekens is a fellow of the Belgian National Fund for Scientific Research. The authors thank Dr Susan Bonner-Weir for providing Fig. 4.

References

Ahrén, B. and Taborsky, G. J., Jr (1986). The mechanism of vagal nerve stimulation of glucagon and insulin secretion in the dog. *Endocrinology* **118,** 1551–7.

Ahrén, B., Taborsky, G. J., Jr, and Porte, D., Jr (1986). Neuropeptidergic versus cholinergic and adrenergic regulation of islet hormone secretion. *Diabetologia* **29,** 827–36.

Alpert, G., Hanahan, D., and Teitelman, G. (1988). Hybrid insulin genes reveal a developmental lineage for pancreatic endocrine cells and imply a relationship with neurons. *Cell* **53,** 295–308.

Baekkeskov, S., Aanstoot, H. J., Christgau, S., Reetz, A., Solimena, M., Cascalho, M., Folli, F., Richter-Olesen, H., and De Camilli, P. (1990). Identification of the 64K autoantigen in insulin-dependent diabetes as the GABA-synthesizing enzyme glutamic acid decarboxylase. *Nature* **347,** 151–6.

Bearer, E. L. and Orci, L. (1985). Endothelial fenestral diaphragms: a quick-freeze, deep-etch study. *J. Cell Biol.* **100,** 418–28.

Bendayan, M. (1987). Presence of endocrine cells in pancreatic ducts. *Pancreas* **2,** 393–7.

Bergeron, J. J. M., Rachubinski, R., Searle, N., Sikstrom, R., Borts, D., Bastian, P., and Posner, B. I. (1980). Radioautographic visualization of *in vivo* insulin binding to the exocrine pancreas. *Endocrinology* **107,** 1069–80.

Bonner-Weir, S. (1988). Morphological evidence for pancreatic polarity of β-cells within islets of Langerhans. *Diabetes* **37,** 616–21.

Bonner-Weir, S. and Orci, L. (1982). New perspectives on the microvasculature of the islets of Langerhans in the rat. *Diabetes* **31,** 883–9.

Bonnevie-Nielsen, V., Skovgaard, L. T., and Lernmark, A. (1983). β-cell function relative to islet volume and hormone content in the isolated perfused mouse pancreas. *Endocrinology* **112,** 1049–56.

Cegrell, L. (1968). The occurrence of biogenic monoamines in the mammalian endocrine pancreas. *Acta Physiol. Scand. Suppl.* **314,** 1–58.

Chen, L., Komiya, I., Imman, L., McCorkle, P., Alam, T., and Unger, R. H. (1989). Molecular and cellular responses of islets during perturbations of glucose homeostasis determined by *in situ* hybridization histochemistry. *Proc. Natl Acad. Sci. USA* **86,** 1367–71.

Cooper, G. J. S., Willis, A. C., Clark, A., Turner, R. C., Sim, R. B., and Reid, K. B. M. (1987). Purification and characterization of a peptide from amyloid-rich pancreases of Type 2 diabetic patients. *Proc. Natl Acad. Sci USA* **84,** 8628–32.

Davidson, H. W. and Hutton, J. C. (1987). The insulin-secretory-granule carboxypeptidase H. Purification and demonstration of involvement in proinsulin processing. *Biochem. J.* **245,** 575–82.

Davidson, H. W., Rhodes, C. J., and Hutton, J. C. (1988). Intraorganellar calcium and pH control proinsulin cleavage in the pancreatic B-cell via two distinct site-specific endopeptidases. *Nature* **333,** 93–6.

Dean, P. M. and Matthews, E. K. (1970). Glucose-induced electrical activity in pancreatic islet cells. *J. Physiol.* **210,** 255–64.

Docherty, K. and Steiner, D. F. (1982). Post translational proteolysis in polypeptide hormone biosynthesis. *Ann. Rev. Physiol.* **44,** 625–38.

Donev, S. R. (1984). Ultrastructural evidence for the presence of a glial sheath investing the islets of Langerhans in the pancreas of mammals. *Cell Tissue Res.* **237,** 343–8.

Falkmer, S., Dafgard, E., El-Salhy, M., Engström, W., Grimelius, L., and Zetterberg, A.

(1985). Phylogenetical aspects of islet hormone families: a minireview with particular reference to insulin as a growth factor and to the phylogeny of PYY and NPY immunoreactive cells and nerves in the endocrine and exocrine pancreas. *Peptides* **6** (Suppl. 3), 315–20.

Farr, A. G. and Anderson, S. K. (1985). *In situ* ultrastructural demonstration of cells bearing Ia antigens in the murine pancreas. *Diabetes* **34,** 987–90.

Faustman, D., Hauptfeld, V., Lacy, P., and Davie, J. (1981). Prolongation of murine islet allograft survival by pretreatment of islets with antibody directed to Ia determinants. *Proc. Natl Acad. Sci. USA* **78,** 5156–9.

Fehmann, H. C., Weber, V., Göke, R., Göke, B., and Arnold, R. (1990). Cosecretion of amylin and insulin from isolated rat pancreas. *FEBS Lett.* **262,** 279–81.

Ferner, H. (1952). *Das Inselsystem des pankreas,* pp. 109–86. Thieme Verlag, Stuttgart.

Fujita, T. (1959). Histological studies on the neuro-insular complex in the pancreas of some mammals. *Z. Zellforsch.* **50,** 94–109.

Fujita, T. and Murakami, T. (1973). Microcirculation of monkey pancreas with special reference to the insulo-acinar portal system. A scanning electron microscope study of vascular casts. *Arch. Histol. Jap.* **35,** 255–63.

Fujita, T., Kobayashi, S., and Serizawa, Y. (1981). Intercellular canalicule system in pancreatic islet. *Biomed. Res.* **2** (Suppl. 1), 115–18.

Garry, D. J., Sorenson, R. L., Elde, R. P., Maley, B. E., and Madsen, A. (1986). Immunohistochemical colocalization of GABA and insulin in β-cells of rat islet. *Diabetes* **35,** 1090–5.

Gerber, J. C. and Hare, T. A. (1979). Gamma-aminobutyric acid in peripheral tissue, with emphasis on the endocrine pancreas. *Diabetes* **28,** 1073–6.

Gomori, G. (1950). A new stain for elastic tissue. *Am. J. Clin. Pathol.* **20,** 665–6.

Gorus, F., Malaisse, W. J., and Pipeleers, D. G. (1984). Differences in glucose handling by pancreatic A- and B-cells. *J. Biol. Chem.* **259,** 1196–200.

Grodsky, G. M., Bennett, L. L., Smith, D. F., and Schmid, F. G. (1967). Effect of pulse administration of glucose or glucagon on insulin secretion *in vitro. Metabolism* **16,** 222–3.

Grube, D., Eckert, I., Speck, P. T., and Wagner, H.-J. (1983). Immunohistochemistry and microanatomy of the islets of Langerhans. In *Brain-gut axis* (ed. T. Fujito, T. Kanno and N. Yanaihara). *Biomed. Res.* **4** (Suppl.), 25–36.

Guest, P. C., Pipeleers, D. G., Rossier, J., Rhodes, C. J., and Hutton, J. C. (1989). Co-secretion of carboxypeptidase H and insulin from isolated rat islets of Langerhans. *Biochem. J.* **264,** 503–8.

Halban, P. A., Powers, S. L., George, K. L., and Bonner-Weir, S. (1987). Spontaneous reassociation of dispersed adult pancreatic islet cells into aggregates with three-dimensional architecture typical of native islets. *Diabetes* **36,** 783–90.

Hellerström, C. (1964). A method for the microdissection of intact pancreatic islets of mammals. *Acta Endocrin.* **45,** 122–32.

Hellman, B. (1959). The volumetric distribution of the pancreatic islet tissue in young and old rats. *Acta Endocrin.* **31,** 91–106.

Hellman, B. (1967). Islet morphology and glucose metabolism in relation to the specific function of the pancreatic B-cells. *Excerpta Med. Int. Congr. Ser.* **172,** 92–109.

Hellman, B. (1970). Methodological approaches to studies on the pancreatic islets. *Diabetologia* **6,** 110.

Hellman, B., Petersson, B., and Hellerström, C. (1964). The growth pattern of the endocrine pancreas in mammals. In *The structure and metabolism of the pancreatic islets,* pp. 45–61. Pergamon Press, Oxford.

Henderson, J. R. and Daniel, P. M. (1979). A comparative study of the portal vessels connecting the endocrine pancreas, with a discussion of some functional implications. *Q. J. Exp. Physiol.* **64,** 267–75.

Hopcroft, D. W., Mason, D. R., and Scott, R. S. (1985). Insulin secretion from perifused rat pancreatic pseudoislets. *In Vitro Cellular and Developmental Biology* **21,** 421–7.

Howell, S. L. (1984). The mechanism of insulin secretion. *Diabetologia* **26,** 319–27.

Hutton, J. C. (1984). Secretory granules. *Experientia* **40,** 1091–8.

Hutton, J. C. (1989). The insulin secretory granule. *Diabetologia* **32,** 271–81.

Hutton, J. C., Peshavaria, M., Johnston, C. F., Ravazzola, M., and Orci, L. (1988). Immunolocalization of betagranin: a chromogranin A-related protein of the pancreatic B-cell. *Endocrinology* **12,** 1014–20.

In't Veld, P. A. and Pipeleers, D. G. (1988). *In situ* analysis of pancreatic islets in rats developing diabetes: appearance of non-endocrine cells with surface MHC class II antigens and cytoplasmic insulin immunoreactivity. *J. Clin. Invest.* **82,** 1123–8.

In't Veld, P., Schuit, F., and Pipeleers, D. (1985). Gap junctions between pancreatic β-cells are modulated by cyclic AMP. *Eur. J. Cell Biol.* **36,** 269–76.

In't Veld, P., Pipeleers, D., and Gepts, W. (1986). Glucose alters configuration of gap junctions between pancreatic islet cells. *Am. J. Physiol.* **251,** C191–C196.

Jaim-Etcheverry, G. and Zieher, L. M. (1968). Electron microscopic cytochemistry of 5-hydroxytryptamine (5-HT) in the beta cells of guinea pig endocrine pancreas. *Endocrinology* **83,** 917–23.

Jansson, L. (1984). The blood flow to the pancreas and the islets of Langerhans during an intraperitoneal glucose load in the rat. *Diabetes Research* **1,** 111–14.

Jansson, L. and Hellerström, C. (1983). Stimulation by glucose of the blood flow to the pancreatic islets of the rat. *Diabetologia* **25,** 45–50.

Jansson, L. and Hellerström, C. (1986). Glucose-induced changes in pancreatic islet blood flow mediated by central nervous system. *Am. J. Physiol.* **251,** E644–E647.

Jörns, A., Barklage, E., and Grube, D. (1988). Heterogeneities of the islets on the rabbit pancreas and the problem of 'paracrine' regulation of islet cells. *Anat. Embryol.* **178,** 297–307.

Kiekens, R., Van De Winkel, M., Ling, Z., In't Veld, P., Schuit, F., and Pipeleers, D. (1989). Glucose recognition determines functional heterogeneity in the pancreatic β-cell population. *Diabetologia* **32,** 503 A.

Kitahara, A. and Adelman, R. C. (1979). Altered regulation of insulin secretion in isolated islets of different sizes in aging rats. *Biochem. Biophys. Res. Comm.* **87,** 1207–13.

Koerken, D. J., Ruch, W., Chideckel, E., Palmer, J., Goodner, C. J., Ensinck, J., and Gale, C. G. (1974). Somatostatin: hypothalamic inhibitor of the endocrine pancreas. *Science* **184,** 482–4.

Kohen, E., Kohen, C., and Rabinovitch, A. (1983). Cell-to-cell communication in rat pancreatic islet monolayer cultures is modulated by agents affecting islet-cell secretory activity. *Diabetes* **32,** 95–8.

Kolb-Bachofen, V., Epstein, S., Kiesel, U., and Kolb, H. (1988). Low-dose streptozotocin-induced diabetes in mice: electron microscopy reveals single-cell insulitis before diabetes onset. *Diabetes* **37,** 21–7.

Korc, M., Oberbach, D., Quinto, C., and Rutter, W. J. (1981). Pancreatic islet-acinar cell interactions: amylase messenger RNA levels are determined by insulin. *Science* **213,** 351–3.

Lacy, P. E. (1959). Electron microscopic and fluorescent antibody studies on islets of Langerhans. *Exp. Cell Res.* **7,** 296–308.

Lacy, P. E. (1961). Electron microscopy of β-cells of pancreas. *Am. J. Med.* **31,** 851–9.
Lacy, P. E. and Greider, M. H. (1972). Ultrastructural organization of mammalian pancreatic islets. In *Handbook of physiology* (ed. S. R. Greiger), Section 7, Vol. 1, pp. 77–90. American Physiological Society, Washington, DC.
Lacy, P. E. and Kostianovsky, M. (1967). Method for the isolation of intact islets of Langerhans from the rat pancreas. *Diabetes* **16,** 35–9.
Le Douarin, N. M. (1988). On the origin of pancreatic endocrine cells. *Cell* **53,** 169–71.
Lernmark, A. and Hellman, B. (1970). Effect of epinephrine and mannoheptulose on early and late phases of glucose-stimulated insulin release. *Metabolism* **19,** 614–18.
Lifson, N., Kramlinger, K. G., Mayrand, R. R., and Lender, E. J. (1980). Blood flow to the rabbit pancreas with special reference to the islets of Langerhans. *Gastroenterology* **79,** 466–73.
Like, A. A. and Orci, L. (1972). Embryogenesis of the human pancreatic islets: a light and electron microscopic study. *Diabetes* **21,** 511–34.
Liljenquist, J. E., Horwitz, D. L., Jennings, A. S., Chiasson, J. L., Keller, U., and Rubenstein, A. H. (1978). Inhibition of insulin secretion by exogenous insulin in normal man as demonstrated by C-peptide assay. *Diabetes* **27,** 563–70.
Lukinius, A., Wilander, E., Westermark, G. T., Engström, U., and Westermark, P. (1989). Co-localisation of islet amyloid polypeptide and insulin in the B-cell-secretory granules of the human pancreatic islets. *Diabetologia* **32,** 240–4.
Lundquist, I., Sundler, F., Hakanson, R., Larsson, L.-I., and Heding, L. G. (1975). Differential changes in the 5-hydroxytryptamine and insulin content of guinea-pig β-cells. *Endocrinology* **97,** 937–47.
McCuskey, R. S. and Chapman, T. M. (1969). Microscopy of the living pancreas *in situ*. *Am. J. Anat.* **126,** 395–408.
McEvoy, R. C. (1981). Changes in the volumes of the A-, B-, and D-cell populations in the pancreatic islets during the postnatal development of the rat. *Diabetes* **30,** 813–17.
Maes, E. and Pipeleers, D. (1984). Effects of glucose and 3′, 5′-cyclic adenosine monophosphate upon reaggregation of single pancreatic β-cells. *Endocrinology* **114,** 2205.
Malaisse, W. J., Malaisse-Lagae, F., and Mayhew, D. (1967). A possible role for the adenylcyclase system in insulin secretion. *J. Clin. Invest.* **46,** 1724–34.
Malaisse-Lagae, F., Ravazzola, M., Robberecht, P., Vandermeers, A., Malaisse, W. J., and Orci, L. (1975). Exocrine pancreas: evidence for topographic partition of secretory function. *Science* **190,** 795–7.
Meda, P., Perrelet, A., and Orci, L. (1979). Increase of gap junctions between pancreatic β-cells during stimulation of insulin secretion. *J. Cell Biol.* **82,** 441–8.
Meda, P., Denef, J.-F., Perrelet, A., and Orci, L. (1980). Nonrandom distribution of gap junctions between pancreatic β-cells. *Am. J. Physiol.* **238,** C114–C119.
Meda, P., Kohen, E., Kohen, C., Rabinovitch, A., and Orci, L. (1982). Direct communication of homologous and heterologous endocrine islet cells in culture. *J. Cell Biol.* **92,** 221–6.
Meda, P., Michaels, R. L., Halban, P. A., Orci, L., and Sheridan, J. D. (1983). *In vivo* modulation of gap junctions and dye coupling between B-cells of the intact pancreatic islet. *Diabetes* **32,** 858–68.
Michaels, R. L. and Sheridan, J. D. (1981). Islets of Langerhans: dye coupling among immunocytochemically distinct cell types. *Science* **214,** 801–2.
Miller, R. E. (1981). Pancreatic neuroendocrinology: peripheral neural mechanisms in the regulation of the islets of Langerhans. *Endocrine Rev.* **2,** 471–94.

Montesano, R. (1986). Cell–extracellular matrix interactions in morphogenesis: an *in vitro* approach. *Experientia* **42,** 977–85.

Mowry, R. W. (1983). Selective staining of pancreatic beta-cell granules. *Arch. Pathol. Lab. Med.* **107,** 464–8.

Murphy, W. A., Fries, J. L., Meyers, C. A., and Coy, D. H. (1981). Human pancreatic polypeptide inhibits insulin release in the rat. *Biochem. Biophys. Res. Comm.* **101,** 189–93.

Nakazato, M., Asai, J., Kangawa, K., Matsukura, S., and Matsuo, H. (1989). Establishment of radioimmunoassay for human islet amyloid polypeptide and its tissue content and plasma concentration. *Biochem. Biophys. Res. Comm.* **164,** 394–9.

Nerup, J., Mandrup-Poulsen, T., Molvig, J., and Spinas, G. (1988). In *The pathology of the endocrine pancreas in diabetes* (ed. P. Lefebvre and D. Pipeleers), pp. 71–84. Springer-Verlag, Berlin.

Ohtani, O. (1983). Microcirculation of the pancreas: a correlative study of intravital microscopy with scanning electron microscopy of vascular corrosion casts. *Arch. Histol. Jap.* **46,** 315–25.

Okada, Y., Taniguchi, H., and Shimada, C. (1976). High concentration of GABA and high glutamate decarboxylase activity in rat pancreatic islets and human insulinoma. *Science* **194,** 620–2.

Orci, L. (1974). A portrait of the pancreatic B-cell. *Diabetologia* **10,** 163–87.

Orci, L. (1982). Macro- and microdomains in the endocrine pancreas. *Diabetes* **31,** 538–65.

Orci, L. (1985). The insulin factory: a tour of the plant surroundings and a visit to the assembly line. *Diabetologia* **28,** 528–46.

Orci, L. and Unger, R. H. (1975). Functional subdivision of islets of Langerhans and possible role of D cells. *Lancet* **2,** 1243.

Orci, L., Amherdt, M., Malaisse-Lagae, F., and Renold, A. E. (1973). Insulin release by emiocytosis: demonstration with freeze-etching technique. *Science* **179,** 82–4.

Orci, L., Malaisse Lagae, F., Ravazolla, M., Rouiller, D., Renold, A. E., Perrelet, A., and Unger, R. (1975). A morphological basis for intercellular communication between α- and β-cells in the endocrine pancreas. *J. Clin. Invest.* **56,** 1066–70.

Orci, L., Amherdt, M., Roth, J., and Perrelet, A. (1979). Inhomogeneity of surface labelling of β-cells at prospective sites of exocytosis. *Diabetologia* **16,** 135–8.

Orci, L., Halban, P., Amherdt, M., Ravazzola, M., Vassalli, J. D., and Perrelet, A. (1984). A clathrin-coated, Golgi-related compartment of the insulin secreting cell accumulates proinsulin in the presence of monensin. *Cell* **39,** 39–47.

Orci, L., Ravazzola, M., Amherdt, M., Madsen, O., Vassalli, J.-D., and Perrelet, A. (1985). Direct identification of prohormone conversion site in insulin-secreting cells. *Cell* **42,** 671–81.

Orci, L., Vassali, J.-D., and Perrelet, A. (1988). The insulin factory. *Sci. Am.* **259,** 50–61.

Orci, L., Thorens, B., Ravazzola, M., and Lodish, H. F. (1989). Localization of the pancreatic beta cell glucose transporter to specific plasma membrane domains. *Science* **245,** 295–7.

Pearse, A. G. E. (1982). Islet cell precursors are neurones. *Nature* **295,** 96–7.

Pictet, R. and Rutter, W. J. (1972). Development of the embryonic endocrine pancreas. In *Handbook of physiology* (ed. S. R. Greiger), Section 7, Vol. 1, pp. 25–66. American Physiological Society, Washington, DC.

Pipeleers, D. (1984). Islet cell interactions with pancreatic B-cells. *Experientia* **40,** 1114–26.

Pipeleers, D. G. (1987). The biosociology of the pancreatic B-cell. *Diabetologia* **30,** 277–91.

Pipeleers, D. G. and Pipeleers-Marichal, M. A. (1981). A method for the purification of single A, B and D cells and for the isolation of coupled cells from isolated rat islets. *Diabetologia* **20,** 654–63.

Pipeleers, D. G., In't Veld, P. A., Maes, E., and Winkel, M. van de (1982). Glucose-induced insulin release depends on functional cooperation between islet cells. *Proc. Natl Acad. Sci. USA* **79,** 7322–5.

Pipeleers, D. G., Schuit, F., In't Veld, P. A., Maes, E., Hooghe-Peters, E. L., Winkel, M. van de, and Gepts, W. (1985). Interplay of nutrients and hormones in the regulation of insulin release. *Endocrinology* **117,** 824–33.

Pipeleers, D. G., In't Veld, P. A., Pipeleers-Marichal, M. A., Gepts, W., and Winkel, M. van de (1987). Presence of pancreatic hormones in islet cells with MHC-class II antigen expression. *Diabetes* **36,** 872–6.

Porte, D., Jr (1967). A receptor mechanism for the inhibition of insulin release by epinephrine in man. *J. Clin. Invest.* **46,** 86–94.

Rahier, J., Goebbels, R. M., and Henquin, J. C. (1983). Cellular composition of the human diabetic pancreas. *Diabetologia* **24,** 366–71.

Reaven, E. P., Gold, G., and Reaven, G. M. (1979). Effect of age on glucose-stimulated insulin release by the B-cell of the rat. *J. Clin. Invest.* **64,** 591–9.

Rorsman, P., Berggren, P.-O., Bokvist, K., Ericson, H., Möhler, H., Östenson, C. G., and Smith, P. A. (1989). Glucose-inhibition of glucagon secretion involves activation of GABAA-receptor chloride channels. *Nature* **341,** 233–6.

Saito, A., Williams, J. A., and Kanno, T. (1980). Potentiation of cholecystokinin-induced exocrine secretion by both exogeneous and endogeneous insulin in isolated and perfused rat pancreata. *J. Clin. Invest.* **65,** 777–82.

Salmon, D. and Meda, P. (1986). Heterogeneity and contact-depending regulation of hormone secretion by individual B-cells. *Exp. Cell Res.* **162,** 507–20.

Samols, E. and Stagner, J. I. (1988). Intra-islet regulation. *Am. J. Med.* **85** (5A), 31–5.

Samols, E., Marri, G., and Marks, V. (1965). Promotion of insulin secretion by glucagon. *Lancet* **2,** 415–16.

Samols, E., Bonner-Weir, S., and Weir, G. C. (1986). Intra-islet insulin–glucagon–somatostatin relationships. *Clin. Endocr. Metab.* **15,** 33–58.

Samols, E., Stagner, J. I., Ewart, R. B. L., and Marks, V. (1988). The order of islet microvascular cellular perfusion is B-A-D in the perfused rat pancreas. *J. Clin. Invest.* **82,** 350–3.

Sanket, T., Bell, G. I., Sample, C., Rubenstein, A. H., and Steiner, D. F. (1988). An islet amyloid peptide derived from an 89-amino acid precursor by proteolytic processing. *J. Biol. Chem.* **263,** 17243–6.

Schmechel, D., Marangos, P. J., and Brightman, M. (1978). Neuron-specific enolase is a molecular marker for peripheral and central neuroendocrine cells. *Nature* **276,** 834–6.

Schuit, F. C. and Pipeleers, D. G. (1985). Regulation of adenosine 3′, 5′-monophosphate levels in the pancreatic B cell. *Endocrinology* **117,** 834–40.

Schuit, F. C. and Pipeleers, D. G. (1986). Differences in adrenergic recognition by pancreatic A and B cells. *Science* **232,** 875–7.

Schuit, F. C., In't Veld, P. A., and Pipeleers, D. G. (1988). Glucose stimulates proinsulin biosynthesis by a dose-dependent recruitment of pancreatic beta cells. *Proc. Natl Acad. Sci. USA* **85,** 3865–9.

Schuit, F. C., Derde, M.-P., and Pipeleers, D. G. (1989). Sensitivity of rat pancreatic A and B cells to somatostatin. *Diabetologia* **32,** 207–12.

Simon, J.-P. and Aunis, D. (1989). Review article: Biochemistry of the chromogranin A protein family. *Biochem. J.* **262,** 1–13.

Stagner, J. I., Samols, E., and Bonner-Weir, S. (1988). b-a-d pancreatic islet cellular perfusion in dogs. *Diabetes* **37,** 1715–21.

Stefan, Y., Orci, L., Malaisse-Lagae, F., Perrelet, A., Patel, Y., and Unger, R. H. (1982). Quantitation of endocrine cell content in the pancreas of non-diabetic and diabetic humans. *Diabetes* **31,** 694–700.

Stefan, Y., Meda, P., Neufeld, M., and Orci, L. (1987). Stimulation of insulin secretion reveals heterogeneity of pancreatic B-cells *in vivo. J. Clin. Invest.* **80,** 175–83.

Steiner, D. F., Kemmler, W., Clark, J. L., Oyer, P. E., and Rubenstein, A. H. (1972). The biosynthesis of insulin. In *Handbook of physiology* (ed. S. R. Greiger), Section 7, Vol. 1, pp. 175–98. American Physiological Society, Washington, DC.

Swenne, I. (1982). Regulation of growth of the pancreatic B-cell: an experimental study in the rat. *Acta Univ. Upsal. Abstr. Upps. Diss. Fac. Med.* **414,** 1–33.

Tatemoto, K., Efendic, S., Mutt, V., Makk, G., Feistner, J., and Barchas, J. D. (1986). Pancreastatin, a novel pancreatic peptide that inhibits insulin secretion. *Nature* **324,** 476–8.

Teitelman, G., Joh, T. H., and Reis, D. J. (1981). Transformation of catecholaminergic precursors into glucagon (A) cells in mouse embryonic pancreas. *Proc. Natl Acad. Sci. USA* **78,** 5225–9.

Teitelman, G., Lee, J. K., and Alpert, S. (1987). Expression of cell type-specific markers during pancreatic development in the mouse: implications for pancreatic cell lineages. *Cell Tissue Res.* **250,** 435–9.

Trimble, E. R., Halban, P. A., Wollheim, C. B., and Renold, A. E. (1982). Functional differences between rat islets of ventral and dorsal pancreatic origin. *J. Clin. Invest.* **69,** 405–13.

Van De Winkel, M. and Pipeleers, D. (1983). Autofluorescence-activated cell sorting of pancreatic islet cells: purification of insulin-containing B-cells according to glucose-induced changes in cellular redox state. *Biochem. Biophys. Res. Comm.* **114,** 835–42.

Van Schravendijk, C. F. H., Foriers, A., Hooghe-Peters, E. L., Rogiers, V., De Meyts P., Sodoyez, J.-C., and Pipeleers, D. G. (1985). Pancreatic hormone receptors on islet cells. *Endocrinology* **117,** 841–8.

Van Schravendijk, C., Foriers, A., Van den Brande, J. L., and Pipeleers, D. (1987). Evidence for the presence of type I-insulin-like growth factor receptors on pancreatic islet cells. *Endocrinology* **121,** 1784–8.

Van Schravendijk, C. F. H., Heylen, L., van den Brande, J. L., and Pipeleers, D. G. (1990). Direct effect of insulin and insulin-like growth factor-I on the secretory activity of rat pancreatic beta cells. *Diabetologia* **33,** 649–53.

Waldhäusl, W. K., Gasic, S., Bratusch-Marrain, P., Korn, A., and Nowotny, P. (1982). Feedback inhibition by biosynthetic human insulin of insulin release in healthy human subjects. *Am. J. Phys.* **243,** E476–E482.

Westermark, P., Wernstedt, C., Wilander, E., and Sletten, K. (1986). A novel peptide in the calcitonin gene related peptide family as an amyloid fibril protein in the endocrine pancreas. *Biochem. Biophys. Res. Comm.* **140,** 827–31.

Westermark, P., Wernstedt, C., O'Brien, T. D., Hayden, D. W., and Johnson, K. H. (1987). Islet amyloid in type 2 human diabetes mellitus and adult diabetic cats contains a novel putative polypeptide hormone. *Am. J. Pathol.* **127,** 414–17.

Wiedenmann, B. and Franke, W. W. (1985). Identification and localization of synaptophysin, an integral membrane glycoprotein of Mr 38,000 characteristic of presynatic vesicles. *Cell* **41,** 1017–28.

Wiedenmann, B., Franke, W. W., Kuhn, C., Moll, R., and Gould, V. E. (1986). Synaptophysin: A marker protein for neuroendocrine cells and neoplasms. *Proc. Natl Acad. Sci. USA* **83,** 3500–4.

PART II INSULIN BIOSYNTHESIS

Editors' introduction

The next two chapters consider the mechanism by which insulin is synthesized and packaged for secretion.

Insulin is a small protein with a molecular weight of about 6000. It consists of two polypeptide chains, the A-chain containing 21 amino acids and the B-chain containing 30 amino acids, joined by two interchain disulphide bridges between cysteine residues. The molecule also contains one intrachain disulphide bridge. The insulin gene specifies the structure of a single-chain precursor of insulin, preproinsulin. This protein consists of a 24 amino acid signal peptide, followed by the insulin B-chain sequence, a connecting peptide (C-peptide) containing about 30 amino acids, and, finally, the A-chain sequence.

The human insulin gene is located towards the end of the short arm of chromosome 11. The structure of the gene is highly conserved. It is split up into three expressed sequences (exons) interrupted by two non-coding sequences (introns) which have to be spliced out in the nucleus to produce transcriptionally active preproinsulin mRNA. The first exon contains 5′ untranslated sequences involved in the initiation of transcription; the second exon contains the sequence that encodes the signal peptide, the insulin B-chain, and a fraction of the C-peptide; and the third exon encodes the remainder of the C-peptide, the insulin A-chain, and 3′ untranslated sequences. Expression of the gene is controlled by DNA sequences upstream of the region that encodes preproinsulin mRNA itself. It is likely that both positive and negative regulatory proteins interact with the DNA to control gene expression.

Interest in the insulin gene stems from the possibility that there may be mutations that produce defective insulins. Such mutations could be in control regions, altering the expression of the gene; in coding regions specifying the structure of insulin; or in coding sequences significant for intracellular processing of preproinsulin. Mutations in all these regions have been identified, but they are comparatively rare and individuals with such mutations only account for a few per cent of diabetics.

Mutations of the insulin gene can be detected by treatment of an individual's DNA with a suitable restriction enzyme that cleaves the DNA at a specific site. Different lengths of DNA will be produced when acted upon by such an enzyme. Thus, variations in DNA sequence will lead to restriction fragment length polymorphisms (RFLPs). The association of a given RFLP with the insulin gene is demonstrated by applying a radioactive probe for the insulin gene to gels on which restriction-enzyme-treated DNA has been separated according to size. The insulin probe will only bind to those restriction fragments that encode part of the insulin gene. The pattern of radioactive bands will be characteristic for a given individual's

DNA but will vary between individuals if polymorphism exists. It is also possible to selectively amplify sequences in the insulin gene using a technique called the polymerase chain reaction (PCR). Variation in the size of the amplified sequence provides evidence for polymorphism.

The newly synthesized preproinsulin is processed in the cell during its voyage through the endoplasmic reticulum (ER) and Golgi complex to the secretory granules. Cells contain two main secretory pathways, the constitutive and the regulated. It is believed that the default pathway for secretory proteins is the constitutive route. Insulin is secreted almost entirely via the regulated pathway, to which it must be directed by specific signals. The 'pre'-region of preproinsulin acts as a signal sequence to direct ribosomes bearing the nascent protein to the ER. This process involves the interaction with a cytosolic signal recognition particle (SRP) which causes association of the complex with the SRP receptor on the ER. The signal sequence then interacts with the signal sequence receptor (SSR) on the ER, and a functional pore is formed which permits transfer of the nascent protein into the lumen of the ER. During this transfer, preproinsulin is converted to proinsulin. Proteins that possess a specific amino-acid sequence (KDEL) are retained by the ER; other proteins, such as proinsulin, move from the ER to the *cis* Golgi complex. Passage from the ER and through the Golgi complex occurs within transport vesicles by repeated cycles of budding and fusion. At the *trans* face of the Golgi, insulin is committed to the regulated secretory pathway. Conversion of proinsulin to insulin and C-peptide takes place in the Golgi complex and immature secretory granules; two endopeptidases responsible for this conversion have been identified. The C-peptide is retained within the secretory granule and is co-secreted with insulin on stimulation. As many as 50 other proteins are also found within the secretory granule, underlining the complexity of this organelle.

2 The insulin gene

ANDREW R. CLARK and KEVIN DOCHERTY

Insulin is a member of a superfamily of structurally related peptides which includes the ovarian hormone relaxin, and insulin-like growth factors I and II (IGF-I and IGF-II) (Steiner *et al.* 1985; Bell and Seino 1990). Insulin is also related to two insect hormones (prothoraciotrophic hormone of the silkworm *Bombyx mori* (Nagasawa *et al.* 1986) and LIRP, or locust insulin-related peptide (Lagueux *et al.* 1990)), and a molluscan insulin-related peptide (MIP), which is synthesized by the cerebral light-green cell of the pond snail (*Lymnea stagnatis*) (Smit *et al.* 1988). It seems likely that the various members of the insulin gene family evolved from a common ancestral gene through a process of gene duplication and diversification (Steiner *et al.* 1985). The structure and evolution of the insulin gene have been previously reviewed in detail (Steiner 1990; Steiner *et al.* 1985; Bell and Seino 1990). However, since this subject was reviewed a number of important advances have been made on the regulation of expression of the insulin gene (Selden *et al.* 1987; Walker 1990), and this will form the major topic of this chapter.

2.1 Structure of the insulin gene

Insulin has been studied extensively at the protein and gene level: primary structure data are available on insulins from about 50 different species, while the gene structure and organization is known for around 10 different insulins. The structure of the insulin gene is highly conserved amongst species and consists of three exons and two introns (Bell *et al.* 1980) (Fig. 1). Exon 1 is located in the 5′ untranslated region of the gene. Exon 2 contains sequences encoding the signal peptide, the insulin B-chain and part of the C-peptide, while exon 3 encodes the remainder of the C-peptide, the insulin A-chain, and 3′ untranslated sequences. The lengths and sequences of the introns are highly variable amongst species. However the relative length (intron 1 is always shorter than intron 2) and position (intron 1 is located in the 5′ untranslated region while intron 2 interrupts the gene between the first and second nucleotides of the codon for amino acid 7 of the C-peptide) are highly conserved. The insulin gene is present as a single copy in all species examined with the exception of the rat and mouse, where there are two non-allelic

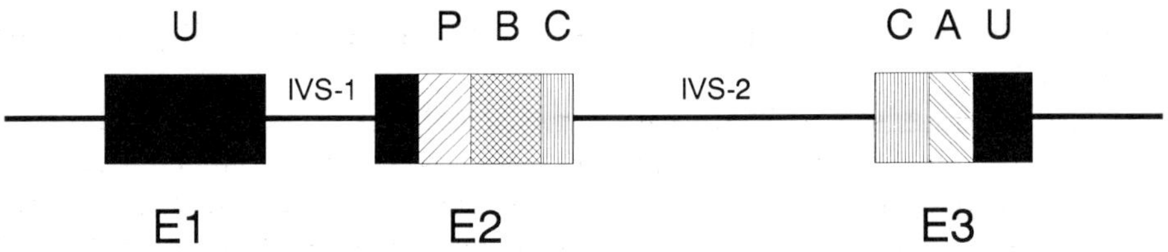

Fig. 1 Structure of the insulin gene. The arrangement of exons E1–E3 and intervening sequences of introns (IVS-1 and IVS-2) is indicated. The human insulin gene is 1430 base pairs long (E1, 42 bp; IVS-1 179 bp; E2, 204 bp; IVS-2, 786 bp; and E3, 219 bp). For other insulin genes, readers are referred to Bell and Seino (1990). The regions of the mRNA and prepoinsulin are noted: U denotes the 5′ and 3′ untranslated region of the mRNA; P, B, C, and A are the pre-peptide, insulin B, chain, C-peptide, and insulin A-chain, respectively

insulin genes (Lomedico *et al.* 1979; Wentworth *et al.* 1986). Intron 2 is absent in the rat and mouse 1 genes. The human insulin gene is located near the end of the short arm of chromosome 11 (Harper *et al.* 1981), flanked on the 5′ side with the gene for tyrosine hydroxylase and on the 3′ side with the gene for IGF-II (Bell *et al.* 1985; O'Malley and Rotwein 1988).

Several abnormal products of the human insulin gene have been described (see Tager 1990); their relevance to the aetiology of diabetes is discussed in Chapter 11. These abnormal insulins include those resulting from point mutations which affect receptor binding potency and which result in a syndrome of mild hyperinsulinaemic diabetes. Thus insulin Chicago has been characterized as having a leucine instead of phenylalanine at position B25 caused by a codon change TTC → TTG; insulin Los Angeles has the change Phe24 → Ser (TTC → TCC); and insulin Wakayama has been identified as Val43 → Leu (GTG → TTG).

Other mutations have been described that affect the processing of proinsulin to insulin and result in hyperproinsulinaemia. For example, proinsulin Tokyo results from a mutation in the codon for residue 65 (CGC → CAC) such that the paired basic amino-acid cleavage site at the C-peptide/A-chain junction changes from Lys64Arg65 to Lys64His65, and this single basic residue is no longer cleaved by the proinsulin converting endoprotease (Shibasaki *et al.* 1985). The amino-acid change in proinsulin Providence is HisB10 → Asp, resulting from a codon change CAC → GAC (Chan *et al.* 1987). The reason why proinsulin Providence is not cleaved (since the mutation is distant from the processing site) is not clear, although it is likely that this mutation may affect the intracellular sorting and processing of the prohormone (see Halban 1990).

In addition to point mutations resulting in abnormal gene products, at least four silent mutations have been identified in the human insulin gene (see Elbein and Permutt 1990). These mutations are of use to geneticists if they occur at a restriction enzyme site and generate a restriction fragment length polymorphism (RFLP). Of these silent point mutations, only one (C to A transversion at position 1628) generates an RFLP; a new *Pst*I site in the 3′ untranslated region of the gene. It is not known whether this mutation has an effect on the production of insulin e.g.

through changes in the efficiency of translation or stability of the preproinsulin mRNA.

In contrast to the 3′ untranslated *Pst*I RFLP, variations in the 5′ untranslated region of the human insulin gene have been extensively studied in diabetes. Varying numbers of a 14- or 15-nucleotide sequence inserted 365 bp upstream of exon 1 of the insulin gene generate an RFLP (Bell *et al.* 1982). When human genomic DNA is digested with restriction enzymes that flank the repetitive sequence on the 5′ side and the insulin gene on the 3′ side, and then probed with an insulin gene probe, restriction fragments are obtained, the length of which depend on the number of repeat units. Population studies have shown that the size of the insulin-linked polymorphic region (ILPR) can vary. The most common alleles contain 40, 95, or 175 repeat units, corresponding to 590, 1320, and 2140 base pairs. These alleles have been referred to as small or class 1, large or class 3, and intermediate sized or class 2 (Bell *et al.* 1984). Racial differences in the frequencies of these alleles have been described, with class 2 alleles restricted to Black populations, and class 3 alleles rare in Japanese populations. In population studies, a weak association of class 3 alleles and non-insulin-dependent diabetes mellitus (NIDDM) in Caucasians has been described (Rotwein *et al.* 1981; Owerbach and Nerup 1982; Bell *et al.* 1984), but no association of this allele with NIDDM was found in either Blacks, Pima Indians, Nauruans, or Japanese (see Elbein and Permutt 1990). However, studies in families have failed to confirm these associations in Caucasians, although other racial groups have yet to be analysed by linkage studies. Associations of the class 1 allele with insulin-dependent diabetes mellitus (IDDM) in Caucasians appeared to be significant (Bell *et al.* 1984), but again linkage studies have failed to confirm these findings (see Hitman and Niven 1989).

An unusual DNA structure has recently been described for the ILPR (see below).

2.2 Regulation of insulin biosynthesis

The insulin gene is transcribed by RNA polymerase II to a pre-mRNA which contains the exon and intron sequences. Within the nucleus of the cell the 5′ end of the RNA is modified by addition of a 5′ methylguanine cap, the RNA is cleaved 10–30 nucleotides beyond an AAUAAA sequence that constitutes a signal for the addition of a poly A tail, and the non-coding introns are excised by an enzymatic mechanism involving small ribonucleoprotein particles (Fig. 2). The mature mRNA is then transferred to the cytoplasm where translation occurs on membrane-bound ribosomes.

There are several stages at which regulation of insulin biosynthesis might occur:

(1) transcription of the gene;

(2) RNA processing and transport into the cytoplasm; and

(3) translation of the preproinsulin mRNA.

The major control point for acute insulin production is at the level of translation; glucose stimulates translation of preproinsulin mRNA over a short time-scale

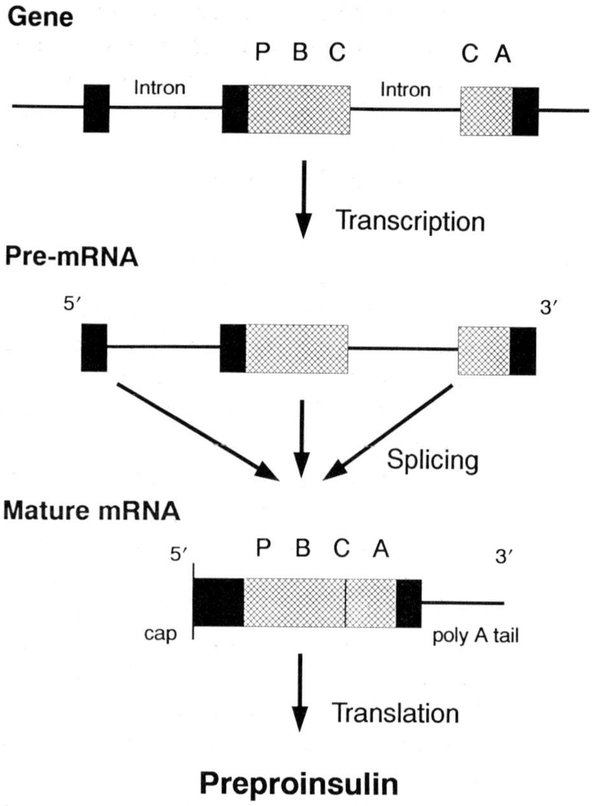

Fig. 2 Schematic demonstrating the stages involved in the production of preproinsulin mRNA. The insulin gene is transcribed to generate a pre-mRNA, which is then processed to the mature mRNA within the nucleus. A 5′ methylguanine cap is added to the 5′ end, the mRNA is polyadenylated, and the introns spliced. The mature mRNA is then transported to the cytoplasm, where translation occurs

(within minutes) (see Chapter 3). This is in keeping with the requirement to replenish insulin stores rapidly, in response to elevated circulating glucose levels following food intake. In addition to the short-term effect on translation, glucose also affects insulin mRNA production and mRNA stability over longer periods (Permutt and Kipnis 1972; Welsh *et al.* 1985).

It is unlikely that nuclear processing of preproinsulin pre-mRNA plays an important role in the control of insulin production. It is worth noting, however, that a defect in the 5′ capping efficiency of rat II insulin mRNA in a rat transplantable tumour is responsible for the approximately tenfold higher levels of rat I than rat II insulin in this tissue (Cordell *et al.* 1982), and an extended poly A tail correlates with glucose responsiveness in growth arrested insulinoma cells (Muschell *et al.* 1986). Also, it is not known whether the 5′ and 3′ untranslated regions of the mRNA have a specific function in controlling insulin production. When intron 1 is spliced from the 5′ untranslated region of the preproinsulin pre-mRNA, an approximately 60 nucleotide long segment is left upstream of the initiation codon (AUG). A comparison between species shows an overall sequence conservation of 70 per cent in this region, which in short stretches amounts to as much as 95 per cent (Bell *et al.* 1980). Preliminary experiments have demonstrated specific binding of proteins to conserved sequences within this region (Knight and Docherty 1991). It will be interesting to determine whether these 5′ and 3′ untranslated regions are involved in the stability or initiation of translation of the mRNA, and whether protein binding is modulated by glucose concentrations.

The third level of control, i.e. transcription of the insulin gene, will form the main topic of this chapter.

2.2.1 Regulation of gene expression in eukaryotes

Before addressing the mechanisms involved in regulating expression of the insulin gene, we will briefly review current knowledge on the regulation of eukaryotic gene expression. Regulation of expression of eukaryotic genes transcribed by RNA polymerase II occurs at the level of initiation; there is little evidence at present that transcription elongation or termination represent major control points. Regulation of transcription is dependent on the interaction of *cis*-acting DNA sequences close to the specific gene with protein factors (Mitchell and Tjian 1989). These sequences have been classified according to their properties as promoters and enhancers, although this distinction is often blurred.

Promoters are located proximal to the transcription initiation site, are typically 100 bp in length, and represent the site of binding and assembly of the initiation complex, which comprises RNA polymerase II and related general transcription factors (i.e. TFIIA, TFIIB, TFIIC, TFIID, TFIIE, TFIIF, and TFIIS; Saltzmann and Weinmann 1989). Most promoters possess a conserved sequence motif, TATAAA (the TATA box), around 20 to 30 nucleotides upstream of the transcription start site (i.e. −20 to −30), which functions primarily to ensure that transcripts are accurately initiated. However, many promoters, particularly those of housekeeping genes, lack TATA boxes and are instead composed of GC-rich elements that are often located within methylation-free islands (Bird 1986). A typical promoter may also contain short (8–12 bp) sequence elements upstream of the TATA box (upstream promoter elements, UPEs), which are thought to increase the rate of transcription. The pivotal event in transcription initiation is the binding of TFIID to a 20 bp sequence close to the transcription start site which normally contains the TATA box (Kadonaga 1990). In promoters that lack TATA boxes, TFIID binding is equally important, although the affinity of TFIID for these promoters is less than that for TATA-box-containing promoters.

Enhancers are composed of discrete sequence elements which act in *cis* with nuclear regulatory proteins to modulate the activity of the promoter (Atchison 1988; Muller *et al*. 1988) in a positive or negative manner (Levine and Manley 1989). Enhancers were first identified in viral genes where they were shown to activate the gene in a position- and orientation-independent manner over distances as great as 6 kb from either the 5′ or 3′ end of the gene. Unlike viral enhancers, which act in a variety of cell types, the activity of enhancers in eukaryotic genes is, for the most part, restricted to particular cell types. Eukaryotic enhancers can function in the temporal and tissue-specific control of gene expression, and in mediating the response to agents that induce or modulate gene expression (e.g. heat shock response elements, heavy metal response elements of metallothionein, and steroid response elements of a number of genes) (Maniatis *et al.* 1987).

Promoters and enhancers are thought to operate by interacting with transcriptional regulatory proteins (Jones *et al.* 1988; Ptashne 1988; Johnson and McKnight 1989; Parker 1989). These proteins can be grouped in structurally related families based on DNA binding/dimerization motifs:

(1) helix-turn-helix;
(2) zinc finger;
(3) leucine zipper; and
(4) helix-loop-helix.

Many of these factors bind DNA as dimers, and different members of a family bind as homodimers or as heterodimers with other members of the same family (Jones 1990). Present ideas on the mode of action of transcriptional regulatory proteins include protein–protein interactions, bringing the regulatory factors into juxtaposition with the transcriptional machinery at the start point. This implies that the DNA or chromatin is flexible enough to allow bending or looping (Ptashne 1986). Even more complex is the observation that some regulatory proteins function by binding to other proteins, called co-activators or adaptors, that in turn bind to the transcription factors (TFIIA, etc.) (Berk and Schmidt 1990; Lewin 1990; Ptashne and Gann 1990).

2.2.2 Methods used in the study of transcriptional regulation

The following steps constitute a rational approach to the elucidation of mechanisms involved in the regulation of transcription of a particular gene:

(1) identify the DNA sequence elements important for the transcriptional activity of the gene;
(2) identify and characterize proteins that bind to these sites;
(3) quantify the amount of these proteins in cells expressing or not expressing the gene in question;
(4) describe the interactions that occur between such proteins;
(5) determine whether these proteins are transcriptionally active alone or in combination with other proteins.

The now classic approach to delineating transcriptional control elements involves joining the putative controlling region to a reporter gene (e.g. chloramphenicol acetyl transferase (CAT) or growth hormone) and introducing this hybrid construction into cells in culture. Approximately 72 h after transfection the cells are harvested and the reporter gene product assayed in cell extracts. The level of reporter gene activity is proportional to the strength of the controlling region cloned upstream.

This transient expression system can be used to define tissue-specific and inducible controlling elements. Thus, for example, a DNA fragment containing a strong viral enhancer will generate high reporter activity in a wide range of cells, while a fragment containing a tissue-specific insulin enhancer will generate high reporter gene activity in β-cell lines but not in non-insulin-expressing cell lines. Most of the work on the insulin enhancer/promoter has been performed on a β-cell line, HIT M 2.2.2 (Edlund *et al.* 1985), which was subcloned from the SV40 transformed hamster cell line HIT T15 (Santerre *et al.* 1981); more recently, βTC cells, which were derived from transgenic mice bearing an insulin genes/SV40 large T antigen construct (Efrat *et al.* 1988), have been used. Other β-cell lines, such as the rat insulinoma cell line RIN m5F (Gazdar *et al.* 1980) and the HIT T15 line, have proved difficult to transfect transiently with foreign DNA.

The mapping of protein binding sites within a fragment of DNA involves three main techniques: DNase footprinting, gel mobility shift, and methylation interference (see Boam *et al.* 1990*b*). In the former two methods, DNA containing the sequence of interest is radiolabelled and incubated with nuclear protein extracts; the binding of protein is monitored by its ability to protect the DNA from nuclease digestion or to alter the mobility of the DNA on polyacrylamide gels. In methylation interference analysis, the radiolabelled DNA is partially methylated before reacting with the protein extract, such that binding will only occur to that population of DNA molecules where the protein-binding site is unmethylated. Protein-bound and free DNA are separated on a polyacrylamide gel, and after treatment with piperidine, which will cleave at methylated Gs and As, the radiolabelled DNA fragments are analysed on a sequencing gel. A gap corresponding to the protein-binding site, where the DNA is uncleaved, can be observed on the protein-bound sample.

2.2.3 Regulation of insulin gene expression

Most of the work on the regulation of the insulin gene has been performed on the rat I and rat II insulin genes, and on the human insulin gene. As mentioned previously, both rat and mouse possess two insulin genes, gene I having arisen from the ancestral gene II by an RNA-mediated duplication–transposition event. The RNA transcript involved was initiated from a site upstream of the normal physiological promoter of the insulin II gene; thus the insulin I gene is homologous to the insulin II gene within about 500 bp upstream of the transcription start site (Soares *et al.* 1985). Duplication of the gene presumably altered the selective pressures operating upon transcriptional regulation, with the result that the rat and mouse genes may not be entirely typical of mammalian insulin genes in this respect. The rat and mouse insulin I and II genes are expressed equally in normal islets (Clark and Steiner 1969; Giddings and Carnaghi 1988; Koranyi *et al.* 1989), though not in fetal yolk sac (Giddings and Carnaghi 1989), nor in some rat insulinoma-derived cells (Fiedorek *et al.* 1990). Since the rat insulin I gene has been studied in greatest detail, we shall begin our survey there, before examining the rat insulin II and human insulin genes.

Rat insulin I gene

Both promoter and enhancer of the rat insulin I gene behave as cell-specific regulatory elements (Edlund *et al.* 1985). When a strong viral enhancer is placed upstream of the insulin promoter, the rate of transcription of this construct is 10 times greater in insulin-secreting than in non-secreting cell lines. Since the viral enhancer displays no cell-specificity, the insulin promoter must be β-cell specific. Likewise, an enhancer-containing DNA fragment, when placed upstream of the herpes simplex virus thymidine kinase promoter, will increase the transcriptional activity of this promoter only in insulin-secreting cell lines.

The enhancer was delineated by means of deletion analysis of a chimeric construct in which expression of the reporter gene chloramphenicol acetyl transferase (CAT) was directed by the 5′ flanking region of the rat insulin I gene. Deletion of sequences upstream of −302 had little effect on CAT activity, whereas deletion of sequences downstream of −302 led to dramatic losses of CAT activity (Walker *et al.* 1983). This is in agreement with the observation that a fragment containing sequences −302 to −103 is able to act as a powerful tissue-specific enhancer (Edlund *et al.* 1985).

Fine mapping of the enhancer was achieved by systematic mutagenesis of the 5′ region (Karlsson *et al.* 1987). No single sequence was absolutely required for transcription, but some mutations showed moderate effects upon reporter gene activity, and mutations within three discrete regions showed drastic effects: −23 to −32 (which spans the TATA box), −104 to −112, and −233 to −241. The latter two regions contained an identical 8 bp sequence, GCCATCTG, at −104 to −112 and −230 to −238. Mutation of either motif alone reduced transcription by roughly 85 per cent, whereas mutation of both essentially abolished transcription. These sequence motifs have been designated, amongst other things, IEB1 or NIR (−104 to −112) and IEB2 or FAR (−230 to −238). Here we will adopt the nomenclature IEB1 and IEB2 for these motifs and their counterparts in other insulin genes (Fig. 3).

When cloned upstream of the thymidine kinase promoter, IEB1 was shown to exhibit 10 per cent of the β-cell-specific enhancer activity of a full-length enhancer fragment (−103 to −346), while IEB2 exhibited 16 per cent of enhancer activity, and in combination IEB1 and IEB2 generated 40 per cent of enhancer activity. Thus it appears that the IEB motifs are critical elements of the rat insulin I enhancer, acting synergistically and in a tissue-specific manner, with other regulatory elements being involved to a lesser extent (Karlsson *et al.* 1989).

Mapping of protein-binding sites by DNAse footprinting identified five binding sites, designated E1 to E5, in the rat insulin I gene 5′ region (Fig. 3). These included IEB1 (footprint E4), IEB2 (footprint E5), and footprints at: −285 to −332 (E1), −201 to −222 (E2), and −153 to −174 (E3) (Ohlsson and Edlund 1986; Ohlsson *et al.* 1988). Under the conditions used, footprints E1, E4, and E5 appeared to be β-cell-specific. E1 is an extensive footprint which remains to be characterized in detail. Although it contains an element similar to the enhancer core TGTGGAAAG (Khoury and Gruss 1983) it is the sequence adjacent to this that is most sensitive to

```
                                          |---------------- E1 ----------------|
Rat I    CAAAGATACCAGGTCCCC-AACAACTGCAACTTTCTGGGAAATGAGGTGGAAAA-TGCTCAGCCAAGGAAAAAGAGGGCCTTACCCTCTCTGGGACAAT-GATT
Rat II   CAAAGTTACTAGGTCCCCCAACAACTGCAGCCTCCTGGGGAATGATGTGGAAAAATGCTCAGCCAAGGACAAAGAAGGCCTCACCCTCTCTGAGACAAT-GTCC
Human    CAAAGA-GCCCCGCCCTGCAGC--CTCCAGCTCTCCTGGTC-TAATGTGGAAAGTGGCCCAGGT---------GAGGGCTTTGCTCTCCTGGAGACATTTGCCC
              Sp 1                                 core                                              NRE

            |------------ E5 ------------| |---- E2 ----|
Rat I    G-TGCTGTGAACTGCTTCA--TCAGGCCATCTGGC--CCCTTGTTAATAATCTAATTACCC--TAGGTCTAAGTA-GAGTTGTTGACGTCCAATGAGCGC
Rat II   CCTGCTGTGAACTGGTTCA--TCAGGCCACCCAGGAGCCCCTCTTAAGACTCTAATTACCC--TAAGGCTAAGTA-GAGGTGTTGTTGTCCAATGAGCAC
Human    CCAGCTGTGAGCAGGGACAGGTCTGGCCACCGGGC--CCCTGGTTAAGACTCTAATGACCCGCTGGTCCTGAGGAAGAGGTGCTGACGACCAAGGAGATC
                                IEB 2                CT II                                      CRE

|--- E3 ---|                                 |-------- E4 --------|
Rat I    TTTCTGCAGACTTAGCACTAGGCAAGTG-TTTGGAAATTACAGCTTCAGCCCCTCTCGCCATCTGCCTACCTACCCCTCCTAGAGCCCTTAATGGGCCAAA
Rat II   TTTCTGCAGACCTAGCACCAGGCAAGTG-TTTGGAAACTGCAGCTTCAGCCCCTCTGGCCATCTGCTGATCCAC-----------CCTTAATGGGACAAA
Human    TTCCCACAGACCCAGCACCAGGGAAATGGTCCGGAAATTGCAGCCTCAGCCCC--CAGCCATCTGCCGACCCCCCCACCCCAGG--CCCTAATGGGCCAGG
                             GG II    GG I                   IEB1                             CT I

Rat I    CGGCAAAGTCCAGGGGGCAGAGAGGAGGTGCTTTG-GACTATAAAGCTAGTGGAGACCCAGTAACTCCC
Rat II   CAGCAAAGTCCAGGGGTCAGGGGGGGGGTGCTTTG-GACTATAAAGCTAGTGGGATTCAGTAACCCCC
Human    CGGCAGGGGTTGACAGGTAGGGGAGATGGGCTCTGAGACTATAAAGCCAGCGGGGGCCCAGCAGCCCTC
                        COUP                   TATA                      -1
```

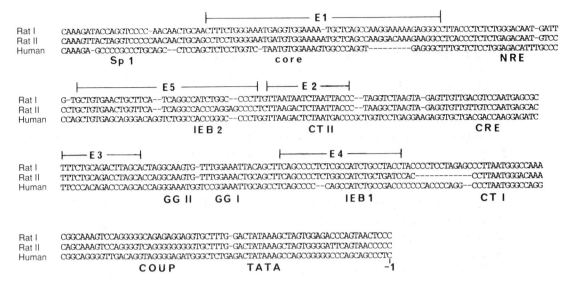

Fig. 3 Comparison of the 5′ regions of the rat and human insulin genes. The nucleotide immediately upstream of the transcription start site is marked −1. Regions protected from DNase digestion in the rat insulin I gene are marked above the sequences. Known or putative protein-binding sites, as described in the text, are marked below the sequences. Note that the sites indicated are not present in all genes

mutation in transcriptional assays (Karlsson *et al.* 1987). Similarly, the results of systematic mutagenesis imply a transcriptional role for footprint E3, which also remains to be characterized. The E2 region contains the sequence TCTAAT (CT II, see human gene); it demonstrated no activity when placed on its own upstream of the thymidine kinase promoter, but would potentiate the activity of IEB2 in this context (Karlsson *et al.* 1989).

A further regulatory element was identified by testing the ability of chimeric insulin/CAT constructs to respond to cAMP, which had previously been shown to elevate insulin mRNA levels to a modest extent in β-cell lines (Welsh *et al.* 1985; Hammonds *et al.* 1987). Chimeric constructs containing 410 bp of insulin 5′ material upstream of the CAT gene showed a rapid, roughly fourfold increase in transcription in response to cAMP, and deletion or mutagenesis of the sequence TGACGTCC (−184 to −177) compromised this responsiveness. The sequence element implicated is similar to the cAMP responsive elements (CREs) of other genes, and binds a nuclear factor similar in size to a previously identified CRE binding factor (Philippe and Missotten 1990). Partially overlapping with the putative CRE is the sequence CCAAT, a potential binding site for CTF/NF-1 or a related transcription factor. It is suggested that such a factor may be involved in the cAMP response, perhaps by interaction with a CRE binding factor.

Although no DNAse protection corresponding to the cAMP response element has been demonstrated, it is broadly true that identified protein-binding sites correspond to functionally identified regulatory regions, some of which remain to be investigated in detail. Two elements of critical importance in regulation of rat insulin I gene expression are the IEB boxes, with other sites playing a lesser role.

Rat insulin II gene

Although the rat insulin II gene is 80 per cent identical to the insulin I gene within the region −370 to +1, its regulation clearly differs in a number of respects. First, in the absence of enhancer sequences the promoter is quite strongly active in both insulin-secreting and non-secreting cell lines (Whelan *et al.* 1989). A β-cell-specific protein binds within the promoter region (Sample and Steiner 1987), and a ubiquitous protein related to the chicken ovalbumin upstream promoter (COUP) transcription factor binds to the promoter at position −54 to −45 (Hwung *et al.* 1988). The role of these proteins, and the nature of the differences between the rat insulin promoters remains unclear.

Mutagenesis of the rat insulin II gene 5′ region reveals up to eight distinct regions in which mutation lowers the rate of transcription by twofold or more (Crowe and Tsai 1989). Again the IEB1 motif is of critical importance, mutation causing a 25-fold decrease in transcription of the linked reporter gene. There is no exact homologue of the IEB2 motif (see Table 1), and deletion through the corresponding region has no significant effect upon transcription. One linker-scanning mutation that disrupts this region causes a twofold decrease in transcription; however two other linker-scanning mutations within the same area show no significant effect. Thus it remains unclear whether this region is transcriptionally active; certainly it does not play the same critical role as the IEB2 box of the insulin I gene and if a transcription factor binds here it may not be the same as the factor that binds to IEB1.

One other element, located between −110 and −129, is highly sensitive to mutation and contains the sequence GGAAA (GGI, see below). This element appears to act co-operatively with the IEB1 box, since a restriction fragment con-

Table 1 Comparison of 'E boxes' of a variety of genes

Generalized E box	CANNTG
IEB1 of mammalian insulin genes	GCCATCTGC
IEB2 of rat insulin I gene	GCCATCTGG
IEB2 homology in rat insulin II gene	GCCACCCAG
IEB2 homology in human insulin gene	GCCACCGGG
μE1 of immunoglobulin μ gene	GCCATCTTG
μE2 of immunoglobulin μ gene	GCCAGCTGC
μE3 of immunoglobulin μ gene	GCCACATGA
κE2 of immunoglobulin κ gene	GGCAGGTGG
Elastase gene	CACAGGTGA
α-Amylase gene	TTCAGCTGT
Mouse muscle creatine kinase (i)	AACACCTGC
(ii)	GACATGTGG

taining both sites is able to stimulate transcription from a heterologous promoter in a β-cell-specific manner (Hwung *et al.* 1990).

Other elements that are sensitive to mutation include: regions of homology to the gene I footprints E1 and E3; a region of homology to the gene I cAMP response element; the COUP factor binding site; a region containing the sequence CTTAAT (CT I site, see human gene), and a further uncharacterized site around −185 (Fig. 3).

Human insulin gene

A preliminary investigation of the human insulin gene indicated that enhancer sequences were located downstream of −258 (Walker *et al.* 1983). More recent deletion analysis reveals the presence of transcriptionally active regions upstream of this position (Boam *et al.* 1989). A negative regulatory element is centred at −270, and will be described in a later section. Immediately upstream of this we have evidence for a positively acting region roughly corresponding to the DNAse I footprint E1. This region antagonizes the adjacent negative regulatory element (NRE), and is active in both β-like and non-β-like cells. Several factors are able to bind within the positively acting region, one of which appears to be a factor that recognizes the CT sites (see below).

Results of both our current transient expression studies and transgenic mice studies (Fromont-Racine *et al.* 1990) suggest that sequences located even further upstream of the transcription start site may also be involved in gene regulation. Putative regulatory sites remain to be identified and characterized, but may include a sequence at about −340 which resembles a binding site for the transcription factor Sp1.

Deletion of the human IEB2 region has a drastic effect upon transcription, suggesting a strong positive role in regulation of gene activity (Boam *et al.* 1990*a*). However, the homology to the rat I IEB2 is incomplete (six residues out of eight), and the two differing residues appear to be important for IEB activity (Whelan *et al.* 1990; and see Table 1). Nevertheless, the human IEB1 and IEB2 regions are able to cross-compete for the binding of nuclear factors, suggesting that at least one protein is able to recognize both sites (Docherty *et al.*, unpublished results). Whether the human IEB1 and IEB2 boxes regulate transcription in the same manner remains unclear at present.

Further deletion to −201 causes another drop in transcriptional activity. This region contains the sequence TCTAAT (CT II), strongly conserved in both rat genes, and corresponding to footprint E2. A related motif CCTAAT (CT I) is present at −83 to −78, and is also quite strongly conserved in the rat genes. The same nuclear factor binds to both CT boxes and to a related motif within the E1 region of the human insulin gene (see Fig. 3): this factor appears to be specific to insulin secreting cell lines (Boam and Docherty 1989; Boam *et al.* 1990*a*).

Other putative regulatory sites, as indicated by protein-binding studies, are as follows:

1. −183 to −179, which contains a homology to the cAMP response element. Nuclear proteins binding to this region are detected in all cell lines tested, as might be expected for an important regulatory factor such as CREB.

2. −153 to −127, which contains two copies of the sequence GGAAAT (GG II and GG I). The DNAse protection pattern within this region clearly differs between extracts from insulin-secreting and non-secreting cell lines.

The exact role of these sequences in regulation of transcription of the human gene must await the outcome of mutagenesis and expression studies which are currently in progress in our laboratory.

Summary

The following conclusions can be made concerning regulatory sequences within the insulin gene:

1. Although the mammalian insulin genes are quite strongly conserved within the 5′ region (Fig. 3) there are clear differences in transcriptional regulation. At least some of the identified regulatory sites are common to all three genes studied, but the utilization of these sites appears to differ between genes.
2. The IEB1 motif is the most strongly conserved sequence within mammalian insulin gene 5′ regions, and is probably of critical importance in all of these genes. The IEB2 motif is not strongly conserved in rat II and human genes, and if the corresponding region does play a role in transcriptional regulation of these genes, it may not be identical to that of the rat I IEB2 motif.
3. The CT motifs are of interest because they apparently bind a factor found only in cells of pancreatic origin, are quite strongly conserved, and contain the sequence TAAT, which is part of the binding site for homeodomain-containing proteins (see later). Although inactive on their own, CT boxes may potentiate other enhancer elements.
4. A cyclic AMP responsive element exists within the 5′ region of the rat I gene, and probably also other insulin genes.
5. Other regions probably involved in transcriptional regulation, but not yet fully characterized, include the E1 and E3 footprints, and the GG motifs.
6. There is some evidence that sequences located up to 1 kb or more upstream of the transcription start site may be involved in transcriptional regulation, at least in the human gene. As yet there is no evidence for involvement of intronic or 3′ sequences.
7. Tissue specificity of insulin gene expression may depend upon the activity of several regulatory elements, only a few (possibly even none) of which bind β-cell-specific transcription factors. Tissue-specific expression may require a β-cell-specific combination of transcription factors, rather than a single β-cell-specific factor. This remains an open question which will be further addressed in a later section.

2.2.4 Negative regulation of insulin genes

The insulin enhancer is able to bind positively acting transcription factors which are present in a variety of cell types (see below), yet is active only in insulin-secreting cell lines. This may be due in part to negative regulation of enhancer activity, as demonstrated by the following experiment (Nir *et al.* 1986). A non-insulin-secreting cell line was transfected with a plasmid containing the reporter gene CAT under the control of the insulin enhancer, in the absence or presence of a second plasmid containing a copy of the enhancer. The presence of the second (competitor) enhancer allowed the first enhancer to become active, presumably by titrating out factors which bind to the enhancer and repress enhancer activity.

A 5′ deletion analysis of the rat insulin II gene indicates that the IEB1 box is a site of negative regulation of transcription in non-β-cells (Whelan *et al.* 1989). However, an internal deletion, which disrupts IEB1 while leaving the remainder of the enhancer intact, does not cause the enhancer to become active in non-β-cells. Therefore, there must be other sites of negative regulation. The adenovirus 5 E1a gene products transcriptionally repress several enhancers, including those of the insulin genes. A site of action of has been mapped to the IEB1 box of the rat insulin II gene (Stein and Whelan 1989), leading to the suggestion that negative regulation of insulin gene expression in non-β-cells may be mechanistically similar to repression by E1A proteins (see Fig. 4). Repression by the E1A proteins is thought not to involve disruption of DNA binding by transcription factors; rather the ability of bound factors to stimulate transcription is affected, as shown schematically in panel (d) of Fig. 4.

Another negative regulatory element is located within a long interspersed repetitive sequence, roughly three kilobases upstream of the rat insulin I gene (Laimins *et al.* 1986). This possesses some of the characteristics of an enhancer, in that its activity is independent of position and orientation. Since it appears to be active in both β- and non-β-cells, its function remains unclear.

As mentioned earlier, 5′ deletion analysis indicates the presence of a powerful negative regulatory element centred at 270 base pairs upstream of the human insulin gene transcription start site (Boam *et al.* 1990*a*). This element is active in both β- and non-β-cell lines; however, its deletion does not allow the enhancer to become active in non-β-cells, and thus it does not appear to play a pivotal role in restricting insulin gene expression to the β-cell. Like the negative regulatory element described by Laimins, it possesses the properties of being at least partially independent of position and orientation. The pattern of binding of nuclear factors within the region of this negative element is highly complex, with three distinct and overlapping protein-binding sites detected. One of the binding sites contains the sequence GAGACAT, which is also present in the corresponding position of the rat insulin II gene. Mutagenesis of this sequence in the rat insulin II gene increases the rate of transcription by more than twofold (Crowe and Tsai 1989), suggesting that both of these genes are negatively regulated by a similar mechanism.

Finally, the polymorphic region of the human insulin gene has been implicated as a target for negative regulation in non-pancreatic cells, where it appears to function as an orientation-independent silencer (Takeda *et al.* 1989). Since this activity was not assayed in insulin-secreting cells, it remains unclear whether negative regulation by the polymorphic region is released in β-cells.

Negative regulation probably plays a role in restricting expression of the insulin

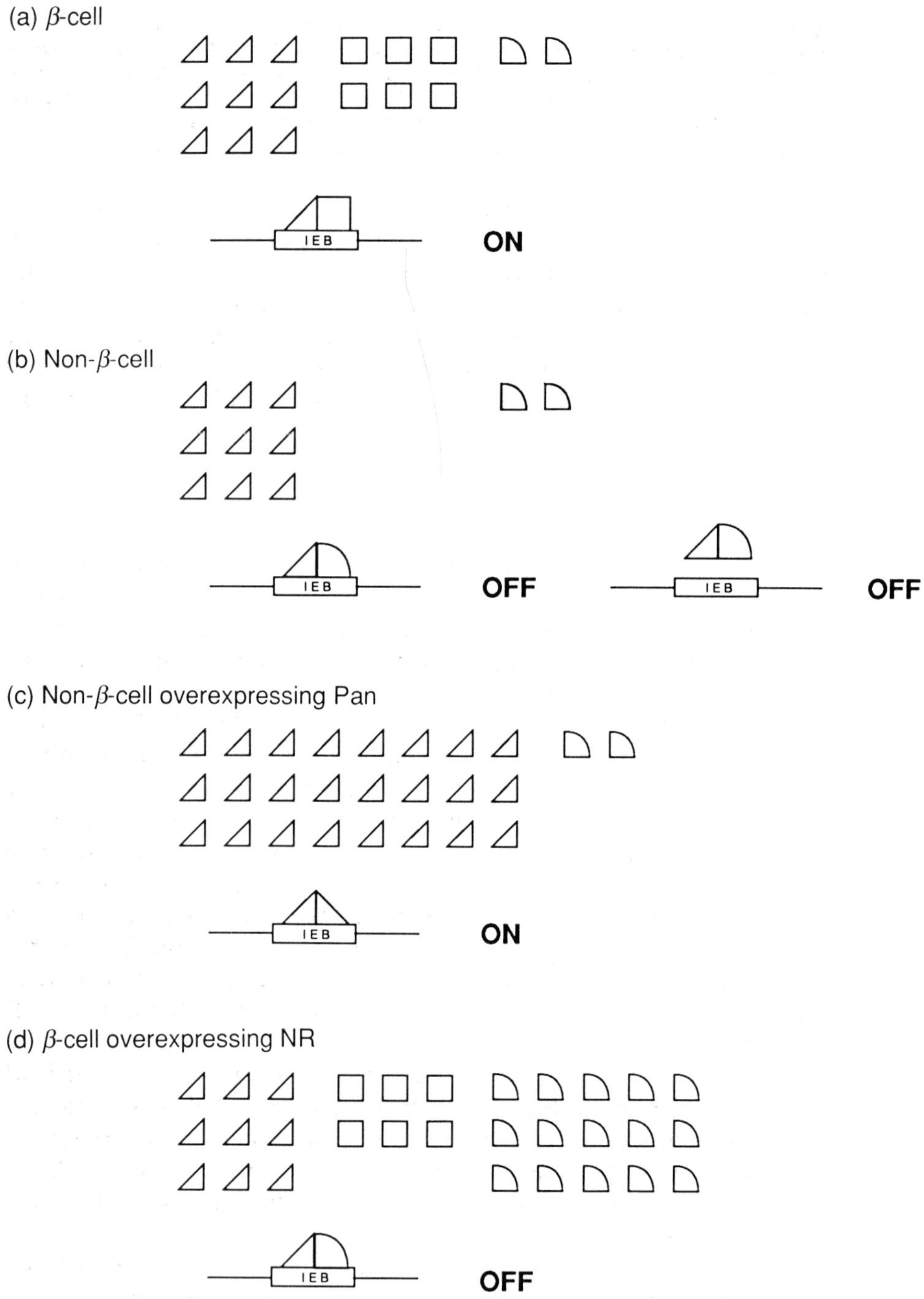
(a) β-cell
IEB
ON
(b) Non-β-cell
IEB
OFF
IEB
OFF
(c) Non-β-cell overexpressing Pan
IEB
ON
(d) β-cell overexpressing NR
IEB
OFF

Fig. 4 A simplified model of positive and negative regulation of transcription by the IEB site. The insulin gene is (a) transcribed in β-cells; (b) switched off in non-β-cells; (c) transcribed in non-β-cells overexpressing Pan (see Nelson *et al.* 1990); and (d) switched off in β-cells overexpressing a negative regulatory protein, e.g. E1A (see Whelan *et al.* 1990). △, Pan (A1, E47, etc); □, βEP (β-specific HLH protein); ◺, NR (negative regulatory protein; might be related to E1A or to Id).

gene to the β-cell. Its function may be to prevent low level, ectopic expression of the gene due to the presence of ubiquitous positively acting transcription factors which are able to bind to the insulin enhancer. For some factors such a function cannot be postulated, since the negative regulatory activity is present in both insulin-secreting and non-secreting cell lines. In such instances, negative regulation may be involved in the modulation of gene expression in response to external signals, or in the programme of developmental switching of pancreatic genes on or off.

2.2.5 Insulin gene regulatory proteins

Binding of proteins to putative regulatory sequences has been extensively investigated by means of the gel mobility shift assay. Analysis of the results is complicated by the fact that virtually any oligonucleotide probe from an insulin gene 5′ region will specifically bind to nuclear protein factors, while it seems unlikely that all of the binding complexes observed are of functional significance (Moss *et al.* 1988). Furthermore, the number of specific protein–DNA complexes detected using a given oligonucleotide probe depends upon the salt conditions used in the binding assay (Whelan *et al.* 1990). Thus a single β-cell-specific protein was shown to bind to IEB boxes at salt concentrations above 100 mM (Ohlsson *et al.* 1988), whereas at lower salt concentrations (50–75 mM) up to four diferent complexes could be detected, at least one of which was not β-cell-specific (Moss *et al.* 1988; Boam *et al.* 1990*a*; Whelan *et al.* 1990). It has been proposed that the low mobility complex detected under high salt conditions is responsible for transcriptional activation, since this complex is β-cell-specific, and mutations that interfere with complex formation also reduce the activity of the IEB site in functional assays (Whelan *et al.* 1990).

Moss *et al.* (1988) noted that the IEB motif was similar to the μE1, μE2, and μE3 motifs of the immunoglobulin enhancer (Ephrussi *et al.* 1985; and see Table 1). By means of competition experiments, they were able to demonstrate that factors recognizing the immunoglobulin regulatory sequences were also able to bind to the IEB boxes, although only the μE2 sequence was able to fully substitute for an IEB box in functional assays.

The IEB and μE boxes belong to a class of regulatory sites known as E boxes (Kingston 1989; and see Table 1). These possess a consensus core, CANNTG, and have been implicated in the regulation of tissue-specific expression of muscle and exocrine pancreas (Cockell *et al.* 1989; Rosenthal 1989) as well as immunoglobulin and insulin genes. Several cDNAs have been isolated for protein factors that recognize E boxes. The factors are homologous to one another and to the *myc* family of oncogenes, and are known as HLH (helix-loop-helix) proteins. They possess a domain capable of forming two amphipathic helices separated by a loop structure, and there is usually an adjacent basic domain. The helix-loop-helix domain is involved in dimerization between HLH proteins, while the basic domain mediates DNA binding. Murre *et al.* (1989*b*) propose that HLH proteins may be divided into classes A (ubiquitous, for example E12 and E47) and B (tissue-specific, for example the protein MyoD which is involved in muscle-specific gene expression). A heterodimer of a class A protein and an appropriate class B protein will bind with high affinity to the cognate E box; for instance a mixture of E12 and MyoD binds strongly to the CAGGTG motif of the muscle creatine kinase gene. Whereas class A proteins may bind as homodimers, class B proteins appear to bind only as heterodimers with class A proteins (Chakraborty *et al.* 1991). A further class of HLH proteins is exemplified by Id (Benezra *et al.* 1990) and extramacrochaetae (Ellis *et al.* 1990). These proteins have HLH domains but lack the basic domains required for DNA binding. They are able to heterodimerize with other HLH proteins and antagonize the function of HLH transcription factors, apparently by interfering with DNA binding.

By screening expression libraries with radiolabelled probes containing tandem repeats of the IEB motif, cDNAs for factors that recognize this motif have been obtained (Shibasaki *et al.* 1990; Walker *et al.* 1990). Allowing for species differences, the factors identified appear to be identical or very closely related, and to correspond to a ubiquitous HLH protein identified using other E boxes as probes (Murre *et al.* 1989*a*,*b*; Nelson *et al.* 1990). Indeed, it appears that a messenger RNA species encoding this factor (E12/47, Pan1/2, A1, IEBP1) is widely distributed. The factor is able to bind to a variety of E boxes, and when overexpressed in HeLa cells will activate a promoter containing multimers of the chymotrypsin E box CACCTG (Nelson *et al.* 1990).

Further evidence for a role of HLH proteins in insulin gene regulation comes from the recent observation (Cordle *et al.* 1991) that expression of Id in β-like cell lines blocks transcriptional activation through the IEB sites. Also, it has been observed that extinction of insulin gene expression in hybrids between β-like cells and fibroblasts is accompanied by decreased binding at the IEB motifs (Leshkowitz and Walker 1991). A possible explanation for the latter observation is that Id or a related factor present in fibroblasts interacts with HLH proteins present in the β-like cells to prevent their DNA binding. From the above discussion it seems reasonable to speculate that a β-cell-specific (i.e. class B) HLH protein remains to be identified, and might play a pivotal role in specification of the β-cell lineage, just as MyoD is critical in muscle cell development. This putative factor (βEP) may possess low DNA affinity in the absence of a class A protein and may therefore prove difficult to isolate by the method of screening expression libraries with the recognition site, as described earlier. Since β-cell-specific DNA binding can be observed using the IEB motif, an alternative strategy might be to attempt partial protein purification by DNA affinity chromatography. Purified protein could be sequenced and, on the basis of this information, appropriate oligonucleotides synthesized and used to screen a cDNA library.

Some theoretical features of transcriptional regulation through the IEB boxes are illustrated in simplified schematic form in Fig. 4. Figure 4a indicates the normal situation in β-cells, in which an interaction of ubiquitous and β-cell-specific HLH proteins leads to activation of transcription through the IEB boxes. Negative regulatory factors may be absent or present. Figure 4b indicates that βEP is absent from non-β-cells, and that negative regulatory factors (related to E1A or to Id?) may also be involved in the inactivation of these sites. Figure 4c indicates a possible mechanism of transcriptional activation in cells overexpressing Pan; this model depends upon the ability of class A HLH proteins to bind to E boxes as homodimers (albeit weakly). Finally, Fig. 4d illustrates that the IEB boxes may be silenced in β-cells by means of overexpressing E1A or Id (Stein and Whelan 1989; Cordle *et al.* 1991).

By screening a library from the insulin-secreting RIN cell line, Karlsson *et al.* (1990) identified a cDNA for a protein Isl-1 which binds to the CT II motif of the rat insulin I gene. The protein has a relative molecular mass of 38 000, and binds within the sequence TTAATAATCTAATTA, which contains three copies of the motif TAAT, known to be important in the DNA binding of homeodomain-containing proteins. Although the equivalent rat insulin II sequence possesses only one TAAT motif, it will compete for binding of Isl-1. Examination of the sequence of Isl-1 reveals three domains of interest:

(1) a Cys–His domain which potentially forms zinc finger-like structures, of the type found in several transcription factors (Evans 1988);
(2) a homeodomain bearing striking resemblance to those of the *C. elegans* mec-3 and lin-11 gene products;
(3) an acid-rich domain, again reminiscent of several transcription factors.

Homeodomain-containing proteins are commonly found to play a role in developmental regulation of gene expression (Levine and Hoey 1988), and Karlsson *et al.* (1990) argue strongly that the Isl-1 factor is involved in the development of islet cells.

Three DNA–protein complexes can be detected using a CT site as a probe in gel retardation assays (Boam and Docherty 1989). The protein factors responsible for these three complexes can be separated by means of DEAE chromatography (Scott *et al.* 1991). We find that the CT binding factors can be shown by South-Western blotting to have relative molecular masses of approximately 115 000 and 46 000. The identity and function of these factors, designated IUF1, remains to be clarified.

2.2.6 Chromatin structure and insulin gene expression

The structure of DNA within chromosomes is also important in regulating the expression of genes. It has been known for several years that actively transcribed genes are in a more open chromatin structure, and are hence more sensitive to DNAse digestion than non-transcribed genes (Weintraub *et al.* 1981). In addition, active promoters coincide with nuclease hypersensitive sites (Gross and Garrard 1988; Patient and Allan 1989). In the insulin gene, regions of DNA 200–300 bp upstream of the start site have been shown to be preferentially sensitive to DNAse, in keeping with the role of these sequences in interacting with regulatory proteins (Wu and Gilbert 1981). The substantial unfolding of DNA which is associated with DNAse hypersensitivity can arise from modification of nucleosomal histones (e.g. histone acetylation), absence of histone H1, presence of non-histone proteins (e.g. high-mobility group proteins), and undermethylation of DNA sequences (see Latchman 1990; Tazi and Bird 1990). It is not clear whether these features occur together or can function separately, but in any case, for insulin expression, only the methylation state of the DNA has been studied. In rat insulinomas and cell lines that make fivefold higher levels of insulin I mRNA than insulin II mRNA, it was found that the general level of methylation of the genes does not correlate with their differential expression (Cate *et al.* 1983). Therefore, at least in the rat, there is no specific control by methylation of the insulin genes, although further studies on other insulin genes may be warranted.

Chromosomes are also endowed with a higher order structure, the molecular determinants of which are not fullly understood. There is some evidence for the existence of a proteinaceous network spanning the interior of the nucleus (nuclear scaffold) that serves to organize the chromosomes into a non-random configuration (Gasser and Laemmli 1987; Cook 1989). One implication of this scaffold is that it can attach chromosomal domains, and it has been claimed that such attachments increase the transcriptional activity and DNAse hypersensitivity of the DNA (see Fisher 1989; Patient and Allan 1989). An explanation for this effect is that attachment induces torsional stress (supercoiling) in the DNA, perhaps involving the participation of a DNA gyrase-like enzyme, although no such activity has been identified unequivocally in eukaryotes (see Patient and Allan 1989). The existence of supercoils in eukaryotic chromatin is highly controversial, but it is well documented in bacteria, where it has profound effects upon gene expression. It is worth noting, therefore, that an unusual structure, which exhibits DNAse hypersensitivity and is dependent upon supercoiling of the DNA, has been identified within the insulin-linked polymorphic region (ILPR) (see above) close to the human insulin gene promoter (Hammond-Kosack *et al.* 1991).

The first evidence that there was an unusual structure in the ILPR was obtained using a small chemical probe, bromoacetaldehyde (BAA), which reacts with single-stranded or unpaired nucleotides in otherwise double-stranded DNA. When a cloned human insulin gene was treated with BAA, and Southern blots probed with labelled DNA fragments corresponding to various regions of the insulin gene, it was observed that BAA reacted at multiple sites within the ILPR. An understanding of the structural basis of this BAA hypersensitivity came first from a close examination of the sequence within the ILPR. The ILPR is composed of tandem repeats of the consensus sequence ACAGGGGT(G/C)(T/C)GGGG (Bell *et al.*

1982). What is of particular interest is that the sequence within the ILPR contains four G residues (G4) separated by three nucleotides, e.g. ACA GGGG TGT GGGG. It next became clear that the two strands within the ILPR behaved very differently. Primer extension sequencing showed that *Taq* polymerase could copy the bottom (C-rich) strand but not the top strand. In fact the polymerase stopped prematurely in front of the G4 stretches on the top strand. This implied a secondary structure within the top strand involving the G4 stretches. In support of this, chemical sequencing of the two strands showed protection of the G4 stretches from dimethyl sulphate during the G-specific chemical reactions. Finally, electron microscopic homoduplex analysis demonstrated directly that one of the two strands formed a compact secondary structure in the ILPR (Hammond-Kosack *et al.* 1992).

Although at this stage we cannot entirely rule out a triplex or double hairpin structure, the most likely interpretation of these data is that the ILPR contains a four-strand intramolecular foldback within the G-strand, with the C-strand largely single stranded or loosely associated with the intramolecular structure of the G-strand (Fig. 5). The tetrastrand structure is dependent on torsional stress within the DNA, and is also present within assembled nucleosomes, tempting the speculation that it may be present in the human insulin gene *in vivo*. It is not clear whether the G4 structure within the ILPR plays a role in the function of the human insulin gene, since such structures are usually present in the genome within the telomeres at chromosomal ends (Bloom and Yeh 1989), and the insulin gene is, to our knowledge, the first gene in which such a structure has been described.

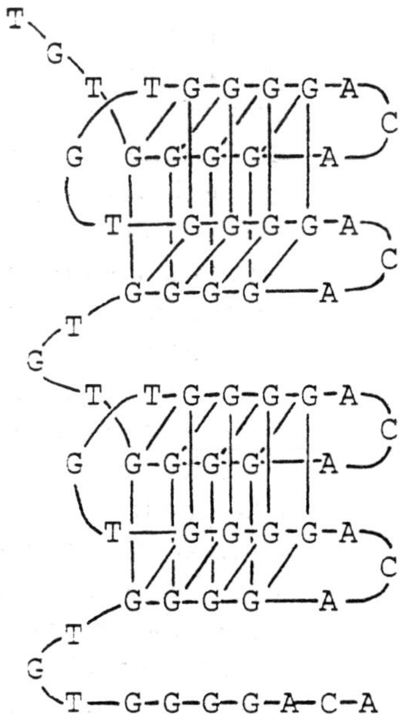

Fig. 5 A proposed model for the G-strand of the ILPR. The model assumes that the oligonucleotide repeats form tetrastrand foldback structures as discussed in the text

2.2.7 Summary and conclusions

The emerging picture of insulin gene regulation is highly complex. Identified regulatory sites can be shown to bind several proteins, not all of which are restricted to insulin-secreting cells, and not all of which are necessarily involved in the regulation of transcription. A major *cis*-acting site (IEB) is active only in β-cell lines, yet is able to bind a protein factor (Pan) which is not restricted to β-cells, and apparently demonstrates positive regulatory activity. At the same time, a site that does bind β-cell-specific proteins (CT-site) does not, on its own, act as a strong activator of transcription. Further, a negative regulatory element, which one might expect to be active only in non-insulin-expressing cells, demonstrates powerful activity in β-cell lines.

Any model of gene regulation must also account for the developmental pattern of expression. In the mouse, the various cells of the endocrine pancreas appear sequentially during embryogenesis (Teitelman and Lee 1987). In transgenic mice, hybrid insulin genes are initially expressed in all cells of the endocrine pancreas (Alpert *et al.* 1988), suggesting a common progenitor cell type. This conclusion is supported by the observation that cell lines of pancreatic endocrine origin are capable of producing more than one hormone (Gazdar *et al.* 1980; Madsen *et al.* 1986). It thus seems likely that the cell lines commonly used in the study of insulin gene regulation resemble dedifferentiated endocrine cells rather than mature β-cells, and may not possess the complete regulatory machinery of terminally differentiated cells. Further transgenic mice studies (Hanahan 1985; Buchini *et al.* 1986; Selden *et al.* 1987) and *in vivo* expression models (Madsen *et al.* 1988; Serup *et al.* 1990) may prove more informative.

Although it is too early to form a comprehensive picture of insulin gene regulation, it is likely that the process involves the combinatorial action of several distinct positively and negatively acting regulatory sites. According to one model, cell-specific gene expression requires a cell-specific repertoire of transcription factors, no one of which need necessarily be restricted to the β-cell. Another model postulates the presence of positively acting transcription factors found exclusively in the β-cell. The latter model would be strongly supported by the discovery of a genuinely β-cell-specific transcription factor which bound to the IEB box. Such a factor might act analogously to the MyoD protein, which is found only in muscle cells and appears to act as a specification factor for muscle cell identity (Davis *et al.* 1987; Sassoon *et al.* 1989). Of course, the situation may be yet more complicated than suggested by either of these models. Cell specificity of gene expression may depend on the relative levels of transcription factors, rather than their simple absence or presence; on tissue-specific modifications of transcription factors; on some feature of higher-order chromatin structure; or, indeed, on some combination of these.

Acknowledgements

We wish to acknowledge several colleagues who have contributed to our work on the insulin gene. These include Drs D. S. W. Boam, M. C. U. Hammond-Kosack, V. Scott, I. Leibiger, and M. W. Kilpatrick. Funding was provided by the British Diabetic Association and the Wellcome Trust.

References

Affolter, M., Schier, A., and Gehring, W. J. (1990). Homeodomain proteins and the regulation of gene expression. *Curr. Opinion in Cell Biol.* **2,** 485–95.

Alpert, S., Hanahan, D., and Teitelman, G. (1988). Hybrid insulin genes reveal a developmental lineage for pancreatic endocrine cells and imply a relationship with neurons. *Cell* **53,** 295–308.

Atchison, M. L. (1988). Enhancers: mechanisms of action and cell specificity. *Ann. Rev. Cell Biol.* **4,** 127–53.

Bell, G. I. and Seino, S. (1990). The organisation and structure of the insulin gene. In *Molecular biology of the islets of Langerhans,* (ed. H. Okamoto), pp. 9–25. Cambridge University Press.

Bell, G. I., Pictet, R. L., Rutter, W. J., Cordell, B., Tischer, E., and Goodman, H. M. (1980). Sequence of the human insulin gene. *Nature* **284,** 26–32.

Bell, G. I., Selby, M. J., and Rutter, W. J. (1982). The highly polymorphic region near the human insulin gene is composed of simple tandemly repeating sequences. *Nature* **295,** 31–5.

Bell, G. I., Horita, S., and Karam, J. H. (1984). A polymorphic locus near the human insulin gene is associated with insulin dependent diabetes mellitus. *Diabetes* **33,** 176–83.

Bell, G. I., Gerhard, D. S., Fong, N. M., Sanchez-Pescador, R., and Rall, L. B. (1985). Isolation of the human insulin like growth factor genes: insulin-like growth factor II and insulin genes are contiguous. *Proc. Natl Acad. Sci. USA* **82,** 6450–4.

Benezra, R., Davis, R. L., Lockshon, D., Turner, D. L., and Weintraub, H. (1990). The protein Id: a negative regulator of helix-loop-helix DNA binding proteins. *Cell* **61,** 49–59.

Berk, A. J. and Schmidt, M. C. (1990). How do transcription factors work? *Genes and Development* **4,** 151–5.

Bird, A. P. (1986). CpG-rich islands and the function of DNA methylation. *Nature* **321,** 209–13.

Bloom, K. and Yeh, E. (1989). Centromeres and telomeres: structural elements of eukaryotic chromosomes. *Curr. Opinion in Cell Biol.* **1,** 526–32.

Boam, D. S. W. and Docherty, K. (1989). A tissue specific nuclear factor binds to multiple sites in the human insulin gene enhancer. *Biochem. J.* **264,** 233–9.

Boam, D. S. W., Clark, A. R., and Docherty, K. (1990*a*). Positive and negative regulation of the human insulin gene by multiple *trans*-acting factors. *J. Biol. Chem.* **265,** 8285–96.

Boam, D. S. W., Clark, A. R., Shennan, K. I. J., and Docherty, K. (1990*b*). The molecular biology of polypeptide hormone expression and biosynthesis. In *Polypeptide hormones, a practical approach* (ed. J. C. Hutton and K. Siddle), pp. 269–305. IRL Press, Oxford.

Buchini, D., Ripoche, M.-A., Stinnakpre, M.-G., Desbois, P., Lores, P., Monthioux, E., Absil, J., Lepesant, J.-A., Pictet, R., and Jami, J. (1986). Pancreatic expression of the human insulin gene in transgenic mice. *Proc. Natl Acad. Sci. USA* **83,** 2511–15.

Cate, R. L., Chick, W., and Gilbert, W. (1983). Comparison of the methylation patterns of the two rat insulin genes. *J. Biol. Chem.* **258,** 6645–52.

Chakraborty, T., Brennan, T., and Olson, E. (1991). Differential *trans*-activation of a muscle-specific enhancer by myogenic helix-loop-helix proteins is separable from DNA binding. *J. Biol. Chem.* **266,** 2878–82.

Chan, S. J., Seino, S., Gruppuso, P. A., Gordon, P., and Steiner, D. F. (1987). A mutation in the B chain coding region is associated with impaired conversion in a family with hyperproinsulinemia. *Proc. Natl Acad. Sci. USA* **84,** 2194–7.

Clark, J. L. and Steiner, D. F. (1969). Insulin biosynthesis in the rat: demonstration of two proinsulins. *Proc. Natl Acad. Sci. USA* **62,** 278–85.

Cockell, M., Stevenson, B. J., Staubin, M., Hagenbuchle, O., and Wellauer, P. K. (1989). Identification of a transcription factor that interacts with a transcriptional activator of genes expressed in the acinar pancreas. *Mol. Cell. Biol.* **9,** 2464–76.

Cook, P. R. (1989). The nucleoskeleton and the topology of transcription. *Eur. J. Biochem.* **185,** 487–501.

Cordell, B., Diamond, D., Smith, S., Punter, J., Schone, H. H., and Goodman, H. M. (1982). Disproportionate expression of the two nonallelic rat insulin genes in a pancreatic tumour is due to translational control. *Cell* **31,** 531–42.

Cordle, S. R., Henderson, E., Masuoka, H., Weil, P. A., and Stein, R. (1991). Pancreatic β-cell-type-specific transcription of the insulin gene is mediated by basic helix-loop-helix DNA-binding proteins. *Mol. Cell. Biol.* **11,** 1734–8.

Crowe, D. T. and Tsai, M. (1989). Mutagenesis of the rat insulin II 5′-flanking region defines sequences important for expression in HIT cells. *Mol. Cell. Biol.* **9,** 1784–9.

Davis, R. L., Weintraub, H., and Lassar, A. B. (1987). Expression of a single transfected cDNA converts fibroblasts to myoblasts. *Cell* **51,** 987–1000.

Edlund, T., Walker, M. D., Barr, P. J., and Rutter, W. J. (1985). Cell specific expression of the rat insulin gene: evidence for the role of two distinct 5′ flanking sequences. *Science* **230,** 912–16.

Efrat, S., Linde, S., Kofod, H., Spector, D., Delannoy, M., Grant, S., Hanahan, D., and Baekkeskov, S. (1988). Beta cell lines derived from transgenic mice expressing a hybrid insulin gene-oncogene. *Proc. Natl Acad. Sci. USA* **85,** 9037–41.

Elbein, S. C. and Permutt, M. A. (1990). The role of the insulin gene in diabetes: use of restriction fragment length polymorphisms in diagnosis. In *Molecular biology of the islets of Langerhans* (ed. H. Okamoto), pp. 251–62. Cambridge University Press.

Ellis, H. M., Spann, D. R., and Posakony, J. W. (1990). Extramacrochaetae, a negative regulator of sensory organ development in *Drosophila*, defines a new class of helix-loop-helix proteins. *Cell* **61,** 27–38.

Ephrussi, A., Church, G. M., Tonegawa, S., and Gilbert, W. (1985). B lineage-specific interactions of an immunoglobulin enhancer with cellular factors *in vivo*. *Science* **227,** 134–40.

Evans, R. M. (1988). The steroid and thyroid hormone receptor superfamily. *Science* **240,** 889–95.

Fiedorek, F. T., Jr, Carnaghi, L. R. and Giddings, S. J. (1990). Selective expression of the insulin I gene in rat insulinoma-derived cell lines. *Mol. Endocrinol.* **4,** 990–9.

Fisher, P. A. (1989). Chromosomes and chromatin structure: the extrachromosomal karyoskeleton. *Curr. Opinion in Cell Biol.* **1,** 447–53.

Fromont-Racine, M., Bucchini, D., Madsen, O., Besbois, P., Linde, S., Neilsen, J. H., Saulnier, C., Ripoche, M.-A., Jami, J., and Pictet, R. (1990). Effect of 5′-flanking sequence deletions on expression of the human insulin gene in transgenic mice. *Mol. Endocrinol.* **4,** 669–77.

Gasser, S. M. and Laemmli, U. K. (1987). A glimpse at chromosomal order. *Trends in Genetics* **3,** 16–21.

Gazdar, A. F., Chick, W. L., Oie, H. K., Sims, H. L., King, D. L., Weir, G. C., and Lau ris, V. (1980). Continuous clonal insulin- and somatostatin-secreting cell lines established from a transplantable rat islet cell tumour. *Proc. Natl Acad. Sci. USA* **77,** 3519–23.

Giddings, S. J. and Carnaghi, L. R. (1988). The two nonallelic rat insulin mRNAs and pre-mRNAs are regulated coordinately *in vivo*. *J. Biol. Chem.* **263,** 3845–9.

Giddings, S. J. and Carnaghi, L. (1989). Rat insulin II gene expression by extraplacental membranes. *J. Biol. Chem.* **264,** 9462–9.

Gross, D. S. and Garrard, W. T. (1988). Nuclease hypersensitivity in chromatin. *Ann. Rev. Biochem.* **57,** 159–97.

Halban, P. A. (1990). Proinsulin trafficking and processing in the pancreatic B cell. *Trends in Endocrinology and Metabolism* **1,** 261–5.

Hammond-Kosack, M. C. U., Dobrinski, B., Lurz, R., Docherty, K., and Kilpatrick, M. W. (1992). The human insulin gene linked polymorphic region exhibits an altered DNA structure. *Nucleic Acids Res.*, in press.

Hammonds, P., Schofield, P. N., and Ashcroft, S. J. H. (1987). Glucose regulates prepro-insulin messenger RNA levels in a clonal cell line of simian virus 40-transformed B cells. *FEBS Lett.* **213,** 149–54.

Hanahan, D. (1985). Hereditable formation of pancreatic β cell tumours in transgenic mice expressing recombinant insulin/simian virus 40 oncogenes. *Nature* **315,** 115–22.

Harper, M. E., Ullrich, A., and Saunders, G. F. (1981). Localisation of the human insulin gene to the distal end of the short arm of chromosome 11. *Proc. Natl Acad. Sci. USA* **78,** 4458–60.

Hitman, G. A. and Niven, M. J. (1989). Genes and diabetes mellitus. *British Medical Bulletin* **45,** 191–205.

Hwung, Y.-P., Crowe, D. T., Wang, L.-H., Tsai, S. Y., and Tsai, M.-J. (1988). The COUP transcription factor binds to a promoter element of the rat II insulin gene. *Mol. Cell. Biol.* **8,** 2070–7.

Hwung, Y-P., Gu, Y-Z., and Tsai, M-J. (1990). Cooperativity of sequence elements mediates tissue specificity of the rat insulin II gene. *Mol. Cell. Biol.* **10,** 1784–8.

Johnson, P. F. and McKnight, S. L. (1989). Eukaryotic transcriptional regulatory proteins. *Ann. Rev. Biochem.* **58,** 799–839.

Jones, N. (1990). Transcriptional regulation by dimerization: two sides of an incestuous relationship. *Cell* **61,** 9–11.

Jones, N. C., Rigby, P. W. J., and Ziff, E. B. (1988). *Trans*-acting protein factors and the regulation of eukaryotic transcription: lessons from studies on DNA tumour viruses. *Genes and Development* **2,** 267–81.

Kadonaga, J. T. (1990). Gene transcription: basal and regulated transcription by RNA polymerase II. *Curr. Opinion in Cell Biol.* **2,** 496–501.

Karlsson, O., Edlund, T., Moss, L. B., Rutter, W. J., and Walker, M. D. (1987). A mutational analysis of the insulin gene transcription control region: expression in beta cells is dependent on two related sequences within the enhancer. *Proc. Natl Acad. Sci. USA* **84,** 8819–23.

Karlsson, O., Walker, M. D., Rutter, W. J., and Edlund, T. (1989). Individual protein-binding domains of the insulin gene enhancer positively activate β cell specific transcription. *Mol. Cell. Biol.* **9,** 823–7.

Karlsson, O., Thor, S., Norberg, T., Ohlsson, H., and Edlund, T. (1990). Insulin gene enhancer binding protein Isl-1 is a member of a novel class of proteins containing both a homeo- and Cys–His domain. *Nature* **344,** 879–82.

Kingston, R. E. (1989). Transcription control and differentiation: the HLH family, c-*myc* and C/EBP. *Curr. Opinion in Cell Biol.* **1,** 1081–7.

Knight, S. W. and Docherty, K. (1991). The identification of specific protein–RNA interactions within the 5′ untranslated region of human insulin mRNA. *Biochem. Soc. Trans.* **19,** 120S.

Koranyi, L., Permutt, M. A., Chirgwin, J. M., and Giddings, S. J. (1989). Proinsulin I and II gene expression in inbred mouse strains. *Mol. Endocrinol.* **3,** 1895–902.

Lagueux, M., Lwoff, L., Meister, M., Goltzené, and Hofmann, J. A. (1990). cDNAs from neurosecretory cells of brains of *Locusta migratoria* (Insecta, Orthoptera) encoding a novel member of the superfamily of insulins. *Eur. J. Biochem.* **187,** 249–54.

Laimins, L., Holmgren-Konig, M., and Khoury, G. (1986). Transcriptional 'silencer' element in rat repetitive sequences associated with the rat insulin I gene locus. *Proc. Natl Acad. Sci. USA* **83,** 3151–5.

Latchman, D. (1990). *Gene regulation: a eukaryotic perspective.* Unwin Hyman, London.

Leshkowitz, D. and Walker, M. D. (1991). Extinction of insulin gene expression in hybrids between β cells and fibroblasts is accompanied by loss of the putative β-cell-specific transcription factor IEF1. *Mol. Cell. Biol.* **11,** 1547–52.

Levine, M. and Hoey, T. (1988). Homeobox proteins as sequence-specific transcription factors. *Cell* **55,** 537–40.

Levine, M. and Manley, J. L. (1989). Transcriptional repression of eukaryotic promoters. *Cell* **59,** 405–8.

Lewin, B. (1990). Commitment and activation at pol II promoters: a tail of protein–protein interactions. *Cell* **61,** 1161–4.

Lomedico, P., Rosenthal, N., Efstradiatis, A., Gilbert, W., Kolodner, R., and Tizard, R. (1979). The structure and evolution of two nonallelic rat preproinsulin genes. *Cell* **18,** 545–58.

Madsen, O. D., Larson, L-I., Rehfeld, J. F., Schwartz, T. W., Lernmark, A., Labrecque, A. D., and Steiner, D. F. (1986). Cloned cell lines from a transplantable islet cell tumour are heterogeneous and express cholecystokinin in addition to islet hormones. *J. Cell Biol.* **103,** 2025–34.

Madsen, O. D., Andersen, L. C., Michelsen, B., Owerbach, D., Larsson, L.-I., Lernmark, A., and Steiner, D. F. (1988). Tissue-specific expression of transfected human insulin genes in pluripotent clonal rat insulinoma lines during passage *in vivo*. *Proc. Natl Acad. of Sci. USA* **85,** 6652–6.

Maniatis, T., Goodbourn, S., and Fischer, J. A. (1987). Regulation of inducible and tissue-specific expression. *Science* **236,** 1237–45.

Mitchell, P. J. and Tjian, R. (1989). Transcriptional regulation in mammalian cells by sequence-specific DNA binding proteins. *Science* **245,** 371–8.

Moss, L. G., Moss, J. B., and Rutter, W. J. (1988). Systematic binding analysis of the insulin gene transcriptional control region: insulin and immunoglobulin enhancers utilize similar transactivators. *Mol. Cell. Biol.* **8,** 2620–7.

Muller, M. M., Gerster, T., and Schaffner, W. (1988). Enhancer sequences and regulation of gene expression. *Eur. J. Biochem.* **176,** 485–95.

Murre, C., McCaw, P. S., and Baltimore, D. (1989*a*). A new DNA binding and dimerization motif in immunoglobulin enhancer binding, daughterless, MyoD, and myc proteins. *Cell* **56,** 777–83.

Murre, C., McCaw, P. S., Uaessin, H., Candy, M., Jan, L. Y., Jan, Y. N., Cabrera, C. V., Buskin, J. N., Hauschka, S. D., Lassar, A. B., Weintraub, H., and Baltimore, D. (1989*b*). Interactions between heterologous helix-loop-helix proteins generate complexes that bind specifically to a common DNA sequence. *Cell* **58,** 537–44.

Muschell, R., Khoury, G., and Reid, L. M. (1986). Regulation of insulin mRNA abundance and adenylation: dependence on hormones and matrix substrata. *Mol. Cell. Biol.* **6,** 337–41.

Nagasawa, H., Kataoka, H., Isogai, A., Tamura, S., Suzuki, A., Mizogichi, A., Fujiwara, Y., Suzuki, A., Takahashi, S. Y., and Ishizaki, H. (1986). Amino acid sequence of a prothoraciotropic hormone of the silkworm. *Bombyx mori. Proc. Natl Acad. Sci. USA* **83,** 5840–3.

Nelson, C., Shen, L.-P., Meister, A., Fodor, E., and Rutter, W. J. (1990). Pan: a transcriptional regulator that binds chymotrypsin, insulin, and AP-4 enhancer motifs. *Genes and Development* **4,** 1035–43.

Nir, U., Walker, M. D., and Rutter, W. J. (1986). Regulation of rat insulin I gene expression: evidence for negative regulation in non-pancreatic cells. *Proc. Natl Acad. Sci. USA* **83,** 3180–4.

Ohlsson, H. and Edlund, T. (1986). Sequence-specific interactions of nuclear factors with the insulin gene enhancer. *Cell* **45,** 35–44.

Ohlsson, H., Karlsson, O., and Edlund, T. (1988). A beta cell specific protein binds to two major regulatory sequences of the insulin gene enhancer. *Proc. Natl Acad. Sci. USA* **85,** 4228–31.

O'Malley, K. L. and Rotwein, P. (1988). Human tyrosine hydroxylase and insulin genes are contiguous on chromosome 11. *Nucleic Acids Res.* **16,** 4437–45.

Owerbach, D. and Nerup, J. (1982). Restriction fragment length polymorphism of the insulin gene in diabetes mellitus. *Diabetes* **31,** 275–7.

Parker, C. S. (1989). Transcription factors. *Curr. Opinion in Cell Biol.* **1,** 512–18.

Patient, R. K. and Allan, J. (1989). Active chromatin. *Curr. Opinion in Cell Biol.* **1,** 454–9.

Permutt, M. A. and Kipnis, D. M. (1972). Insulin biosynthesis. I. On the mechanism of glucose stimulation. *J. Biol. Chem.* **247,** 1194–9.

Philippe, J. and Missotten, M. (1990). Functional characterisation of a cAMP-responsive element of the rat insulin I gene. *J. Biol. Chem.* **265,** 1465–9.

Ptashne, M. (1986). Gene regulation by proteins acting nearby and at a distance. *Nature* **322,** 697–701.

Ptashne, M. (1988). How eukaryotic transcriptional activators work. *Nature* **333,** 683–9.

Ptashne, M. and Gann, A. A. F. (1990). Activators and targets. *Nature* **346,** 329–31.

Rosenthal, N. (1989). Muscle cell differentiation. *Curr. Opinion in Cell Biol.* **1,** 1094–101.

Rotwein, P. S., Chirgwin, J., Cordell, B., Goodman, H. M., and Permutt, M. A. (1981). Polymorphism in the 5′ flanking region of the human insulin gene and its possible relationship to type 2 diabetes. *Science* **213,** 1117–20.

Saltzmann, A. G. and Weinmann, R. (1989). Promoter specificity and modulation of RNA polymerase II transcription. *FASEB J.* **3,** 1723–33.

Sample, C. E. and Steiner, D. F. (1987). Tissue-specific binding of a nuclear factor to the insulin gene promoter. *FEBS Lett.* **222,** 332–6.

Santerre, R. F., Cook, R. A., Crisel, R. M. D., Sharp, J. D., Schmidt, R. J., Williams, D. C., and Wilson, C. P. (1981). Insulin synthesis in a clonal cell line of simian virus 40-transformed hamster pancreatic beta cells. *Proc. Natl Acad. Sci. USA* **78,** 4339–43.

Sassoon, D., Lyons, G., Wright, W. E., Lin, V., Lassar, A., Weintraub, H., and Buckingham

(1989). Expression of two myogenic regulatory factors myogenin and MyoD1 during mouse embryogenesis. *Nature* **341,** 303–9.

Scott, V., Clark, A. R., Hutton, J. C. and Docherty, K. (1991). Two proteins act as the IUF1 insulin gene enhancer binding factor. *FEBS Letts,* **290,** 27–30.

Selden, R. F., Skoskiewicz, M. J., Howie, K. B., Russell, P. S., and Goodman, H. M. (1986). Regulation of human insulin gene expression in transgenic mice. *Nature* **32,** 525–8.

Selden, R. F., Skoskieicz, M. J., Russell, P. S., and Goodman, H. M. (1987). Regulation of insulin-gene expression. Implications for gene therapy. *New Engl. J. Med.* **317,** 1067–76.

Serup, P., Michelsen, B., Lund, K., and Madsen, O. D. (1990). Induction of insulin enhancer binding factors during β-cell differentiation. *Diabetologia* **33** (Suppl.), A34.

Shibasaki, Y., Kawakami, T., Kanazawa, Y., Akanuma, Y., and Takaku, F. (1985). Postranslational cleavage of proinsulin is blocked by a point mutation in familial hyperproinsulinemia. *J. Clin. Invest.* **76,** 378–80.

Shibasaki, Y., Sakura, H., Takaku, F., and Kasuga, M. (1990). Insulin enhancer binding protein has helix-loop-helix structure. *Biochem. Biophys. Res. Comm.* **170,** 314–21.

Smit, A. B., Vreugdenhill, E., Ebberink, R. H. M., Garaerts, W. P. M., Klootwijk, J., and Joose, J. (1988). Growth-controlling mulloscan neurons produce the precursor of an insulin related peptide. *Nature* **331,** 535–8.

Soares, M. B., Schon, E., Henderson, A., Karathanasis, S. K., Cate, R., Zeitlin, S., Chirgwin, J., and Efstradiatis, A. (1985). RNA-mediated gene duplication: the rat preproinsulin I gene is a functional retroposon. *Mol. Cell. Biol.* **5,** 2090–103.

Stein, R. W. and Whelan, J. (1989). Insulin gene enhancer activity is inhibited by adenovirus 5 E1a gene products. *Mol. Cell. Biol.* **9,** 4531–4.

Steiner, D. F. (1990). The biosynthesis of insulin. In *Handbook of experimental pharmacology,* Vol. 92, *Insulin* (ed. P. Cautrecasas and S. Jacobs), pp. 67–92. Springer Verlag, Berlin.

Steiner, D. F., Chan, S. J., Welsh, J. M., and Kwok, S. C. M. (1985). Structure and evolution of the insulin gene. *Ann. Rev. Genet.* **19,** 463–8.

Tager, H. S. (1990). Abnormal insulin gene products. In *Molecular biology of the islets of Langerhans* (ed. H. Okamoto), pp. 263–86. Cambridge University Press.

Takeda, J., Ishii, S., Seino, Y., Immamoto, F., and Imura, H. (1989). Negative regulation of human insulin gene expression by the 5′ flanking region in non-pancreatic cells. *FEBS Lett.* **247,** 41–5.

Tazi, J. and Bird, A. (1990). Alternative chromatin at CpG islands. *Cell* **60,** 909–20.

Teitelman, G. and Lee, J. K. (1987). Cell lineage analysis of pancreatic islet cell development: glucagon and insulin cells arise from catecholaminergic precursors present in the pancreatic duct. *Developmental Biol.* **121,** 454–66.

Walker, M. D. (1990). Insulin gene regulation. In *Handbook of experimental pharmacology,* Vol. 92, *Insulin* (ed. P. Cautrecasas and S. Jacobs), pp. 93–112. Springer Verlag, Berlin.

Walker, M. D., Edlund, T., Boulet, A. M., and Rutter, W. J. (1983). Cell-specific expression controlled by the 5′ flanking region of insulin and chymotrypsin genes. *Nature* **306,** 557–61.

Walker, M. D., Park, C. W., Rosen, A., and Aronheim, A. (1990). A cDNA from a mouse pancreatic β cell encoding a putative transcription factor of the insulin gene. *Nucleic Acids Res.* **18,** 1159–66.

Weintraub, H., Larsen, A., and Groudine, M. (1981). Alpha globin gene switching during the development of chicken embryos: expression and chromosome structure. *Cell* **24,** 333–44.

Welsh, M., Scherberg, N., Gilmore, R., and Steiner, D. F. (1985). Translational control of insulin biosynthesis: evidence for regulation of elongation, initiation, and signal recognition particle-mediated translational arrest by glucose. *Biochem. J.* **235,** 459–67.

Wentworth, B. M., Schaefer, I. M., Villa-Komaroff, L., and Chirgwin, J. M. (1986). Characterisation of two nonallelic genes encoding mouse preproinsulin. *J. Mol. Evolution* **23,** 305–12.

Whelan, J., Poon, D., Weil, P. A., and Stein, R. (1989). Pancreatic β-cell-specific expression of the rat insulin II gene is controlled by positive and negative cellular transcriptional elements. *Mol. Cell. Biol.* **9,** 3253–9.

Whelan, J., Cordle, S. R., Henderson, E., Weil, P. A., and Stein, R. (1990). Identification of a pancreatic β-cell insulin gene transcription factor that appears to activate cell-type-specific expression: its possible relationship to other cellular factors that bind to a common insulin gene sequence. *Mol. Cell. Biol.* **10,** 1564–72.

Wu, C. and Gilbert, W. (1981). Tissue-specific exposure of chromatin at the 5′ terminus of the rat preproinsulin II gene. *Proc. Natl Acad. Sci. USA* **78,** 1577–80.

3 Insulin synthesis

ELAINE M. BAILYES, PAUL C. GUEST, and
JOHN C. HUTTON

3.1 Insulin structure

The structure of insulin has been highly conserved throughout higher vertebrate evolution. The hormone consists of an A-chain (21 amino acids) and a B-chain (30 amino acids), connected by two intermolecular disulphide bonds (A7–B7, A20–B19), with an intramolecular bond between A6 and A11. Invariant features include the positions of the three disulphide bonds, the N- and C-terminal regions of the A-chain, and the hydrophobic residues in the C-terminal region of the B-chain. In the presence of zinc, at acid pH, most insulins form crystals composed of insulin hexamers. The hexamer consists of three dimers arranged roughly in a plane. The central axis perpendicular to the plane passes through two zinc ions; each is co-ordinated with the imidazole groups of three B10 histidine residues with one located above and one below the plane of the hexamer. The insulin dimers are held together in the crystal by hydrogen bonds between amino acid B24 and amino acid B26, forming an antiparallel β-pleated sheet structure (Steiner *et al.* 1989).

Insulin is synthesized initially as a single-chain precursor, preproinsulin. Human preproinsulin (Fig. 1) comprises an N-terminal signal sequence of 24 amino acids, followed by the insulin B-chain, an Arg-Arg sequence, the connecting C-peptide of 31 amino acids, a Lys-Arg sequence, and the insulin A-chain. Signal sequences are a common feature of secretory, plasma membrane, lysosomal, and resident endoplasmic reticulum (ER) and Golgi proteins, and function to direct the translating ribosomal/mRNA complex from the cytoplasm to the rough ER (Pfeffer and Rothman 1987). The signal sequence of preproinsulin contains the three main structural features characteristic of such peptides (Briggs and Gierasch 1986):

(1) a central core of strongly hydrophobic residues

(2) a more polar but usually uncharged region C-terminal to the hydrophobic core

(3) an N-terminal segment containing one or more charged (usually basic) residues.

Removal of the preproinsulin signal sequence to generate proinsulin, occurs during translation as the molecule is translocated into the lumen of the ER. The three-dimensional structure of proinsulin, once the disulphide linkages have been established, resembles that of insulin, except for the presence of C-peptide. Both molecules aggregate to form dimers and Zn^{2+} hexamers, and share common

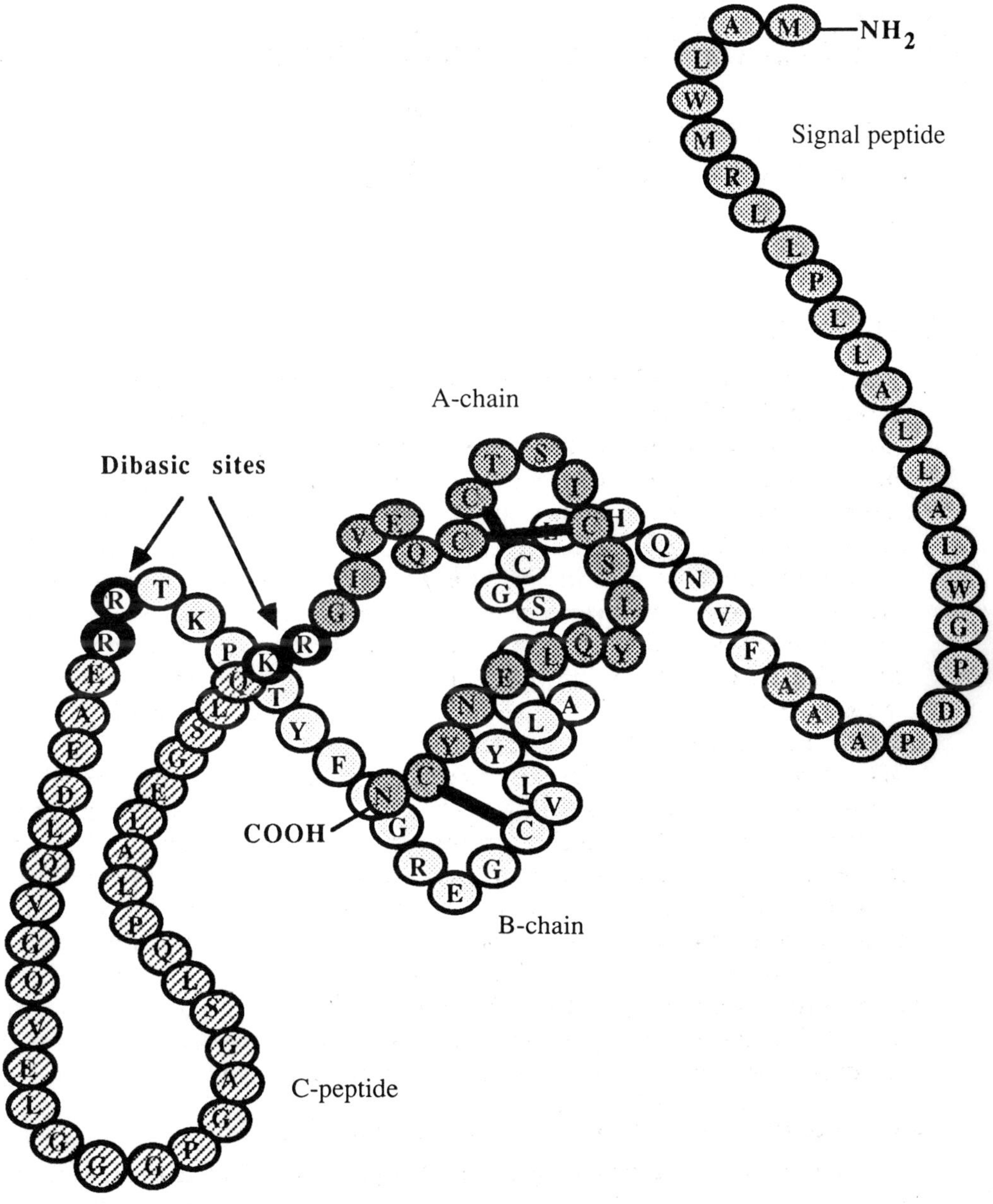

Fig. 1 The structure of human preproinsulin. The four domains of preproinsulin are shown. The dibasic sites excised by proteolytic cleavage at the C-peptide junctions are outlined in black. The A- and B-chains are folded as in insulin, the C-peptide is unstructured

epitopes (Derewenda *et al.* 1989). Proinsulin is more soluble than insulin, however, owing to the presence of the charged residues in C-peptide. The length of C-peptide in higher vertebrates varies between 26 and 38 residues (Gross *et al.* 1989*a*) and shows a greater sequence variation than insulin itself. The dibasic amino-acid residues at the B–C junction (Arg31, Arg32) and C–A junction (Lys64, Arg65) are always present and are essential for the proteolytic processing of proinsulin into insulin and C-peptide. Dibasic amino-acid sequences at the sites of controlled proteolysis in precursor proteins also occur in a wide variety of proteins and species (Docherty and Steiner 1982; Loh *et al.* 1984). Prediction of the secondary structure around prohormone cleavage sites suggests that most dibasic processing sites occur in a β-turn adjacent to a region of high helical content or propensity to form a β-sheet (Rholam *et al.* 1986). Omega (Ω) loops (6–16 amino acids, of no ordered secondary structure, whose ends are separated by no more than 1nm) may provide an additional or alternative recognition feature (Bek and Berry 1990). In the latter analysis, proinsulin residues 19–31 and 53–66 were predicted to form β-turns, and residues 46–54 and 69–76 to form Ω loops. The structure of proinsulin has been deduced from comparison of the known crystal structure of insulin, from far-UV circular dichroism, and from NMR and photochemically induced dynamic nuclear polarization studies (Derewenda *et al.* 1989; Weiss *et al.* 1990). The A- and B-chains are folded as in insulin, with the two dibasic sites within 1 nm of each other on the surface of the molecule, and C-peptide has a mobile, flexible structure capable of interaction with the surface of the insulin chains. However, a difference was detected between proinsulin and insulin in the two-dimensional NMR resonances of the hydrophobic core. The intermediate generated by cleavage of proinsulin at the C–A junction was insulin-like in that regard, whereas the intermediate generated by cleavage at the B–C junction had a conformation similar to that of proinsulin. This suggests that there is a stable local structure at the C–A junction (termed the CA knuckle) which may form a specific recognition structure for the endopeptidase responsible for cleavage of the molecule at this site (Weiss *et al.* 1990).

3.2 Translational control

3.2.1 Translation and translocation

Translation of preproinsulin mRNA begins in the cytosol with the binding of free ribosomes to the molecule. Transfer to the ER membrane occurs co-translationally and is initiated by the binding of the emerging signal sequence to the 54 kDa subunit of the signal recognition particle (SRP), an 11S ribonucleoprotein complex (Eskridge and Shields 1983; Kurzchalia *et al.* 1986) (Fig. 2). This retards further elongation of the nascent polypeptide chain (Meyer *et al.* 1982). The SRP-mediated translational arrest of preproinsulin appears to be possible after the translation of around 50 residues, when as few as 10 residues of signal peptide protrude from the ribosomal complex (Okun *et al.* 1990). Association of the complex with the ER membrane is mediated by the interaction of the SRP with an integral ER membrane protein, the SRP receptor (docking protein) (Meyer *et al.* 1982), which occurs when 70–80 amino acids of preproinsulin have been polymerized (Eskridge and Shields 1983). SRP is then released from the complex (Gilmore and Blobel 1983; Wiedmann *et al.* 1987) and the signal sequence is transferred to the signal sequence receptor (SSR), a glycosylated, integral ER membrane protein, which may also be a com-

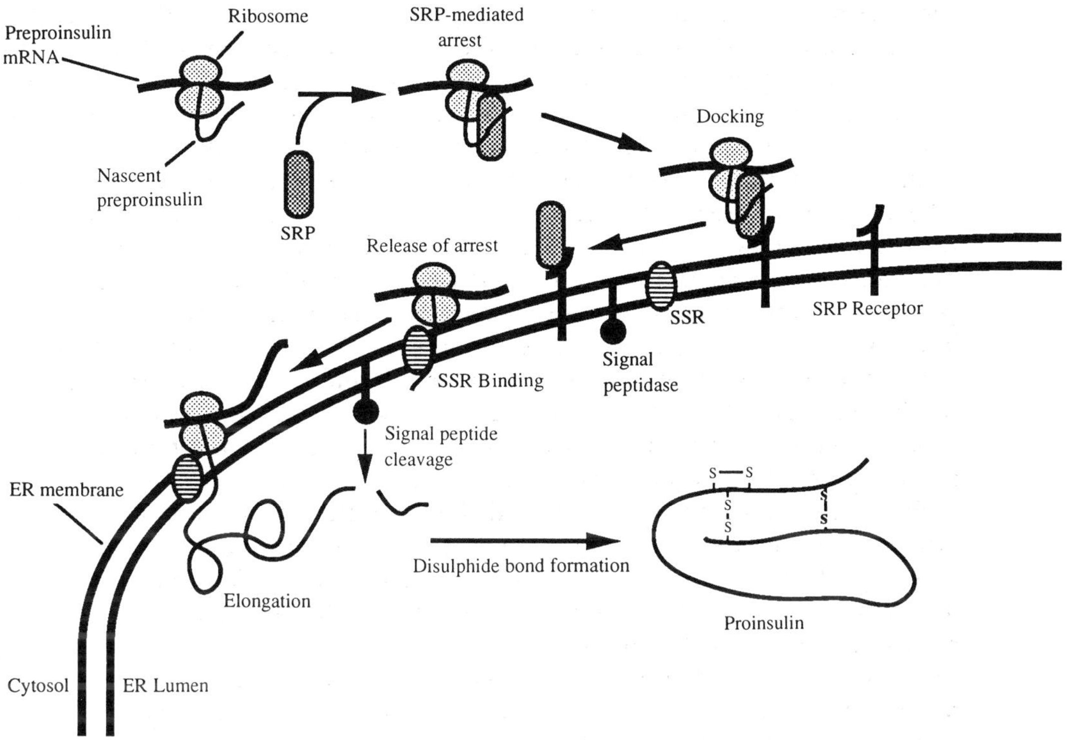

Fig. 2 Translation of insulin mRNA begins in the cytoplasm. Interaction of the nascent chain with the signal recognition particle (SRP) stops further translation and allows the binding of the translation complex to the ER membrane via interaction with the SRP receptor. SRP is released, the signal sequence transfers to the signal sequence receptor (SSR), and translation continues with translocation of the elongating chain across the membrane into the ER lumen. Signal peptidase removes the signal sequence and the mature structure of proinsulin is formed

ponent of the putative proteinaceous translocation pore in the ER membrane (Wiedmann *et al.* 1989). The dissociation of SRP and the signal sequence is mediated by the SRP receptor and requires receptor-bound GTP (Connolly and Gilmore 1989). The dissociation of SRP from its receptor is triggered by hydrolysis of GTP (Rapoport 1990). The attachment of the ribosomes to the ER membrane is mediated by another integral ER membrane protein of M_r 180 000, the ribosome receptor (Savitz and Meyer 1990).

3.2.2 Regulation of preproinsulin mRNA translation

Insulin biosynthesis is regulated by many circulating factors, including glucose, some amino acids, neurotransmitters, and hormones (Hedeskov 1980; Permutt 1981). The metabolism of glucose is generally required for the stimulation of synthesis since only metabolizable analogues of glucose and other sugars or amino acids are effective (Ashcroft *et al.* 1978). Thus, insulin synthesis is stimulated by mannose and leucine, but not by fructose, xylitol, or ribose. Furthermore, the glucose-induced stimulation of insulin biosynthesis is inhibited by mannoheptulose, which blocks glucose metabolism in islets by inhibiting its phosphorylation to glucose 6-phosphate.

Insulin biosynthesis and secretion are not obligatorily coupled. Glucose-stimulated insulin secretion, for example, is inhibited in a Ca^{2+}-free medium, whereas under these conditions biosynthesis is still activated (Pipeleers *et al.* 1973*a*). Tolbutamide stimulates insulin secretion but not insulin biosynthesis (Ashcroft *et al.* 1978). Furthermore, the threshold for the glucose-induced activation of insulin secretion (4.2–5.6 mM) is higher than that for insulin synthesis (2.5–3.9 mM) (Pipeleers *et al.* 1973*b*; Maldonato *et al.* 1977).

Although glucose increases islet preproinsulin mRNA levels over longer periods of exposure (2–72 h), blot hybridization studies with cDNA probes have shown that there is no significant increase after exposure of 1 h to the sugar (Itoh and Okamoto 1980). Over the same time period, however, incorporation of radiolabelled amino acids into (pro)insulin is increased ten- to twentyfold whereas incorporation into TCA-precipitable total protein only increases 2.5- to threefold (Ashcroft *et al.* 1978). This effect does not require the synthesis of new mRNA, as indicated by the rapidity of the response and the lack of effect on it of transcriptional inhibitors such as actinomycin D (Permutt and Kipnis 1972). Thus, the initial stimulation of (pro)insulin biosynthesis, which occurs within 20 min of exposure to glucose (Permutt and Kipnis 1972; Ashcroft *et al.* 1978), must use pre-existing mRNA and involve translational regulation.

3.2.3 Regulation of initiation of preproinsulin mRNA translation

Analysis of the subcellular distribution of preproinsulin mRNA in isolated rat islets suggests that glucose increases the rate of translational initiation (Welsh *et al.* 1986). Exposure of islets to glucose concentrations above 3.3 mM resulted in an increased rate of transfer of cytoplasmic preproinsulin mRNA to subcellular fractions containing ribosomes and larger polysomes (Welsh *et al.* 1986). Preincubation of islets with theophylline (phosphodiesterase inhibitor), or with a combination of 10 mM L-leucine and 3.3 mM glucose, gave the same response (Welsh *et al.* 1987). Similar effects have been observed for the distribution of total islet mRNA, which suggests that altered rates of initiation of translation may be a general response to glucose (Permutt and Kipnis 1975) and not specific to insulin.

3.2.4 Regulation of SRP-mediated nascent preproinsulin translocation

Glucose also increases the transfer of initiated preproinsulin mRNA from free to membrane-bound polysomes (Welsh *et al.* 1986), a finding that is consistent with an increased rate of SRP-mediated transfer of ribosomes bearing nascent preproinsulin from the cytoplasm to the ER. This effect could be mimicked by 10 mM L-leucine in combination with 3.3 mM glucose, although, in contrast to effects on initiation of translocation, not by theophylline (Welsh *et al.* 1987). This indicates that the substrate, leucine, stimulates preproinsulin translation by a mechanism similar to that of glucose, whereas agents that increase the β-cell cyclic AMP content, such as theophylline, only stimulate translational initiation rates.

Investigations on *in vitro* translation in islet homogenates indicate that glucose stimulation of preproinsulin synthesis might be the result of an increased association of the SRP initiation complex with the SRP receptor (Welsh *et al.* 1986). Addition of purified dog pancreas SRP receptor to the islet homogenates increased the run-off incorporation of [^{125}I]tyrosine into preproinsulin. This response was even greater when islets were preincubated in 16.7 mM glucose, indicating that in glucose-stimulated islets the SRP may be altered structurally, enhancing its interaction with the SRP receptor.

3.2.5 Regulation of nascent preproinsulin elongation

The rate of translational elongation of preproinsulin may be regulated specifically by glucose in the range of 0–5.6 mM. This has been assessed in experiments on (pro)insulin biosynthesis under conditions in which elongation is the rate-limiting step, i.e. in the presence of low concentrations of cycloheximide (Welsh *et al.* 1986). Proinsulin synthesis was stimulated by glucose at concentrations up to 5.6 mM; the synthesis of other proteins was not significantly affected by glucose. Stimulation took place without any change in the intracellular distribution of preproinsulin mRNA and was suggestive of an increase in the rate of peptide chain elongation.

3.2.6 Other constituents of insulin secretory granules

In addition to insulin, the secretory granules contain C-peptide, in amounts equimolar with insulin. These two peptides alone account for approximately 75 per cent of the granule protein mass. Two-dimensional gel electrophoresis of purified insulin granules, however, reveal the presence of an additional 100 or more different polypeptides (Hutton *et al.* 1982). Many of these are related in as much as they are differentially glycosylated forms of the same protein or various intermediates and products of post-translational proteolysis of precursor proteins. Nevertheless, independent gene products in the granule probably number around 50. These include the proteases involved in pro-polypeptide conversion (endopeptidases, carboxypeptidases and amidating enzymes), co-secreted regulatory factors (islet amyloid polypeptide, chromogranin A, and thyrotropin-releasing hormone), ion-translocating proteins that regulate the intragranular environment (e.g. proton-translocating ATPase), and extrinsic membrane proteins involved in intracellular granule movement and exocytosis.

3.2.7 Regulation of the biosynthesis of other insulin secretory granule constituents

Recent investigations of insulin secretory granule biogenesis in isolated rat islets of Langerhans have shown that the biosyntheses of several other granule constituents are co-ordinated with that of insulin (Grimaldi *et al.* 1987; Guest *et al.* 1989). These include a granule membrane glycoprotein of 110 kDa, designated SGM 110, and the relatively abundant granule matrix constituent, chromogranin A. Glucose stimulation produced an increase in the synthesis of chromogranin A and SGM 110 to an extent similar to that observed for proinsulin, as determined by studies involving specific immunoprecipitation of radiolabelled islet proteins (Fig. 3). In contrast, glucose stimulation produced little or no effect on the biosynthesis of carboxypeptidase H, a proprotein-converting enzyme of the secretory

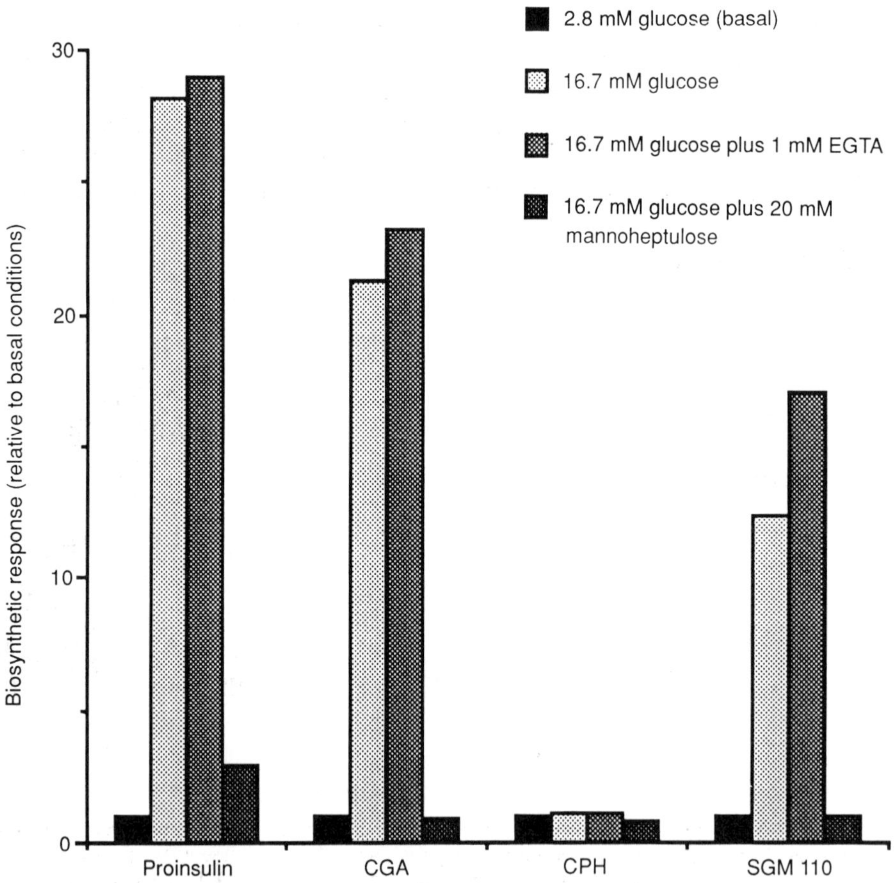

Fig. 3 Glucose stimulation of islet protein biosynthesis. Isolated rat islets were labelled with [^{35}S]methionine and the indicated proteins immunoprecipitated and electrophoresed (CGA, chromogranin-A; CPH, carboxypeptidase H). The results are expressed relative to the incorporation under basal conditions

granule matrix. The stimulatory effect of glucose on the biosynthesis of proinsulin, chromogranin A, and SGM 110, required the metabolism of glucose, as indicated by the inhibitory effect of mannoheptulose. The finding that removal of Ca^{2+} from the medium did not affect the biosynthetic responses of chromogranin A and SGM 110 to glucose, indicated that activation of the synthesis of these proteins, like proinsulin, was not strictly coupled to exocytosis but occurred as a primary response to a signal generated by the secretagogue.

Investigations using two-dimensional electrophoretic analysis of radiolabelled islets have shown that glucose exerts a similar stimulatory effect on the biosynthesis of more than 100 islet proteins (Guest *et al.* 1991). Subcellular fractionation of these islets revealed that the majority of glucose-stimulated proteins were localized to fractions enriched in secretory granules (Fig. 4), although a small proportion were also present in cytoplasmic fractions. It is unlikely that release of SRP-mediated arrest is the only mechanism involved in the biosynthetic response of islet proteins to glucose since not all proteins that are co-translationally inserted into the ER lumen, such as secretory and plasma membrane proteins, responded in an equivalent manner. In addition, a small number of cytoplasmic proteins, which are synthesized on free ribosomes, exhibited a large stimulation.

Since the biosynthetic regulation of this glucose-stimulated subset of proteins paralleled that of insulin and occurred within 2 h, it would appear that control is exerted at the translational level. It is conceivable that the mRNAs encoding these

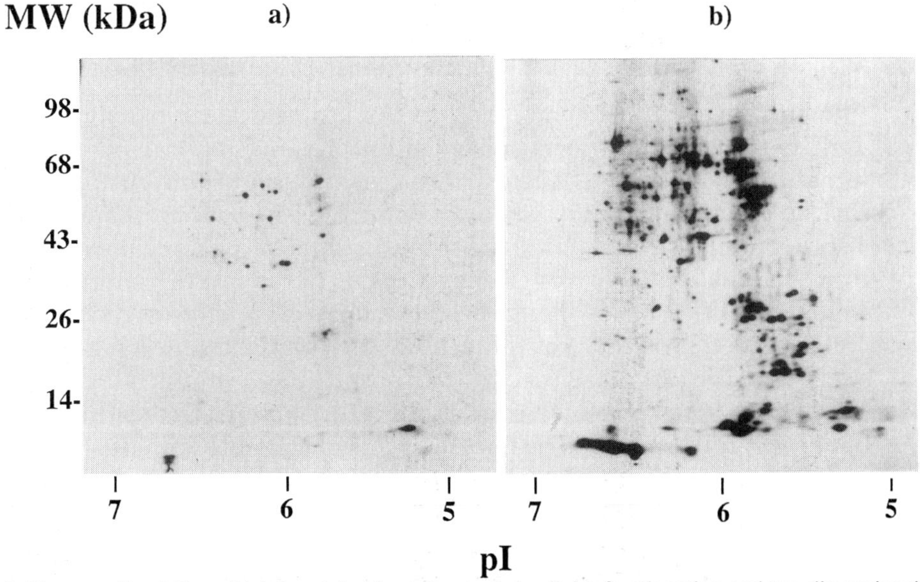

Fig. 4 Glucose stimulation of islet protein biosynthesis: subcellular fractionation and two-dimensional gel analysis. Isolated rat islets were labelled for 1 h with [^{35}S]methionine in the presence of either (a) 2.8 mM glucose or (b) 16.7 mM glucose, then incubated for a further 3 h in non-radioactive medium containing 2.8 mM glucose to permit conversion of proproteins to their mature forms. Subcellular fractions were prepared on Nycodenz density gradients, and those enriched in insulin (secretory granules) were subjected to two-dimensional gel electrophoresis

proteins contain common structural elements which allow their selective recognition by the translational machinery. Regulation of translation via the 5′ non-coding region of eukaryotic mRNAs is thought to occur via the ability of these sequences to form secondary structures and modulate the accessibility of the 5′ cap (reviewed by Clemens 1989). However, such elements are not readily discernible by direct comparison of the linear cDNA sequences encoding insulin and chromogranin A (Hutton *et al.* 1990). Future studies involving the application of gene-fusion techniques may provide an experimental means of demonstrating the existence of such regulatory sequences.

3.3 Post-translational events

3.3.1 Formation of proinsulin within the endoplasmic reticulum (ER)

The signal sequence is removed from preproinsulin by signal peptidase and degraded in the lumen of the ER, either during or shortly after translocation (Lively 1989). Dog pancreatic signal peptidase is a complex of five subunits (12, 18, 21, 22/23, and 25 kDa). The 22/23 kDa subunits are differently glycosylated forms of the same transmembrane polypeptide, and the 18 and 21 kDa subunits are highly homologous but distinct gene products which span the ER membrane (Shelness and Blobel 1990). Homologous proteins have been identified in other species and include the hen oviduct signal peptidase glycoprotein and a yeast gene product essential for signal peptide processing. The signal peptidases appear to belong to a class of protease hitherto unidentified. No specific inhibitors of catalysis have yet been identified (Lively 1989).

In the case of transmembrane proteins, translocation is halted if a sufficiently long hydrophobic section is present in the protein (stop transfer membrane anchor signal). The absence of such a signal in preproinsulin ensures that the whole molecule is delivered into the lumen of the ER. The nascent chain then folds into its stable configuration, forming the three disulphide bonds of insulin. The process is probably catalysed by two resident proteins of the ER, BiP and protein disulphide isomerase (PDI). BiP is thought to bind transiently to newly synthesized, incompletely folded proteins, thus preventing their aggregation. It binds permanently to aberrant proteins whose further transport through the secretory pathway is thus blocked (Pelham 1989). PDI catalyses the isomerization of protein disulphide bonds (Freedman 1984). The enzyme's role in protein synthesis has been shown in an *in vitro* protein translation system composed of a reticulocyte lysate and dog pancreatic microsomes. Normal co-translational disulphide bond formation was disrupted in such a system when the microsomes were PDI-deficient. Addition of purified PDI restored normal function (Bulleid and Freedman 1988).

3.3.2 Exit from the endoplasmic reticulum (ER)

The mechanism by which resident ER proteins are sorted from proteins destined for elsewhere appears to be one of specific retention of soluble ER proteins, signalled by possession of a C-terminal sequence, Lys–Asp–Glu–Leu (KDEL in the single-letter amino acid code). Secretory and membrane proteins which are not retained appear to leave by passive bulk flow (Pelham 1989). Deletion of KDEL (HDEL in yeast) from an ER protein results in secretion of the truncated protein. Conversely, addition of the sequence to a secretory protein results in the retention of the protein within the ER. Candidate receptors for proteins bearing this motif have been identified in mammalian and yeast systems, located in a 'salvage' compartment intermediate between the ER and Golgi complex (Lewis and Pelham 1990; Semenza *et al.* 1990; Vaux *et al.* 1990). Since proinsulin lacks the C-terminal KDEL sequence, it travels through this salvage compartment with the other secretory and membrane proteins to the Golgi complex.

3.3.3 Transport through the Golgi complex

Proteins are thought to leave the ER and travel to the Golgi in transport vesicles, and pass through the *cis, medial* and *trans* Golgi cisternae by repeated cycles of vesicle budding and fusion (Fig. 5). Studies on vesicle transfer between the endoplasmic reticulum and Golgi using reconstituted membrane transport systems and semi-intact, perforated cells, have demonstrated a requirement for GTP and calcium (Beckers and Balch 1989). Incubation of isolated Golgi fractions with cytosol and ATP, under conditions that reconstitute intercisternal transport, released transfer vesicles bearing a cytoplasmic protein coat which was distinct from clathrin. Vesicle transfer required the action of a peripheral Golgi membrane protein, the NEM (*N*-ethylmaleimide)-sensitive protein, and was inhibited by the GTP analogue, GTPγS. Treatment with GTPγS caused the accumulation of coated vesicles, whereas treatment with NEM caused the accumulation of uncoated vesicles. NEM and GTPγS together resulted in the accumulation of coated vesicles (Orci *et al.* 1989). Coated vesicles thus give rise to uncoated vesicles in the transfer between the cisternae. The molecular basis for the movement of the carrier vesicles between these subcellular organelles is not fully understood, but the requirement for GTP hydrolysis appears to involve low molecular weight GTP-binding proteins (~20–25 kDa) of the *ras/rab* oncogene family. In the yeast *Saccharomyces cerevisiae*, two small GTP-binding proteins, YPT1 and SEC4, are required in the secretory pathway for Golgi transport and exocytosis respectively (Burgoyne 1989). Cloned sequences highly homologous to the yeast genes have been isolated from the Madin–Darby canine kidney cell line (MDCK cells) and shown to code for proteins located to a compartment intermediate between the ER and *cis* Golgi cisterna and to compartments of the endocytic pathway (Chavrier *et al.* 1990). How these proteins function in membrane traffic is not clear. One model predicts that each step of membrane transfer requires a distinct GTP-binding protein (Bourne 1988). It has been proposed that the GTP-binding protein (containing bound GDP) binds a protein exposed on the cytoplasmic face of the donor organelle and catalyses the assembly of a protein-coated pit. Replacement of GDP by GTP allows the vesicle to bud off and bind to the target cisterna, where GTP hydrolysis releases the GTP-binding protein and catalyses coat removal to allow membrane fusion (Fine 1989). GTP hydrolysis is also required in the formation of constitutive vesicles and secretory granules from the *trans* Golgi network in the neuroendocrine cell line PC12 (Tooze *et al*. 1990).

The Golgi is a site of major post-translational modifications of proteins, such as terminal *N*-glycosylation, *O*-glycosylation, sulphation, and the addition of mannose-6-phosphate residues to lysosomal enzymes, the latter modification being a signal for sorting to lysosomes (Farquhar 1985). None of these modifications appear to apply to proinsulin, although they may affect other insulin secretory granule constituents.

3.3.4 Secretory granule formation

Sorting of granule constituents

The secretory granules are formed from the *trans* Golgi network (TGN), a tubular reticulum on the *trans* face of the Golgi stack, which sorts proteins into different vesicles depending on their final destination (Griffiths and Simons 1986). The granules bud from dilated regions of the TGN that contain condensing secretory material. Both the TGN dilations and the newly formed secretory granules have partial cytoplasmic coats of clathrin which are lost on maturation. This has been demonstrated not only in studies on the formation of granules in pancreatic β-cells but also in other cells of the endocrine pancreas, the anterior pituitary (Orci *et al*. 1985*a*), and a pituitary cell line (Tooze and Tooze 1986).

Experimental evidence that proinsulin is separated at the TGN from constitutively secreted proteins and proteins destined for insertion into the plasma membrane has been provided by a study of pancreatic β-cells infected with influenza virus. This provided an easily identifiable constitutive protein marker — the viral membrane protein, haemagglutinin (Orci *et al*. 1987). Haemagglutinin and proinsulin co-localized throughout the Golgi as far as the TGN. Haemagglutinin, however, was excluded from the membrane directly surrounding condensing proinsulin. Haemagglutinin in post-Golgi compartments was found on non-coated vesicles which did not contain insulin. Conversely, immature clathrin-coated granules contained proinsulin but not haemagglutinin. Further evidence for the formation of immature secretory granules and constitutive vesicles from the TGN has been provided in a cell-free system derived from an adrenal medullary phaeochromocytoma cell line, PC12 (Tooze and Huttner 1990).

Sorting signals

It has been proposed that the sorting of secretory proteins into the regulated pathway requires a positive signal and that constitutive release occurs by default (Burgess and Kelly 1987). The signal must be relatively independent of cell type and species, since many secretory cells can target transfected foreign secretory proteins to their own secretory granules. The nature of the signal is not known;

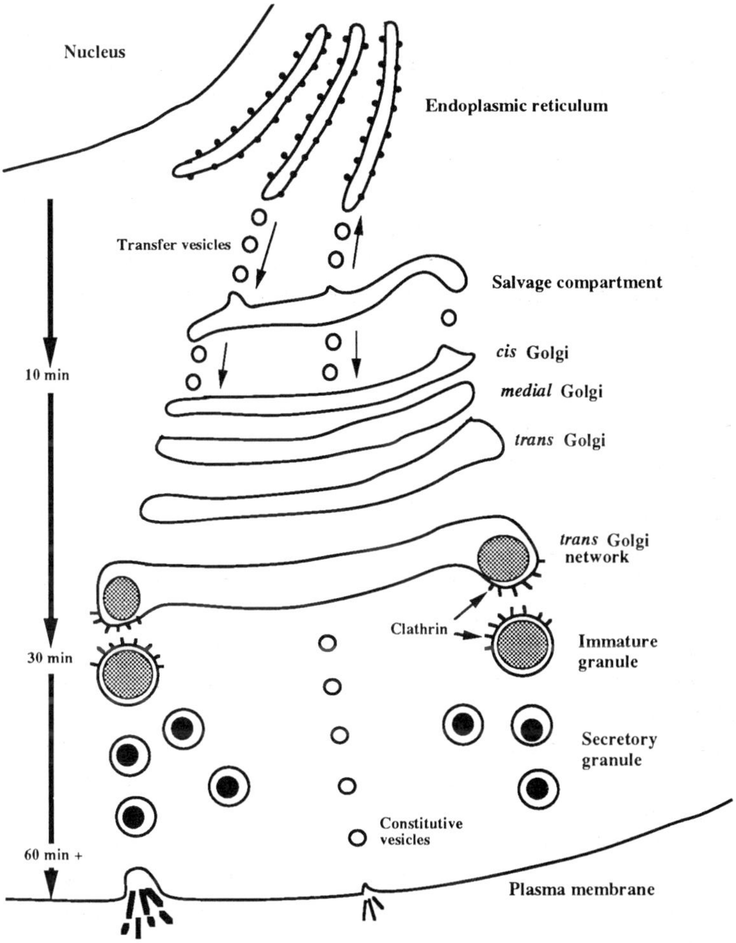

Fig. 5 The secretory pathway in the pancreatic β-cell

there is no obvious common linear motif analogous to the C-terminal KDEL sequence of soluble ER proteins. The process of segregation of insulin into secretory granules is remarkably efficient in normal tissue; less than 1 per cent of the hormone is released constitutively (Rhodes and Halban 1987). However, a greater proportion of insulin is released constitutively in transplantable rat insulinoma cells (Sopwith *et al.* 1981). Similar findings are observed in the mouse pituitary corticotrophic cell line, AtT20, in which only about 15–30 per cent of the ACTH (adrenocorticotrophin) is targeted to the secretory granules (Moore and Kelly 1986). These findings suggest that regulated secretion is a specialized cellular function which has been partially lost in the tumour cell lines.

Studies on the intracellular targeting of hybrid proteins have provided evidence that the sorting signal for the regulated pathway is dominant. A chimeric protein composed of a regulated secretory protein (growth hormone) and the ectodomain of a membrane protein (VSV-G protein) is thus segregated into secretory granules in transfected AtT20 cells (Moore and Kelly 1986). Similar conclusions have been reached in a study using PC12 cells which avoided the use of engineered, unnatural proteins (Rosa *et al.* 1989). In these experiments, microinjection of mRNA encoding a monoclonal antibody to an antigen not present in the cells, resulted in constitutive secretion of the antibody. Substitution of a monoclonal antibody to a secretory granule component (secretogranin 1) resulted in the storage of antigen–antibody complexes in the secretory granules (Rosa *et al.* 1989). Thus, the targeting information present on secretogranin 1 was dominant and not obscured by reaction with the antibody. Furthermore, these results indicate that the constitutive and regulated proteins follow a common biosynthetic pathway which enables the antibody to bind to its antigen, which then diverts the antibody to the secretory granules.

Sorting of proinsulin

The presence of C-peptide and proteolytic conversion are not required for correct sorting to the secretory granules. Mutant proinsulins with the C-peptide domain deleted, altered, or replaced by the much shorter (12 amino acids) and non-homologous C-peptide domain of insulin-like growth factor, were sorted correctly to the secretory granules of transfected AtT20 cells, even though they were not converted to insulin (Powell *et al.* 1988; Gross *et al.* 1989*a*). Further evidence that processing is not required for sorting is provided by the observation that >99 per cent of the proinsulin released by normal islets (accounting for about 10 per cent of the secreted hormone) is secreted by the regulated pathway (Rhodes and Halban 1987).

In contrast to the C-peptide alterations, a single point mutation changing the B10 His to an Asp residue in the insulin B-chain significantly disrupts both the processing of proinsulin to insulin and the sorting to the secretory granules. Experimental introduction of proinsulin genes bearing this mutation into mice and AtT20 cells showed that the substitution led to a high level of constitutive secretion of the mutant proinsulin (Carroll *et al.* 1988; Gross *et al.* 1989*b*). Since B10 histidine is involved in the formation of proinsulin and insulin hexamers by its co-ordination with zinc, it suggests that the sorting information resides in the secondary or tertiary structure of the molecule and that the selection procedure may require hexamer formation. However, although B10 His is highly conserved throughout the higher vertebrates, it cannot be essential for proinsulin sorting to granules in all cases, since New World hystricomorph rodents, such as the guinea pig (B10 Asn) and coypu (B10 Gln), do not form Zn–insulin crystals but still segregate insulin into secretory granules. The failure of processing to insulin observed was probably a secondary consequence of the lack of exposure to the prohormone-converting enzymes of the regulated pathway.

Sorting mechanism

The mechanism involved in the sorting of proproteins such as proinsulin into dense-cored secretory granules is at present unclear. Currently, two models receive considerable attention (Burgess and Kelly 1987). The first model is based on the concept that the TGN contains receptor molecules that concentrate the prohormone to the exclusion of other molecules, which therefore pass into the constitutive pathway by default. The second postulates that proteins destined for the regulated pathway undergo selective aggregation by virtue of their physicochemical properties of isoelectric point and Ca^{2+} binding (Huttner and Tooze 1989).

Evidence consistent with a receptor-mediated process has come from a study on hormone-binding proteins in the dog exocrine pancreas (Chung *et al.* 1989). A group of 25 kDa proteins were isolated from a Golgi membrane fraction which bound regulated secretory proteins such as prolactin, growth hormone, and insulin but not constitutively secreted proteins (serum albumin, haemoglobin, immunoglobulin). The hormone-binding proteins (HBP25s) bound at pH 7.4 and dissociated at pH 5.0. The HBP25s could bind simultaneously to insulin and prolactin. Immunolocalization studies showed that the HBP25s were found in the Golgi but not the secretory granules in both the pancreas and AtT20 cells. The molecules were not integral membrane proteins; only about 50 per cent remained associated with the membranes after freeze-thawing and these could be removed by a high pH wash. These proteins can be envisaged either as part of a receptor-mediated process or they could provide possible nucleation sites in an aggregation mechanism, since they appear to be bivalent.

The role of receptor molecules in the selective endocytosis of plasma-membrane proteins and in the targeting of lysosomal enzymes has been well established. Clathrin is an important part of this sorting mechanism which concentrates plasma-membrane receptor proteins, such as the low-density lipoprotein receptor, into coated pits and endocytic vesicles. Clathrin is also involved in the *trans* Golgi formation of vesicles containing the mannose 6-phosphate receptors which transport lysosomal enzymes containing this sugar to lysosomes. The coating of the pits and vesicles consists of an outer lattice of clathrin and an inner shell of adaptor molecules, termed adaptins. Adaptins are bivalent in that they interact both with clathrin and with the cytoplasmic domain of the various receptors in the membrane of the vesicle (Pearse and Robinson 1990).

A clathrin coat also forms around TGN dilations and newly formed secretory granules, but not around constitutive vesicles. Some indication of the role that clathrin may play in secretory granule formation has come from a study of yeast mutants lacking the clathrin heavy-chain gene (Payne and Schekman 1989). In these mutants, the transmembrane KEX2 protease, which is normally found in the Golgi, is expressed at the cell surface. Clathrin is not, therefore, an essential component of vesicular traffic. One interpretation of these studies is that clathrin normally anchors the KEX2 protease to the Golgi membrane. Another possibility is that clathrin may function as a salvage receptor which retrieves membrane constituents, such as the protease, from TGN-derived vesicles or from the cell surface after exocytosis of vesicle contents. In higher eukaryotes, by an analogous process, clathrin might be used to remove and recycle constitutents of the immature granule which are not required in the mature granule, or to target proteins to the immature granule. Data consistent with a recycling model come from studies on AtT20 cells in which clathrin-coated vesicles have been observed apparently budding from forming secretory granules (Tooze and Tooze 1986). However, it is unlikely that clathrin is used to recycle proinsulin-converting proteases since these enzymes are found in the mature granule and do not appear to be membrane associated (Davidson *et al.* 1988). The yeast model is also not directly applicable to endocrine cells, since yeasts do not possess a regulated pathway of secretion and do not form dense-cored storage granules.

Support for the aggregation model of sorting has come from studies of the physical characteristics of the granule constituents (Huttner and Tooze 1989). Many granule proteins co-precipitate or aggregate with each other, particularly at the acid pH characteristic of the secretory granules of endocrine cells. These include sulphated proteoglycans, sulphated glycoproteins, secretogranins, and chromogranins (Burgess and Kelly 1987). The secretory granule protein secretogranin II of the PC12 cell line, in the presence of 10 mM Ca^{2+} at pH 5.2 but not at pH 7.4, forms aggregates that exclude constitutive secretory proteins (Gerdes *et al.* 1989). A similar finding has been observed for the chromogranin A of bovine adrenal chromaffin granules (Yoo and Albanesi 1990). In this case, aggregation was favoured by high protein concentrations and required ~20–30 mM Ca^{2+} at pH 5.5. At pH 7.5, aggregation was only about 30 per cent of that at acid pH and required 40–60 mM Ca^{2+}.

Although the chromogranins and secretogranins are widely distributed in neuroendocrine granules, their concentrations are usually far less than that of the stored hormone and therefore could only function as catalysts of aggregation or as nucleation sites. It is likely, however, that the physicochemical properties of the segregated molecules are important in the sorting process. It is noteworthy in this regard that insulin, like many other hormones, is highly insoluble at pH values close to its isoelectric point (5.3).

3.3.5 Proteolytic conversion to insulin

Proteolytic processing of proinsulin in pancreatic tissue produces insulin, C-peptide and free amino acids derived from the flanking dibasic sequences. This can be achieved *in vitro* by two enzymes of the exocrine pancreas, trypsin, which cleaves on the carboxyl side of basic residues, and carboxypeptidase B, which removes basic C-terminal amino acids (Docherty and Steiner 1982). However, these enzymes cannot be responsible *in vivo* since they are not constituents of the β-cell. Nevertheless, it is likely that the endogenous proteases operate by a similar mechanism except that the endopeptidase must be specific for dibasic amino-acid sequences and both enzymes should exhibit an acidic pH optimum.

Intracellular site of proinsulin conversion

Studies using antibodies that recognize the B–C junction (Orci *et al.* 1985*b*) and C–A junction (Steiner *et al.* 1987) of proinsulin have shown that the prohormone is found predominantly in the Golgi and clathrin-coated granules. Proinsulin disappeared rapidly from these sites when protein synthesis was inhibited. Cellular depletion of ATP prevented movement of proinsulin from the Golgi to secretory granules. Under these conditions, proinsulin in preformed coated granules was still converted to insulin, as indicated by the loss of immunoreactivity, whereas that in the Golgi was not (Orci *et al.* 1985*b*). Ultrastructural studies based on the accumulation of a basic congener of dinitrophenol, DAMP (3-(2,4-dinitroanilino)-3′-amino-*N*-methyldipropylamine) to visualize compartmental differences in pH in the β-cell have demonstrated a lower pH in mature granules than in the TGN (Orci *et al.* 1985*b*). The proinsulin-rich immature granules were 15-fold more heavily labelled by anti-dinitrophenol antibodies than was the *trans* Golgi, and the labelling of the mature insulin granules was a further fivefold higher. This suggested a difference between the immature and mature granules of 0.7 pH units if there was a direct relationship between DAMP accumulation and pH. The pH of intact, isolated insulin granules has been measured, by the accumulation of a permeant base (methylamine), as being of the order of 5.0–5.5 (Hutton 1982). The experiments on DAMP accumulation suggest that the TGN has a near neutral pH and the immature granule a pH value intermediate between that of the TGN and insulin granules. Since this correlates with the localization of proinsulin and insulin in the cell, it follows that acidification may be an important event in the initiation of proteolytic conversion of proinsulin (Davidson *et al.* 1988).

Endopeptidase activity

Identification of the endopeptidase activity converting proinsulin to insulin has proved a difficult task. One approach involved the use of a radiolabelled active-site-directed protease inhibitor containing a dibasic amino-acid sequence ([^{125}I]Tyr–Ala–Lys–Arg–CH_2Cl), to probe rat insulin secretory-granule-enriched fractions under acidic conditions. Two proteins, of M_r 39 000 and 31 500, were labelled by this technique and shown to be immunologically related to the rat liver lysosomal thiol protease, cathepsin B. It was thought unlikely that cathepsin B itself was involved, given its substrate specificity *in vitro*, but it was suggested that the enzyme responsible for the conversion of proinsulin to insulin might be a granule cathepsin of M_r 39 000 (Docherty *et al.* 1983, 1984). In these experiments, the granule fraction degraded proinsulin in a manner suggestive of the participation of a thiol protease. However, the products of proteolysis were not identified, and the extracts would have contained lysosomal proteases which may have degraded proinsulin non-specifically.

Other investigations of the conditions necessary for the conversion of proinsulin to insulin by rat insulin granule extracts *in vitro* argue against the involvement of cathepsins, since the process was not affected by thiol protease inhibitors such as iodoacetic acid and E-64 (*trans*-epoxysuccinyl-L-leucylamido-(4-guanidino)butane), or activators such as dithiothreitol (Davidson *et al.* 1987). These studies suggested that the lysed granules contained two distinct endopeptidase activities, which could be separated by ion-exchange chromatography (Davidson *et al.* 1988). The enzymes, designated type 1 and type 2 endopeptidase, were also insensitive to group-specific inhibitors of thiol proteases, serine proteases (1 mM PMSF, phenylmethylsulphonyl fluoride; 1 mM DFP, di-isopropylfluorophosphate), and aspartyl proteases (1 mM pepstatin A). Both enzymes were inhibited by the chelating agents EDTA and CDTA (1,2-cyclohexanediaminetetraacetic acid) but not by the heavy-metal chelator 1,10-phenanthroline, which is diagnostic of metalloproteases. The inhibition by metal chelation was reversed by the addition of calcium. The two enzymes differed in Ca^{2+} sensitivity; type 2 endopeptidase was reactivated by micromolar Ca^{2+} ($K_{0.5}$ = 0.1 mM), but type 1 endopeptidase required millimolar Ca^{2+} ($K_{0.5}$ = 2.5 mM). The enzymes also had different substrate specificities; type 1 endopeptidase only cleaved proinsulin on the C-terminal side of Arg31, Arg32, whereas type 2 endopeptidase cleaved predominantly on the carboxyl side of Lys64, Arg65 but with some activity (~10 per cent) on the Arg31, Arg32 site (Fig. 6) (Davidson *et al.* 1988). The site specificity of the endopeptidases was also

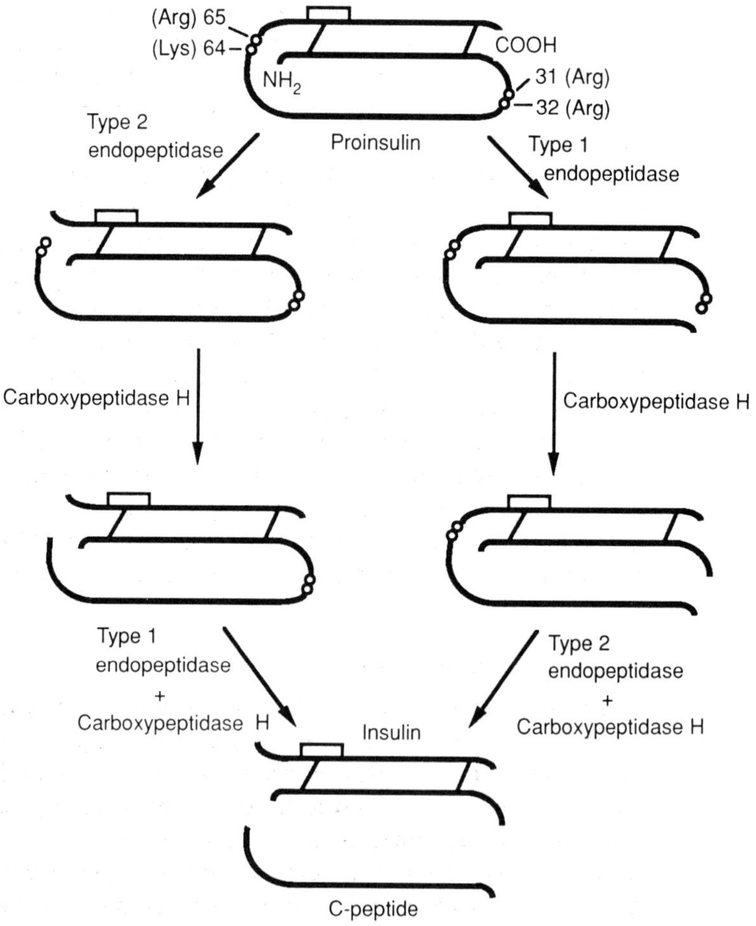

Fig. 6 Proteolytic conversion of proinsulin to insulin by insulin granule proteases

reflected in their sensitivity to protease inhibitors containing different dibasic amino-acid sequences (tripeptidyl sulphonium salts ($X_nCH_2S^+(CH_3)_2$) and a chloromethylketone). Type 1 endopeptidase was more sensitive to Ala–Arg–Arg–$CH_2S^+(CH_3)_2$ ($K_{0.5} = 6.1$ μM) than to Ala–Lys–Arg–$CH_2S^+(CH_3)_2$ ($K_{0.5} = 106$ μM); type 2 endopeptidase was more sensitive to Ala–Lys–Arg–$CH_2S^+(CH_3)_2$ ($K_{0.5} = 0.065$ μM) than to Ala–Arg–Arg–$CH_2S^+(CH_3)_2$ ($K_{0.5} = 9.6$ μM). Similarly, the tripeptide chloromethylketone, Ala–Lys–Arg–CH_2Cl, produced a greater inhibition with the type 2 endopeptidase ($K_{0.5} = 0.8$ μM) than with the type 1 endopeptidase ($K_{0.5} = 5$ μM). Both endopeptidases were much less sensitive to a chloromethylketone with a single basic residue (Ala–Norleu–Arg–CH_2Cl), 100 μM gave 40 per cent inhibition (Rhodes *et al.* 1989*a*).

Both enzymes had an acidic pH optimum of 5.5 but, whereas type 1 endopeptidase had a sharp pH optimum, that of type 2 endopeptidase was broader, with about 30 per cent of the maximum activity remaining at pH 7.5 (Davidson *et al.* 1988). The pH profile and calcium requirements of the two enzymes are consistent with the intracellular environment of the secretory granule. The granule is markedly more acidic than the other organelles of the secretory pathway, with an internal pH of around 5.5, and the free calcium concentration of the granule is calculated to be of the order of 1–10 mM (Hutton *et al.* 1983). This indicated that the action of the type 1 enzyme, in particular, is probably restricted to the intragranular environment.

Pulse-chase studies of the biosynthesis of insulin in rat islets showed that the first intermediate formed was des-64,65 proinsulin, i.e. proinsulin from which the basic residues at positions 64 and 65 have been removed. This intermediate was produced at a time when most of the newly synthesized secretory protein was within the Golgi; the time of appearance of des-31,32 proinsulin and insulin coincided with the transfer of newly synthesized protein to secretory granules (Davidson *et al.* 1988). These findings are consistent with the ability of type 2 endopeptidase (Lys–Arg-directed cleavage) to be active at a more neutral pH and at lower calcium concentrations than type 1 endopeptidase (Arg–Arg-directed cleavage). Most of the Lys–Arg-directed cleavage of proinsulin occurred post Golgi, in agreement with morphological studies using anti-proinsulin antibodies to detect conversion (Orci *et al.* 1985*b*; Steiner *et al.* 1987), which showed that insulin formation was confined to the maturing secretory granules.

Substrate specificity

The dibasic specificities of the two proinsulin endopeptidases have been further confirmed by their action on proalbumin. The processing of proalbumin to albumin in liver requires a single endopeptidase cleavage on the C-terminal side of an Arg–Arg sequence. Type 1 endopeptidase converted proalbumin to albumin with the same pH and calcium dependency shown towards proinsulin. Type 2 endopeptidase was only about a tenth as active as the type 1 enzyme, consistent with its Arg–Arg-directed catalysis being only a minor activity. A natural mutant proalbumin in which Arg–Arg was replaced by Arg–Gln was not cleaved by either insulin granule endopeptidase (Rhodes *et al.* 1989*b*).

Transfection of mouse POMC (pro-opiomelanocortin) into the rat insulinoma cell line RIN m5F resulted in processing of this multi-peptide precursor at several of the Lys–Arg cleavage sites. Mutation of the sites to Lys–Lys, Arg–Lys, Lys–Arg, and Arg–Arg showed strong preference for cleavage at Lys–Arg and Arg–Arg sites, consistent with the proinsulin processing sites (Thorne and Thomas 1990).

Mutant proinsulins were constructed and tested as substrates for the endopeptidases (Docherty *et al.* 1989). Des-38-62 proinsulin, in which all but six of the C-peptide residues were removed, was not processed by either enzyme. The mutant (Lys64–Arg65 to Thr64–Arg65) was not cleaved by the type 2 enzyme, but its Arg31–Arg32 site was still a substrate for the type 1 enzyme. The mutant (Arg31–Arg32 to Arg31–Gly32), however, was not processed at either C-peptide junction by either enzyme. Thus, disruption of the proinsulin structure at the B–C junction prevented processing at the C–A junction.

Alteration of the C-peptide sequence adjacent to the B–C junction also prevented proinsulin conversion. Comparison of different proinsulin sequences revealed the presence of a highly conserved tetrapeptide sequence adjacent to the Arg–Arg pair at the B–C junction: Glu/Asp–X–Glu/Asp (X being a hydrophobic amino acid, alanine, valine, or leucine), followed in most mammalian C-peptide sequences by aspartate. Transfection studies in AtT20 cells showed that deletion of this sequence (Glu–Val–Glu–Asp) from rat proinsulin II DNA did not affect targeting to the secretory granules, but did prevent processing to insulin (Gross *et al.* 1989*a*).

3.4 Cloning of the genes of proprotein-processing endopeptidases

3.4.1 Proteolytic processing in yeast

Saccharomyces cerevisiae, in common with higher eukaryotic organisms, uses dibasic amino-acid sequences to signal proteolytic processing of precursor proteins. Pro-pheromone conversion in yeast has been well characterized; genetic manipulation has provided positive identification of the proteases involved and the relevant genes have been isolated and sequenced. Processing of pro-alpha mating factor is initiated by endopeptidase cleavage on the carboxyl side of four Lys–Arg sites (Fuller *et al.* 1988). The enzyme responsible, the product of the KEX2 gene, can cleave foreign proproteins at dibasic sites; proinsulin is cleaved at the C–A and B–C junction and insulin is secreted from the cell (Thim *et al.* 1986). The KEX2 protein has been partially purified (to ~10 per cent purity) and its gene has been sequenced (Mizuno *et al.* 1988; Fuller *et al.* 1989*a,b*). KEX2 protease is a glycosylated membrane-bound protein of apparent M_r 135 000, 90 000 of which is contributed by the single polypeptide chain. The enzyme is located within the Golgi, requires micromolar calcium for activity, and is active at neutral pH. It has been classified as a calcium-dependent serine protease, although its inhibitor profile is rather mixed (Fuller *et al.* 1989*a*). KEX2 protease was inhibited by the chelating agents EDTA and EGTA, but not by the heavy-metal chelator 1,10-phenanthroline. The enzyme was not inhibited by millimolar concentrations of the serine protease inhibitor phenylmethylsulphonyl fluoride (PMSF), but was inhibited by di-isopropylfluorophosphate (DFP), albeit at rather high concentrations (50 per cent inhibition at 10 mM). The chloromethylketone class of compounds ($X–CH_2Cl$), which can inhibit serine and thiol proteases and whose specificity depends on the nature of the X group, had a mixed effect. TPCK (1-1-tosylamido-2-phenylethylchloromethyl-ketone) and TLCK (N^a-(*r*-tosyl)Lys-chloromethyl-ketone) were not inhibitory at millimolar concentrations. Ala–Lys–Arg–chloromethylketone, an inhibitor of the serine protease, trypsin, and the thiol protease, cathepsin B, (Docherty *et al.* 1983), was inhibitory at micromolar concentrations. The effect of reagents acting on thiol residues was inconclusive. Heavy-metal ions (Hg^{2+}, Zn^{2+}, Cu^{2+}), alkylating agents, dithiothreitol, and 4-(chloromercuri)phenylsulphonate were inhibitory, but 2-mercaptoethanol and the *trans*-epoxysuccinate inhibitors (e.g. E-64) of thiol proteases were not.

Convincing evidence for the classification of KEX2 as a serine protease came from its deduced amino-acid sequence (Mizuno *et al.* 1988; Fuller *et al.* 1989*b*). The KEX2 gene encodes 814 amino acids, containing an N-terminal signal sequence (~20 amino acids), a putative membrane-spanning domain (residues 679–699), and an acidic C-terminal tail (115 amino acids). Residues 152–410 of KEX2 were 50 per cent homologous with residues 9–246 of the bacterial serine protease, subtilisin BFN. The charge-relay catalytic mechanism of serine proteases involves a triad of amino acids at the active site (aspartate, histidine, and serine), with a further asparagine residue acting to stabilize an intermediate state during catalysis. This triad of amino acids in subtilisin (Asp32, His64, Ser221) and the Asn155 were conserved in KEX2 (Asp175, His213, Ser385, and Asn314); see Table 1 for comparison of the amino-acid residues around the four active-site residues.

3.4.2 Furin

Examination of the DNA data-bases revealed a human homologue with KEX2 and subtilisin, the *fur* gene, which is located immediately upstream of the *fps/fes* oncogene (Roebroek *et al.* 1986; Van den Ouweland *et al.* 1990). *fur* Encodes a protein of 794 amino acids, furin, containing an N-terminal signal peptide and a putative membrane-spanning domain. A subtilisin-like domain of ~270 amino acids, containing the active-site triad of aspartate, histidine, and serine, plus asparagine, was 45 per cent homologous with the corresponding region in KEX2. Furin mRNA was also detected in rats and African Green Monkey cells; its expression was high in the liver and kidney; low in the brain, spleen, and thymus; and very low in heart muscle, lung, and testes (Roebroek *et al.* 1986; Schalken *et al.* 1987). Furin enzymic activity has been demonstrated by co-transfecting COS-1 cells with furin and prepro-von Willebrand factor (vWF). Pro-vWF was converted to vWF by cleavage

Table 1 Sequences around the active-site amino acids of KEX2 protease and related proteins. The amino-acid sequences around the active-site amino acids of serine proteases (the essential charge-relay triad of Asp, His, and Ser, plus the catalytically important Asn/Asp residue) are shown for the yeast protease KEX2, the bacterial subtilisin BFN and the related proteins, human furin, human PC2 and PC3, and mouse PC1. Gaps were introduced into the KEX2 sequence to improve the alignment. The active-site amino acid is shown in bold type. Sequence data was obtained from Mizuno *et al.*, (1988); Seidah *et al.*, (1990); Smeekens and Steiner (1990); Smeekens *et al.* (1991), and van den Ouweland *et al.* (1990)

Protein	**Asp region**	**His region**
KEX2	VVAAIV**D**DGLDYENED	DDY---**H**GTRCAGEI
Subtilisin BFN	VKVAVI**D**SGIDSSHPD	FQDNNS**H**GTHVAGTV
Furin	IVVSIL**D**DGIEKVHPD	QMNDNR**H**GTRCAGEV
hPC2	VTIGIM**D**DGIDYLHPD	DDWFNS**H**GTRCAGEV
hPC3	VVITVL**D**DGLEWNHTD	LTNENK**H**GTRCAGEI
	Asn/Asp region	**Ser region**
KEX2	GAIYVFASG**N**GGTRGD	HGGT**S**AAAPLAAGVY
Subtilisin BFN	GVVVVAAAG**N**EGTSGS	YNGT**S**MASPHVAGAA
Furin	GSIFVWASG**N**GGREHD	HTGT**S**ASAPLAAGII
hPC2	GSIFVWASG**D**GGSYDD	HSGT**S**AAAPEAAGVF
mPC1	GSIFVWASG**N**GGRQGD	HTGT**S**ASAPLAAGIF
hPC3	GSIFVWASG**N**GGRQGD	HTGT**S**ASAPLAAGIF

at the Lys–Arg site (Van de Ven *et al.* 1990). Mutation of the cleavage site to Lys–Gly prevented processing, suggesting that furin is specific for dibasic amino-acids sites.

3.4.3 Serine protease homologues in endocrine tissue

Discovery of a mammalian homologue to KEX2 protease prompted searches for similar sequences in endocrine tissues that might encode the prohormone-processing endopeptidases. Two such sequences have been identified, PC2 and PC1 (PC3). PC2 was cloned from a human insulinoma library (Smeekens and Steiner 1990) and a mouse insulinoma library (Seidah *et al.* 1990). PC3 was first identified as a partial sequence (PC1, 826–1621 bp of PC3) in a mouse pituitary library (Seidah *et al.* 1990), and cloned in full from a mouse AtT20 library (Smeekens *et al.* 1991). Both groups used a similar strategy to obtain the probes used to isolate the clones. For human PC2, degenerate nucleotide primers were designed from the sequences surrounding the active-site aspartate and histidine residues of KEX2. These were used to identify and amplify the intervening sequence from a sample of human insulinoma RNA by the polymerase chain reaction. The 150 nucleotide probe thus generated was used to isolate PC2 from a human insulinoma cDNA library. Human PC2 comprised 2223 base pairs and encoded a protein of 638 amino acids; mouse PC3 comprised 2259 base pairs and encoded 758 amino acids. Both proteins contained an N-terminal signal peptide and a subtilisin-like domain. PC2 and PC3 were closely related; of the first 598 amino acids of PC3, 82 per cent were identical or similar in PC2. The serine protease domain of PC2 (Gly155 to Asp424) showed 27 per cent identity and 49 per cent conservative substitution compared to subtilisin BFN, and 49 per cent identity, 35 per cent similar compared to KEX2. The active-site catalytic triad of serine proteases was present in both PC2 and PC3, but the asparagine residue was substituted in PC2, but not PC3, by Asp310. The homology between PC2 and KEX2 extended outside that of the subtilisin domain; the first 594 amino acids of PC2 were 34 per cent identical with the corresponding KEX2 sequence, with an additional 41 per cent being similar. The main structural difference between KEX2 and PC2 and PC3 was the absence of a transmembrane-spanning domain and cytoplasmic tail. However, the C-terminal regions of PC2 and PC3 have a possible amphipathic helix, similar to that proposed for the membrane anchor of carboxypeptidase H (Fricker *et al.* 1990), which raises the possibility that they may be membrane-associated under certain circumstances.

PC2 and PC3 mRNAs were found in rat pancreatic islets and brain, mouse and human insulinomas, and in mouse AtT20 cells. Both clones had two sizes, 2.8 and 5 kb for PC2, and 3 and 5 kb for PC3. The abundance of the PC2 and PC3 mRNAs appeared to be inversely related: brain, islets, and insulinomas contained high levels of PC2 mRNA but low levels of PC3 mRNA; AtT20 cells contained high levels of PC3 and low levels of PC2 (Smeekens *et al.* 1991). PC2 and PC3 mRNAs were undetectable in liver, kidney, heart, intestine, skeletal muscle, and spleen. The 2.8 kb form of PC2 predominated in the insulinoma cell line (Seidah *et al.* 1990).

3.4.4 The cellular roles of furin, PC2, and PC3

The tissue distribution of furin suggests that it is a protease involved in the processing of constitutively secreted proteins and membrane proteins. The question of identity between the PC sequences and the processing enzyme activities of neuroendocrine tissues awaits resolution. There is circumstantial evidence that PC2 and PC3 correspond to the two rat insulinoma endopeptidases that process proinsulin to insulin (Davidson *et al.* 1988). The neuroendocrine location of the PC sequences is appropriate, and the inhibitor profile and calcium requirement of the KEX2 protease is similar, though not identical, to those of type 1 and 2 endopeptidases. Both endopeptidases are soluble proteins, which fits with the absence of a transmembrane domain on the PC sequences. Our unpublished studies show that the size of type 1 endopeptidase (~60 kDa by gel filtration) is similar to that predicted by the PC sequences. One interesting feature of the PC2 sequence is the substitution of Asn by Asp, which might, by analogy with the subtilisins, considerably decrease its catalytic activity. It may be possible, however, that the aspartate residue could substitute with reasonable efficiency for asparagine at the acidic pH optima of the endopeptidases (Smeekens *et al.* 1991) and be an important determinant of the acidic pH optimum of the enzyme.

Expression of furin and the PC sequences in enzymically active forms in appropriate cell types is obviously necessary to enable their full characterization and correspondence with the processing enzymes identified by the methods of protein biochemistry. How the proteases are targeted to the secretory granules and whether they are synthesized in a precursor form and, if so, how they are activated, are questions that should soon be addressed.

3.4.5 Carboxypeptidase H

The basic amino-acid residues exposed at the B–C and C–A junctions by endopeptidase activity are removed by carboxypeptidase H (Davidson and Hutton 1987). This enzyme corresponds to carboxypeptidase E ('enkephalin convertase'; EC 3.4.17.10) which is widespread in many endocrine and neural tissues. The enzyme activity is localized to the secretory granules, where it exists in soluble and membrane-bound forms, which differ in molecular mass by 2–3 kDa (Fricker 1988). The membrane-bound form can be extracted from the membrane at pH 9. Solubilization at low pH (5–6) requires the addition of detergent and a high salt concentration. In the pancreatic β-cell, carboxypeptidase H comprises 2–5 per cent of the total granule protein, and is present mostly in its soluble form, with an apparent M_r of 55 000 (Davidson and Hutton 1987). The soluble enzyme has a narrow, acidic pH optimum between 5 and 6, consistent with its acidic intragranular location. Carboxypeptidase H activity is stimulated fivefold by millimolar $CoCl_2$, and is not affected by calcium ions. It is inhibited by metal chelators such as EDTA, CDTA, and 1,10-phenanthroline due to the presence of bound zinc which is involved in the catalytic mechanism. This is also reflected in the ability of the active-site-directed inhibitor GEMSA (guanidinomercaptosuccinic acid) to inhibit the enzyme at low concentrations (K_i = 9 nM).

The cDNA for the enzyme (1989 nucleotides) has been cloned from rat hippocampal and hypothalamic libraries (Fricker *et al.* 1989) and also, in our own unpublished work, from rat islet and insulinoma libraries. It encodes a protein of 476 amino acids. Northern blot analysis of the mRNA distribution in different rat tissues showed a single species of mRNA, with the highest levels found in endocrine and neural tissues.

The rat, bovine, and human carboxypeptidase H sequences are highly conserved, with >90 per cent homology being observed (Manser *et al.* 1990). The homology with the exocrine pancreatic enzymes, carboxypeptidase B and A is 17 per cent and 20 per cent respectively, and is highest in the regions of the molecule thought to be important for zinc binding and catalytic function.

The first 16 amino acids at the N-terminus form a typical hydrophobic signal peptide sequence which directs translocation into the rough endoplasmic reticulum during protein synthesis. The molecule contains five dibasic amino-acid sites, including a sequence of five arginine residues at amino-acid position 38–42, which might be a signal for processing by endopeptidase activity during biosynthesis. Furthermore, Arg455, Lys456 precedes a sequence of 11 amino-acid residues at the

C-terminus, which is predicted to form an amphiphilic alpha-helix, capable of interaction with the membrane. The function of this sequence has been examined in bovine pituitary carboxypeptidase E in two separate studies using antibodies raised to the C-terminal region. One study found that antibodies raised to the last nine residues recognized the soluble enzyme (Parkinson 1990), whereas Fricker and co-workers found that only the membrane-bound form contained the 11 amino-acid C-terminal epitope(s). It was postulated that this structure formed a pH-dependent membrane anchor which is proteolytically removed to generate the soluble form (Fricker *et al.* 1990).

References

Ashcroft, S. J. H., Bunce, J., Lowry, M., Hansen, S. E., and Hedeskov, C. J. (1978). The effect of sugars on (pro)insulin biosynthesis. *Biochem. J.* **174,** 517–26.

Bek, E. and Berry, R. (1990). Prohormone cleavage sites are associated with omega loops. *Biochemistry* **29,** 178–83.

Beckers, C. J. M. and Balch, W. E. (1989). Ca^{2+} & GTP: Essential components in vesicular trafficking between the ER and Golgi apparatus. *J. Cell Biol.* **108,** 1245–56.

Bourne, H. R. (1988). Do GTPases direct membrane traffic in secretion? *Cell* **53,** 669–71.

Briggs, M. S. and Gierasch, L. M. (1986). Molecular mechanisms of protein secretion: the role of the signal sequence. *Adv. Prot. Chem.* **38,** 109–80.

Bulleid, N. J. and Freedman, R. B. (1988). Defective co-translational formation of disulphide bonds in protein disulphide isomerase-deficient microsomes. *Nature* **335,** 649–51.

Burgess, T. L. and Kelly, R. B. (1987). Constitutive and regulated secretion of proteins. *Ann. Rev. Cell Biol.* **3,** 243–94.

Burgoyne, R. D. (1989). Small GTP-binding proteins. *TIBS* **14,** 394–6.

Carroll, R. J., Hammer, R. E., Chan, S. J., Swift, H. H., Rubenstein, A. H., and Steiner, D. F. (1988). A mutant human proinsulin is secreted from the islets of Langerhans in increased amounts via an unregulated pathway. *Proc. Natl Acad. Sci. USA* **85,** 8943–7.

Chavrier, P. C., Paron, R. G., Hauri, H. P., Simons, K., and Zerial, M. (1990). Localization of low M_r GTP-binding proteins to the exocytic and endocytic compartments. *Cell* **62,** 317–29.

Chung, K.-N., Walter, P., Aponte, G. W., and Moore, H. H.-P. (1989). Molecular sorting in the secretory pathway. *Science* **243,** 192–7.

Clemens, M. J. (1989). Regulatory mechanisms in translational control. *Curr. Opinion in Cell Biol.* **1,** 1160–7.

Connolly, T. and Gilmore, R. (1989). The signal recognition particle receptor mediates the GTP-dependent displacement of SRP from the signal sequence of the nascent polypeptide. *Cell* **57,** 599–610.

Davidson, H. W. and Hutton, J. C. (1987). The insulin secretory-granule carboxypeptidase H. *Biochem. J.* **245,** 575–62.

Davidson, H. W., Peshavaria, M., and Hutton, J. C. (1987). Proteolytic conversion of proinsulin into insulin. Identification of a Ca^{+}-dependent endopeptidase in isolated insulin-secretory granules. *Biochem. J.* **246,** 279–86.

Davidson, H. W., Rhodes, C. J., and Hutton, J. C. (1988). Intraorganellar calcium and pH control proinsulin cleavage in the pancreatic B cell via two distinct site-specific endopeptidases. *Nature* **333,** 93–6.

Derewenda, U., Derewenda, Z., Dodson, G. G., Hubbard, R. E., and Korber, F. (1989). Molecular structure of insulin: The insulin monomer and its assembly. *Br. Med. Bull.* **45,** 4–18.

Docherty, K. and Steiner, D. F. (1982). Post-translational proteolysis in polypeptide hormone biosynthesis. *Ann. Rev. Physiol.* **44,** 625–38.

Docherty, K., Carroll, R., and Steiner, D. F. (1983). Identification of a 31,500 molecular weight islet cell protease as cathepsin B. *Proc. Natl Acad. Sci. USA* **80,** 3245–9.

Docherty, K., Hutton, J. C., and Steiner, D. F. (1984). Cathepsin B-related proteases in the insulin secretory granules. *J. Biol. Chem.* **259,** 6041–4.

Docherty, K., Rhodes, C. J., Taylor, N. A., Shennan, K. I. J., and Hutton, J. C. (1989). Proinsulin endopeptidase substrate specificities defined by site-directed mutagenesis of proinsulin. *J. Biol. Chem.* **264,** 18335–9.

Eskridge, E. M. and Shields, D. (1983). Cell-free processing and segregation of insulin precursors. *J. Biol. Chem.* **258,** 11487–91.

Farquhar, M. G. (1985). Progress in unraveling pathways of Golgi traffic. *Ann. Rev. Cell Biol.* **1,** 447–88.

Fine, R. E. (1989). Vesicles without clathrin: intermediates in bulk flow exocytosis. *Cell* **56,** 609–10.

Freedman, R. B. (1984). Native disulphide bond formation in protein biosynthesis: evidence for the role of protein disulphide isomerase. *TIBS* **9,** 438–41.

Fricker, L. D. (1988). Carboxypeptidase E. *Ann. Rev. Physiol.* **50,** 309–21.

Fricker, L. D., Adelman, J. P., Douglass, J., Thompson, R. C., von Strandman, R. P., and Hutton, J. C. (1989). Isolation and sequence analysis of cDNA for rat carboxypeptidase E [EC 3.4.17.10], a neuropeptide processing enzyme. *Mol. Endocrinol.* **3,** 666–73.

Fricker, L. D., Das, B., and Hogue-Angeletti, R. (1990). Identification of the pH dependent membrane anchor of carboxypeptidase E [EC 3.4.17.10.]. *J. Biol. Chem.* **265,** 2476–82.

Fuller, R. S., Sterne, R. E., and Thorner, J. (1988). Enzymes required for yeast prohormone processing. *Ann. Rev. Physiol.* **50,** 345–62.

Fuller, R. S., Brake, A., and Thorner, J. (1989*a*). Yeast prohormone processing enzyme (KEX2 gene product) is a Ca^{++} dependent serine protease. *Proc. Natl Acad. Sci. USA* **86,** 1434–8.

Fuller, R. S., Brake, A. J., and Thorner, J. (1989*b*). Intracellular targeting and structural conservation of a prohormone processing endopeptidase. *Science* **246,** 482–6.

Gerdes, H.-H., Rosa, P., Phillips, E., Baeuerle, P. A., Frank, R., Argos, P., and Huttner, W. B. (1989). The primary structure of human secretogranin II, a widespread tyrosine-sulphated secretory granule protein that exhibits low pH and calcium-induced aggregation. *J. Biol. Chem.* **264,** 12009–15.

Gilmore, R. and Blobel, G. (1983). Transient involvement of signal recognition particle and its receptor in the microsomal membrane prior to protein translocation. *Cell* **35,** 677–85.

Griffiths, G. and Simons, K. (1986). The *trans* Golgi network: sorting at the exit site of the Golgi complex. *Science* **234,** 438–43.

Grimaldi, K. A., Siddle, K., and Hutton, J. C. (1987). Biosynthesis of insulin secretory granule membrane proteins: control by glucose. *Biochem. J.* **245,** 567–73.

Gross, D. J., Villa-Komaroff, L., Kaln, C. R., Weir, G. C., and Halban, P. A. (1989*a*). Deletion of a highly conserved tetrapeptide sequence of the proinsulin connecting peptide (C-peptide) inhibits proinsulin to insulin conversion by transfected pituitary corticotroph (AtT20) cells. *J. Biol. Chem.* **264,** 21486–90.

Gross, D. J., Halban, P. A., Kaln, C. R., Weir, G. C., and Villa-Komaroff, L. (1989*b*). Partial diversion of a mutant proinsulin (B10 aspartic acid) from the regulated to the constitutive secretory pathway in transfected AtT-20 cells. *Proc. Natl Acad. Sci. USA* **86,** 4107–11.

Guest, P. C., Rhodes, C. J., and Hutton, J. C. (1989). Regulation of the biosynthesis of insulin-secretory-granule proteins: Co-ordinate translational control is exerted on some, but not all, granule matrix constituents. *Biochem. J.* **257,** 431–7.

Guest, P. C., Bailyes, E. M., Rutherford, N. G., and Hutton, J. C. (1991). Insulin secretory granule biogenesis: Co-ordinate regulation of the biosynthesis of the majority of constituent proteins. *Biochem. J.*, **274,** 73–8.

Hedeskov, C. J. (1980). Mechanisms of glucose induced insulin secretion. *Physiol. Rev.* **60,** 442–509.

Huttner, W. B. and Tooze, S. A. (1989). Biosynthetic protein transport in the secretory pathway. *Curr. Opinion in Cell Biol.* **1,** 648–54.

Hutton, J. C. (1982). The internal pH and membrane potential of the insulin-secretory granule. *Biochem. J.* **204,** 171–8.

Hutton, J. C., Penn, E. J., and Peshavaria, M. (1982). Isolation and characterisation of insulin secretory granules from a rat islet cell tumour. *Diabetologia* **23,** 365–73.

Hutton, J. C., Penn, E. J., and Peshavaria, M. (1983). Low molecular weight constituents of isolated insulin secretory granules: divalent cations, adenine nucleotides and inorganic phosphate. *Biochem. J.* **210,** 297–305.

Hutton, J. C., Bailyes, E. M., Rhodes, C. J., Rutherford, N. G., Arden, S. A., and Guest, P. C. (1990). Biosynthesis and storage of insulin. *Biochem. Soc. Trans.* **18,** 122–4.

Itoh, N. and Okamoto, H. (1980). Translational control of proinsulin synthesis by glucose. *Nature* **283,** 100–2.

Kurzchalia, T. V., Wiedmann, M., Girshovich, A. S., Bochkareva, E. S., Bielka, H., and Rapoport, T. A. (1986). The signal sequence of nascent preprolactin interacts with the 54kD polypeptide of the signal recognition particle. *Nature* **320,** 634–6.

Lewis, M. J. and Pelham, H. R. B. (1990). A human homologue of the yeast HDEL receptor. *Nature* **348,** 162–3.

Lively, M. O. (1989). Signal peptidases in protein biosynthesis and intracellular transport. *Curr. Opinion in Cell Biol.* **1,** 1188–93.

Loh, Y. P., Brownstein, M. J., and Gainer, H. (1984). Proteolysis in neuropeptide processing and other neural functions. *Ann. Rev. Neurosci.* **7,** 189–222.

Maldonato, A., Renold, A. E., Sharp, G. W. G., and Cerasi, E. (1977). Glucose-induced proinsulin biosynthesis: role of cyclic AMP. *Diabetes* **26,** 538–45.

Manser, E., Fernandez, D., Loo, L., Goh, P. Y., Monfries, C., Hall, C., and Lim, L. (1990). Human carboxypeptidase E. Isolation and characterization of the cDNA sequence conservation, expression and processing *in vitro. Biochem. J.* **267,** 517–25.

Meyer, D. I., Krause, E., and Dobberstein, B. (1982). Secretory protein translocation across membranes – the role of the 'docking protein'. *Nature* **297,** 647–50.

Mizuno, K., Nakamura, T., Oshima, T., Tanaka, S., and Matsuo, H. (1988). Yeast KEX2 gene encodes an endopeptidase homologous to subtilisin-like serine proteases. *Biochem. Biophys. Res. Comm.* **156,** 246–54.

Moore, H.-P. H. and Kelly, R. B. (1986). Rerouting of a secretory protein by fusion with human growth hormone sequence. *Nature* **321,** 443–6.

Okun, M. M., Eskridge, E. M., and Shields, D. (1990). Truncations of a secretory protein define minimum lengths required for binding to signal recognition particle and translocation across the endoplasmic reticulum membrane. *J. Biol. Chem.* **265,** 7478–84.

Orci, L., Ravazzola, M., Amherdt, M., Louvard, D., and Perrelet, A. (1985*a*). Clathrin-immunoreactive sites in the Golgi apparatus are concentrated at the *trans* pole in polypeptide hormone-secreting cells. *Proc. Natl Acad. Sci. USA* **82,** 5385–9.

Orci, L., Ravazzola, M., Amherdt, M., Madsen, O., Vassalli, J.-D., and Perrelet, A. (1985*b*). Direct identification of the prohormone conversion site in insulin-secreting cells. *Cell* **42,** 671–91.

Orci, L., Ravazzola, M., Amherdt, M., Madsen, O., Perrelet, A., Vassalli, J.-D., and Anderson, R. G. W. (1986). Conversion of proinsulin to insulin occurs coordinately with acidification of maturing secretory vesicles. *J. Cell Biol.* **103,** 2273–81.

Orci, L., Ravazzola, M., Amherdt, M., Perrelet, A., Powell, S. K., Quinn, D. L., and Moore, H.-P. H. (1987). The *trans*-most cisternae of the Golgi complex: a compartment for the sorting of secretory and plasma membrane proteins. *Cell* **51,** 1039–51.

Orci, L., Malhotra, V., Amherdt, M., Serafini, T., and Rothman, J. E. (1989). Dissection of a single round of vesicular transport: sequential intermediates for intercisternal movement in the Golgi stack. *Cell* **56,** 357–68.

Parkinson, D. (1990). Two soluble forms of bovine carboxypeptidase H have different NH_2-terminal sequences. *J. Biol. Chem.* **265,** 17101–5.

Payne, G. S. and Schekman, R. (1989). Clathrin: a role in the intracellular retention of a Golgi membrane protein. *Science* **245,** 1358–65.

Pearse, B. M. F. and Robinson, M. S. (1990). Clathrin, adaptors and sorting. *Ann. Rev. Cell Biol.* **6,** 151–71.

Pelham, H. R. B. (1989). Control of protein exit from the endoplasmic reticulum. *Ann. Rev. Cell Biol.* **5,** 1–24.

Permutt, M. A. (1981). In *Islets of Langerhans* (ed. S. Cooperstein and D. T. Watkins), pp. 75–95. Academic Press, New York.

Permutt, M. A. and Kipnis, D. M. (1972). 1. On the mechanism of glucose stimulation. *J. Biol. Chem.* **247,** 1194–9.

Permutt, M. A. and Kipnis, D. M. (1975). Insulin biosynthesis and secretion. *Fed. Proc.* **34,** 1549–55.

Pfeffer, S. R. and Rothman, J. E. (1987). Biosynthetic protein transport and sorting by the endoplasmic reticulum and Golgi. *Ann. Rev. Biochem.* **56,** 829–52.

Pipeleers, D. G., Marichal, M., and Malaisse, W. J. (1973*a*). The stimulus–secretion coupling of glucose-induced insulin release. XIV. Glucose regulation of insular biosynthetic activity. *Endocrinology* **93,** 1001–11.

Pipeleers, D. G., Marichal, M., and Malaisse, W. J. (1973*b*). The stimulus–secretion coupling of glucose-induced insulin release. XV. Participation of cations in the recognition of glucose by the beta-cell. *Endocrinology* **93,** 1012–18.

Powell, S. K., Orci, L., Craik, C. S., and Moore, H. H.-P. (1988). Efficient targeting to storage granules of human proinsulins with altered propeptide domains. *J. Cell Biol.* **106,** 1843–51.

Pugsley, A. P. (1990). Translocation of proteins with signal sequences across membranes. *Curr. Opinion in Cell Biol.* **2,** 609–16.

Rapoport, T. A. (1990). Protein transport across the ER membrane. *TIBS* **15,** 355–8.

Rhodes, C. J. and Halban, P. A. (1987). Newly synthesised proinsulin/insulin and stored insulin are released from pancreatic B cells predominantly via a regulated rather than a constitutive pathway. *J. Cell Biol.* **105,** 145–53.

Rhodes, C. J., Zumbrunn, A., Bailyes, E. M., Shaw, E., and Hutton, J. C. (1989*a*). The inhibition of proinsulin-processing endopeptidase activities by active-site-directed peptides. *Biochem. J.* **258,** 305–8.

Rhodes, C. J., Brennan, S. O., and Hutton, J. C. (1989*b*). Proalbumin to albumin conversion by a proinsulin processing endopeptidase of insulin secretory granules. *J. Biol. Chem.* **264,** 14240–5.

Rholam, M., Nicolas, P., and Cohen, P. (1986). Precursors for peptide hormones share common secondary structures forming features at the proteolytic processing site. *FEBS Lett.* **207,** 1–6.

Roebroek, A. J. M., Schalken, J. A., Bussemakers, M. J. G., Heerikhuizen, H. van, Onnekink, C., Debruyne, F. M. J., Bloemers, H. P. J., and Ven, W. J. M. van de (1986). Characterization of human c fes/fps reveals a new transcription unit (fur) in the immediately upstream region of the proto-oncogene. *Molec. Biol. Rep.* **11,** 117–25.

Rosa, P., Weiss, U., Pepperkok, R., Ansorge, W., Niehrs, C., Stelzer, E. H. K., and Huttner, W. B. (1989). An antibody against secretogranin 1 (Chromogranin B) is packaged into secretory granules. *J. Cell Biol.* **109,** 17–34.

Savitz, A. J. and Meyer, D. I. (1990). Identification of a ribosome receptor in the rough endoplasmic reticulum. *Nature* **346,** 540–4.

Schalken, J. A., Roebroek, A. J. M., Oomen, P. P. C. A., Wagenaar, S. S., Debruyne, F. M. J., Bloemers, H. P. J., and Van de Ven, W. J. M. (1987). *fur* gene expression as a discriminating marker for small cell and non-small cell lung carcinomas. *J. Clin. Invest.* **80,** 1545–9.

Seidah, N. G., Gaspar, L., Mion, P., Marcinkiewicz, M., Mbikay, M., and Chretien, M. (1990). cDNA sequence of two distinct pituitary proteins homologous to Kex2 and furin gene products: tissue-specific mRNAs encoding candidates for prohormone processing proteinases. *DNA Cell Biol.* **9,** 415–24.

Semenza, J. C., Hardwick, K. G., Dean, N., and Pelham, H. R. B. (1990). ERD2, a yeast gene required for the receptor-mediated retrieval of luminal ER proteins from the secretory pathway. *Cell* **61,** 1349–57.

Shelness, G. S. and Blobel, G. (1990). Two subunits of the canine signal peptidase complex are homologous to yeast SEC11 protein. *J. Biol. Chem.* **265,** 9512–19.

Smeekens, S. P. and Steiner, D. F. (1990). Identification of a human insulinoma cDNA encoding a novel mammalian protein structurally related to the yeast dibasic processing protease KEX2. *J. Biol. Chem.* **265,** 2997–3000.

Smeekens, S. P., Avruch, A. S., LaMendola, J., Chan, S. J., and Steiner, D. F. (1991). Identification of a cDNA encoding a second putative prohormone convertase related to PC2 in AtT20 cells and islets of Langerhans. *Proc. Natl Acad. Sci. USA* **88,** 340–344.

Sopwith, A. M., Hutton, J. C., Naber, S. P., Chick, W. L., and Hales, C. N. (1981). Insulin secretion by a transplantable rat islet cell insulinoma. *Diabetologia* **21,** 224–9.

Steiner, D. F., Michael, J., Houghten, R., Mathieu, M., Gardner, P. R., Ravazzola, M., and Orci, L. (1987). Use of a synthetic peptide antigen to generate antisera reactive with a proteolytic processing site in native human proinsulin: Demonstration of cleavage within clathrin-coated (pro)secretory vesicles. *Proc. Natl Acad. Sci. USA* **84,** 6184–8.

Steiner, D. F., Bell, G. I., and Tager, H. S. (1989). Chemistry and biosynthesis of pancreatic proteins. In *Endocrinology* (ed. L. J. De Groot) Chapter 15. Saunders Co., Harcourt Brace Jovanovich Inc., Philadelphia.

Thim, L., Hansen, M. T., Norris, K., Hoegh, I., Boel, E., Forstrom, J., Ammerer, G., and Fiil, N. P. (1986). Secretion and processing of insulin precursors in yeast. *Proc. Natl Acad. Sci. USA* **83,** 6766–70.

Thorne, B. A. and Thomas, G. (1990). An *in vivo* characterization of the cleavage site specificity of the insulin cell prohormone processing enzymes. *J. Biol. Chem.* **265,** 8436–43.

Tooze, J. and Tooze, S. A. (1986). Clathrin-coated vesicular transport of secretory proteins during the formation of ACTH-containing secretory granules in AtT20 cells. *J. Cell Biol.* **103,** 839–50.

Tooze, S. A. and Huttner, W. B. (1990). Cell-free protein sorting to the regulated and constitutive pathways. *Cell* **60,** 837–47.

Tooze, S. A., Weiss, U., and Huttner, W. B. (1990). Requirement for GTP hydrolysis in the formation of secretory vesicles. *Nature* **347,** 207–8.

Van de Ven, W. J. M., Voorberg, J., Fontijn, R., Pannekoek, H., Ans, M. W., Van den Ouweland, A. M. W., Duijhoven, H. L. P., Roebroek, A. J. M., and Siezen, R. J. (1990). Furin is a subtilisin-like proprotein processing enzyme in higher eucaryotes. *Molec. Biol. Rep.* **14,** 265–75.

Van den Ouweland, A. M. W., Van Duijnhoven, H. L. P., Keizer, G. D., Dorssers, C. J., and Van de Ven, W. J. M. (1990). Structural homology between the human *fur* gene product and the subtilisin-like protease encoded by yeast KEX2. *Nucleic Acids Res.* **18,** 664.

Vaux, D., Tooze, J., and Fuller, S. (1990). Identification by anti-idiotypic antibodies of an intracellular membrane protein that recognizes a mammalian endoplasmic reticulum retention signal. *Nature* **345,** 495–502.

Weiss, M. A., Frank, B. H., Khait, I., Pekar, A., Heiney, R., Shoelson, S. E., and Neuringer, L. J. (1990). NMR and photo-CIDNP studies of human proinsulin and prohormone processing intermediates with application to endopeptidase recognition. *Biochemistry* **29,** 8389–401.

Welsh, M., Scherberg, N., Gilmore, R., and Steiner, D. F. (1986). Translational control of insulin biosynthesis. Evidence for regulation of elongation, initiation and signal-recognition-particle-mediated translational arrest by glucose. *Biochem. J.* **235,** 459–67.

Welsh, N., Welsh, M., Steiner, D. F., and Hellerstrom, C. (1987). Mechanisms of leucine- and theophylline-stimulated insulin biosynthesis in isolated rat pancreatic islets. *Biochem. J.* **246,** 245–8.

Wiedmann, M., Kurzchalia, T. V., Bielka, H., and Rapoport, T. A. (1987). Direct probing of the interaction between the signal sequence of nascent preprolactin and the signal recognition particle by specific cross-linking. *J. Cell Biol.* **104,** 201–8.

Wiedmann, M., Goerlich, D., Hartmann, E., Kurzchalia, T. V., and Rapoport, T. A. (1989). Photocrosslinking demonstrates proximity of a 34 kDa membrane protein to different portions of preprolactin during translocation through the endoplasmic reticulum. *FEBS Lett.* **257,** 263–8.

Yoo, H. S. and Albanesi, J. P. (1990). Ca^{2+}-induced conformational change and aggregation of chromogranin A. *J. Biol. Chem.* **265,** 14414–21.

PART III INSULIN SECRETION

Editors' introduction

The principal determinant of the rate of insulin release is the circulating level of blood glucose: hormones and neurotransmitters act to modulate the secretory response to glucose.

Insulin secretagogues may be divided into two groups, the initiators and the potentiators. The former are substances that are capable of stimulating insulin release on their own and include nutrients (such as glucose) which are metabolized by the β-cell; substances that stimulate metabolism of endogenous nutrients; and drugs such as the sulphonylureas tolbutamide and glibenclamide. Potentiators of insulin release comprise a number of hormones such as glugacon, gastrointestinal peptide (GIP), vasoactive intestinal peptide (VIP), and transmitters such as acetylcholine (ACh). These substances are able to potentiate insulin secretion in the presence of glucose but cannot initiate release in the absence of the sugar. Insulin release can be inhibited by drugs such as diazoxide and by physiological agents such as somatostatin, galanin, calcitonin gene-related peptide (CGRP), and α_2-adrenergic agents.

It has been known for a long time that the β-cell is electrically excitable and that electrical activity and its associated Ca^{2+} influx into the β-cell is a key event in initiating secretion. The last few years have seen a dramatic advance in our understanding of the molecular mechanisms by which glucose and sulphonylureas elicit electrical activity and insulin secretion. It is now clear that both these agents act by inhibiting a specific type of potassium channel in the plasma membrane of the β-cell, the ATP-regulated K-channel (K-ATP channel) so called because it is blocked by intracellular ATP. Thus the common denominator for these and other initiators of insulin secretion is their ability to close K-ATP channels and depolarize the β-cell. In contrast, the effects of potentiators involve other mechanisms such as stimulation of β-cell protein kinases. Inhibitors may act at the level of metabolism, the plasma membrane, or exocytosis itself.

The key elements in our current understanding of β-cell stimulus-secretion coupling are as follows. The K-ATP channel controls the β-cell resting potential and as a consequence of its inhibition, the β-cell membrane depolarizes. This opens voltage-dependent Ca-channels and the resulting Ca^{2+} influx stimulates insulin secretion. Whereas sulphonylureas act as direct channel blockers, glucose and other nutrient secretagogues have to be metabolized by the β-cell in order to inhibit the K-ATP channel and thereby stimulate insulin secretion.

The metabolism of glucose and other nutrient secretagogues is an essential step in stimulus-secretion coupling in β-cells. In the cytosol, glucose is converted to pyruvate by glycolysis in a series of nine enzymic steps, some of which are shown in Fig. 6. Although ATP is synthesized during this process, the bulk of the energy derived from glucose is produced during the subsequent oxidation of pyruvate within the mitochondrion. Pyruvate is first converted to acetyl CoA which is then oxidized to CO_2 and water by the citric acid cycle. The NADH produced by these reactions is oxidized by the respiratory chain, which is coupled to the synthesis of ATP in a process known as oxidative phosphorylation. Under anaerobic conditions, pyruvate is converted into lactate within the cytosol. Other routes for the generation of ATP and NAD(P)H from glucose are also known but play relatively minor roles in the β-cell. Fatty acids are converted to their acyl CoA derivatives, before being transported into the mitochondria where they are converted to acetyl CoA and oxidized via the citric acid cycle and oxidative phosphorylation. Metabolism of amino acids involves their transamination to produce a 2-keto acid, which is then converted to a citric acid cycle precursor or intermediate, and subsequently oxidized.

The β-cell is enzymically equipped to respond to an increase in bood glucose by an increased rate of metabolism of the sugar. The key elements in this response are:

(1) the non-rate-limiting glucose transporter (GLUT-2);

(2) the presence of high K_m glucokinase;

(3) the presence of glucose 6-phosphatase.

In these respects β-cell glucose metabolism closely resembles that of the liver which, of course, also serves as a glucose-sensing organ. It is envisaged that an increase in the rate at which glucose is metabolized by the β-cell increases the intracellular ATP/ADP ratio and so inhibits K-ATP channel activity.

Secretion occurs by exocytosis as a consequence of a rise in the intracellular concentration of calcium ions. In this poorly understood process, the secretory granules move to, and fuse with, the plasma membrane, discharging their contents to the extracellular milieu. There is evidence that Ca^{2+}-dependent protein phosphorylation plays a role in the initiation of exocytosis, but the molecular details are not understood. Potentiatory effects of hormones and neurotransmitters may also involve activation of protein phosphorylation via effects on protein kinases A and C. Release of calcium from intracellular stores may be involved in certain potentiatory effects; however, a major mechanism for potentiation of secretion seems to be a sensitization of the secretory system to calcium ions.

4 | Mechanism of insulin secretion

FRANCES M. ASHCROFT and STEPHEN J. H. ASHCROFT

Insulin release *in vivo* is regulated principally by the circulating glucose level together with extensive modulatory effects exerted by hormones, neurotransmitters, and other nutrients. The main agents capable of influencing insulin release, both physiologically and experimentally, are summarized in Fig. 1. These include initiators of secretion, such as metabolic substrates and the sulphonylureas, and a variety of substances that act to potentiate or inhibit the secretory response to an initiator. Of major importance are paracrine influences from adjacent non-β-cells of the islet; these include glucagon, released from α-cells, and somatostatin, released from D-cells. Both neural inputs and gastrointestinal hormones may underlie the enhanced secretory response to oral, as opposed to intravenous, glucose loading. Pharmacological agents, in particular the sulphonylureas, also influence insulin secretion.

In this chapter, we consider in detail the molecular mechanisms involved in regulating insulin release.

4.1 Methods

The following basic techniques have been used widely to study the secretory function of the pancreatic β-cell.

4.1.1 Measurement of insulin

Insulin levels in plasma or in incubation media can be quantified by radioimmunoassay, immunoradiometric assay, or enzyme-linked immuno-absorption assay (ELISA). These methods rely on the specific high-affinity binding of insulin by antibodies to the hormone. Figure 2 illustrates the principles. In the radioimmunoassay (Hales and Randle 1963), radioactively labelled insulin is incubated with a limiting concentration of anti-insulin antibodies. The bound and free radioactivity is then separated. The amount of radioactive insulin bound will be reduced in the simultaneous presence of unlabelled insulin. Thus it is possible to construct a standard curve, using known concentrations of unlabelled insulin, which can then be used to determine the concentration of (unlabelled) insulin in the test sample.

The immunoradiometric assay (Hales 1972) differs in that the antibody is the

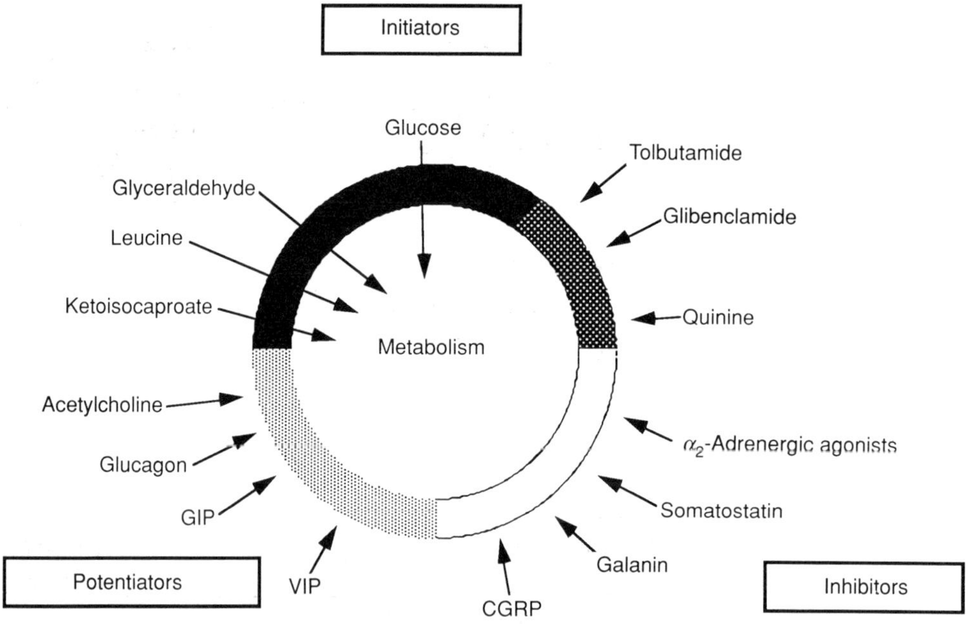

Fig. 1 Modulators of insulin secretion. Initiators of secretion are able to elicit insulin release by themselves. They fall into two classes—metabolic substrates and drugs—with a common mode of action to depolarize the β-cell membrane. Potentiators of secretion act via classical second messenger pathways to augment the secretory response to an initiator. Inhibitors may act at various sites in the release process. VIP = vasoactive-intestinal peptide, GIP = gastric inhibitory peptide, CGRP = calcitonin-gene-related peptide. Modified from Panten (1987)

labelled species. Insulin (standard or unknown) is first absorbed to an excess of insulin antibody which is immobilized on a solid support such as a microtitre plate. It is then bound to a radiolabelled second insulin antibody, which reacts with a different epitope. This technique results in a considerable gain in sensitivity compared with the radioimmunoassay method, since all the insulin, rather than only a fraction of it, binds to the antibody.

In ELISA methods, a colorimetric assay replaces the use of radioactivity. In one variant of this technique used to measure insulin (Kekow *et al.* 1988), an insulin antibody is first immobilized to a microtitre plate. Addition of a mixture of a constant amount of peroxidase-linked insulin and a varying amount of unknown or standard insulin results in a variable amount of binding of peroxidase-linked insulin to the plate. Subsequent addition of peroxidase substrates allows this to be visualized.

4.1.2 Reverse haemolytic plaque assay

This method (Fig. 3) allows measurement of insulin secretion from single β-cells (Salomon and Meda 1986; Lewis *et al.* 1988). Dispersed β-cells are surrounded by a

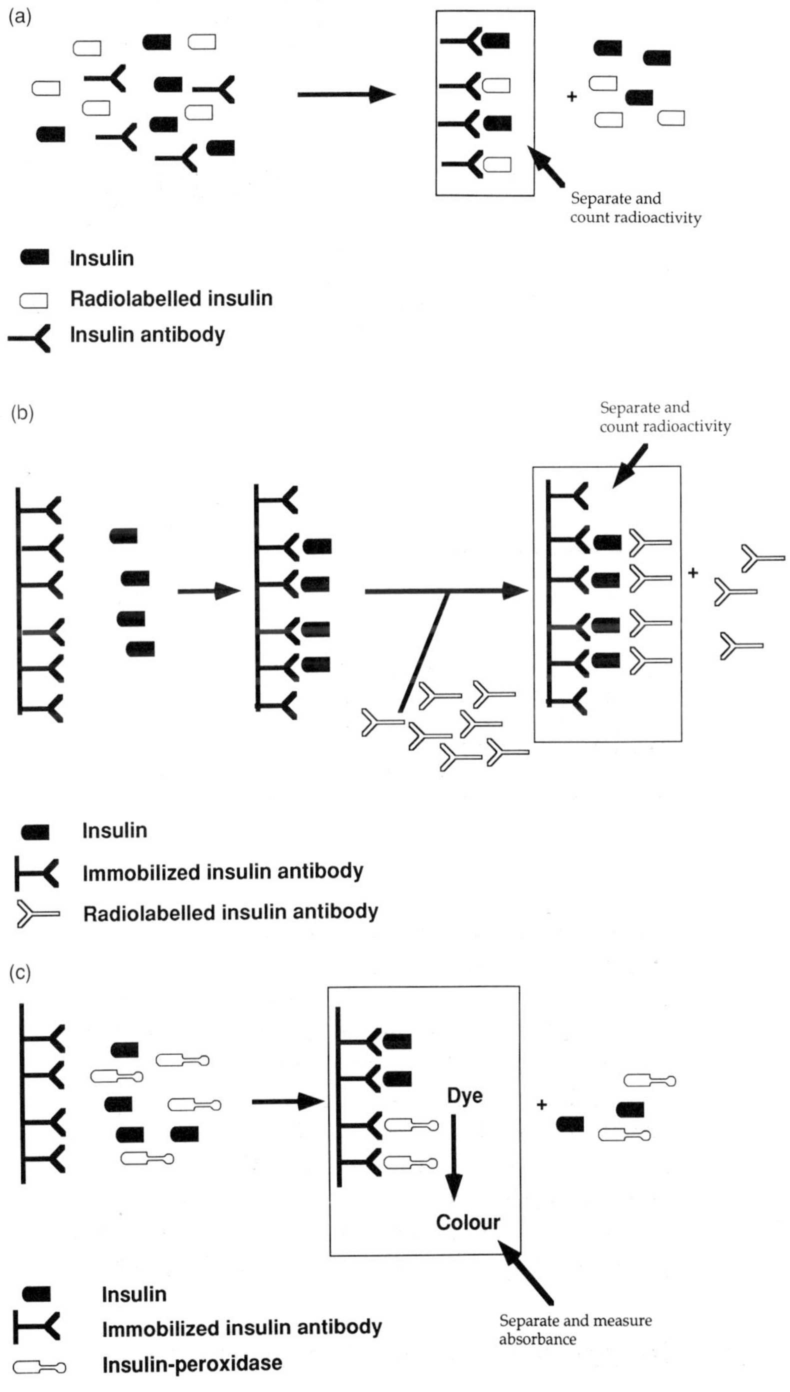

Fig. 2 Insulin assay. Insulin can be assayed by (a) radioimmunoassay, (b) immunoradiometric assay, (c) ELISA

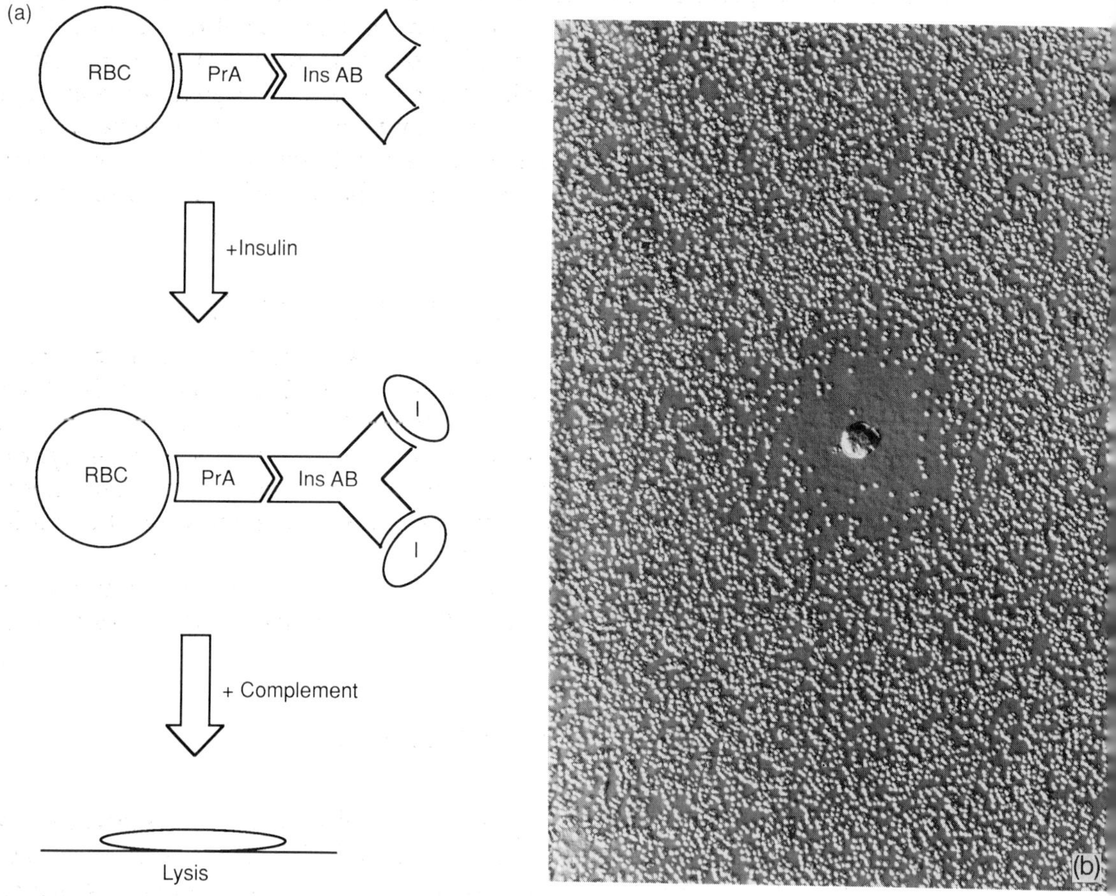

Fig. 3 Haemolytic plaque assay. (a) Insulin secretion can be measured from single β-cells using the reverse haemolytic plaque method. Red blood cells (RBC) are conjugated with protein A (PrA) to which insulin antibody (Ins AB) binds. Insulin (I), secreted from a β-cell, binds to insulin antibody and the presence of complement induces red cell lysis. (b) Haemolytic plaques (dark area) surrounding a secreting β-cell which is lying in a lawn of red blood cells. Red cell ghosts can be seen surrounding the β-cell. Secretion was stimulated with 20 mM glucose. (M. Faehling and F.M. Ashcroft, unpublished)

lawn of erythrocytes which have been conjugated with protein A. Anti-insulin antibody is then added and binds to the protein A. Insulin secretion results in the binding of insulin to the anti-insulin antibody. Subsequent addition of complement promotes lysis of those erythrocytes binding insulin and is visualized as a clear area surrounding an individual β-cell. The number and size of the plaques is related to the number of secreting cells and the extent of secretion, respectively.

4.1.3 Permeabilized cells

Plasma membrane permeabilization allows the composition of the intracellular solution to be precisely controlled and has been used widely for studying the secretory process itself. Strategies employed for β-cell membrane permeabilization have included high-voltage discharge, the use of toxins (such as staphylococcal α-toxin), incubation in low Ca^{2+} media, treatment with detergents such as digitonin and saponin, and hypotonic shock. The relative advantages and disadvantages of these methods have been reviewed by Hersey and Perez (1990). It is worth emphasizing that the choice of method determines the size of the pore produced and hence may influence the results obtained.

4.1.4 Single-cell Ca^{2+} measurements and imaging

Fluorescent probes, such as fura II and indo-1, can be used to monitor intracellular Ca^{2+} levels in the whole islet (Valdeolmillos *et al.* 1989), in single β-cells (Pralong *et al.* 1990) and in suspensions of β-cells (Gylfe 1988). These probes are most often used in an esterified form which passes across the cell membrane and is hydrolysed by intracellular esterases to the impermeant cationic form. Fluorescent emission is monitored using dual wavelength fluorimetry, which enables specific monitoring of intracellular free Ca^{2+} concentrations.

Spatial resolution of the fluorescence signals can be achieved using an imaging system which allows local variation of $[Ca^{2+}]_i$ to be visualized as a pseudo-colour image (Santos *et al.* 1991).

4.1.5 Electrophysiological measurements

Microelectrode recording

In this technique, the potential across the β-cell membrane is measured by the insertion of a fine glass microelectrode into the cell. To date, this method has been exclusively used with β-cells within the intact islet. It has produced considerable information about the β-cell resting potential and the electrical activity initiated by secretagogues.

Patch-clamp recording

The patch-clamp technique allows the current that flows through a single ion channel when it is open to be measured: it can also be used to measure the summed activity of all the ion channels present in the cell membrane (the whole-cell current). There are four main configurations used for recording single-channel currents and two for recording whole-cell currents (Hamill *et al.* 1981; Horn and Marty 1988). Each of these has its own particular advantages and disadvantages.

All configurations derive from the cell-attached patch, which is obtained by forming a high-resistance seal between the cell membrane and a glass recording pipette pressed against the cell surface (Fig. 4a). Cell-attached recording has the advantage that the cell is intact and ion channels may therefore be studied under physiological conditions. It can also be used to determine whether nutrient secreta-

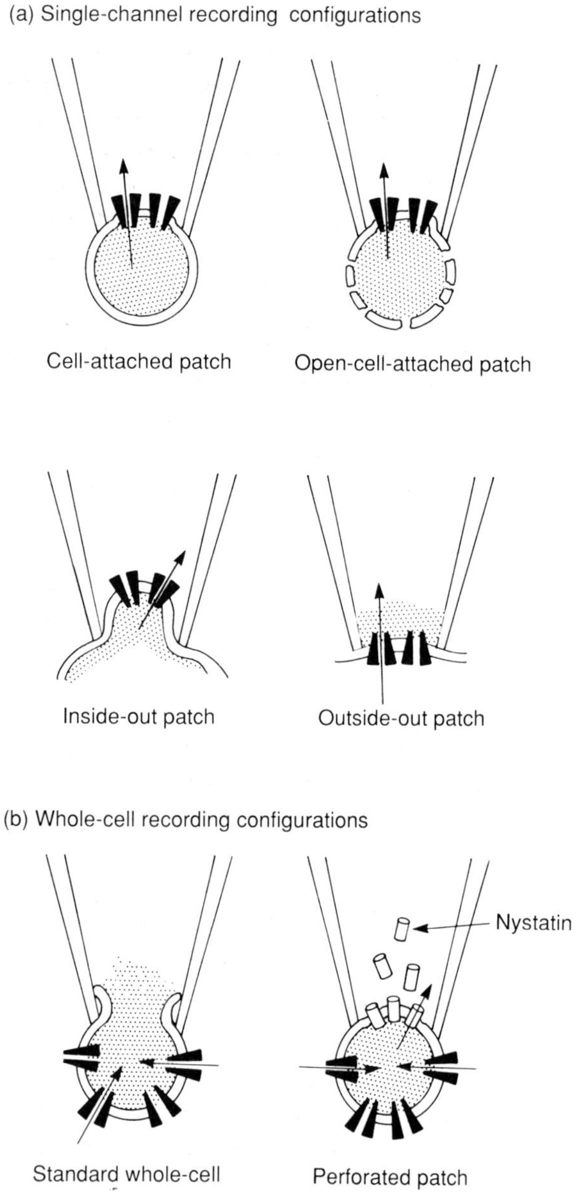

Fig. 4 The various patch clamp configurations used for recording single-channel and whole-cell currents. The stippled area indicates the cytoplasm (reproduced from F.M. Ashcroft and Rorsman 1989)

gogues or hormones mediate their effects by metabolism or intracellular second messengers. This is because the glass-membrane seal is too tight to permit the diffusion of substances between the pipette and bath solutions: thus any effect of a substance added to the bath solution on the activity of ion channels in the patch of membrane beneath the pipette tip must be mediated via an intracellular route.

The open-cell attached patch (Fig. 4a) is a variant of the cell-attached patch in which the rest of the cell membrane is permeabilized by substances such as saponin. This allows the intracellular solution to be modified and putative channel modulators to be introduced. It has the advantage that cytosolic second messengers, which may be required for channel function, are lost less rapidly than in the excised patch configurations (see below).

Two other single-channel recording configurations (Fig. 4a) are possible because the high mechanical stability of the seal results in the membrane breaking before the seal. Thus if the pipette is withdrawn from the cell surface, an isolated membrane patch is produced, spanning the pipette tip, which has its intracellular surface exposed to the bath solution (inside-out patch). An outside-out patch is produced when the pipette is withdrawn from the whole-cell configuration. With these two configurations, channel properties can be measured under controlled ionic conditions and measurements made in different solutions without detachment of the patch pipette. In addition, inside-out patches can be used to test the effects of putative intracellular modulators on channel activity.

The two whole-cell configurations (Fig. 4b) allow the summed activity of many ion channels (the whole-cell current) to be investigated. The standard whole-cell method (Fig. 4b, left) is obtained by forming a cell-attached patch and then destroying the patch membrane with strong suction. It has the advantage that the intracellular solution can be manipulated, but the concomitant disadvantage that soluble cytosolic constituents are lost from the cell. By contrast, the perforated patch method (Fig. 4b, right) preserves cellular metabolism and intracellular second messenger systems (Horn and Marty 1988), as in this configuration the patch membrane is permeabilized by a pore-forming antibiotic, such as nystatin, rather than destroyed.

4.2 Metabolism of the β-cell

4.2.1 Substrate-site hypothesis

Early suggestions that β-cell metabolism was in some way linked to insulin release (Grodsky *et al.* 1963; Coore and Randle 1964) prompted intensive study of islet metabolic pathways. The idea that the effects on insulin release of glucose and other metabolic substrates were mediated by some product of their intracellular transformation in the β-cell was first explicitly formulated as the 'substrate-site hypothesis' by Randle *et al.* in 1968 (Fig. 5). This concept was intended to contrast with the alternative 'regulator-site' model in which the β-cell was envisaged as being equipped with specific cell-surface receptors for glucose and other stimulants. Attempts to discriminate between these models provided the conceptual framework for many years of investigation, which have given overwhelming evidence in favour of the substrate-site hypothesis. The key observations (reviewed by Ashcroft 1980; Hedeskov 1980; Meglasson and Matschinsky 1986) which have led to general acceptance of this hypothesis are as follows. The dependence on glucose concentration of effects of the sugar on insulin release

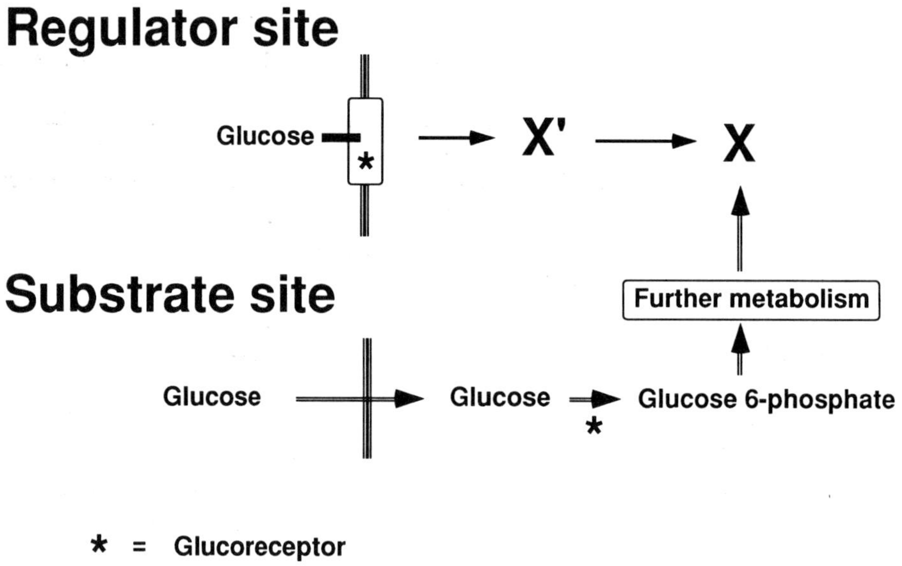

Fig. 5 Models for glucose recognition. In the regulator-site model, glucose binds to a cell-membrane receptor. This initiates events leading to production of X, an intracellular trigger to secretion, from a precursor (X′). In the substrate-site model, it is the metabolism of glucose within the β-cell that generates the intracellular trigger. Current evidence supports the latter model

parallels that for the rate of glucose utilization by islets. Both processes display a sigmoidal concentration dependence, being half-maximal at around 10^{-2} M glucose. Extensive studies on the specificity of secretory responses to sugars has established that only those sugars that are well metabolized by islets are able to elicit secretion. Other metabolites capable of entering the glycolytic pathway, such as glyceraldehyde (at the triose level) or inosine (via the pentose cycle), are also effective secretagogues. The effects on metabolism of various inhibitors invariably paralleled their effects on insulin secretion.

It followed from these considerations that the specificity and concentration-dependence of the β-cell secretory response to sugars must reflect that of the enzymic step exerting regulatory control over their entry into metabolism. The β-cell glucose sensor responsible for monitoring and appropriately responding to changes of blood glucose concentration can therefore be identified as the enzyme(s) catalysing this rate-limiting step. Figure 6 summarizes some of the β-cell metabolic pathways that have been studied.

4.2.2 Control of β-cell glucose metabolism

Detailed accounts of β-cell glucose metabolism can be found in Hedeskov (1980), Ashcroft (1981), and Meglasson and Matschinsky (1986).

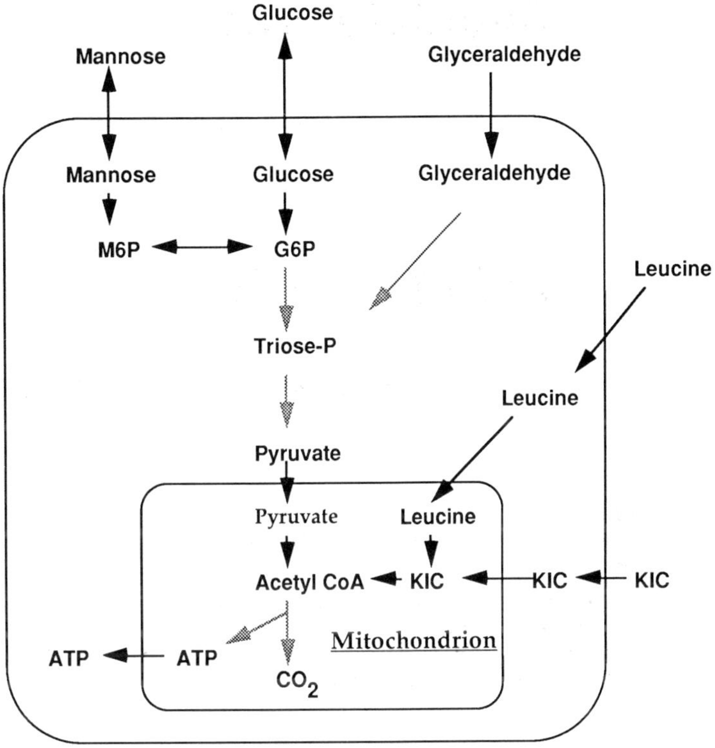

Fig. 6 β-Cell metabolic pathways. The main classes of substrates that can be metabolized by the β-cell. The heavy arrows indicate single metabolic steps. The light arrows indicate metabolic pathways containing several steps

Metabolic fate of glucose

The main route for glucose metabolism in islets is glycolysis. Glycogen levels are low and glycogen synthesis accounts for less than 7 per cent of glucose uptake. The sorbitol pathway and conversion of glucose to lipid and amino acids are also of only minor quantitative importance. Flux via the pentose phosphate pathway is slow and does not vary substantially with glucose concentration; indeed, the fraction of glucose metabolized by the pentose phosphate pathway actually declines as the glucose concentration is raised. Other routes for generation of NADPH are present, however, since islets contain both 'malic enzyme' and substantial amounts of NADPH-isocitrate dehydrogenase. Therefore glycolysis and the subsequent oxidation of pyruvate via the Krebs cycle are of major functional importance for control of insulin secretion.

Glucose utilization by isolated islets demonstrates two separate kinetic components. A high-affinity component shows hyperbolic kinetics with a K_m of 0.2 mM glucose. The second component has sigmoid kinetics with a Hill coefficient of 1.5, an apparent K_m of 11 mM glucose and a V_{max} almost tenfold higher than the high-affinity component. Thus, at a low physiological concentration of glucose (5 mM) the low-affinity component constitutes 68 per cent of glucose utilization, rising to 84 per cent at a high glucose concentration (15 mM).

Glucose transport

Entry of glucose into the β-cell occurs by facilitated diffusion. The intracellular glucose concentration in the β-cell is similar to the extracellular, i.e. glucose is equilibrated across the β-cell plasma membrane and transport is thus non-rate-limiting. Consistent with these observations, flux studies showed that the β-cell glucose transporter is of high capacity with a V_{max} some tenfold greater than measured rates of glucose utilization; and marked inhibition of transport (for example by phloretin) can occur without significant alteration in the rate of utilization. The K_m of the transporter for glucose is 50 mM. In these respects β-cell glucose uptake resembles that of the liver. Cloning of glucose transporters has confirmed this identity (Thorens *et al.* 1988; Johnson *et al.* 1990). The facilitative glucose transporters have been found to be a family of structurally related proteins coded for by distinct genes (Gould and Bell 1990). Both the liver and the β-cell preferentially express the same isoform, GLUT-2, which shows 55 per cent homology with the erythrocyte-type (GLUT-1) glucose transporter and similarly has 12 predicted transmembrane helices. In man, the GLUT-2 gene is located on chromosome 3 and codes for a protein of 524 amino acids.

Immunocytochemical techniques have localized the β-cell glucose transporter preferentially to microvilli facing adjacent endocrine cells (Orci *et al.* 1989), suggesting some interaction with the underlying cytoskeleton and, incidentally, providing evidence that β-cells are polarized.

Glucose phosphorylation

The first non-equilibrium reaction for glucose metabolism capable of exerting significant control of overall flux is phosphorylation of the sugar (glucose → glucose-6-phosphate). This step is enzymically complex in the β-cell (Ashcroft and Randle, 1970; Meglasson *et al.* 1983). Measurement of glucose phosphorylation rates in islet homogenates reveals both a low and a high K_m component of glucose phosphorylation, with K_m values of approximately 0.1 and 10 mM, respectively. The low K_m component, identified as Type 1 hexokinase, will be saturated with substrate at all physiological glucose concentrations and, moreover, must be substantially inhibited in the intact cell, since the V_{max} is well in excess of observed rates of glycolytic flux. The main inhibitor is glucose 6-phosphate (G6P) (non-competitive with glucose, K_i 25 μM). The kinetics and elution profile in several chromatography systems suggested that the low-affinity glucose phosphorylating activity was similar to liver glucokinase (Type 4 hexokinase isoenzyme). This identity was also supported by immunoblotting (Iynedjian *et al.* 1986), and the demonstration that there is probably only one copy of the glucokinase gene in the rat further indicated that the β-cell protein was likely to be identical to that of the liver. However, using cDNAs to rat liver glucokinase it has been shown that the glucokinase mRNA is actually larger in β-cells than in liver (Iynedjian *et al.* 1989; Magnuson *et al.* 1989) — the two mRNAs showed differences at the 5′ end upstream from amino acid 19. These differences arise both from differential tissue splicing and from the use of alternative promoters.

The maximal activity of glucokinase in islets is similar to that of the maximal rates of the low-affinity glucose utilization measured in intact cells. Moreover, the sigmoidicity of the low-affinity component of glucose usage is identical to the co-operativity of islet glucokinase. Thus the low-affinity glucose usage by intact islets closely fits the activity of glucokinase at physiological glucose and ATP concentrations.

The discrimination of glucose anomers by glucokinase may also account for the more rapid metabolism of the alpha anomeric form of glucose as compared with the beta, and hence the anomeric specificity of glucose-stimulated insulin release.

A decrease in β-cell glucokinase activity is observed in several situations associated with decreased insulin release; the rat insulinoma β-cell line RIN m5F secretes little or no insulin in response to glucose and is correspondingly deficient in glucokinase; normal islets treated with alloxan have lowered glucokinase activity and secretory response to glucose (Meglasson *et al*. 1986; Lenzen *et al*. 1987; Lenzen and Panten 1988*a*); and starvation induces a reduction in islet glucokinase and insulin secretion. Note, however, that in the case of starvation the reduced insulin response is not specific for glucose-induced release and is apparent before a reduction in glucokinase can be measured. Hence additional regulatory mechanisms are involved here. The effect of alloxan is also difficult to interpret unequivocally since the drug also induces a marked inhibition of a Ca-calmodulin-dependent protein kinase suggested to be important for insulin secretion (Harrison *et al*. 1986).

Glucose 6-phosphatase

The control of glucose phosphorylation is complicated by the presence in islets of glucose 6-phosphatase (G6Pase) (Ashcroft and Randle 1970; Waddell and Burchell 1988). The latter is a microsomal enzyme that has proved difficult to purify. In the liver it has been shown that the G6Pase active site is on the lumenal surface of the endoplasmic reticulum membrane and functions in concert with transport systems responsible for uptake of G6P into the endoplasmic reticulum (T1), the equilibration of the product phosphate (T2), and the release of glucose from this compartment (T3). Using an antibody to the liver G6Pase, it has been demonstrated that the same 36 500 Da G6Pase enzyme protein is present in both liver and islets from several different species (Waddell and Burchell 1988). Strikingly, the activity of G6Pase in islets is some tenfold higher than in liver. This may have great importance for regulation of glucose and G6P concentrations. In addition, the G6Pase complex functions to liberate phosphate within the endoplasmic reticulum. One may speculate that this could have importance in complexing Ca^{2+} to replenish Ca^{2+} stores. It is also of considerable significance that the combined operation of glucokinase and G6Pase can constitute a substrate cycle (Fig. 7).

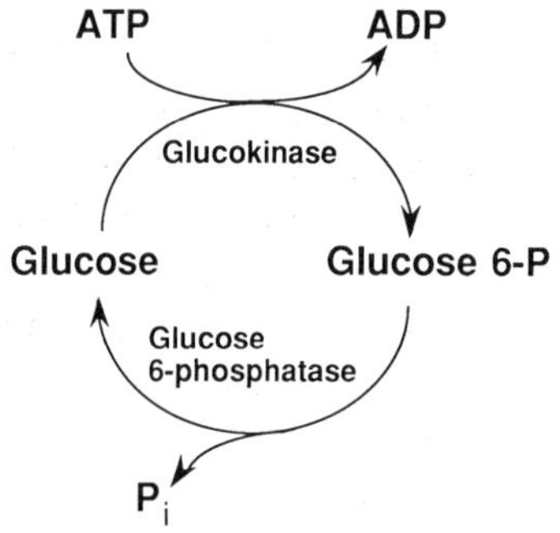

Fig. 7 Substrate cycling in the β-cell. The presence of glucokinase and glucose 6-phosphatase within the β-cell allows substrate cycling at the initial step for entry of glucose into metabolism

Glucose cycling

Direct evidence that substrate cycling between glucose and G6P does occur has been provided in ob/ob mouse islets and, to a lesser extent, in normal islets (Khan *et al*. 1989, 1990*a*, *b*). This could have two important consequences. Sensitivity of control at this step will be increased as a function of the cycling rate. Secondly, the cycle consumes ATP. In view of the importance of ATP in coupling glucose metabolism to ionic fluxes, as discussed below, the rate of cycling may be of considerable importance for the strength of the connection between metabolic and ionic fluxes. It is noteworthy that cycling was increased in islets from mildly diabetic rats. Unfortunately no data are available for normal or diabetic human islets, but the possibility that an increased rate of cycling may contribute to the impaired secretory response of diabetic islets to glucose is intriguing.

Additional regulatory steps

In islets the maximum extractable activity of phosphofructokinase is some 35-fold greater than glycolytic flux. Regulation of the enzyme is likely to be complex and the precise role of potential regulators, such as fructose 2,6-bisphosphate, is controversial. The reported absence of fructose bisphosphatase from islets indicates that substrate cycling does not occur at this stage and also precludes the islet from carrying out gluconeogenesis. In many tissues the operation of the Pasteur effect, in which glycolysis is speeded up under conditions that lower ATP, is mediated by changes in phosphofructokinase activity. Unusually, the islet does not show this phenomenon; rates of glycolysis are positively correlated with islet ATP levels (Ashcroft *et al*. 1973).

Further controls not currently understood must restrain the velocity of the pyruvate kinase reaction. This enzyme has a very high maximal activity in islet extracts and the velocity predicted from the kinetic parameters of the enzyme at the measured concentrations of substrates and products present in the β-cell is vastly in excess of glycolytic flux (Sugden and Ashcroft 1977). One possibility not fully explored is the presence of an additional substrate cycle at this step, since islets contain both pyruvate carboxylase and phosphoenol-pyruvate carboxykinase (PEPCK) (Ashcroft and Randle 1970).

Pyruvate metabolism

Lactate is a major end-product of glucose metabolism, accounting for 76 per cent of glucose usage at low and 47 per cent at high glucose concentrations. The activity of lactate dehydrogenase is high, so the [lactate]/[pyruvate] ratio is likely to be at equilibrium with cytosolic [NAD^+]/[NADH]; the latter ratio has thus been estimated from measurements of [lactate]/[pyruvate] and shown to decrease with increasing glucose concentration (Ashcroft and Christie 1979). Pyruvate enters mitochondria via a transporter inhibitable with cyanocinnamate. Since islet pyruvate levels are around 0.5 mM, the transporter (K_m 0.15 mM) is likely to be operating near to its maximal velocity. Measurements of maximal activities of islet Krebs cycle enzymes and actual rates of oxygen uptake suggest that the Krebs cycle normally operates at well below its maximal capacity. Although there are insufficient data on β-cell mitochondrial metabolite levels available to establish the mechanism of regulation of Krebs cycle activity in the β-cell, factors important for control of pyruvate oxidation may be inferred from work on other tissues. In particular, the importance of intramitochondrial levels of Ca^{2+} has been stressed (Denton and McCormack 1990). Pyruvate dehydrogenase (PDH) is one of three intramitochondrial dehydrogenases whose activity can be increased by Ca^{2+} in the range 0.05–2 μM. Ca^{2+} activates the phosphatase responsible for dephosphorylation and hence activation of PDH; activities of NAD^+-isocitrate dehydrogenase and 2-oxoglutarate dehydrogenase are increased by Ca^{2+} by causing a marked decrease in their K_m for isocitrate and 2-oxoglutarate, respectively. It has been shown (McCormack *et al.* 1990) that at low glucose concentrations only 16 per cent of rat islet PDH is in the active dephosphorylated form but, in the presence of 15 mM glucose, the proportion of active form was increased to 50 per cent. Evidence for activation of isocitrate and 2-oxoglutarate dehydrogenases by high glucose has also been presented (Sener *et al.* 1990). It should be pointed out, however, that it is by no means clear that these effects are mediated by increased mitochondrial Ca^{2+}. It seems also plausible, for example, that the increased activity of PDH could result from inhibition of PDH kinase by pyruvate. This point is important conceptually; increased generation of ATP is currently postulated to lead to increased cytosolic Ca^{2+} levels via inhibition of K-ATP channels (see below), whereas the view that activation of PDH by Ca^{2+} leads to increased ATP generation reverses this order of events. We favour the view that increased provision of substrate to the Krebs cycle is primarily responsible for increased oxidation of glucose and ATP generation at high glucose concentrations. Subsequent increases in cytosolic Ca^{2+} may result in increased mitochondrial Ca^{2+}, which would further activate mitochondrial oxidation and the rate of oxidative phosphorylation. This would permit enhanced oxidative phosphorylation, providing ATP for the secretory process to be sustained despite an increased [ATP]/[ADP] ratio.

Role of malonyl-CoA

Stimulation of islets by glucose is associated with a switch from lipid oxidation to lipid synthesis (Berne 1975). The finding that there is an increased concentration of malonyl-CoA in β-cells exposed to high glucose concentrations may provide an explanation for the mechanism, since malonyl-CoA is a potent inhibitor of carnitine acyl transferase 1, which catalyses the first specific step in the metabolism of lipid. It is less clear what the function of this switch might be. It has been suggested that increased generation of diacylglycerol (DAG) and acyl-CoA is of functional importance for insulin release (Corkey *et al.* 1989). This remains speculative at present. DAG might activate protein kinase C but current evidence suggests that activation of protein kinase C by DAG is not important for glucose-induced insulin release; a role for DAG in inhibition of K-ATP channels was suggested from data obtained with RIN m5F cells (Wollheim and Regazzi 1990) but such a mechanism appears unlikely in normal β-cells. Acyl-CoA might conceivably be involved in acylation of proteins of relevance to secretion, but there is no direct evidence to support such a mechanism.

Metabolism of amino and keto acids

The possibility has been considered that a signal for secretion could arise during glycolysis and evidence that phosphoenolpyruvate (PEP) might influence mitochondrial Ca-handling has been obtained (Sugden and Ashcroft 1977). However, there is abundant evidence that the stimulatory effects on insulin secretion of the deamination product of leucine, 2-ketoisocaproate (KIC), are, as in the case of glucose, causally linked to its metabolism (Hutton *et al.* 1980; Lenzen *et al.* 1982, 1986). Since KIC metabolism occurs entirely intramitochondrially, it is clear that mitochondrial metabolism is of fundamental importance for initiation of insulin secretion.

Studies on the metabolism of leucine have revealed another facet to stimulation of insulin release. The non-metabolizable analogue of leucine, BCH, itself produces marked stimulation of insulin release. Although initially regarded as evidence against a metabolic basis for stimulation of insulin release by amino acids, this effect was later found to be explained by stimulation by BCH of the oxidation by glutamate dehydrogenase (GDH) of endogenous glutamate with, presumably, increased generation of NADH (Panten *et al.* 1984; Panten and Lenzen 1985). At least part of the action of leucine on insulin release is therefore attributed to activation of GDH. The actions of a number of other agents are also suggested to be attributable, directly or indirectly, to effects on GDH activity (Fahien *et al.* 1988). For example, the insulinotropic properties of esters of succinate are suggested to result from increasing succinyl-CoA and decreasing GTP, which are activators and inhibitors, respectively, of GDH. Although these findings are of interest, they do not alter the main thrust of the conclusion from a vast area of data that insulin secretion in response to glucose and other metabolites is dependent on their ability to increase β-cell oxidative metabolism.

Importance of ATP

The common denominator for the metabolic effects described is an enhanced rate of oxidative phosphorylation. Since the closure of K-ATP channels by ATP represents an attractive mechanism for linking metabolic to ionic fluxes (see below), the key question is whether changes in β-cell [ATP]/[ADP] ratio occur with sufficient rapidity and magnitude to constitute the linking factor. In isolated islets a rapid increase in [ATP]/[ADP] occurs in response to elevation of glucose (Meglasson *et al.* 1989). The increase occurs over the range of glucose concentrations most effective in closing K-ATP channels, i.e. up to and just above the threshold for insulin secretion. Whether the absolute values for the ATP and ADP concentrations are consistent with the sensitivity of K-ATP channels in isolated patches is less clear (see below). However, there is evidence (Niki *et al.* 1989*b*) from measurement of activities of K-ATP channels and of Na-K-ATPase in β-cells that the effective ATP concentrations in the vicinity of the membrane may be rather different from those in the bulk cytosol as measured in cell extracts. Thus, lowering bulk ATP from 5 to 2 mM elicited marked changes in activity of both these membrane proteins, despite the fact that their sensitivity to ATP in isolated membrane systems is in the μM range. Studies using the isolated perfused rat pancreas (Ghosh *et al.* 1991) have, however, questioned the primacy of changes in the ATP/ADP ratio in mediating the effect of glucose on insulin secretion. The levels of ATP and related metabolites were determined in β-cells from freeze-dried samples of pancreas which had been frozen at various time-intervals during first phase glucose-stimulated insulin release. There was no change in the ATP/ADP ratio despite a ten-fold increase in insulin release. These data suggest that there may exist alternative or complementary factors responsible for the metabolic regulation of K-ATP channel activity, which remain to be identified. However, other candidates proposed for linking metabolic and membrane events, such as changes in the redox state, pH and free Ca^{2+} levels, have received little experimental support.

In addition to its proposed signal function, ATP may also be required as a source of energy for the secretory process itself. Although there is some evidence for an ATP-requiring step in exocytosis, the quantitative importance of this requirement is unknown. In the basal state at least 75 per cent of the calculated ATP turnover in the β-cell is accounted for by the activity of Na-K-ATPase (Hedeskov 1980). How much of the increased ATP generation found on raising the glucose to stimulatory levels is actually consumed by exocytosis itself is not known.

Pyridine nucleotide fluorescence

Early studies showed that there was an increase in the fluorescence of pyridine nucleotides in intact islets exposed to glucose or KIC (Panten *et al.* 1973). The time-lag for reduction of NAD(P) is similar to that for increased oxygen consumption (Hutton and Malaisse 1980); therefore increases in NAD(P)H closely reflect increased ATP generation. Simultaneous measurement of NAD(P)H fluorescence and Ca^{2+} in single β-cells (Pralong *et al.* 1990) showed that the reduction of NAD(P) preceded nutrient-induced changes in $[Ca^{2+}]_i$. The NAD(P)H responses were usually biphasic, although some cellular heterogeneity was observed. The delay before there was an increase in NAD(P)H was identical for glucose and for KIC. In contrast, the latency of the rise in $[Ca^{2+}]_i$ was greater for glucose than for KIC. This is explicable in terms of the metabolism of these two agents. During glucose stimulation the initial consumption of ATP by glucokinase and phosphofructokinase may delay the increase at the cell-membrane of [ATP]/[ADP] relative to KIC, whose metabolism leads immediately and exclusively to mitochondrial generation of ATP.

There is little evidence to link changes in NAD(P)H directly to the activity of β-cell ion channels.

4.3 β-Cell electrical activity

There is considerable evidence that electrical activity of the β-cell membrane plays a central role in stimulus–secretion coupling. Much of this evidence has been obtained from microelectrode recordings of membrane potential from β-cells within intact islets of Langerhans (reviewed by Henquin and Meissner 1984). Such studies have shown that when the glucose concentration is below that required to elicit insulin secretion, the β-cell is electrically silent; the resting potential is around −70 mV in 0 mM glucose. Glucose induces a slow depolarization of the β-cell membrane. At glucose concentrations that elicit insulin release (>7 mM), this depolarization is sufficient to bring the membrane to the threshold potential at which electrical activity is initiated. A characteristic pattern of electrical activity then ensues (Fig. 8). In 10 mM glucose this consists of slow oscillations in membrane potential (known as slow waves) between a depolarized plateau, on which Ca-dependent action potentials are superimposed, and a more negative interburst interval. As the glucose concentration is increased, the duration of the plateau increases and that of the intervals between them decreases. Continuous electrical activity finally occurs above about 20 mM glucose.

There is a good correlation between the time spent at the plateau firing action potentials and insulin secretion, obtained at different glucose concentrations (Meisser and Preissler 1979). Both have a threshold at about 7 mM glucose, are half maximal at around 10 mM, and saturate above 20 mM glucose. This close relationship suggests that electrical activity plays an important role in regulating insulin release, a view supported by the findings that:

(1) Ca^{2+}-channel blockers, or removal of extracellular calcium, inhibits both electrical activity and insulin release; and
(2) other initiators of insulin release, such as the sulphonylureas, also depolarize the β-cell and stimulate electrical activity.

Application of the patch-clamp technique (Hamill *et al.* 1981) to single β-cells has clarified our understanding of the way in which secretagogues modulate β-cell electrical activity (see Fig. 4 for methodology). The technique enables identification of the different ion channels that underlie β-cell electrical activity, and determination of the effects of nutrient secretagogues on the properties of these channels. Patch-clamp studies have led to the model of stimulus–secretion coupling shown

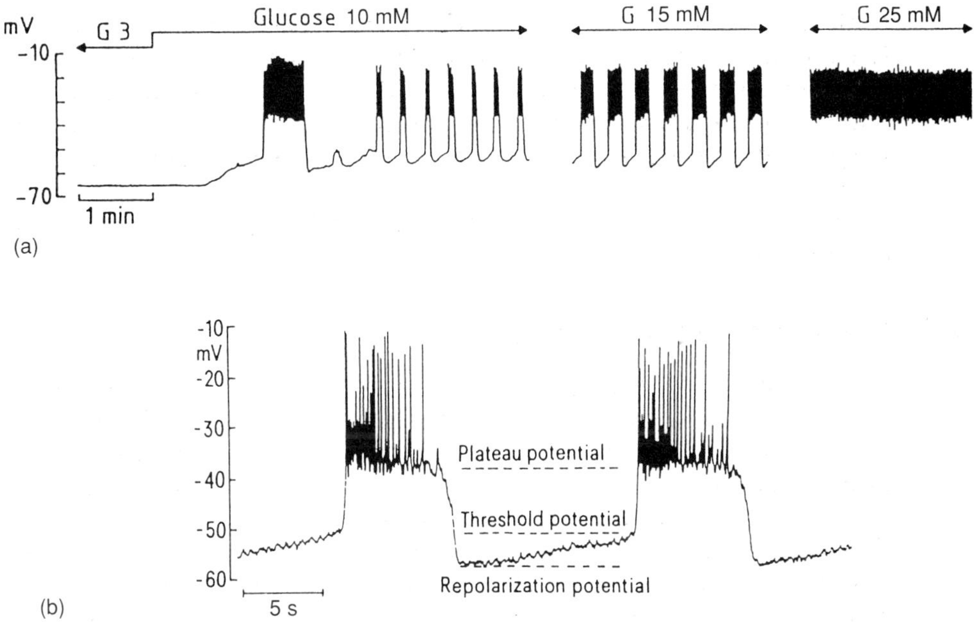

Fig. 8 β-Cell electrical activity. Microelectrode recording of the β-cell membrane potential. (a) Effect of increasing glucose from 3 mM to 10 mM, 15 mM, and finally 25 mM. All traces come from the same cell but an interval separates the recordings at the higher glucose concentrations (from Henquin *et al.* 1990 with permission). (b) Expanded section of two successive bursts recorded in 10 mM glucose (from Henquin and Meissner 1984, with permission)

in Fig. 9, which ascribes a key role to the ATP-regulated K-channel (K-ATP channel). This channel constitutes the link between the metabolic events described above and β-cell electrical activity. The resting membrane potential of the β-cell is governed by the activity of the K-ATP channel. Closure of this channel, either by glucose metabolism (Ashcroft *et al.* 1984, 1988; Rorsman and Trube 1985; Misler *et al.* 1986) or by sulphonylureas (Sturgess *et al.* 1985) reduces the membrane K^+-permeability and leads to membrane depolarization. This opens voltage-dependent Ca^{2+} channels (Rorsman *et al.* 1988) and so elicits β-cell electrical activity. The increased Ca^{2+} influx that takes place through the open Ca^{2+} channels produces a rise in intracellular calcium, which stimulates insulin secretion (Prentki and Matchinsky 1987). β-Cell electrical activity thus provides a mechanism for linking glucose metabolism to the rise in intracellular calcium.

4.3.1 Ion channels present in the β-cell membrane

The different types of ion channel that have been identified in β-cells and the various cell lines derived from them, are illustrated in Fig. 10. There are at least five different types of K-selective channels: the ATP-regulated K-channel (K-ATP channel), the delayed rectifier (K-DR channel), the Ca-activated K-channel (K-Ca channel), a K-channel activated by inhibitory transmitters (K-I channel), and a

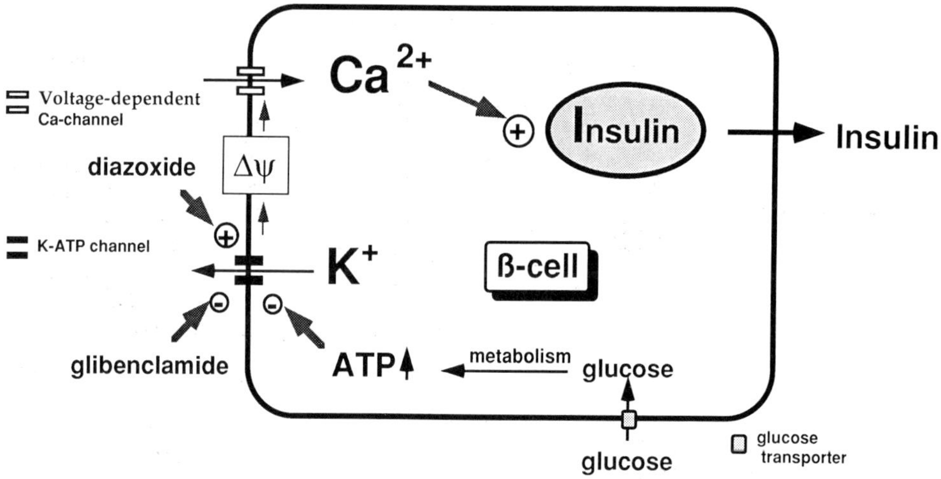

Fig. 9 Role of K-ATP channels in insulin secretion. The closure of K-ATPP channels in response to a rise in intracellular ATP or extracellular glibenclamide causes β-cell depolarization and subsequent opening of voltage-sensitive Ca^{2+}-channels. The resulting Ca^{2+}-influx triggers insulin secretion

transient K-channel (A-channel). Two types of Ca-selective channel (T-type and L-type), a single type of sodium channel (Na-channel), and a non-selective cation channel (NS-channel) have also been observed. A full account of the properties of these ion channels and their importance for insulin secretion may be found in the review by Ashcroft and Rorsman (1989).

The K-ATP channel

This channel is inhibited by a rise in the intracellular ATP concentration (Fig. 11). In excised membrane patches, half maximal inhibition of channel activity is produced by between 10 and 50 μM ATP and complete block by 1 mM ATP (Cook and Hales 1984; Rorsman and Trube 1985). A much lower ATP sensitivity is estimated for the intact cell (1–2 mM; Niki *et al.* 1989*b*, and see also above). This difference has been attributed in part to the loss of cytosolic constituents in the excised patch, and in part to a lower ATP concentration below the plasma membrane than that of average cytosol; the latter concentration is measured in experiments used to determine the ATP sensitivity of the channel in intact cells. The inhibitory effect of ATP does not involve a phosphorylation event, since non-hydrolysable ATP analogues are also effective (Ohno-shosaku *et al.* 1987; Ashcroft and Kakei 1989); thus ATP is considered to interact directly with a site on the channel or an associated control protein. There is evidence, however, that phosphorylation may play a role in maintaining channel activity (Ohno-shosaku *et al.* 1987). Thus ATP appears to have a dual modulatory effect on the K-ATP channel.

In addition to ATP, the K-ATP channel is regulated by a considerable number of other cytosolic constituents (reviewed by Ashcroft and Rorsman 1989). How far

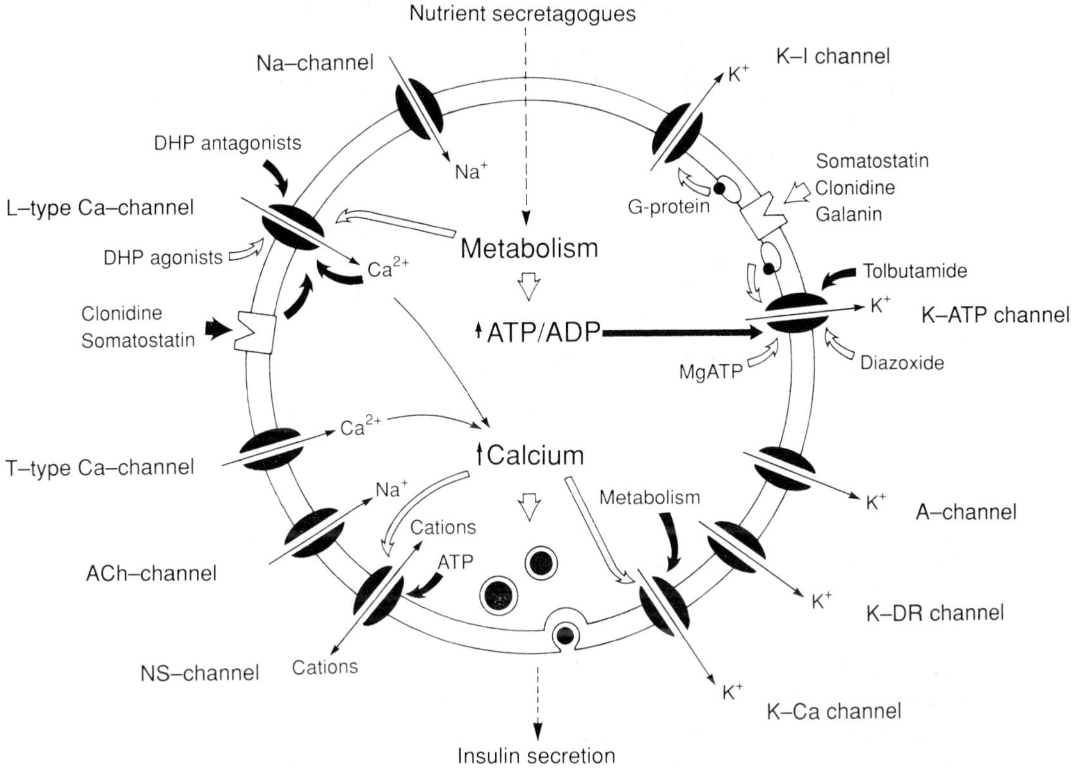

Fig. 10 Types of ion channels found in the β-cell membrane. Activatory pathways are indicated by open arrows and inhibitory pathways by closed arrows. Channels are indicated by closed symbols and receptors by open symbols. Abbreviations: K-I channel, K-channel activated by inhibitors such as clonidine; K-ATP channel, ATP-regulated K-channel; A-channel, transient outward K-channel; K-DR channel, delayed rectifying K-channel; K-Ca channel, Ca-activated K-channel; NS-channel, non-selective cation channel; ACh channel, acetylcholine-activated channel; G-protein, GTP-activated protein. Modified from F.M. Ashcroft and Rorsman (1989)

this modulation is of physiological relevance remains unclear. However, it is thought that the ability of ADP to decrease the ATP sensitivity of the channel is of particular importance and that the ATP/ADP ratio is a major determinant of channel activity in the β-cell (Kakei *et al.* 1986; Dunne *et al.* 1988). Metabolism of nutrient secretagogues will increase the ATP/ADP ratio, and so close K-ATP channels. The changes in the cytosolic ATP/ADP, are greater than those in ATP, further amplifying the effect of metabolism on channel activity.

The pharmacological properties of the K-ATP channel are discussed below (section 4.3.5).

The delayed rectifier K-channel

This channel is voltage-dependent, channel activity increasing with depolarization. Its functional role is to repolarize the action potential.

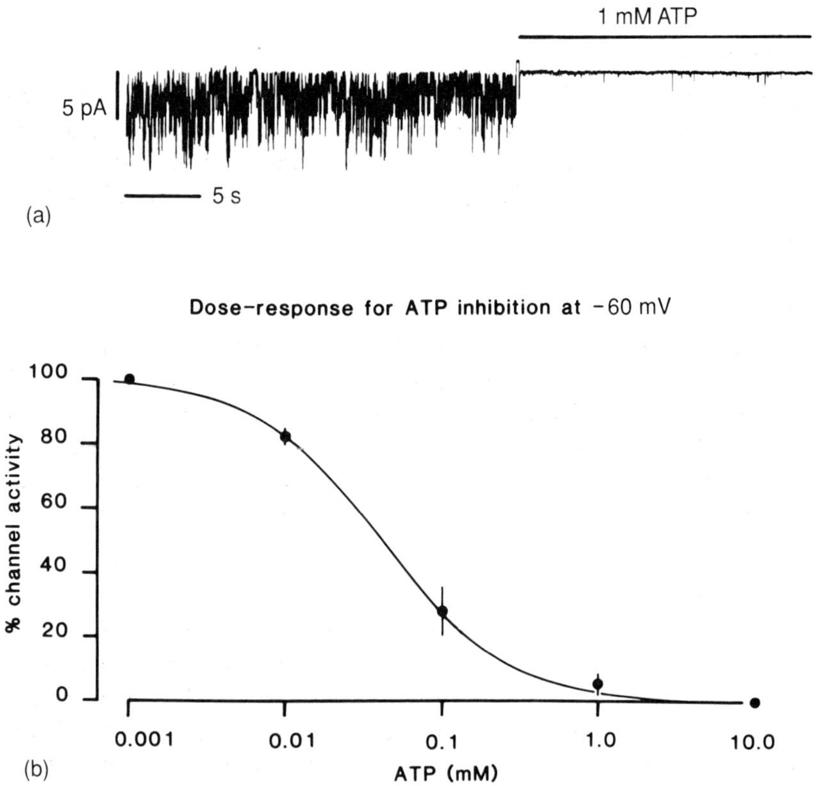

Fig. 11 Inhibition of K-ATP channel activity by ATP. (a) Effect of 1 mM ATP on single K-ATP channel currents recorded from an inside-out patch at a membrane potential of −60 mV. (b) Relationship between the ATP concentration and K-ATP channel activity normalized to its value in 1 μM ATP). (F.M. Ashcroft and M. Kakei, unpublished data)

K-Ca channels

These channels are activated both by depolarization and by an increase in the intracellular Ca^{2+} concentration (Cook *et al.* 1984; Tabcharini and Misler 1989). The reported Ca^{2+} sensitivity is very variable, but in the rat >30 μM Ca^{2+} is needed to stimulate channel activity at the resting potential, and 10–22 μM at potentials achieved during the β-cell action potential (Tabcharini and Misler 1989). This may explain why K-Ca channel activity is not observed at the resting potential in cell-attached patches. Ca^{2+} influx during the action potential, however, may lead to channel activation.

K-Ca channels are blocked by charybdotoxin, and their sensitivity to external tetraethylammonium ions (TEA) is higher than that of other β-cell K-channels ($K_i \sim 0.2$ mM; Bokvist *et al.* 1990), so that it is possible to block this channel relatively selectively with charybdotoxin or low concentrations of TEA. Such studies suggest that K-Ca channels contribute to action potential repolarization but not to the slow waves (Henquin 1990; Kukuljan *et al.* 1991).

A-current

The A-current has been observed in some, but not all, preparations of mouse β-cells and its functional role remains obscure. It constitutes a transient outward K-current which is about 50 per cent inactivated at the resting potential and largely inactivated at potentials that occur during β-cell electrical activity (Smith *et al.* 1989*a*).

Non-selective cation channel

Such a channel, which is permeable to both sodium and potassium ions, has been seen in inside-out patches from human β-cells (Sturgess *et al.* 1987*a*) and GRI-G1 insulinoma cells (Sturgess *et al.* 1986*b*, 1987*b*). This channel is activated by intracellular Ca^{2+} (>100 μM) and blocked by adenine nucleotides, with AMP being more effective than ATP. Its functional importance is uncertain as these properties would suggest that the channel is closed in the intact cell.

Na^+ channels

These do not contribute significantly to the β-cell action potential depolarization in rodents, since they are largely inactivated at the plateau potential (Hirriat and Matteson 1988; Plant 1988*a*). This explains why the Na^+ channel blocker tetrodotoxin has no effect on electrical activity (Meissner and Schmelz 1974). Canine β-cells, however, are an exception where Na-channels contribute significantly to action potential depolarization (Pressel and Misler 1990).

Ca^{2+} channels

Two types of voltage-dependent single Ca^{2+} channel currents have been described in rat β-cells (Ashcroft *et al.* 1990*b*), with properties resembling those of L-type and T-type Ca^{2+} channels (Nowycky *et al.* 1985). It seems unlikely that the T-type channel plays a substantial role in β-cell electrical activity, since this channel is not present in mouse β-cells (Rorsman *et al.* 1988; Plant 1988*b*).

L-type Ca^{2+} currents are activated by depolarization to potentials more positive than −60 mV (in the presence of glucose, see section 4.3.2). Ca^{2+} currents are maximal around −20 mV and of sufficient magnitude to account for action potential depolarization. Inactivation of the currents during a maintained voltage step probably involves two processes: Ca^{2+}-dependent inactivation due to preceding Ca^{2+} entry (Plant 1988*b*) and a slow, voltage-dependent inactivation with time constants in the range of seconds (Satin and Cook 1987). These properties of the L-type Ca^{2+} current suggest that it underlies both action potential depolarization and the plateau potential, implying that most of the Ca^{2+} influx required for insulin secretion flows through L-type Ca^{2+} channels. Consistent with this idea, dihydropyridine antagonists, which specifically inhibit these channels, block insulin secretion (Malaisse-Lagae *et al.* 1984).

4.3.2 Regulation of β-cell ion channels by cellular metabolism

As described above, glucose modulates β-cell electrical activity and thus insulin release. There is evidence that the metabolic regulation of K-ATP channel activity is the principal mechanism underlying this effect of glucose, and that metabolic regulation of Ca^{2+} channel activity is also of importance. In addition, K-Ca channels may be modulated by cellular metabolism, although the physiological role of this regulation is currently unclear (Ribalet *et al.* 1988; Eddlestone *et al.* 1989).

K-ATP channels

In the absence of glucose, the resting potential of the β-cell is dominated by the activity of the K-ATP channel (Ashcroft *et al.* 1984; Ashcroft and Rorsman 1990). Interestingly, less than 25 per cent of the total K-ATP conductance is activated at rest, indicating that the channel is subject to considerable inhibition, even in the absence of exogenous substrate.

Substrates for β-cell metabolism produce a rapid, reversible, and dose-dependent decrease of K-ATP channel activity (Fig. 12). Cell-attached patch recordings indicate that 50 per cent of the channels active in the absence of glucose are inhibited by around 2 mM glucose, and more than 90 per cent at the physiological (fasting) glucose concentration of 5 mM (Misler *et al.* 1986; Ashcroft *et al.*

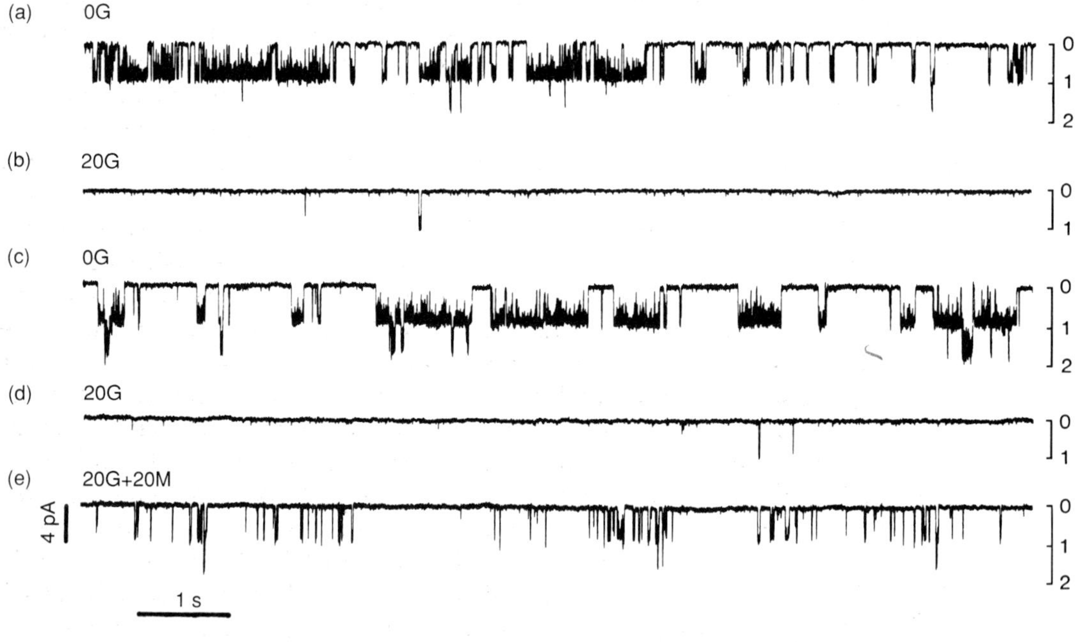

Fig. 12 Inhibition of K-ATP channel activity by glucose. Single-channel currents recorded from a cell-attached patch at a pipette potential of 0 mV: (a) in glucose-free solution; (b) 7 min after changing to 20 mM glucose; (c) after return to glucose-free solution; (d) 4 min after addition of 20 mM glucose; and (e) 2 min after perfusion with 20 mM glucose plus 20 mM mannoheptulose. The scale on the right indicates the number of simultaneous channel openings (from F.M. Ashcroft *et al.* 1984)

1988; Rorsman and Trube 1990). When whole-cell K-ATP currents are studied using the perforated patch method, however, it is clear that there is, in fact, considerable variability in the glucose-sensitivity of different β-cells (Ashcroft *et al.* 1990*b*). Thus, whereas some β-cells show a sensitivity similar to that reported for the single-channel recordings, other β-cells are considerably less sensitive to glucose, with more than 8 mM being required to produce 50 per cent inhibition. In almost all cells, however, complete inhibition is found with 15–20 mM glucose. Differences in β-cell metabolism probably underlie the observed heterogeneity in the glucose sensitivity of the K-ATP current, since a similar variability has been reported for glucose-induced changes in NAD(P)H fluorescence in single β-cells, which provides an index of cellular metabolism (Pralong *et al.* 1990). Variation in the glucose dose–response curve for the K-ATP currents may account for the heterogeneity in the glucose sensitivity of insulin secretion found for single β-cells (Salomon and Meda 1986; Hirriat and Matteson 1988).

In addition to glucose, glycolytic intermediates and metabolizable (but not non-metabolized) sugars inhibit K-ATP channel activity. For example, glyceraldehyde almost completely blocks K-ATP channel activity in rat β-cells ($K_{0.5} \sim 5$ mM) whereas galactose and 3-*O*-methyl glucose are ineffective (Misler *et al.* 1986). It is clear that these nutrient secretagogues must be metabolized to close K-ATP channels, since the specificity and the effects of metabolic inhibitors closely resemble those described above for glucose utilization. Since leucine and α-ketoisocaproate also close K-ATP channels, it appears that oxidative phosphorylation also generates an intracellular mediator of channel inhibition (Ashcroft *et al.* 1987; Ribalet *et al.* 1988; Eddlestone *et al.* 1989). There is even a possibility that oxidative metabolism may be a more important regulator of K-ATP channel activity than glycolysis, because inhibitors of mitochondrial metabolism (such as rotenone, azide, and oligomycin) are very effective at rapidly increasing channel activity blocked by glucose (Ashcroft *et al.* 1985; Misler *et al.* 1986; Ribalet and Ciani 1987). Indeed, these agents are able to promote channel activity in the absence of exogenous fuels (Ashcroft *et al.* 1985).

The effects of all these nutrients are believed to be mediated by an increase the intracellular ATP/ADP ratio. As discussed above, changes in K-ATP channel activity and in cytosolic ATP/ADP occur over a similar time course and range of glucose concentrations. Furthermore, all substances that lower intracellular ATP stimulate channel activity, whereas those that increase ATP levels produce channel inhibition. Although it has been postulated that *de novo* synthesis of diacylglycerol, and subsequent activation of protein kinase C, may couple glucose metabolism to channel inhibition in RIN m5F cells (Wollheim *et al.* 1988), this pathway does not seem to be important in regulating channel activity in normal β-cells (Henquin *et al.* 1989). Other candidates for linking metabolic events to K-ATP channel activity have been proposed (e.g. pH changes), but have received little experimental support.

L-type Ca-channels

Theoretically, there are two ways in which nutrient secretagogues may influence Ca^{2+} influx through voltage-dependent Ca^{2+} channels: by altering the β-cell membrane potential or by biochemical modulation of the Ca^{2+} channel itself. Both of these mechanisms operate in β-cells. Thus, as discussed above, closure of K-ATP channels depolarizes the β-cell, thereby activating voltage-dependent Ca^{2+} channels; in addition, metabolism modulates L-type Ca^{2+} channels directly.

Smith *et al.* (1989*b*) found that increasing glucose from 0 to 20 mM produced an approximate doubling of the whole-cell Ca^{2+} current in mouse β-cells, as a result of an increase in L-type Ca^{2+} channel activity. L-type Ca^{2+} channel openings occur in bursts separated by longer closed intervals; the principal effect of the sugar was to increase the burst duration and decrease the duration of the long closures between bursts. A similar increase in Ca^{2+} channel activity is produced by glyceraldehyde in RIN m5F cells (Velasco *et al.* 1988), which, in addition to increasing the frequency of channel openings, also increases channel lifetime and shifts the activation curve to more negative potentials.

Several pieces of evidence support the idea that the effect of glucose is mediated by an intracellular second messenger (Smith *et al.* 1989*b*). First, the effect is slow, requiring several minutes to develop. Secondly, the sugar was added only to the bath solution; since the glass–membrane seal prevents diffusion between bath and pipette solutions, the effect of glucose must have been mediated by an intracellular route. Thirdly, inhibition of glucose metabolism reverses the effect of the sugar on Ca^{2+} channel activity. The identity of the second messenger which links glucose metabolism to Ca^{2+} channel activation has not yet been established. In the case of glyceraldehyde, however, it has been argued that activation of protein kinase C may be involved (Velasco and Petersen 1989).

4.3.3 The contribution of the different ion channels to β-cell electrical activity

A hypothetical model for the generation of β-cell electrical activity and its regulation by glucose, based on the data currently available, is given in Fig. 13.

In the absence of glucose, the resting potential of the β-cell is determined principally by the activity of the K-ATP channel. Closure of these channels, either by glucose metabolism or by sulphonylureas, depolarizes the β-cell. This implies the presence of a background inward current, which has not yet been identified (one candidate might be the NS-channel).

If the depolarization is sufficient, electrical activity is initiated which, as described above, consists of action potentials riding on the top of slow waves. The precise mechanism by which the slow waves are generated has not been established. However, the evidence favours the idea that both the depolarizing phase of the action potentials that arise from the plateau, and that of the slow waves themselves, result from activation of L-type Ca^{2+} channels. It also seems probable that the slow waves are principally determined by a balance between the L-type Ca^{2+} current, the K-DR and K-Ca currents, and the K-ATP current. Thus, we hypothesize that the slow wave is initiated by opening the voltage-dependent Ca^{2+} channels.

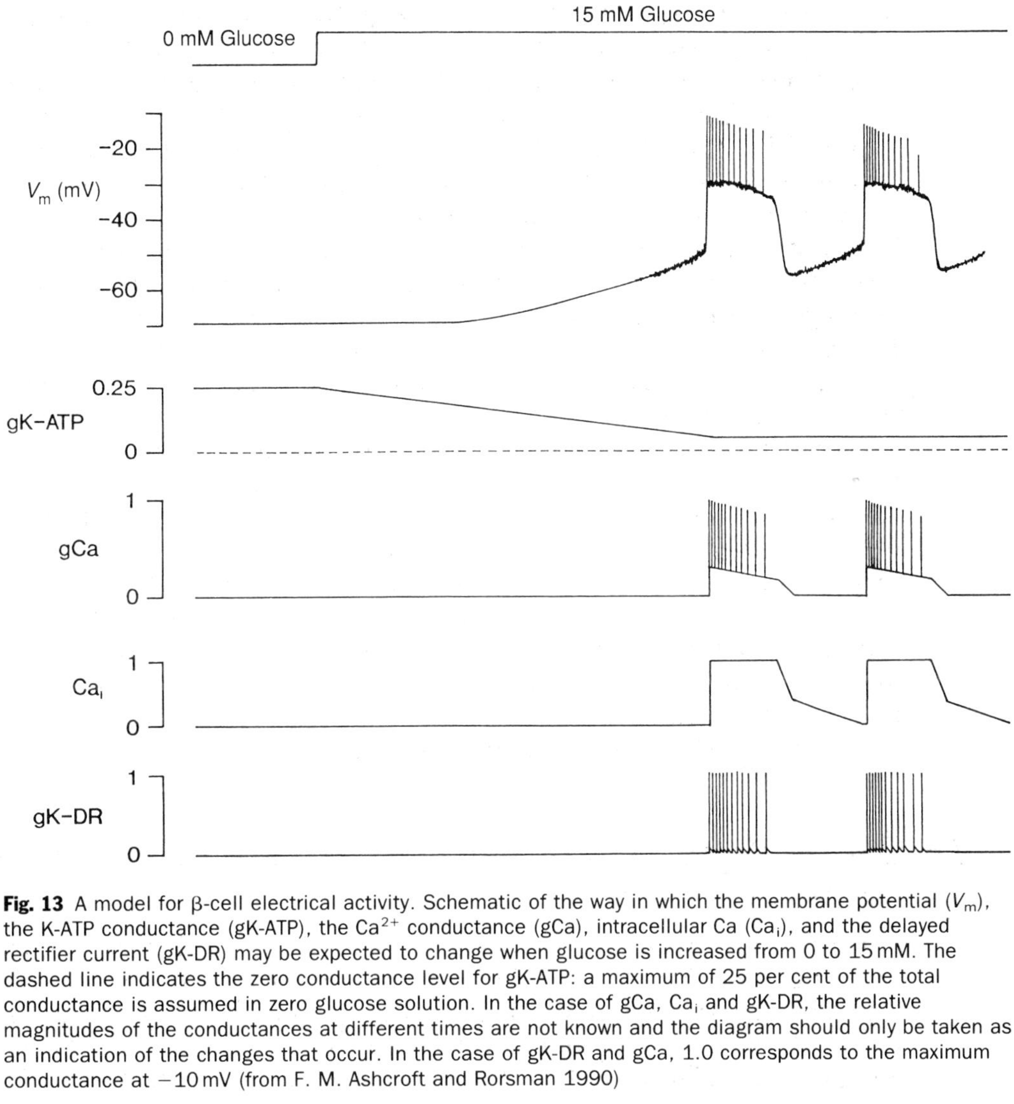

Fig. 13 A model for β-cell electrical activity. Schematic of the way in which the membrane potential (V_m), the K-ATP conductance (gK-ATP), the Ca^{2+} conductance (gCa), intracellular Ca (Ca_i), and the delayed rectifier current (gK-DR) may be expected to change when glucose is increased from 0 to 15 mM. The dashed line indicates the zero conductance level for gK-ATP: a maximum of 25 per cent of the total conductance is assumed in zero glucose solution. In the case of gCa, Ca_i and gK-DR, the relative magnitudes of the conductances at different times are not known and the diagram should only be taken as an indication of the changes that occur. In the case of gK-DR and gCa, 1.0 corresponds to the maximum conductance at −10 mV (from F. M. Ashcroft and Rorsman 1990)

Subsequent activation of K-DR and K-Ca channels repolarizes the membrane to the plateau potential, where these channels slowly close (deactivate). Since Ca^{2+} channels are still open at this potential, the decrease in the K-Ca and K-DR currents produces a net inward current which depolarizes the membrane and initiates another action potential. Repolarization at the end of the burst may be a consequence of a slow inactivation of the Ca^{2+} current, which finally becomes too small to balance the hyperpolarizing effect of the K-ATP current. The interburst interval may therefore reflect the rate at which the Ca^{2+} channels recover from inactivation. Throughout the slow wave the K-ATP current is roughly constant, as observed in patch-clamp studies. A fuller discussion of this model for β-cell electrical activity may be found in the review by Ashcroft and Rorsman (1989).

Oscillations in glycolysis and in the ATP/ADP ratio have been described in a number of systems. It has therefore been proposed that a similar phenomenon, if present in the β-cell, might lead to oscillations in K-ATP channel activity and so produce the slow waves. This hypothesis seems unlikely to be correct, however, since no difference in K-ATP current amplitude was found between the plateau phase of the slow wave and the interburst interval (Smith *et al.* 1990*b*). Instead, the K-ATP current appears to constitute a time-invariant, outward K-current. There is also good evidence that the K-Ca channel does not play a role in the generation of the slow waves, since they are unaffected by selective inhibition of this channel (Henquin 1990; Kukuljan *et al.* 1991).

Glucose modulates β-cell electrical activity, the duration of the slow waves becoming longer as the sugar concentration is increased until, finally, electrical activity is continuous. Current evidence favours the idea that glucose regulates the duration of the slow waves by further inhibiting K-ATP channels. In most β-cells, glucose concentrations that initiate electrical activity do not completely block the whole-cell K-ATP current, and the remaining current is further inhibited as glucose is increased. Inhibition is almost complete at glucose concentrations of ~20 mM, where electrical activity is continuous (Ashcroft *et al.* 1990*b*). We have also found a good correlation between the pattern of β-cell electrical activity and the magnitude of the whole-cell K-ATP current in perforated patch recordings; that is, the duration of the slow waves varies inversely with the magnitude of the K-ATP current. Finally, tolbutamide can modulate the duration of the slow waves in a manner similar to glucose, and diazoxide can convert continuous electrical activity to slow waves (Henquin 1988; Cook and Ickeuchi 1989). Taken together, these results indicate that closure of K-ATP channels (in addition to depolarizing the β-cell and initiating electrical activity), contributes to the changes in electrical activity that occur as glucose is increased. The high resistance of the β-cell membrane in the presence of glucose means that only very small changes in K-ATP current are needed to markedly affect β-cell electrical activity.

The physiological role of glucose modulation of the L-type Ca^{2+} channel is difficult to assess because no data concerning the concentration-dependence of the effect are available. Since glucose increases the Ca^{2+} current at a given membrane potential, the threshold potential for activation of both the action potential and the slow wave will be lowered. Furthermore, if the latter effect is graded with glucose concentration, it might contribute to the glucose-dependent decrease in the interburst interval (by reducing the threshold for initiation of the next burst). Likewise, a glucose-dependent increase in Ca^{2+} current might contribute to the prolongation of the slow wave produced by progressively raising glucose.

In summary, therefore, we suggest that glucose prolongs the burst by reducing the K-ATP current and thus maintaining the net inward (depolarizing) current for longer. Modulation of the Ca^{2+} current produced by glucose may also contribute to both the increase in slow wave duration and the reduction of the interburst interval.

4.3.4 Effects of pharmacological agents on β-cell electrical activity

A variety of pharmacological agents influence β-cell electrical activity as a consequence of their ability to activate or inhibit ion-channel activity. A full account of the effects of these agents may be found in the review by Ashcroft and Rorsman (1989).

Of clinical importance are the sulphonylureas, such as tolbutamide and glibenclamide, which have been used for many years to stimulate insulin secretion in non-insulin-dependent diabetics. These drugs act as potent and specific blockers of K-ATP channel activity (Fig. 14a; Sturgess *et al.* 1985; Trube *et al.* 1986). They interact directly with a specific sulphonylurea receptor in the β-cell membrane, and thereby inhibit channel activity; whether the sulphonylurea receptor is also the K-ATP channel is unknown. The order of potency is glibenclamide ($K_i = 4$ nM) > glipizide (6 nM) > meglitinide ($K_i = 2$ μM) > tolbutamide ($K_i = 7$ μM) (Zunkler *et al.* 1988*a*). Albumin and plasma proteins bind to sulphonylureas and reduce the free drug concentration. Providing that this binding is taken into account, there is a very good correlation between the potency of the various sulphonylureas (Zunkler *et al.* 1988*a*; Panten *et al.* 1989)

(1) in binding to β-cell membranes;

(2) in blocking K-ATP channels;

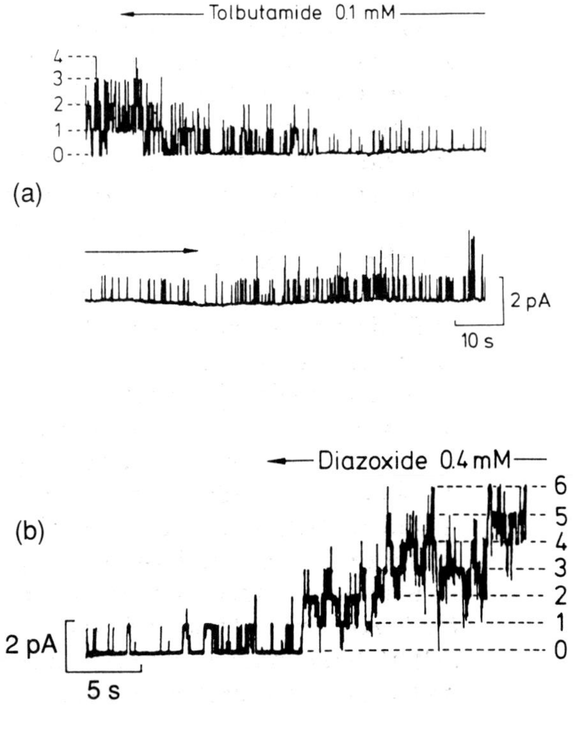

Fig. 14 Effects of tolbutamide and diazoxide on K-ATP channel activity. Effect of tolbutamide (a) and diazoxide (b) on K-ATP channel activity in an outside-out patch. The bath contained a standard extracellular solution. The pipette was filled with intracellular solution, which was supplemented with 0.3 mM MgATP in (b). Channel activity was recorded at a membrane potential of −20 mV in (a) and at 0 mV in (b). The numbers adjacent to the current traces indicate the number of simultaneous channel openings (from Trube *et al.* 1986, with permission)

(3) in inhibiting insulin secretion from mouse β-cells; and
(4) in the therapeutic free plasma levels of the drug in diabetic patients

This strongly supports the idea that inhibition of the K-ATP channel is the sole mechanism by which sulphonylureas initiate insulin release. In this context, it is important that human β-cells possess K-ATP channels with similar ATP- and tolbutamide-sensitivity to that found for rodents (Ashcroft *et al.* 1989*a*; Misler *et al.* 1989).

The related compound, diazoxide, a sulphonamide, has the opposite effect to sulphonylureas and increases K-ATP channel activity (Fig. 14b), which accounts for the inhibitory effect of diazoxide on insulin secretion. Interestingly, the action of diazoxide requires cytosolic MgATP, suggesting that protein phosphorylation may be an essential step in the activation process (Kozlowski *et al.* 1989).

In addition to the sulphonylureas, a number of other drugs (such as quinine) stimulate insulin secretion as a consequence of their ability to inhibit K-ATP channel activity. These agents, however, are not useful for treating NIDDM, as their action is not confined to the β-cell. Finally, drugs which influence Ca^{2+} channel activity, such as the dihydropyridines, modulate insulin secretion.

4.3.5 The sulphonylurea receptor

The high potency and specificity of the effects of sulphonylureas on K-ATP channels has prompted the use of these drugs as ligands for channel isolation. High-affinity sulphonylurea-binding sites have been described in β-cells (Schmid-Antomarchi *et al.* 1987; Gaines *et al.* 1988; Niki *et al.* 1989*a*; Panten *et al.* 1989). It is not yet clear, however, whether the sulphonylurea receptor is itself the K-ATP channel or whether it is a separate protein.

Exposure to ultraviolet light of β-cell membranes incubated with [^{3}H]glibenclamide results in the covalent labelling of proteins ranging in size from 30 to 140 kDa on SDS-PAGE (Kramer *et al.* 1988). A peptide with a molecular weight of 140 kDa, photo-labelled with a [^{125}I]glibenclamide derivative, has also been solubilized and partially purified from HIT T15 cells (Aguilar-Bryan *et al.* 1990). No sequence data have yet been reported, nor has the protein been functionally reconstituted and shown to possess ionophoretic activity.

The mechanism of interaction of sulphonylureas with their receptor is unknown, but there is both biochemical (Schwanstecher *et al.* 1990) and electrophysiological (Kozlowski *et al.* 1989) evidence suggesting that the properties of glibenclamide binding are modified by phosphorylation. The inhibitory action of tolbutamide on K-ATP channel activity is potentiated by cytosolic ADP (Zunckler *et al.* 1988*b*), but whether this is also true for other sulphonylureas has not been established.

The high sensitivity of the K-ATP channel to glibenclamide raises the interesting possibility that an endogenous ligand for the sulphonylurea receptor may exist. An endogenous peptide capable of displacing glibenclamide binding from brain membranes has been extracted from rat brain (Virsolvy-Vergine *et al.* 1988). It is not yet known whether this peptide is capable of interacting with the β-cell.

4.4 Protein phosphorylation and insulin secretion

Insulin secretagogues such as glucose or sulphonylureas, which lead to an increase in intracellular Ca^{2+} via depolarization of the β-cell, are able to initiate insulin release. However, hormones and neurotransmitters which elevate cyclic AMP or activate turnover of inositol phospholipids produce little or no stimulation of insulin release unless glucose or other initiator is present. In this section we review evidence suggesting that activation of a Ca^{2+}/calmodulin-dependent protein kinase may be central to initiation of insulin release whereas a modulatory role in mediating the potentiatory effects of hormones and neurotransmitters is assigned to protein kinases A and C.

4.4.1 Protein kinase A

It is well established that an increase in β-cell cyclic AMP, in response to glucagon and other hormones, magnifies the secretory response to an initiator of secretion such as glucose (reviewed by Malaisse and Malaisse-Lagae 1984). This protentiation of insulin secretion is accompanied by rapid phosphorylation of specific islet substrates for protein kinase A (Christie and Ashcroft 1985) supporting a modulatory role for protein kinase A in regulation of insulin secretion.

Protein kinase A has been detected and characterized in various β-cell preparations (for review see Harrison *et al.* 1984). Type I and Type II isoenzymes have been demonstrated in rat islets with properties similar to those found in other tissues (Sugden *et al.* 1979). Substrates for protein kinase A have been studied in intact islets of Langerhans prelabelled with $^{32}P_i$ and then stimulated with forskolin, an activator of adenyl cyclase (Christie and Ashcroft 1985), and in permeabilized islets exposed to cyclic AMP in the presence of [γ-^{32}]ATP (Jones *et al.* 1988). Both these approaches demonstrated rapid changes in phosphorylation of several β-cell peptides in both cytosolic and membrane fractions. What is currently lacking, however, is identification of such endogenous substrates for protein kinase A and characterization of their roles.

Although glucose itself causes an increase in β-cell cyclic AMP (Christie and Ashcroft 1984) it is unlikely that this effect contributes significantly to the mechanism whereby glucose initiates insulin release. Thus at low glucose concentrations marked elevation of islet cyclic AMP can be elicited without stimulation of insulin secretion (Christie and Ashcroft 1984). Moreover, in islets exposed to a cyclic AMP analogue that abolishes the activity of protein kinase A, glucose is still able to stimulate insulin secretion (Persaud *et al.* 1990).

4.4.2 Protein kinase C

Ca^{2+} is necessary for the activity of protein kinase C. However, the main physiological regulator of this enzyme is believed to be diacyglycerol, liberated by the action of phospholipase C on phosphatidylinositol bisphosphate, which modulates the sensitivity of protein C to Ca^{+2} (Nishizuka 1988). Protein kinase C has been purified to homogeneity from β-cells (Lord and Ashcroft, 1984). As in other tissues, the enzyme is a monomer of about 80 kDa. Western blot analysis has demonstrated that adult rat islets contain only the β-isoform of protein kinase C (Onoda *et al.* 1990; Fletcher and Ways 1991). Endogenous substrates for protein kinase C have been demonstrated in extracts of rat (Harrison *et al.* 1984) or mouse (Thams *et al.* 1984) islets; in rat insulinoma (Brocklehurst and Hutton 1984); in permeabilized islets (Jones *et al.* 1988); and in intact islets (Hughes and Ashcroft 1988) in which a major substrate was found to be a particulate protein of 37 kDa. As for protein kinase A, however, the identity of these endogenous substrates for protein kinase C remains unknown.

Evidence for the involvement of protein kinase C in insulin secretion has been sought by seeking parallels between the effects of various agents on activity of the enzyme and on insulin release. Tumour-promoting phorbol esters such as 12-O-tetradecanoyl phorbol 13-acetate (TPA), which activate protein kinase C by substituting for diacylglycerol, are potent potentiators of insulin release (Harrison *et al.* 1984; Howell *et al.* 1990); this effect has been shown to be accompanied by enhanced phosphorylation of islet proteins (Hughes and Ashcroft 1988). Clomiphene, a potent inhibitor of protein kinase C, blocks effects of TPA on insulin release and protein phosphorylation in intact islets (Hughes and Ashcroft, 1988). Similarly, the protein kinase C inhibitor H-7 inhibits TPA-stimulated insulin release (Metz 1988).

There is, however, growing evidence to suggest that, as for protein kinase A, protein kinase C is not of major importance for the β-cell secretory reponse to glucose itself (Hii *et al.* 1987; Metz 1988; Howell *et al.* 1990; Hughes *et al.* 1990). Thus when β-cells are incubated for a prolonged period with phorbol esters there is a dramatic loss (down-regulation) of protein kinase C activity. Such down-regulated β-cells nevertheless retain a secretory response to glucose. This suggests that protein kinase C may be more involved in responses to hormones or neurotransmitters acting via breakdown of phosphatidylinositol bisphosphate—the impaired ability of acetylcholine to potentiate insulin secretion from down-regulated islets is consistent with this view and suggests that cholinergic regulation of insulin secretion may be mediated by protein kinase C (Hughes *et al.* 1990; Persaud *et al.* 1991).

4.4.3 Calcium/calmodulin-dependent kinases

Several protein kinases activated by Ca^{+2}-calmodulin have been described in the β-cell. Myosin light chain kinase (MLCK) is present and could plausibly be postulated to be involved in granule translocation. However, although increased phosphorylation of a 20 kDa protein was observed in β-cells incubated with depolarizing concentrations of K^{+} (Oberwetter and Boyd 1987), rigorous identification of myosin light chain has not been reported and there is thus no direct evidence for involvement of MLCK in insulin release; other modes of granule movement (e.g. via an interaction between microtubules and kinesin) seem equally plausible.

β-Cells also contain a calcium/calmodulin-dependent protein kinase (P53 kinase) that phosphorylates an endogenous protein of 53 kDa. The properties of this kinase have been documented in detail (Harrison and Ashcroft 1982) and its activity reported in various insulin-secreting tissues, including human islets (Harrison *et al.* 1985). Phosphorylation is stimulated by micromolar concentrations of calcium in the presence of calmodulin and is blocked by calmodulin antagonists. It has been suggested that the kinase is a component of the β-cell cytoskeleton (Harrison *et al.* 1984; Harrison and Ashcroft 1982; Schubart and Fields 1984); a role in mediating granule-cytoskeleton interaction seems plausible.

Since the kinase activity has not been separated from the phosphorylated peptide it is likely that the observed phosphorylation represents an autophosphorylation. In this respect and others, P53 kinase resembles the multifunctional calmodulin kinase II which is of widespread occurrence and has been implicated in neurotransmitter release (Colbran *et al.* 1989). However, this identity has not yet been established.

Pharmacological evidence implicating P53 kinase in insulin secretion has been obtained. Pretreatment of islets with alloxan at concentrations that inhibit insulin secretion was shown to lead to inhibition of P53 kinase; alloxan also inhibited the kinase in islet homogenates (Colca *et al.* 1983). Dehydrouramil (DHU), a stable analogue of alloxan, also inhibited P53 kinase in extracts of islets but did not inhibit protein kinase A and caused only minor inhibition of protein kinase C (Harrison *et al.* 1986). However, in intact islets, pre-exposure to DHU impaired the insulin-secretory response to glucose and also blocked the potentiatory effects on insulin release of forskolin and TPA. These data suggest that P53 kinase may play a central role in the initiation of insulin release.

4.4.4 Modulation of insulin secretion by protein phosphorylation

There are two distinct mechanisms by which protein kinase activation might potentiate glucose-stimulated insulin release. Protein kinase activation may lead to an additional increase in intracellular Ca^{2+} via net influx of Ca^{2+} across the plasma membrane and/or efflux from intracellular compartments. Alternatively, protein kinase activation may sensitize the secretory mechanism to the existing level of intracellular Ca^{2+}.

Much evidence favours the latter mechanism. Activation of either protein kinase A or protein kinase C reduces the threshold extracellular Ca^{2+} concentration at which glucose-stimulated insulin release can take place and increases the maximal secretory response at the highest extracellular Ca^{2+} concentration (Hughes and Ashcroft 1987). That this altered sensitivity to extracellular Ca^{2+} reflects an altered dependence on intracellular Ca^{2+} has been demonstrated using intracellular Ca^{2+}-binding indicators in intact β-cells (Hughes *et al.* 1989) and in permeabilized islets (Jones *et al.* 1988). In intact β-cells, for example, TPA potentiates insulin release without causing any increase in intracellular Ca^{2+}. Indeed TPA actually tends to lower intracellular Ca^{2+} in the presence of glucose, and blocks the increase in β-cell intracellular Ca^{2+} evoked by vasopressin (Hughes *et al.* 1992). Although forskolin can increase β-cell intracellular Ca^{2+}, this effect is small and does not correlate with the potentiatory effect of forskolin on secretion (Hughes *et al.* 1989).

It can be concluded that stimulation of β-cell protein kinases in response to hormones and neurotransmitters primarily leads to potentiation of secretory responses by enhancing the sensitivity of the secretory mechanism to Ca^{2+}. This does not rule out the possibility of additional effects on β-cell Ca-fluxes. The observation, for example, that forskolin increased electrical activity of mouse islets depolarized by tolbutamide (Henquin *et al.* 1987) is consistent with the findings of a modest rise in cytosolic free Ca^{2+} induced by forskolin (Hughes *et al.* 1989). In general, however, effects of potentiators on Ca-fluxes are of less importance than the ability of protein kinase activation to lead to an increased sensitivity to intracellular Ca^{2+}.

4.5 Role of intracellular calcium

Detailed reviews of the role of Ca^{2+} in insulin secretion are given by Wollheim and Sharp (1981) and Prentki and Matchinsky (1987). Figure 15 summarizes the main mechanisms involved in regulation of intracellular Ca^{2+} concentration ($[Ca^{2+}]_i$).

4.5.1 Sources of Ca^{2+} of importance for secretion

There is no doubt that the influx of Ca^{2+} ions across the plasma membrane is essential for the rise in $[Ca]_i$ elicited by glucose or by sulphonylureas. The evidence is that:

(1) the extent of the rise in $[Ca]_i$ is dependent on the concentration of extracellular Ca^{2+};

(2) drugs that block L-type Ca^{2+} channels abolish the rise in $[Ca]_i$.

What then is the role of intracellular Ca^{2+} stores? One possibility is that Ca^{2+} influx triggers the release of stored Ca^{2+}, thereby amplifying the Ca^{2+} signal. Such Ca^{2+}-induced Ca^{2+} release has been demonstrated in other cells, but its occurrence in the β-cell has not been studied. A second role for which considerable evidence exists is in the Ca^{+2} response to potentiators of release acting via breakdown of phosphatidylinositides (Nilsson *et al.* 1987). Mobilization of stored Ca^{2+} by inositol triphosphate has been demonstrated in permeabilized β-cells; levels of IP_3 have been shown to be increased by muscarinic agonists; muscarinic and α_1-adrenergic agonists have been shown to elevate β-cell Ca^{2+} in the absence of external Ca^{2+} and presence of EGTA. Most evidence favours the view that elevation of cAMP, however, does not mobilize intracellular Ca^{2+} (Rorsman and Abrahamsson 1985).

An initial decrease in $[Ca]_i$ upon stimulation with glucose has been observed in β-cell suspensions (Nilsson *et al.* 1988) and single β-cells (Pralong *et al.* 1990). This

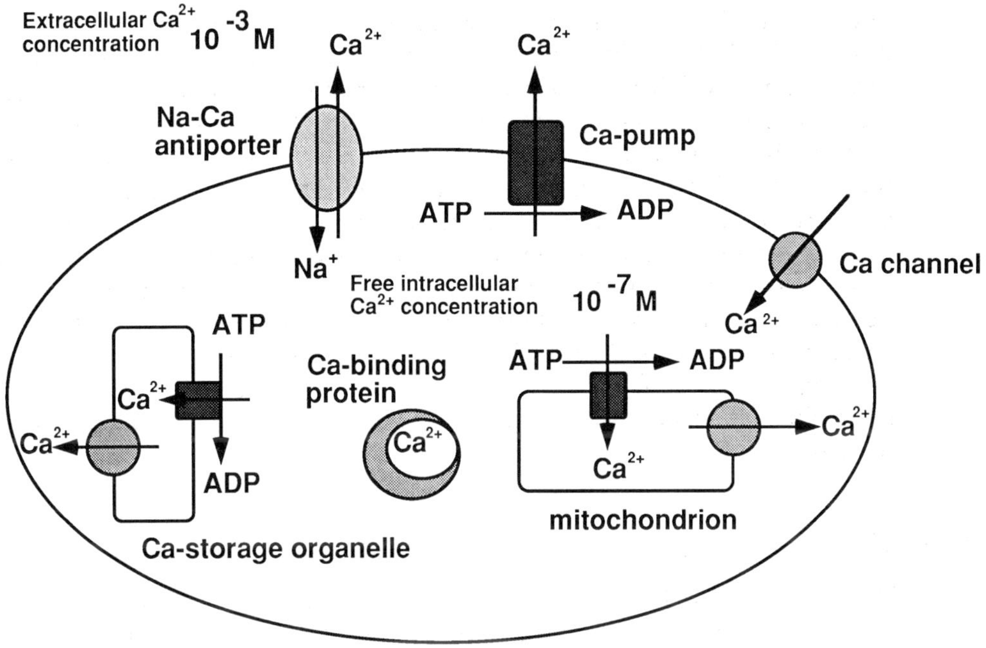

Fig. 15 Intracellular calcium homeostasis. The figure shows the major mechanisms regulating β-cell $[Ca^{2+}]_i$

has been attributed to activation of Ca^{2+} uptake into cellular stores (Rorsman *et al.* 1984). It was suggested that the phenomenon represents simply fuel-depletion prior to elevation of glucose (Prentki and Matschinsky 1987; Nilsson *et al.* 1988) but subsequent studies have supported its physiological significance (Yada *et al.* 1992).

4.5.2 Ca^{2+} oscillations

Measurements of $[Ca]_i$ in single β-cells have shown that the regulation of calcium homeostasis is more complex than had been thought from studies on β-cell suspensions. In particular, oscillations of $[Ca]_i$ were observed in single β-cells (Grapengiesser *et al.* 1990; Pralong *et al.* 1990). In single-cell studies a marked heterogeneity in this Ca^{2+} response is observed. Nevertheless, it is likely that these oscillations correspond to the slow wave oscillations in β-cell electrical activity. Good evidence in support of this idea is provided by the demonstration that oscillations in $[Ca]_i$ in an intact islet are synchronized with β-cell electrical activity (Fig. 16; Valdeolmillos *et al.* 1989). These studies further suggested that the Ca^{2+}

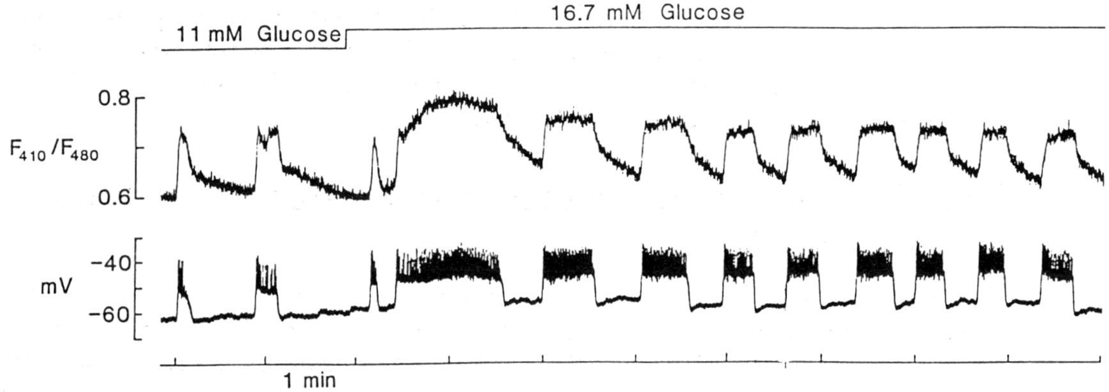

Fig. 16 Oscillations in intracellular Ca^{2+} in an intact islet correlates with β-cell activity. Simultaneous measurement of membrane potential in a single β-cell and intracellular Ca^{2+} concentration measured in the whole islet. The free intracellular Ca^{2+} is given by the indo-1 fluorescence ratio F_{410}/F_{480} (from Santos *et al.* 1991)

signal may be frequency modulated since the effect of glucose was to increase the duration of the Ca^{2+} transients and reduce the intervals between without altering the amplitude.

4.5.3 Actions of intracellular calcium

Many cellular actions of Ca^{2+} are mediated by calmodulin. There is persuasive, although not conclusive, evidence that this intracellular Ca^{2+} receptor may be of importance for insulin secretion. The β-cell contains a high content (50 μM) of calmodulin (Sugden *et al.* 1979) and the ability of a number of phenothiazine derivatives to inhibit calmodulin could be correlated with their inhibitory effects on glucose-stimulated insulin release (Gagliardino *et al.* 1980; Henquin 1981). The main limitation to this conclusion derives from the fact that these inhibitors are not completely specific for calmodulin. However, inhibition of protein kinase C, a known additional site of action for phenothiazines, is unlikely to explain inhibition of glucose-stimulated secretion in view of the lack of importance of protein kinase C in the secretory response to glucose. Of considerable interest is the demonstration (Epstein *et al.* 1989) that overexpression of calmodulin in the β-cells of transgenic mice led to severe diabetes within hours of birth. There was evidence of both secretory defects and of structural damage.

Stimulatory effects of Ca^{2+}, mediated by calmodulin, have been documented on both adenyl cyclase and on cAMP phosphodiesterase (PDE) in islets. Although these effects may form a part of the complex control network for regulating β-cell secretion, it is clear that they are not the basis for the stimulatory effect of Ca^{2+} on exocytosis. Thus activation of PDE would lower cAMP and hence tend to reduce insulin secretion. Activation of adenyl cyclase is capable only of potentiating, but not initiating, insulin release (Christie and Ashcroft 1985).

As discussed in Section 4.4, Ca^{2+}/calmodulin-dependent protein phosphorylation may play a key role in initiating secretion. However our knowledge of exactly how Ca^{2+} modulates secretion is very incomplete; this is because of our limited knowledge of the mechanism of exocytosis.

4.6 Regulation of insulin secretion by hormones and neurotransmitters

A number of hormones and neurotransmitters are capable of modulating the response of the β-cell to glucose and other nutrient secretagogues. These may be either stimulatory or inhibitory in their action. Whereas hormones reach the β-cell via the circulation or act in a paracrine manner, neurotransmitters are released from autonomic nerve fibres terminating within the pancreatic islet (see Chapter 1).

4.6.1 Physiological significance

The modulation of insulin release by hormones and neurotransmitters is of major physiological significance. The cephalic phase of insulin release, which begins prior to glucose absorption and shortly after injesting food, is believed to result from acetylcholine released from parasympathetic nerves terminating within the islet, since it is abolised by vagotomy or atropinization and is absent in rats whose insulin comes entirely from transplanted islets. Stimulation of the sympathetic nervous system results in adrenaline, and possibly also galanin, release with consequent inhibition of insulin secretion (Dunning and Taboborsky 1988). This mechanism may serve to maintain plasma glucose levels during exercise or in response to stress. A number of polypeptides released by the gastrointestinal tract (such as cholecystokinin (CCK), vasoactive intestinal peptide (VIP) and gastric inhibitory peptide (GIP), potentiate insulin release. This explains why glucose taken orally has a more potent effect on secretion than intravenously injected glucose. Insulin release is also subject to paracrine influences from peptides released from other islet cell types. For example, glucagon released from islet α-cells stimulates insulin release whereas somatostatin (released from islet D-cells) inhibits secretion (see Chapter 1). Although immunocytochemical studies have demonstrated peptidergic nerve fibres within the islet containing VIP, CCK and bombesin (which stimulate release), and Substance P and calcitonin-gene-related peptide (CGRP) (which inhibit secretion), the mechanisms responsible for peptide release remain obscure (Ahren *et al.* 1986).

4.6.2 Potentiators of secretion

Potentiators of glucose-induced insulin secretion include glucagon, acetylcholine (acting on muscarinic receptors), vasopressin, vasoactive intestinal peptide (VIP), gastric inhibitory peptide (GIP), bombesin, and purinergic agonists. These agents are often referred to as secondary secretagogues, since they are unable to initiate insulin release, but instead augment secretion induced by primary secretagogues such as glucose (Fig. 1). Although they may induce a small depolarization, secondary secretagogues are unable to initiate β-cell action potential activity and concomitant Ca^{2+} influx. One explanation for the dependence of potentiators of secretion on the presence of a primary secretagogue, therefore, is that the latter is required for triggering electrical activity and Ca^{2+} influx.

In most cases it has been shown that potentiators of secretion act at more than one site; for example, they may stimulate β-cell electrical activity, alter the intracellular concentrations of second messengers such as Ca^{2+}, diacylglycerol or cyclic AMP, or sensitize the secretory machinery to Ca^{2+}. The relative importance of these actions has not been fully established. Secondary secretagogues comprise two main groups: those which elevate cyclic AMP thus activating protein kinase A (PKA), and those which stimulate the phosphoinositide pathway and thereby elevate intracellular Ca^{2+} and activate protein kinase C.

Effects on electrical activity

Potentiators of insulin secretion such as acetylcholine, GIP, and VIP increase β-cell electrical activity (Ashcroft and Rorsman 1990). It is clear that a multiplicity of mechanisms are involved in these effects since the various hormones and neurotransmitters change the pattern of β-cell electrical activity in different ways. The details, however, remain elusive. Evidence exists that the stimulatory effects of acetylcholine on electrical activity result from an increase in membrane Na^+ permeability (Henquin *et al.* 1988) and it has been suggested that glucagon may amplify a Ca^{2+} current (Henquin and Meissner 1983). However, most studies have so far failed to demonstrate any effect of secondary secretagogues on either single channel or whole-cell currents in normal β-cells. Whether this is a consequence of the cell dissociation method or of the patch-clamp configuration employed, or is due to another reason, is unknown. Arginine-vasopressin, however, has been shown to close K-ATP channels (Martin *et al.* 1989) and to activate Ca^{2+} channels (Thorn and Petersen 1991) in RINm5F cells, thereby stimulating electrical activity and increasing Ca^{2+} influx.

Effects on intracellular second messengers (I): activation of phospholipase C

Acetylcholine, cholecystokinin (CCK) and purinergic agonists are believed to stimulate insulin secretion by activation of the phosphoinositide pathway. The evidence is currently strongest for acetylcholine (Prentki and Matchinsky 1987). Activation of muscarinic receptors (Wollheim and Biden 1986), and thus phospholipase C, results in the generation of diacylglycerol and inositol trisphosphate. The latter mobilizes Ca^{2+} from intracellular stores, but the effect is transient and thus probably only important for the initial response to acetylcholine (Hughes *et al.* 1990). By contrast, activation of protein kinase C (PKC) by diacylglycerol appears to play a central role in mediating acetylcholine action, since the transmitter failed to potentiate secretion from islets in which PKC was down-regulated (Hughes *et al.*

1990; Persaud *et al.* 1991). Sensitization of the secretory machinery to Ca^{2+} is thought to mediate the action of PKC and is discussed in Section 4.4.4.

There is evidence that CCK (Zwalich *et al.* 1987) and bombesin (Swope and Schonbrunn 1988) also stimulate insulin release by activation of the phosphoinositide pathway.

Effects on intracellular second messengers (II): activation of adenylate cyclase

Glucagon is thought to act by stimulating adenylate cyclase. In support of this hypothesis, the hormone produces an increase in β-cell cyclic AMP levels (Malaisse and Malaisse-Lagae, 1984) which results in activation of protein kinase A (Christie and Ashcroft 1985). There is some evidence that GIP (Siegel and Creutzfeldt 1985) may also elevate cyclic AMP. Section 4.4.4 discusses the evidence that PKA amplifies insulin secretion by sensitizing the secretory machinery to Ca^{2+}.

4.6.3 Inhibitors of secretion

Little is known of the mechanisms by which CGRP, opioids, Substance P and pancreastatin inhibit insulin secretion. Adenaline, somatostatin and galanin, however, have been shown to act at several different levels in the secretory process. These include inhibition of β-cell electrical activity and ion fluxes, a decrease in the concentrations of intracellular second messengers such as Ca^{2+} and cyclic AMP, and inhibition of a late stage of secretion distal to the increase in intracellular Ca^{2+}. The relative importance of these effects has not been clearly established, although there is evidence that at low adrenaline concentrations, decreased β-cell electrical activity and Ca^{2+} influx are of more significance, whereas at higher concentrations distal effects on secretion dominate (Debuyser *et al.* 1991).

Pharmacological evidence suggests that the inhibitory effect of adrenaline is mediate by α_2-adrenoreceptors. However, the stimulation of insulin release, and the reversal of adrenaline-induced inhibition, by α-adrenoreceptors antagonists may also involve the ability of these drugs to directly block K-ATP channels (Plant and Henquin 1990). We confine our discussion to the actions of adenaline, somatostatin and galanin, because of the paucity of data on other inhibitory hormones. The similarities in the action of these agents suggest that they may act by a common mechanism(s).

Effects on electrical activity

Adrenaline, galanin and somatostatin all reduce β-cell electrical activity, producing an initial transient hyperpolarization which is accompanied by cessation of electrical activity. This is followed by the reappearance of long slow waves of reduced frequency (Ashcroft and Rorsman 1989). These effects on β-cell electrical activity are consistent with the activation of a K-conductance. The apparently contradictory finding that galanin and adrenaline decrease ^{86}Rb efflux can be explained by the reduction in ^{86}Rb efflux through voltage-gated K-channels, which results from the decrease in β-cell electrical activity (Drews *et al.* 1990).

There is evidence from patch-clamp studies that both activation of an outward current and inhibition of an inward current may be involved in mediating the actions of somatostatin, galanin and α_2-adrenergic agonists. However, the precise mechanisms appear to differ between mouse β-cells and β-cell lines. Thus, in the latter, there is evidence that all three agents activate K-ATP channels, thereby hyperpolarizing the membrane and inhibiting electrical activity (Dunne *et al.* 1989*a*; De Weille *et al.* 1988). By contrast, in mouse β-cells, the K-ATP channels are unaffected and instead clonidine activates a different type of K-channel which is insensitive to sulphonylureas (Rorsman *et al.* 1991). The conductance of this channel (the K-I channel) is very low; estimates from analyses of membrane current noise yield values around 1pS with physiological ionic gradients, and single channel currents cannot be resolved. A pertussis toxin-sensitive G-protein appears to mediate the effect of receptor activation on K-I channel activity. Inhibition of Ca^{2+} currents in β-cell lines by somatostatin (Hsu *et al.* 1989), galanin (Homaidan *et al.* 1991) and clonidine has also been reported (Keahey *et al.* 1989) but again it has not been possible to demonstrate such an effect in mouse β-cells (Bokvist *et al.* 1990).

Effects on intracellular second messengers

Galanin, somatostatin, and α_2-adrenergic agonists lower intracellular free Ca^{2+} levels because of their ability to partially repolarize the β-cell and so decrease Ca^{2+} influx (Hurst and Morgan 1989; Nilsson *et al.* 1988; Nilsson *et al.* 1989). All three agents also decrease adenylate cyclase activity and lower cyclic AMP levels (Amiranoff *et al.* 1988). However several studies have established that these effects on Ca^{2+} and the adenylate cyclase system cannot completely account for inhibition of insulin secretion (Jones *et al.* 1987; Ullrich and Wollheim 1988; Sharp *et al.* 1989).

Distal effects on secretion

Studies on permeabilized islets have led to the conclusion that galanin also inhibits insulin secretion at a step distal to the generation of cytosolic second messengers, possibly by interfering with the exocytotic process itself (Ullrich and Wollheim 1989). This inhibition is attenuated by pertussis-toxin, suggesting that a pertussis-toxin sensitive G-protein is involved in the action of the peptide. Similar results are found for both somatostatin and adrenaline (Ullrich and Wollheim 1988; Ullrich *et al.* 1990).

4.7 Mechanism of exocytosis

During exocytosis the secretory granule moves to the plasma membrane; fusion between the membrane surrounding the granule and the cell membrane leads to the development of a hiatus through which insulin is released into the extracellular space. This is a very frequent event: in man, it has been estimated that during rapid insulin secretion around 1.8×10^8 granules are secreted per second (Howell 1984).

4.7.1 Granule translocation

Evidence for the involvement of microtubules and microfilaments in granule movement to the cell membrane has been reviewed by Howell (1984). A well-developed cytoskeletal system in the β-cell has been demonstrated by electron microscopy. Dynin and kinesin power the translocation of secretory vesicles along microtubules in neural cells, but their presence in β-cells has not been documented. Because of the presence of myosin light-chain kinase, it has been envisaged that myosin may provide the motive force for granule movement in the β-cell, presumably by interaction with the actin-containing microfilaments.

It has proven difficult to establish by pharmacological means the role of either microtubules or microfilaments in insulin secretion, for two reasons. One is that the drugs are not specific; the other, that any effect on insulin secretion when these structures are disturbed does not prove that they play a physiological role in secretion.

4.7.2 Role of GTP

Recent studies suggest that Ca^{2+} is not required for secretion in many non-excitable cells, but that GTP fulfils an essential role. A novel GTP-binding protein, G_e, has been postulated to mediate such Ca^{2+}-independent secretion (Gomperts 1990). It is clear that in the β-cell, elevation of $[Ca]_i$ is a sufficient stimulus for secretion (Wollheim and Sharp 1981). However, GTP-induced Ca^{2+}-independent secretion of insulin has also been demonstrated, both in RIN m5F cells (Valler *et al.* 1987) and in electrically permeabilized islet cells (Wollheim *et al.* 1987). GTP also appears to mediate inhibition of insulin secretion at a late stage following the action of α_2-adrenergic, somatostatin, and galanin receptors (Ullrich and Wollheim 1988, 1989; Nilsson *et al.* 1989). This may suggest the involvement of more than one G-protein in the excocytotic process.

4.7.3 Fusion events

Nothing is known of the fusion event in β-cells. The exocytotic event is probably initiated by a pre-assembled fusion pore (Almers 1990). An unknown macromolecule present in the vesicle membrane is believed to dock the vesicle at specialized sites on the plama membrane. Fusion begins when this ion channel opens, forming a fusion pore; subsequent gradual dilation of the pore is due to lipid molecules inserting themselves between the subunits of this structure. A dilation equivalent to the addition of one phospholipid head-group to the pore has been calculated to take only 300 μs.

A possible role for Ca^{2+} is in triggering the phosphorylation of a membrane protein whose phosphorylation state determines whether vesicles can dock at the plasma membrane. In synaptic terminals, there is evidence that synapsin I, which is associated with the vesicle, plays such a role (Nestler and Greengard 1984).

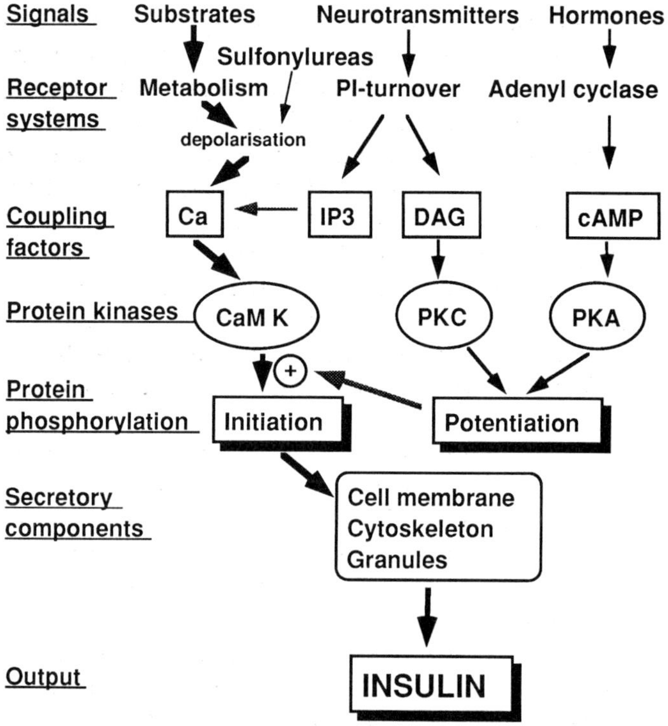

Fig. 17 A model for the control of insulin secretion. The model emphasizes the major role of protein phosphorylation in the control of insulin release. The main initiatory pathway is suggested to involve activation of calmodulin-dependent protein kinase(s) to phosphorylate components of the release system. The magnitude of the secretory response can be modulated by hormones or neurotransmitters via activation of inositol phospholipid turnover and of adenyl cyclase. These potentiatory effects involve primarily a sensitization of the secretory system to calcium

4.8 Model for the control of secretion

Figure 17 outlines a general model for the control of insulin release based on the considerations outlined above. The β-cell is depicted as responding to three classes of physiological stimulants—metabolites, hormones, and neurotransmitters. Initiation of secretion by glucose involves metabolism of the sugar and a rise in $[Ca^{2+}]_i$ via inhibition of K-ATP channels. Activation of calmodulin-dependent protein kinase leads to phosphorylation of key components of the release system and exocytotic discharge of insulin. In addition, in response to hormones and neurotransmitters, the classical second messenger systems involving inositol phospholipid turnover and adenyl cyclase activation generate signals which modulate the initiatory pathway. The additional phosphorylation events involved on activation of protein kinase A or C appear to function by sensitizing the initiatory pathway to calcium.

4.9 Possible defects in the regulation of secretion and their importance for type II diabetes

There is good evidence that type II diabetes is associated with disorders in the secretory function of the β-cell which result in a failure of glucose to stimulate insulin release (Vague and Moulin 1982). Although type II diabetes undoubtedly has a multiple aetiology, the ability of sulphonylureas to stimulate insulin release in type II diabetics suggests that events subsequent to closure of the K-ATP channel remain functional. This raises the intriguing possibility that type II diabetes results from a defect in the K-ATP channel protein or in its regulation: for example, defective glucose uptake or metabolism, or a defect in the channel protein such that it no longer responds to intracellular ATP. The decreased ATP sensitivity seen for the K-ATP channel of the hypertrophied heart provides a precedent for the last suggestion (Cameron *et al.* 1988). Overproduction of an endogenous agent that activates K-ATP channels (and thereby inhibits electrical activity and insulin release) is also possible. Clearly, studies on β-cells isolated from type II diabetics are required to distinguish between these possibilities, but in the absence of such studies it is worthwhile considering the case of glucose-insensitive cell lines and fetal β-cells.

First, alterations in the properties of the glucose transporter are known to influence glucose-stimulated insulin release. A good example is provided by the HIT T15 cell line. Although this cell line retains the ability to respond to glucose, the K_m for glucose-stimulated insulin release is lower than in normal β-cells (Ashcroft *et al.* 1986). This is because glucose metabolism is limited by the rate of the transport of sugars into the cell (Ashcroft and Stubbs 1987). The glucose transporter therefore acts as the glucose sensor in HIT T15 cells, in contrast to normal rodent islets in which glucokinase is thought to serve this function. The glucose transporter of the normal β-cell has been cloned and shown not to be present in RIN m5F cells (Thorens *et al.* 1988), another cell line with a defect in glucose transport (Giroix *et al.* 1986). Instead, RIN m5F cells inappropriately express the erythrocyte glucose transporter (Thorens *et al.* 1988). It would be of interest to ascertain whether the same is true for HIT T15 cells. More importantly, it is now feasible to determine whether the β-cells of type II diabetics have a lower expression of the normal β-cell glucose transporter, or if an inappropriate glucose transporter is expressed.

There is considerable evidence that defects in β-cell metabolism impair insulin secretion. Fetal β-cells manifest an impaired secretory response to nutrient secretagogues but insulin release can be elicited when intracellular Ca^{2+} levels are increased by depolarization with extracellular K^+ (Rorsman *et al.* 1989). Patch-clamp studies have shown that the ATP sensitivity of the K-ATP channel in fetal β-cells is similar to that of the adult β-cell, but that the channel activity does not alter in response to glucose; the failure of glucose to increase cytosolic ATP content probably accounts for this result (Rorsman *et al.* 1989). An immature glucose metabolism is thus the cause of the poor secretory response of the fetal β-cell. In general, insulin-secreting cell lines also respond poorly to glucose, although K-ATP channel closure and insulin secretion can be stimulated by other nutrient secretagogues. For example, glyceraldehyde stimulates RIN m5F cells (Dunne *et al.* 1986) and leucine elicits secretion from CRI-G1 cells (Carrington *et al.* 1987). In both these cell lines, the K-ATP channel has a similar ATP sensitivity to normal β-cells (Sturgess *et al.* 1986*a*; Ribalet and Ciani 1987) and the inability to respond to glucose resides in a defect in glycolytic metabolism. Finally, as discussed earlier, decreased glucokinase activity is observed in several situations associated with decreased insulin release. For example, when compared with normal rat islets, glucose utilization is decreased in islets isolated from rats made diabetic by neonatal injection of streptozotocin (a model for type II diabetes; Portha *et al.* 1988).

One mechanism that might reduce the ability of glucose to increase the cytosolic ATP:ADP ratio is glucose cycling, as this involves an increase in ATP hydrolysis (see Fig. 7). Glucose cycling is fourfold higher in islets isolated from neonatally streptozotocin-induced diabetic rats compared with normal rat islets (Khan *et al.* 1990*a*). Although glucose cycling in human islets has not been studied, hepatic glucose cycling has been measured in man. In normal subjects, hepatic glucose cycling was not detected, but it was markedly increased in type II diabetics (Efendic *et al.* 1985). Whether a similar increase in glucose cycling occurs in the β-cells of type II diabetics is not known but, if it were present, it might impair the ability of glucose to increase cellular ATP levels.

The hypothesis that type II diabetes results from defective β-cell metabolism would be strengthened if it could be shown that known metabolic defects lead to the development of the disease. It might, therefore, be worth investigating whether patients with such disorders (for example, with mitochondrial myopathies) have an increased incidence of type II diabetes.

The answer to the question of whether a defect in the K-ATP channel protein itself is responsible for the failure of the β-cell to respond to nutrient secretagogues must await the cloning of the K-ATP channel from human β-cells.

Finally, is there any evidence that K-ATP channel activity (and thus insulin release) is reduced in type II diabetics because of the enhanced secretion of an endogenous inhibitor? In this context it is relevant that calcitonin gene-related peptide (CGRP) activates K-ATP channels in arterial smooth muscle (Nelson *et al.* 1990). CGRP shows close structural homology with amylin (Cooper *et al.* 1987), a peptide that accumulates within the islets of type II diabetics and has been postulated to be a causal factor in the development of type II diabetes (Clark *et al.* 1987; Cooper *et al.* 1989). Although CGRP acts as an inhibitor of insulin release (Lewis *et al.* 1988), the ability of amylin to inhibit insulin release is controversial. Studies using physiological concentrations of the peptide have failed to demonstrate an inhibition of insulin release, either in isolated islets or *in vivo* (Ghatei *et al.* 1990; Petterson and Ahren 1990). Thus there is little evidence that amylin has a direct inhibitory action on the β-cell. Given that amylin accumulates between the capillaries and the β-cells (Clark *et al.* 1987), an alternative possibility is that amyloid accumulation results in impaired blood flow, and thereby metabolically compromises the β-cell. Such an effect might lead to K-ATP channel activation and decreased insulin release.

The results described here favour the hypothesis that impaired β-cell metabolism may occur in type II diabetics, with the result that K-ATP channels do not close in response to changes in plasma glucose. But since this conclusion is based simply on analogy with animal models of diabetes and glucose-insensitive cell lines, it must be treated with caution. Studies on diabetic human β-cells are required urgently.

Acknowledgements

Work in our own laboratories has been made possible by funding from the British Diabetic Association, the Wellcome Trust, the Medical Research Council, the Royal Society, the British Heart Foundation, Glaxo Inc., and the EP Abraham Trust. We are grateful to them all for their support.

References

Aguilar-Bryan, L., Nelson, D. A., Vu, Q. A., Humphrey, M. B., and Boyd, A. E., III (1990). Photoaffinity labeling and partial purification of the beta cell sulfonylurea receptor using a novel, biologically active glyburide analog. *J. Biol. Chem.* **265,** 8218–24.

Ahren, B., Taborsky, G. J., and Porte, D. (1986). Neuropeptidergic versus cholinergic and adrenergic regulation of islet hormone secretion. *Diabetologia* **29,** 827–36.

Amiranoff, B., Lorinet A.-M., Lagney-Pourmir, I., and Laburthe, M. (1988). Mechanism of galanin-inhibited insulin release. Occurrence of a pertussis-toxin sensitive inhibition of adenylate cyclase. *Eur. J. Biochem.* **177,** 147–52

Almers, W. (1990). Exocytosis. *Ann. Rev. Physiol.* **52,** 607–24.

Ashcroft, F. M. and Kakei, M. (1989). ATP-sensitive K channels in rat pancreatic β-cells; modulation by ATP and Mg^{2+} ions. *J. Physiol.* **416,** 349–67.

Ashcroft, F. M. and Rorsman, P. (1989). Electrophysiology of the pancreatic β-cell. *Prog. Biophys. Molec. Biol.* **54,** 87–145.

Ashcroft, F. M. and Rorsman, P. (1990). ATP-sensitive K^+ channels: A link between B-cell metabolism and insulin secretion. *Biochem. Soc. Trans.* **18,** 109–11.

Ashcroft, F. M., Harrison, D. E., and Ashcroft, S. (1984). Glucose induces closure of single potassium channels in isolated rat pancreatic β-cells. *Nature* **312,** 446–8.

Ashcroft, F. M., Ashcroft, S. J. H., and Harrison, D. E. (1985). The glucose-sensitive potassium channel in rat pancreatic β-cells is inhibited by intracellular ATP. *J. Physiol.* **369,** 101P.

Ashcroft, F. M., Ashcroft, S. J. H., and Harrison, D. E. (1987). Effects of 2-keto-isocaproic acid on insulin release and single potassium channel activity in dispersed rat pancreatic β-cells. *J. Physiol.* **385,** 517–29.

Ashcroft, F. M., Ashcroft, S. J. H., and Harrison, D. E. (1988). Properties of single potassium channels modulated by glucose in rat pancreatic β-cells. *J. Physiol.* **400,** 501–27.

Ashcroft, F. M., Kakei, M., Gibson, J. S., Gray, D. W., and Sutton, R. (1989). The ATP- and tolbutamide-sensitivity of the ATP-sensitive K-channel from human pancreatic β cells. *Diabetologia* **32,** 591–8.

Ashcroft, F. M., Faehling, M., Fewtrell, C. M. S., Rorsman, P., and Smith, P. A. (1990*a*). Perforated patch recordings of ATP-regulated K-currents and electrical activity in murine pancreatic β-cells. *Diabetologia* **33,** A78.

Ashcroft, F. M., Kelly, R. P., and Smith, P. A. (1990*b*). Two types of Ca channel in rat pancreatic beta-cells. *Pflügers Arch.* **415,** 504–6.

Ashcroft, S. J. H. (1980). Glucoreceptor mechanisms and the control of insulin release and biosynthesis. *Diabetologia* **18,** 5–15.

Ashcroft, S. J. H. (1981). Metabolic control of insulin secretion. In *The islets of Langerhans: biochemistry, physiology and pathology* (ed. S. J. Cooperstein and D. Watkins), pp. 117–48. Academic Press, London.

Ashcroft, S. J. H. and Ashcroft, F. M. (1990). Properties and functions of ATP-sensitive K-channels. *Cellular Signalling* **2,** 197–214.

Ashcroft, S. J. H. and Christie, M. R. (1979). Effects of glucose on the cytosolic ratio of reduced/oxidized $NADP^+$ in rat islets of Langerhans. *Biochem. J.* **184,** 697–700.

Ashcroft, S. J. H. and Hughes, S. J. (1990). Protein phosphorylation in the regulation of insulin secretion and biosynthesis. *Biochem. Soc. Trans.* **18,** 116–18.

Aschroft, S. J. H. and Randle, P. J. (1970). Enzymes of glucose metabolism in normal mouse islets. *Biochem. J.* **119,** 5–15.

Ashcroft, S. J. H. and Stubbs, M. (1987). The glucose sensor in HIT cells is the glucose transporter. *FEBS Lett.* **219,** 311–15.

Ashcroft, S. J. H., Weerasinghe, L. C. C., and Randle, P. J. (1973). Inter-relationships of islet metabolism, adenosine triphosphate content and insulin release. *Biochem. J.* **132,** 223–31.

Ashcroft, S. J. H., Hammonds, P., and Harrison, D. E. (1986). Insulin secretory responses of a clonal cell line of simian virus 40-transformed β-cells. *Diabetologia* **29,** 727–33.

Berne, C. (1975). The metabolism of lipids in mouse pancreatic islets. The biosynthesis of triacylglycerols and phospholipids. *Biochem. J.* **152,** 667–73.

Bokvist, K., Rorsman, P., and Smith, P. A. (1990). Block of ATP-regulated and Ca^{2+}-activated K^+ channels in mouse pancreatic beta-cells by external tetraethylammonium and quinine. *J. Physiol.* **423,** 327–42.

Brocklehurst, K. W. and Hutton, J. C. (1984). Involvement of protein kinase C in the phosphorylation of an insulin granule membrane protein. *Biochem.* **220,** 283–90.

Cameron, J. S., Kimuraa, S., Jackson-Burns, D. A., Smith, D. B., and Bassett, A. L. (1988). ATP-sensitive K^+ channels are altered in hypertrophied ventricular myocytes. *Am. J. Physiol.* **255,** H1254–H1258.

Carrington, C. A., Rubery, E. D., Pearson, E. C., and Hales, C. N. (1987). Five new insulin-producing cell lines with differing secretory properties. *J. Endocrinol.* **109,** 193–200.

Christie, M. R. and Ashcroft, S. J. H. (1984). Cyclic AMP-dependent protein phosphorylation and insulin secretion in intact islets of Langerhans. *Biochem. J.* **218,** 87–99.

Christie, M. R. and Ashcroft, S. J. H. (1985). Substrates for cyclic AMP-dependent protein kinase in islets of Langerhans: studies with forskolin and catalytic subunit. *Biochem. J.* **227,** 727–36.

Clark, A., Matthews, D. R., Naylor, B. A., Wells, C. A., Hosker, J. P., and Turner, R. C. (1987). Pancreatic islet amyloid and elevated proinsulin secretion in familial maturity-onset diabetes. *Diabetes Res. Clin. Exp.* **4,** 51–9.

Colbran, R. J., *et al.* (1989). Calcium/calmodulin-dependent protein kinase II. *Biochem. J.* **258,** 313–25.

Colca, J. R., Kotagal, N., Brooks, C. L., Lacy, P. E., Landt, M., and McDaniel, M. L. (1983). Alloxan inhibition of a Ca^{2+} and calmodulin-dependent protein kinase activity in pancreatic islets. *J. Biol. Chem.* **258,** 7260–63.

Cook, D. L. and Hales, C. N. (1984). Intracellular ATP directly blocks K^+ channels in pancreatic β-cells. *Nature* **311,** 271–3.

Cook, D. L. and Ikeuchi, M. (1989). Tolbutamide as mimic of glucose on β-cell electrical activity: ATP-sensitive K^+-channels as common pathway for both stimuli. *Diabetes* **38,** 416–21.

Cook, D. L., Ikeuchi, M., and Fujimoto, W. Y. (1984). Lowering of pH_i inhibits Ca^{2+}-activated K^+ channels in pancreatic β-cells. *Nature* **311,** 269–71.

Cooper, G. J. S., Willis, A. C., Clark, A., Turner, R. C., Sim, R. B., and Reid, K. B. M.

(1987). Purification and characterization of a peptide from amyloid-rich pancreases of type 2 diabetic patients. *Proc. Natl Acad. Sci. USA* **84,** 8628–32.

Cooper, G. J. S., Day, A. J., Willis, A. C., Roberts, A. N., Reid, K. B. M., and Leighton, B. (1989). Amylin and the amylin gene: Structure, function and relationship to islet amyloid and to diabetes mellitus. *Biochim. Biophys. Acta* **1014,** 247–58.

Coore, H. G. and Randle, P. J. (1964). Regulation of insulin secretion studied with pieces of rabbit pancreas incubated *in vitro*. *Biochem. J.* **93,** 66–78.

Corkey, B. E., Glennon, M. C., Chen, K. S., Deeney, J. T., Matschinsky, F. M., and Prentki, M. (1989). A role for malonyl-CoA in glucose-stimulated insulin secretion from clonal pancreatic beta-cells. *J. Biol. Chem.* **264,** 21608–12.

Debuyser, A., Drews, G. and Henquin, J-C (1991). Adrenaline inhibition of insulin release: role of repolarization of the β-cell membrane. *Pflugers Arch.* **419,** 131–7.

Denton, R. M. and McCormack, J. G. (1990). Ca^{2+} as a second messenger within mitochondria of the heart and other tissues. *Ann. Rev. Physiol.* **52,** 451–66.

De Weille, J., Schmid-Antomarchi, H., Fosset, M., and Lasdunski, M. (1988). Pharmacology and regulation of ATP-sensitive K-channels. *Pflügers Arch.* **414,** S80–S87.

Drews, G., Debuyser, A., Nenquin, N., and Henquin, J. C. (1990). Galanin and epinephrine act on distinct receptors to inhibit insulin release by the same mechanisms including an increase in K^+ permeablity of the β-cell membrane. *Endocrinology* **126,** 1646–53.

Dunning. B. E. and Taborsky, G. J. Jr. (1988). Galanin—sympathetic neurotransmitter in the endocrine pancreas? *Diabetes* **37,** 1157–62.

Dunne, M. J., Findlay, I., Petersen, O. H., and Wollheim, C. B. (1986). ATP-sensitive K^+-channels in an insulin secreting cell line are inhibited by glyceraldehyde and activated by membrane permeabilization. *J. Membr. Biol.* **93,** 271–3.

Dunne, M. J., West-Jordan, J. A., Abraham, R. J., Edwards, R. H. T., and Peteresen, O. H. (1988). The gating of ATP-sensitive K-channels in insulin-secreting cells can be modulated by changes in the ratio of ATP^{4-}/ADP^{3-} and by non-hydrolysable analogues of both ATP and ADP. *J. Membr. Biol.* **104,** 165–77.

Dunne, M. J., Bullett, M. J., Li, G., Wollheim, C. B., and Petersen, O. H. (1989). Galanin activates nucleotide-dependent K^+ channels in insulin-secreting cells via a pertussis toxin-sensitive G-protein. *EMBO J.* **8,** 413.

Eddlestone, G. T., Ribalet, B., and Ciani, S. (1989). Comparative study of K channel behavior in beta cell lines with different secretory responses to glucose. *J. Membr. Biol.* **109,** 123–34.

Efendic, S., Wajngot, A., and Vranic, M. (1985). Increased activity of the glucose cycle in the liver: early characteristic of type 2 diabetes. *Proc. Natl Acad. Sci. USA* **82,** 2965–9.

Epstein, P. N., Overbeek, P. A., and Means, A. R. (1989). Calmodulin-induced early-onset diabetes in transgenic mice. *Cell* **58,** 1067–73.

Fahien, L. A., MacDonald, M. J., Kmiotek, E. H., Mertz, R. J., and Fahien, C. M. (1988). Regulation of insulin release by factors that also modify glutamate dehydrogenase. *J. Biol. Chem.* **263,** 13610.

Gagliardino, J. J., Harrison, D. E., Christie, M. R., Gagliardino, E. E., and Ashcroft, S. J. H. (1980). Evidence for the participation of calmodulin in stimulus-secretion coupling in the pancreatic β-cell. *Biochem. J.* **192,** 919–27.

Gaines, K. L., Hamilton, S., and Boyd, A. E., III (1988). Characterization of the sulfonylurea receptor on β-cell membranes. *J. Biol. Chem.* **263,** 2589–92.

Ghatei, M. A., Datta, H. K., Zaidi, M., Bretherton-Watt, D., Wimalawansa, S. J., MacIntyre,

I., and Bloom, S. R. (1990). Amylin and amylin-amide lack an acute effect on blood glucose and insulin. *J. Endocrinol.* **124,** R9–R11.

Ghosh, A., Ronner, P., Cheong, E., Khalid, P. and Matschinsky, F. M. (1991). The role of ATP and free ADP in metabolic coupling during fuel-stimulated insulin release from islet β-cells in the isolated perfused rat pancreas. *J. Biol. Chem.* **266,** 22887–92.

Giroix, M. H., Sener, A., and Malaisse, W. J. (1986). D-glucose transport and concentration in tumoral insulin-producing cells. *Am. J. Physiol.* **251,** C847–C851.

Gomperts, B. D. (1990). G_e: A GTP-binding protein mediating exocytosis. *Ann. Rev. Physiol.* **52,** 591–606.

Gomperts, B. D. and Fernandez, J. M. (1985). Techniques for membrane permeabilization. *TIBS* **10,** 414–17.

Gould, G. W. and Bell, G. I. (1990). Facilitative glucose transporters: an expanding family. *TIBS* **15,** 18–23.

Grapengiesser, E., Gylfe, E., and Hellman, B. (1990). Sulfonylurea mimics the effect of glucose in inducing large amplitude oscillations of cytoplasmic Ca^{2+} in pancreatic beta-cells. *Mol. Pharmacol.* **37,** 461–7.

Grodsky, G. M., Batts, A. A., Bennett, L. L., Vcella, C., McWilliams, N. B., and Smith, D. F. (1963). Effects of carbohydrates on secretion of insulin from isolated rat pancreas. *Am. J. Physiol.* **205,** 638–44.

Gylfe, E. (1988). Glucose-induced early changes in cytoplasmic calcium of pancreatic β-cells studied with time-sharing dual wavelength fluorometry. *J. Biol. Chem.* **263,** 5044–8.

Hales, C. M. (1972). Immunological techniques in diabetes research. *Diabetologia* **8,** 229–35.

Hales, C. N. and Randle, P. J. (1963). Immunoassay of insulin with insulin-antibody precipitate. *Biochem. J.* **88,** 137–46.

Hamill, O. P., Marty, A., Neher, E., Sakmann, B., and Sigworth, F. J. (1981). Improved patch clamp techniques for high resolution current recording from cells and cell-free membrane patches. *Pflügers Arch.* **391,** 85–100.

Harrison, D. E. and Ashcroft, S. J. H. (1982). Effects of Ca^{2+}, calmodulin and cyclic AMP on the phosphorylation of endogenous proteins by homogenates of rat islets of Langerhans. *Biochim. Biphys. Acta* **714,** 313–19.

Harrison, D. E., Ashcroft, S. J. H., Christie, M. R., and Lord, J. M. (1984). Protein phosphorylation in the pancreatic β-cell. *Experientia* **40,** 1057–84.

Harrison, D. E., Christie, M. R., and Gray, D. W. R. (1985). Properties of isolated human islets of Langerhans: insulin secretion, glucose oxidation and protein phosphorylation. *Diabetologica* **28,** 99–103.

Harrison, D. E., Poje, M., Rocic, B., and Ashcroft, S. J. H. (1986). Effects of dehydrouramil on protein phosphorylation and insulin secretion in rat islets of Langerhans. *Biochem. J.* **237,** 191–6.

Hedeskov, C. J. (1980). Mechanism of glucose-induced insulin secretion. *Physiol. Rev.* **60,** 442–509.

Henquin, J-C. (1981). Effects of trifluoperazine and primozide on stimulus-secretion coupling in pancreatic β-cells. Suggestion for a role of calmodulin. *Biochem. J.* **196,** 771–80.

Henquin, J-C. (1985). The interplay between cyclic AMP and ions in the stimulus-secretion coupling in pancreatic β-cells. *Archiv. Int. Physiol. Biochim.* **93,** 37–48.

Henquin, J-C. (1988). ATP-sensitive K^+ channels may control glucose-induced electrical activity in pancreatic β-cells. *Biochem. Biophys. Res. Comm.* **156,** 769–75.

Henquin, J-C. (1990). Role of voltage- and Ca^{2+}-dependent K^+ channels in the control of glucose-induced electrical activity in pancreatic B-cells. *Pflügers Arch.* **416,** 568–72.

Henquin, J-C., Bozem, M., Schmeer, W., and Nenquin, M. (1987). Distinct mechanisms for two amplification systems of insulin release. *Biochem. J.* **246,** 393–9.

Henquin, J-C., Garcia, M. C., Bozem, M., Hermans, M. P. and Nenquin, M. (1988). Muscarinic control of pancreatic β-cell function involves sodium-dependent depolarization and calcium influx. *Endocrinol.* **122,** 2134–42.

Henquin, J-C., Meissner, H. P. (1984). Significance of ionic fluxes and changes in membrane potential for stimulus-secretion coupling in pancreatic β-cells. *Experientia* **40,** 1043–52.

Henquin, J-C., Schmeer, W., Nenquin, M., and Plant, T. D. (1989). Does protein kinase C link glucose metabolism to β-cell membrane depolarization? *Diabetologia* **32,** 496A.

Hersey, S. J. and Perez, A. (1990). Permeable cell models in stimulus-secretion coupling. *Ann. Rev. Physiol.* **52,** 345–61.

Hii, C. S. T., Jones, P. M., Persaud, S. J., and Howell, S. L. (1987). A re-assessment of the role of protein kinase C in glucose-stimulated insulin secretion. *Biochem. J.* **246,** 489–93.

Hirriat, M. and Matteson, D. R. (1988). Na channels and two types of Ca channels in rat pancreatic B-cells identified with the reverse hemolytic plaque assay. *J. Gen. Physiol.* **91,** 617–39.

Homaidan, F. R., Sharp, G. W. G., and Nowak, L. M. (1991). Galanin inhibits a dihydropyridine-sensitive Ca^{2+} current in the RINm5F cell line. *Proc. Natl. Acad. Sci. USA.* **88,** 8744–8.

Horn, R. and Marty, A. (1988). Muscarinic activation of ionic currents measured by a new whole-cell recording method. *J. Gen. Physiol.* **92,** 145–59.

Howell, S. L. (1984). The mechanism of insulin secretion. *Diabetologia* **26,** 319–27.

Howell, S. L., Jones, P. M., and Persaud, S. J. (1990). Protein kinase C and the regulation of insulin secretion. *Biochem. Soc. Trans.* **18,** 114–16.

Hsu, W. H., Xiang, H. D., Kunze, D. L., Rajan, A., and Boyd, A. E., III (1989). Somatostatin decreases insulin secretion by inhibiting Ca influx through voltage-dependent Ca-channels in an insulin-secreting cell line (HIT-cells). *J. Cell Biol.* **12,** 355A.

Hughes, S. J. and Ashcroft, S. J. H. (1988*a*) Effect of secretagogues on cytosolic free Ca^{2+} and insulin release in the hamster clonal β-cell line HIT-T15. *J. Mol. Endocrinol.* **1,** 13–7.

Hughes, S. J. and Ashcroft, S. J. H. (1988*b*). Effects of a phorbol ester and clomiphene on protein phosphorylation and insulin secretion in rat pancreatic islets. *Biochem. J.* **249,** 825–30.

Hughes, S. J., Carpinelli, A., and Ashcroft, S. J. H. (1991). Negative control by protein kinase C of β-cell $[Ca^{2+}]_i$ responses to vasopressin and glucose. *Diabetologia* **34** (Suppl 2), A82.

Hughes, S. J., Christie, M. R., and Ashcroft, S. J. H. (1987). Potentiators of insulin secretion modulate Ca^{2+} sensitivity in rat pancreatic islets. *Mol. Cell Endocrinol.* **50,** 231–6.

Hughes, S. J., Chalk, J. G., and Ashcroft, S. J. H. (1990). The role of cytosolic free Ca^{2+} and protein kinase C in acetylcholine-induced insulin release in the clonal beta-cell line, HIT-T15. *Biochem. J.* **267,** 227–32.

Hughes, S. J., Chalk, J. G., and Ashcroft, S. J. H. (1989). Effect of secretagogues on cytosolic free Ca^{2+} and insulin release at different extracellular Ca^{2+} concentrations in the hamster clonal beta-cell line HIT-T15. *Mol. Cell. Endocrinol.* **65,** 35–41.

Hurst, R. D. and Morgan, N. G. (1989). Intracellular events responsible for the inhibition of insulin secretion by somatostatin. *Biochem. Soc. Trans.* **17,** 1085–6.

Hutton, J. C. and Malaisse, W. J. (1980). Dynamics of O_2 consumption in rat pancreatic islets. *Diabetologia* **18,** 395–405.

Hutton, J. C., Sener, A., and Malaisse, W. J. (1980). Interactions of branched chain amino acids and keto acids upon pancreatic islet metabolism and insulin secretion. *J. Biol. Chem.* **255,** 7340–6.

Iynedjian, P. B., Möbius, G., Seitz, H. J., Wollheim, C. B., and Renold, A. E. (1986). Tissue-specific expression of glucokinase: identification of the gene product in liver and pancreatic islets. *Proc. Natl Acad. Sci. USA* **83,** 1998–2001.

Iynedjian, P. B., Pilot, P-R., Nouspikel, T., Milburn, J. L., Quaade, C., Hughes, S., Ucla, C., and Newgard, C. B. (1989). Differential expression and regulation of the glucokinase gene in liver and islets of Langerhans. *Proc. Natl Acad. Sci. USA* **86,** 7838–42.

Johnson, J. H., Newgard, C. B., Milburn, J. L., Lodish, H. F., and Thorens, B. (1990). The high K_m glucose transporter of islets of Langerhans is functionally similar to the low affinity transporter of liver and has an identical primary sequence. *J. Biol. Chem.* **265,** 6548–51.

Jones, P. M., Salmon, D. M. W., and Howell, S. L. (1988). Protein phosphorylation in electrically permeabilised islets of Langerhans. Effects of Ca^{2+}, cyclic AMP, a phorbol ester and noradrenaline. *Biochem. J.* **254,** 397–403.

Jones, P. M., Fyles, J. M., Persaud, S. J., and Howell, S. L. (1987). Catecholamine inhibition of insulin secretion from electrically permeabilized islets of Langerhans. *FEBS Lett.* **219,** 139.

Kakei, M., Kelly, R. P., Ashcroft, S. J. H., and Ashcroft, F. M. (1986). The ATP-sensitivity of K^+ channels in rat pancreatic β-cells is modulated by ADP. *FEBS Lett.* **208,** 63–6.

Keahey, H. H., Boyd, A. E., III, and Kunze, D. L. (1990). Catecholamine modulation of calcium currents in clonal pancreatic β-cells. *Am. J. Physiol.* **357,** C1171–C1176.

Kekow, J., Ulrichs, K., Müller-Ruchholtz, W., and Gross, W. L. (1988). Measurement of rat insulin. Enzyme-linked immunosorbent assay with increased sensitivity, high accuracy, and greater practicability than established radioimmunoassay. *Diabetes* **37,** 321–6.

Khan, A., Chandramouli, V., Östenson, C.-G., Ahrén, B., Schumann, W. C., Löw, H., Landau, B. R., and Efendic, S. (1989). Evidence for the presence of glucose cycling in pancreatic islets of the ob/ob mouse. *J. Biol. Chem.* **264,** 9732–3.

Khan, A., Chandramouli, V., Östenson, C.-G., Löw, H., Landau, B. R., and Efendic, S. (1990*a*). Glucose cycling in islets from healthy and diabetic rats. *Diabetes* **39,** 456–9.

Khan, A., Chandramouli, V., Östenson, C.-G., Berggren, P.-O., Löw, H., Landau, B. R., and Efendic, S. (1990*b*). Glucose cycling is markedly enhanced in pancreatic islets of obese hyperglycemic mice. *Endocrinology* **126,** 2413–16.

Kozlowski, R. Z., Hales, C. N., and Ashford, M. L. J. (1989). Dual effects of diazoxide on ATP-K currents recorded from an insulin-secreting cell line. *Br. J. Pharmacol.* **97,** 1039–50.

Kramer, W., Oekonimopoulos, R., Punter, J., and Summ, H.-D. (1988). Direct photoaffinity labelling of the putative sulphonylurea receptor in rat β-cell tumor membranes by [^{3}H]glibenclamide. *FEBS Lett.* **229,** 355–9.

Kukuljan, M., Goncalves, A. A., and Atwater, I. (1991). Charybdotoxin-sensitive $K_{(Ca)}$ channel is not involved in glucose-induced electrical activity in pancreatic β-cells. *J. Membr. Biol.* **119,** 187–95.

Lenzen, S. and Panten, U. (1988*a*). Alloxan: history and mechanism of action. *Diabetologia* **31,** 337–42.

Lenzen, S. and Panten, U. (1988*b*). Signal recognition by pancreatic β-cells. *Biochem. Pharmacol.* **37,** 371–8.

Lenzen, S., Formack, H., and Panten, U. (1982). Signal function of metabolism of neutral amino acids and 2-ketoacids for initiation of insulin secretion. *J. Biol. Chem.* **257,** 6631–55.

Lenzen, S., Schmidt, W., Rustenbeck, I., and Panten, U. (1986). 2-Ketoglutarate generation in pancreatic B-cell mitochondria regulates insulin-secretory action of amino acids and 2-keto acids. *Biosci. Rep.* **6,** 163–9.

Lenzen, S., Tiedge, M., and Panten, U. (1987). Glucokinase in pancreatic β-cells and its inhibition by alloxan. *Acta Endocrin.* **115,** 21.

Lewis, C. E., Clark, A., Ashcroft, S. J. H., Cooper, G. J. S., Morris, J. F. (1988). Calcitonin-gene-related peptide and somatostatin inhibit insulin release from individual rat B-cells. *Mol. Cell. Endocrinol.* **57,** 41–9.

Lord, J. M. and Ashcroft, S. J. H. (1984). Identification and characterization of Ca^{2+}-phospholipid-dependent protein kinase in rat islets and hamster β-cells. *Biochem. J.* **219,** 547–51.

McCormack, J. G., Longo, E. A., and Corkey, B. E. (1990). Glucose-induced activation of pyruvate dehydrogenase in isolated rat pancreatic islets. *Biochem. J.* **267,** 527–30.

Magnuson, M. A., Andreone, T. L., Printz, R. L., Koch, S., and Granner, D. K. (1989). Rat glucokinase gene: Structure and regulation by insulin. *Proc. Natl Acad. Sci. USA* **86,** 4838–42.

Malaisse, W. J. and Malaisse-Lagae, F. (1984). The role of cyclic AMP in insulin release. *Experientia* **40,** 1068–74.

Malaisse-Lagae, F., Mathias, P. C. F., and Malaisse, W. J. (1984). Gating and blocking of calcium channels by dihydropyridines in the pancreatic β-cell. *Biochem. Biophys. Res. Comm.* **123,** 1062–8.

Martin, S. C., Yule, D. I., Dunne, M. J., Gallacher, D. V., and Petersen, O. H. (1989). Vasopressin directly closes ATP-sensitive potassium channels evoking membrane depolarizatio and an increase in the free intracellular Ca^{2+} concentration in insulin-secreting cells. *EMBO J.* **8,** 3595–9.

Meglasson, M. D. and Matschinsky, F. M. (1986). Pancreatic islet glucose metabolism and regulation of insulin secretion. *Diabetes/Metabolism Rev.* **2,** 163–214.

Meglasson, M. D., Burch, p. T., Berner, D. K., Najafi, H., Wogin, A. P., and Matschinsky, F. M. (1983). Chromatographic resolution and kinetic characterization of glucokinase from islets of Langerhans. *Proc. Natl Acad. Sci. USA* **80,** 5–9.

Meglasson, M. D., Burch, P. T., Berner, D. K., Najafi, H., and Matchinsky, F. M. (1986). Identification of glucokinase as an alloxan-sensitive glucose sensor of the pancreatic beta-cell. *Diabetes* **35,** 1163–73.

Meglasson, M. D., Nelson, J., Nelson, D., and Erecinska, M. (1989). Bioenergetic response of pancreatic islets to stimulation by fuel molecules. *Metabolism* **38,** 1188–95.

Meissner, H. P. and Preissler, M. (1979). Possible ionic mechanisms of the electrical activity induced by glucose and tolbutamide in pancreatic β-cells. In *Diabetes* (ed. W. K. Waldhausl). International Congress Series 500, pp. 169–72. Exerpta Medica, Amsterdam.

Meissner, H. P. and Schmelz, H. (1974). Membrane potential of β-cells in pancreatic islets. *Pflügers Arch.* **351,** 195–206.

Metz, S. A. (1988). Is protein kinase C required for physiologic insulin release? *Diabetes* **37,** 3–7.

Misler, D. S., Falke, L. C., Gillis, K., and McDaniel, M. L. (1986). A metabolite regulated potassium channel in rat pancreatic β-cells. *Proc. Natl Acad. Sci. USA* **83,** 7119–23.

Misler, S., Gee, W. M., Gillis, K. D., Scharp, D. W., and Falke, L. C. (1989). Metabolite-regulated ATP-sensitive K^+-channel in human pancreatic islet cells. *Diabetes* **38,** 422–7.

Nelson, M. T., Huang, Y., Brayden, J. E., Hescheler, J., and Standen, N. B. (1990). Arterial dilations in response to calcitonin gene-related peptide involve activation of K^+ channels. *Nature* **344,** 770–3.

Nestler, E. J. and Greengard, P. (1984). *Protein phosphorylation in the nervous system.* Wiley-Interscience, New York.

Niki, I., Kelly, R. P., Ashcroft, S. J. H., and Ashcroft, F. M. (1989*a*). ATP-sensitive K-channels in HIT T15 β-cells studied by patch-clamp methods, ^{86}Rb efflux and glibenclamide binding. *Pflügers Arch.* **415,** 47–55.

Niki, I., Ashcroft, F. M., and Ashcroft, S. J. H. (1989*b*). The dependence on intracellular ATP concentration of ATP-sensitive K-channels and of Na,K-ATPase in intact HIT T15 β-cells. *FEBS Lett.* **257,** 361–4.

Nilsson, T., Arkhammer, P., Hallberg, A., Hellman, B., and Berggren, P.-O. (1987). Characterisation of the inositol 1,4,5-triphosphate-induced Ca^{2+} release in pancreatic β-cells. *Biochem. J.* **248,** 329–36.

Nilsson, T., Arkhammer, P., and Berggren, P.-O. (1988). Dual effect of glucose on cytoplasmic free Ca^{2+} concentration and insulin release reflects the beta cell being deprived of fuel. *Biochem. Biophys. Res. Comm.* **153,** 984–91.

Nilsson, T., Arkhammer, P., Rorsman, P., and Berggren, P.-O. (1988). Inhibition of glucose-stimulated insulin release by α_2 adrenoreceptor activation is paralleled by both a repolarization and a reduction in free cytoplasmic Ca^{2+} concentration. *J. Biol. Chem.* **263,** 1855.

Nilsson, T., Arkhammer, P., Rorsmann, P., and Berggren, P.-O. (1989). Suppression of insulin release by galanin and somatostatin is mediated by a G-protein. *J. Biol. Chem.* **264,** 973.

Nishizuka, Y. (1988). The molecular heterogeneity of protein kinase C and its implications for cellular regulation. *Nature* **334,** 661.

Nowycky, M. C., Fox, A. P., and Tsien, R. W. (1985). Three types of neuronal calcium channel with different calcium agonist sensitivity. *Nature* **316,** 440–3.

Oberwetter, J. M. and Boyd III, A. E. (1987). High K^+ rapidly stimulates Ca^{2+}-dependent phosphorylation of three proteins concomitant with insulin secretion from HIT cells. *Diabetes* **36,** 864–71.

Ohno-Shosaku, T., Zunckler, B., and Trube, G. (1987). Dual effects of ATP on K^+ currents of mouse pancreatic β-cells. *Pflügers Arch.* **408,** 133–8.

Onoda, K., Hagiwara, M., Hachiya, T., Usada, N., Nagata, T., and Hidaka, H. (1990). Different expression of protein kinase C isozymes in pancreatic islet cells. *Endocrinology* **126,** 1235–40.

Orci, L., Thorens, B., Ravazzola, M., and Lodish, H. F. (1989). Localization of the pancreatic beta cell glucose transporter to specific plasma membrane domains. *Science* **245,** 295–7.

Panten, U. (1987). Rapid control of insulin secretion from pancreatic islets. *Isi Atlas Sci-Pharmacol.* **1,** 307–9.

Panten, U. and Lenzen, S. (1985). Regulation of glutamate dehydrogenase activity in pancreatic islets by leucine analogues. *IRCS Med. Sci-Biochem.* **13,** 472–3

Panten, U., Christians, J., Kriegstein, E. V., Poser, W., and Hasselblatt, A. (1973). Effect of carbohydrates upon fluorescence of reduced pyridine nucleotides from perifused isolated pancreatic islets. *Diabetologia* **9,** 477–82.

Panten, U., Zielmann, S., Langer, J., Zunkler, B-J., and Lenzen, S. (1984). Regulation of insulin secretion by energy metabolism in pancreatic B-cell mitochondria. Studies with a non-metabolizable leucine analogue. *Biochem. J.* **219,** 189–96.

Panten, U., *et al.* (1989). Control of insulin secretion by sulfonylureas, meglitinide and diazoxide in relation to their binding to the sulfonylurea receptor in pancreatic islets. *Biochem. Pharmacol.* **38,** 1217–29.

Persaud, S. J., Jones, P. M., Sugden, D., and Howell, S. L. (1989). The role of protein kinase C in cholinergic stimulation of insulin secretion from rat islets of Langerhans. *Biochem. J.* **264,** 753–8.

Persaud, S. J., Jones, P. M., and Howell, S. L. (1990). Glucose-stimulated insulin secretion is not dependent on activation of protein kinase A. *Biochem. Biophys. Res. Comm.* **173,** 833–9.

Persaud, S. J., Jones, P. M., Howell, S. L. (1991). Activation of protein kinase C is essential for sustained insulin secretion in response to cholinergic stimulation. *Biochim. Biphys. Acta* **1091,** 120–2.

Pettersson, M. and Ahrén, B. (1990). Failure of islet amyloid polypeptide to inhibit basal and glucose-stimulated insulin secretion in model experiments in mice and rats. *Acta Physiol. Scand.* **138,** 389–94.

Plant, T. D. (1988*a*). Na^+ currents in cultured mouse pancreatic β-cell. *Pflügers Arch.* **411,** 429–35.

Plant, T. D. (1988*b*). Properties of calcium-dependent inactivation of calcium channels in cultured mouse pancreatic B-cells. *J. Physiol.* **404,** 731–47.

Portha, B., *et al.* (1988). Insulin production and glucose metabolism in isolated pancreatic islets of rats with NIDDM. *Diabetes* **37,** 1226–33.

Pralong, W.-F., Bartley, C., and Wollheim, C. B. (1990). Single islet beta-cell stimulation by nutrients: Relationship between pyridine nucleotides, cytosolic Ca^{2+} and secretion. *EMBO J.* **9,** 53–60.

Prentki, M. and Matchinsky, F. M. (1987). Ca^{2+}, cAMP and phospholipid-derived messengers in coupling mechanisms of insulin secretion. *Physiol. Rev.* **67,** 1185.

Pressel, D. and Misler, S. (1991). Sodium channels contribute to action potential generation in canine and human β-cells. *J. Memb. Biol.* **116.**

Randle, P. J., Ashcroft, S. J. H., and Gill, J. R. (1968). Carbohydrate metabolism and the release of hormones. In *Carbohydrate metabolism and its disorders,* Vol. 1, pp. 427–37. Academic Press, London.

Ribalet, B. and Beigelman, P. M. (1980). Effects of sodium on β-cell electrical activity. *Am. J. Physiol.* **242,** C296–C303.

Ribalet, B. and Ciani, S. (1987). Regulation by cell metabolism and adenosine nucleotides of a K channel in insulin-secreting B-cells (RINm5F). *Proc. Natl Acad. Sci. USA* **84,** 1721–5.

Ribalet, B., Eddlestone, G. T., and Ciani, S. (1988). Metabolic regulation of the K(ATP) and a Maxi-K(V) channel in the insulin-secreting RINm5F cell. *J. Gen. Physiol.* **92,** 219–37.

Rorsman, P. and Abrahamsson, H. (1985). Cyclic AMP potentiates glucose-induced insulin release from mouse pancreatic islets without increasing cytosolic free Ca^{2+}. *Acta Physiol. Scand.* **125,** 639–47.

Rorsman, P. and Trube, G. (1985). Glucose dependent K^+-channels in pancreatic beta-cells are regulated by intracellular ATP. *Pflügers Arch.* **405,** 305–9.

Rorsman, P. and Trube, G. (1990). Biophysics and physiology of ATP-regulated K^+ channels. In *Potassium channels: structure function and therapeutic potential* (ed. N. S. Cook), pp. 300–26. Ellis Horwood, Chichester, UK.

Rorsman, P., Abrahamsson, H., Gylfe, E., and Hellman, B. (1984). Dual effects of glucose on the cytosolic Ca^{2+} activity of mouse pancreatic β-cells. *FEBS Lett.* **170,** 196–200.

Rorsman, P., Ashcroft, F. M., and Trube, G. (1988). Single calcium channel currents in mouse pancreatic β-cells. *Pflügers Arch.* **412,** 597–603.

Rorsman, P., *et al.* (1989). Failure of glucose to elicit a normal secretory response in fetal pancreatic beta cells results from glucose insensitivity of the ATP-regulated K^+ channels. *Proc. Natl Acad. Sci. USA* **86,** 4505–9.

Rorsman, P., *et al.* (1991). Activation by adrenaline of a low conductance G protein-dependent K^+ channel in mouse pancreatic β-cells. *Nature* **349,** 77–9.

Salomon, D. and Meda, P. (1986). Heterogeneity and contact-dependent regulation of hormone secretion by individual B-cells. *Exp. Cell Res.* **162,** 507–20.

Santos, R. M., Rosario, L. M., Nadel, A., Garcia-Sancho, J., Soria, B., and Valdeolmillos, M. (1991). Widespread synchronous $[Ca^{2+}]_i$ oscillations due to bursting electrical activity in single pancreatic islets. *Pflügers Arch.* **418,** 417–22.

Schmid-Antomarchi, H., DeWeille, J., Fosset, M., and Lazdunski, M. (1987). The receptor for antidiabetic sulfonylureas controls the activity of the ATP-modulated K^+ channel in insulin secreting cells. *J. Biol. Chem.* **262,** 15840–4.

Schubart, U. K. and Fields, K. L. (1984). Identification of a calcium-regulated insulinoma cell phosphoprotein as an islet cell keratin. *J. Cell. Biol.* **98,** 1001–9.

Schwanstecher, M., Löser, S., Rietze, I., and Panten, U. (1990). MgATP controls glibenclamide- and diazoxide-binding to their receptor in pancreatic B-cells. *Diabetologia* **33,** A78.

Sener, A., Rasschaert, J., and Malaisse, W. J. (1990). Hexose metabolism in pancreatic islets. Participation of Ca^{2+}-sensitive 2-ketoglutarate dehydrogenase in the regulation of mitochondrial function. *Biochim. Biophys. Acta Bio-Energetics* **1019,** 42–50.

Sharp, G. W. G., Le Marchard-Brustel, Y., Yada, T., Russo, L., Bliss, C. R. Cormont, M., Monge, L., and Van Obberghen, E. (1989). Galanin can inhibit insulin release by a mechanism other than membrane hyperpolarization or inhibition of adenylate cyclase. *J. Biol. Chem.* **264,** 7302.

Siegel and Creutzfeldt (1985). Stimulation of insulin release in isolated rat islets by GIP in physiological concentrations and its relation to islet cyclic AMP. *Diabetologia* **28,** 857–61.

Smith, P. A., Bokvist, K., and Rorsman, P. (1989*a*). Demonstration of A-currents in pancreatic islet cells. *Pflügers Arch.* **413,** 441–3.

Smith, P. A., Rorsman, P., and Ashcroft, F. M. (1989*b*). Modulation of dihydropyridine-sensitive Ca^{2+} channels by glucose metabolism in mouse pancreatic beta-cells. *Nature* **342,** 550–3.

Smith, P. A., Bokvist, K., Arkhammar, P., Berggren, P.-O., and Rorsman, P. (1990*a*). Delayed rectifying and calcium-activated K^+ channels and their significance for action potential repolarization in mouse pancreatic beta-cells. *J. Gen. Physiol.* **95,** 1041–59.

Smith, P. A., Rorsman, P., and Ashcroft, F. M. (1990*b*). Simultaneous recording of β-cell electrical activity and ATP-sensitive K-currents in mouse pancreatic β-cells. *FEBS Lett.* **261,** 187–90.

Sturgess, N. C., Ashford, M. L. I., Cook, D. L., and Hales, C. N. (1985). The sulphonylurea receptor may be an ATP-sensitive potassium channel. *Lancet* **2,** 474–5.

Sturgess, N. C., Ashford, M. L., Carrington, C. A., and Hales, C. N. (1986*a*). Single channel recordings of potassium currents in an insulin-secreting cell line. *J. Endocrinol.* **109,** 201–7.

Sturgess, N. C., Hales, C. N., and Ashford, M. L. J. (1986*b*). Inhibition of a calcium-activated, non-selective cation channel, in a rat insulinoma cell line, by adenine derivatives. *FEBS Lett.* **208,** 397–400.

Sturgess, N. C., Ashford, M. L. J., Carrington, C. A., and Hales, C. N. (1987*a*). Nucleotide-sensitive ion channels in human insulin-producing tumor cells. *Pflügers Arch.* **410,** 169–72.

Sturgess, N. C., Hales, C. N., and Ashford, M. L. J. (1987*b*). Calcium and ATP regulate the activity of a non-selective cation channel in a rat insulinoma cell line. *Pflügers Arch.* **409,** 607–15.

Sugden, M. C. and Ashcroft, S. J. H. (1977). Phosphoenol-pyruvate in rat pancreatic islets: a possible intracellular trigger of insulin release. *Diabetologia* **13,** 481–6.

Sugden, M. C., Ashcroft, S. J. H., and Sugden, P. H. (1979). Protein kinase activities in rat pancreatic islets of Langerhans. *Biochem. J.* **180,** 219–29.

Sugden, M. C., Christie, M. R., and Ashcroft, S. J. H. (1979). Presence and possible role of calcium-dependent regulator (calmodulin) in rat islets of Langerhans. *FEBS Lett.* **105,** 95–100.

Swope, S. L. and Schonbrunn, A. (1988). The biphasic stimulation of insulin secretion by bombesin involves both cytosolic free calcium and protein kinase C. *Biochem. J.* **253,** 193–202.

Tabcharini, J. A. and Misler, S. (1989). Ca^{2+}-activated K^+ channel in rat pancreatic islet β-cells: permeation, gating and block by cations. *Biochim. Biophys. Acta* **982,** 67–72.

Thams, P., Capito, K. and Hedeskov, C. J. (1984). Endogenous substrate proteins for Ca^{2+}-calmodulin-dependent, Ca^{2+}-phospholipid-dependent and cyclic AMP-dependent protei kinases in mouse pancreatic islets. *Biochem. J.* **221,** 247–253.

Thorens, B., Sarkar, H. K., Kaback, H. R., and Lodish, H. F. (1988). Cloning and functional expression in bacteria of a novel glucose transporter present in liver, intestine, kidney, and beta-pancreatic islet cells. *Cell* **55,** 281.

Thorn, P. and Petersen, O. H. (1991). Activation of voltage-sensitive Ca^{2+} channels by vasopressin in an insulin-secreting cell line. *J. Memb. Biol.,* **124,** 63–71.

Trube, G., Rorsman, P., and Ohno-Shosaku, T. (1986). Opposite effects of tolbutamide and diazoxide on the ATP-dependent K^+ channel in mouse pancreatic beta-cells. *Pflügers Arch.* **407,** 493–9.

Ullrich, S. and Wollheim, C. B. (1988). GTP-dependent inhibition of insulin secretion by epinephrine in permeabilised RIN m5F cells. Lack of correlation between insulin secretion and cyclic AMP levels. *J. Biol. Chem.* **263,** 8615–20.

Ullrich, S. and Wollheim, C. B. (1989). Galanin inhibits insulin secretion by direct interference with exocytosis. *FEBS Lett.* **247,** 401–4.

Ullrich, S., Prentki, M., and Wollheim, C. B. (1990). Somatostatin inhibition of Ca^{2+}-induced insulin secretion in permeabilized HIT T15 cells. *Biochem. J.* **270,** 273–6.

Vague, P. and Moulin, J.-P. (1982). The defective glucose sensitivity of the B-cell in non insulin dependent diabetes. Improvement after twenty hours of normoglycaemia. *Metabolism* **31,** 139–42.

Valdeolmillos, M., Santos, R. M., Contreras, D., Soria, B., and Rosario, L. M. (1989). Glucose-induced oscillations of intracellular Ca^{2+} concentration resembling bursting electrical activity in single mouse islets of Langerhans. *FEBS Lett.* **259,** 19–23.

Valler, L., Biden, T. J., and Wollheim, C. B. (1987). Guanine nucleotides induce Ca^{2+}-independent insulin secretion from permeabilised RINm5F cells. *J. Biol. Chem.* **262,** 5049–56.

Velasco, J. M. (1988). Calcium channels in rat insulin-secreting RINm5 cell line. *J. Physiol.* **398,** 15P.

Velasco, J. M. and Peterson, O. H. (1989). The effects of a cell-permeable diacylglycerol analogue on Ca^{2+} (Ba^{2+}) channel currents in insulin-secreting cell line RINm5F. *Q. J. Exp. Physiol.* **74,** 367–70.

Velasco, J. M., Petersen, J. U. H., and Petersen, O. H. (1988). Single channel Ba^{2+} currents in insulin secreting cells are activated by glyceraldehyde stimulation. *FEBS Lett.* **213,** 366–70.

Virsolvy-Vergine, A., Bruck, M., Dufour, M., Cauvin, A., Lupo, B., and Bataille, D. (1988). An endogenous ligand for the central sulfonylurea receptor. *FEBS Lett.* **242,** 65–9.

Waddell, I. D. and Burchell, A. (1988). The microsomal glucose-6-phosphatase enzyme of pancreatic islets. *Biochem. J.* **255,** 471–6.

Wollheim, C. B. and Sharp, G. W. G. (1981). Regulation of insulin release by calcium. *Physiol. Rev.* **61,** 914–73.

Wollheim, C. B. and Regazzi, R. (1990). Protein kinase C in insulin releasing cells: Putative role in stimulus secretion coupling. *FEBS Lett.* **268,** 376–80.

Wollheim, C. B., Ullrich, S., Meda, P., and Vallar, L. (1987). Regulation of exocytosis in electrically permeabilised insulin-secreting cells—evidence for Ca^{2+} dependent and independent secretion. *Biosci. Rep.* **7,** 443–54.

Yada, T., Kakai, M., and Tanaka, H. (1992). Single pancreatic β-cells from normal rats exhibit an initial decrease and subsequent increase in cytosolic free Ca^{2+} in response to glucose. *Cell Calcium* **13,** 69–76.

Zunkler, B. J., Lenzen, S., Manner, K., Panten, U., and Trube, G. (1988*a*). Concentration-dependent effects of tolbutamide, meglitinide, glipizide, glibenclamide and diazoxide on ATP-regulated K^+ currents in pancreatic B-cells. *Naunyn-Schmeid. Arch. Pharmacol.* **337,** 225–30.

Zunkler, B. J., Lins, S., Ohno-Shosaku, T., Trube, G., and Panten, U. (1988*b*). Cytosolic ADP enhances the sensitivity to tolbutamide of ATP-dependent K^+ channels from pancreatic B-cells. *FEBS Lett.* **239,** 241.

Zawalich, W., Takuwa, N., Takuwa, Y., Diaz, V. A., and Rasmussen, H. (1987). Interactions of cholecystokinin and glucose in rat pancreatic islets. *Diabetes* **36,** 426–33.

PART IV INSULIN ACTION

Editors' introduction

The following three chapters outline the physiological actions of insulin at three levels. First, the role that insulin plays in integrating metabolic pathways in the intact animal is described in Chapter 5. The mechanism by which insulin affects its target tissues is via interaction with a specific cellular receptor in the plasma membrane. Following the cloning of the insulin receptor, the molecular mechanism of this interaction is now known in considerable detail and is discussed in Chapter 6. Finally, the intracellular events triggered by the binding of insulin to its receptor are considered in Chapter 7.

The importance of insulin stems from the fact that it is the only hormone capable of reducing blood glucose concentration. This is why a deficiency in its secretion (or action) gives rise to the metabolic defects associated with uncompensated diabetes mellitus (see Chapter 8). The main target tissues are adipose tissue, muscle, and liver. In these tissues insulin has a primarily anabolic role. It signals the fed state by promoting the synthesis of carbohydrate, fat, and protein and reducing the rate of fuel degradation.

Central to understanding blood glucose homeostasis is the fact that there is a relatively constant requirement for the provision of glucose, approximately 180 g per day, to the brain. Since food intake, however, occurs sporadically, during starvation blood glucose levels must be maintained by supply from existing stores. The available stores are blood glucose itself (which is very limited) and the stores of glycogen in the liver and muscle. It is important to note that the considerably larger stores of fat in adipose tissue cannot be used for *de novo* synthesis of glucose because of the irreversibility of the pyruvate dehydrogenase reaction. As glycogen stores become depleted, tissue protein must be catabolized to amino acids in order to form glucose by gluconeogenesis. The strategy of the body, therefore, during starvation is to minimize the use of glucose by those tissues, such as muscle, which are able to use other fuels, in particular those derived from fat stores. The reciprocal relationship between fat and carbohydrate metabolism was formulated by Randle and colleagues in 1962 as the 'glucose-fatty acid cycle'. Thus under metabolic conditions favouring fatty-acid mobilization there is a concomitant decrease in the catabolism of glucose as a direct result of the fatty-acid catabolism; conversely high blood glucose favours lipogenesis. Insulin affects muscle glucose utilization directly by increasing glucose uptake (and glycogen synthesis) but also has an important indirect effect as a consequence of its inhibitory action on lipolysis in adipose tissue. Whether the liver functions to take up glucose from the blood or to release it is determined both by the availability of gluconeogenic precursors and by the relative concentrations of insulin and glucagon—other hormones such as cortisol also contribute to the control of gluconeogenesis.

There is evidence that *in vivo* insulin secretion is pulsatile. This may be a consequence of a phase lag between insulin secretion and glucose release by the liver. There are theoretical advantages to phasic release, which has been observed for many hormones, but the physiological relevance for glucose homeostasis remains to be established.

Perhaps the most important advance in recent years in understanding the molecular basis for insulin action has been the cloning of the insulin receptor. This glycoprotein is a heterotetramer consisting of two α- and two β-subunits linked by disulphide bridges. The α-subunit is entirely extracellular and contains the insulin-binding site. The β-subunit is a transmembrane protein which contains within the intracellular portion a tyrosine-specific protein kinase. The receptor is synthesized as a single-chain precursor. Binding of insulin to the receptor is followed by autophosphorylation of β-chain tyrosine residues, which leads to activation of the tyrosine kinase. Many experimental approaches, including site-directed mutagenesis, have converged in indicating a central role for activation of tyrosine kinase in transducing the action of insulin.

The mechanisms by which insulin causes changes in metabolic fluxes are well-established at the cellular level. Tissue-specific increases in the activity of key intracellular enzymes, including glycogen synthase, acetyl-CoA carboxylase, and pyruvate dehydrogenase, occur in response to the hormone. Insulin also facilitates the uptake of glucose and amino acids by increasing the number of transporters in the plasma membrane. These effects account for the anabolic actions of the hormone. In addition, insulin has a number of other effects, such as increasing the activity of the Na/K-ATPase.

A major lacuna in our understanding remains the precise way in which activation of receptor tyrosine kinase is linked to the changes in activity of intracellular enzymes and the translocation of membrane transporters, which underlie the intracellular actions of the hormone. A putative insulin mediator derived from a novel inositol phospholipid glycan has been suggested. However, as discussed in Chapter 7, convincing evidence in favour of such a second messenger is lacking.

The changes in enzymatic activity involve changes in the phosphorylation state of the protein. There is evidence that increased receptor tyrosine kinase activity leads to activation of serine kinase(s). The increased *dephosphorylation*, found for example with glycogen synthase, is due to enhanced activity of a protein phosphatase as a consequence of its phosphorylation by a serine kinase.

5 | Physiological aspects of insulin action

E. A. NEWSHOLME, S. J. BEVAN, G. D. DIMITRIADIS, and R. P. KELLY

The wealth of information about metabolism that has accrued over the past 50 years and the large number of wall charts describing increasingly complex metabolic pathways have led to the view that most of general metabolism is solved and little further work needs to be done. Consequently, there is a feeling that current investigations in the metabolic field are old-fashioned and will not provide any novel information about how insulin controls metabolism, the cause of diabetes mellitus, or lead to an effective and efficient means for regulation of the blood glucose level; only the modern approach of molecular biology will do this. And yet, we are still uncertain as to which is the major metabolic process that controls the blood glucose level in response to a change in the insulin concentration. It is even possible to question whether changes in plasma insulin concentration actually play a physiological role in changing blood glucose concentration.

Because of the vast amount of information available on the effects of insulin, our discussion is restricted to those aspects which we consider play a key role in the integration of metabolism. Use will be made of the principles of metabolic control to select more objectively those areas of metabolism that may play a major role.

First, a list of the effects of insulin on carbohydrate, fat, and protein metabolism is given, followed by presentation of some principles of metabolic control logic. These will then be used to integrate the effects of insulin into a physiological profile of the means by which the hormone is suggested to play a role in the integration of metabolism.

5.1 Summary of the major effects of insulin on metabolism

5.1.1 Effects on carbohydrate metabolism

- Insulin increases the rate of transport of glucose across the cell membrane in adipose tissue and muscle.
- Insulin increases the rate of glycolysis in muscle and adipose tissue.

- Insulin stimulates the rate of glycogen synthesis in a number of tissues, including adipose tissue, muscle, and liver. It also decreases the rate of glycogen breakdown in muscle and liver.
- Insulin inhibits the rate of glycogenolysis and gluconeogenesis in the liver.
- Insulin increases the rate of glucose oxidation by the pentose phosphate pathway in liver and adipose tissue.

It must be emphasized that, although certain major tissues are largely or totally insensitive to the direct action of insulin (e.g. kidney, brain, intestine), insulin can influence the rate of glucose utilization in these tissues indirectly, by effects on mobilization of fuels from other tissues, especially adipose tissue (see below).

5.1.2 Effects on lipid metabolism

- Insulin inhibits the rate of lipolysis in adipose tissue.
- Insulin stimulates fatty acid and triacylglycerol synthesis in adipose tissue and liver.
- It increases the rate of very low density lipoprotein (VLDL) formation in the liver.
- It increases lipoprotein lipase activity in adipose tissue and this increases the uptake of triglyceride from the blood into adipose tissue.
- It may decrease the rate of fatty acid oxidation in liver.
- It increases the rate of cholesterolgenesis in liver.

5.1.3 Effects on protein metabolism

- Insulin increases the rate of transport of some amino acids into muscle, adipose tissue, liver, and other cells.
- Insulin increases the rate of protein synthesis in muscle, adipose tissue, liver, and other tissues.
- Insulin decreases the rate of protein degradation in muscle (and perhaps other tissues).
- Insulin decreases the rate of urea formation.

These effects of insulin serve to encourage protein synthesis rather than amino-acid oxidation, so that the hormone produces positive nitrogen balance in the normal subject, and it can therefore be considered to be an anabolic hormone.

5.2 Some principles of metabolic control logic

Consideration of the importance of steady-state flux through a metabolic pathway has led to the development of the concept of a physiological pathway or a 'metabolic flux', which extends the biochemical interpretation of metabolic pathways and is important in appreciating how the flux through a pathway can be changed, especially by hormones (Newsholme and Crabtree 1981, 1991). Conventional pathways are not always able, by themselves, to generate or maintain a steady-state flux, and hence their significance in cells cannot easily be understood. In order to appreciate how a steady-state flux is established and controlled, it is necessary to understand the difference between near-equilibrium and non-equilibrium reactions, and the meaning of the term 'flux-generating step'.

5.2.1 Near-equilibrium, non-equilibrium, and flux-generating reactions

Reactions in a metabolic pathway, in general, can be divided into two classes: those very close to equilibrium (near-equilibrium) and those far removed from equilibrium (non-equilibrium). A reaction in a metabolic pathway is non-equilibrium if the activity of the enzyme that catalyses the reaction is low in comparison to the activities of other enzymes in the pathway, so that the concentration of substrate(s) of the reaction is maintained high, whereas that of the product is maintained low. Consequently, the rate of the reverse component of the reaction is very much less than the rate of the forward component. In the following example, the rate in the forward direction is 1000-fold greater than the rate in the reverse direction:

$$\longrightarrow A \underset{0.01}{\overset{10.01}{\rightleftarrows}} B \longrightarrow$$

hence the reaction is non-equilibrium.

A reaction is near-equilibrium if the catalytic activity of the enzyme is high in relation to the activities of other enzymes in the pathway, so that the rates of the forward and the reverse components of the reaction are much greater than the overall flux. In the following example, the difference between the forward and reverse components is only 10 per cent, and the rate of the forward component is tenfold greater than the flux:

$$\longrightarrow A \underset{90}{\overset{100}{\rightleftarrows}} B \longrightarrow$$

Note that in both examples the flux is the same, 10 units.

If an enzyme catalyses a non-equilibrium reaction in a metabolic pathway and approaches saturation with its pathway-substrate (that substrate which represents the flow of matter through the pathway), and consequently the catalytic rate is almost independent of the substrate concentration, then the reaction is known as the flux-generating step for the pathway. In other words, in the steady state this reaction initiates a flux to which all other reactions in the pathway must adjust. (Such a reaction must approach saturation with its pathway-substrate, otherwise, as the reaction proceeds, the substrate concentration would decrease and this would decrease the rate of the reaction and hence the flux through the pathway; a steady state would then be impossible.) One important development from the concept of flux-generating step is that it provides a physiologically useful definition of a metabolic pathway. (It should be noted that biochemists, in general, have steered clear of providing a precise definition, perhaps because of the inherent difficulties.) A pathway is here defined as *a series of reactions that is initiated by a flux-generating step and ends with the loss of end-product to a metabolic sink (storage form) or to the environment, or in a reaction that precedes another flux-generating step.* The significance of this definition in relation to understanding the effects of insulin will be made clear below.

One additional advantage of identifying the equilibrium nature of reactions is that a non-equilibrium reaction is more likely than a near-equilibrium reaction to be regulated by allosteric effectors: its importance in the current context is that this includes hormones or second messengers to hormones (see below). Searching for a direct effect of a hormone or a second messenger of that hormone on a near-equilibrium process is unlikely to be profitable! This principle must also apply to those enzymes that are phosphorylated by protein kinase(s): phosphorylation of an enzyme at a serine or tyrosine residue does not automatically indicate that the phosphorylation plays a role in regulation of the flux in the pathway to which the enzyme belongs. The kinetic structure of the pathway must be known (Newsholme and Board 1991).

5.2.2 Control of the transmission of a flux

In the following hypothetical linear pathway, the flux (*J*) is generated at reaction E1 (the flux-generating step which is denoted by the sign (⧣→) from substrate S.

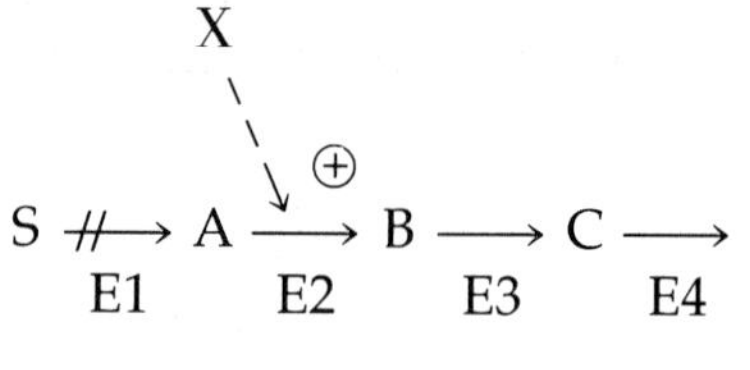

(*J*)

This flux is transmitted along the pathway by the response of the subsequent reactions to the metabolic intermediates A, B, and C, so that these metabolites link all the component reactions of the pathway and help to produce the overall steady-state flux. The role of the metabolic intermediates can best be explained by reference to the effect of an activator of one of the component reactions. Compound X stimulates enzyme E2 so that if the concentration of X were increased, the activity of this enzyme would increase. However, this would result in a decrease in the concentration of substrate A, which would lower the activity of enzyme E2. The decrease in A will continue until the activity of E2 reaches the original activity, so that the overall steady-state flux would not change. The only difference would be a lowered concentration of A, so that the increase in activity of E2 is only transient. This shows that, in this example, the concentration of A is determined by the flux (i.e. the activity of the enzyme that catalyses the flux-generating step, E1) together with the kinetic properties of enzyme E2. Similarly, the concentrations of B and C are determined by the flux and the kinetic properties of E3 and E4, respectively. Metabolic intermediates whose concentrations are determined solely by the flux (and which, therefore, help to maintain the steady-state flux) are termed internal effectors for that flux. From this definition it should be clear that totally internal effectors cannot change the flux: at least, this is the case for a linear pathway. In a branched pathway, however, such intermediates can be partial internal and partial external regulators and hence can influence a flux: the latter may be particularly important in understanding how hormones influence fluxes *in vivo*.

5.2.3 Regulators and regulatory reactions

The term 'regulation' can be described and defined in terms of metabolic communication, which is another way of describing a stimulus–response system. Any effector of an enzyme can be termed a regulator of that enzyme's activity. Of considerable importance is that effectors of an enzyme will not necessarily regulate the flux through the pathway *in vivo*. For example, in the system given above, both A and X are regulators of the enzyme E2 (since they both communicate with the enzyme) but, as explained above, X cannot change the rate of E2 *in vivo* in the steady state. (Any change in the rate of E2 is compensated for by the inverse change in the concentration of A.) On the other hand, the internal effector A adjusts the rate of E2 to the flux and hence is a regulator (an internal effector or regulator) of the rate of E2. It is therefore important to have some knowledge of the structure of the pathway prior to the interpretation of its control.

If, however, the hypothetical system given above is changed, so that Y is now a regulator of E1, a comparison of the effects of X and Y is instructive.

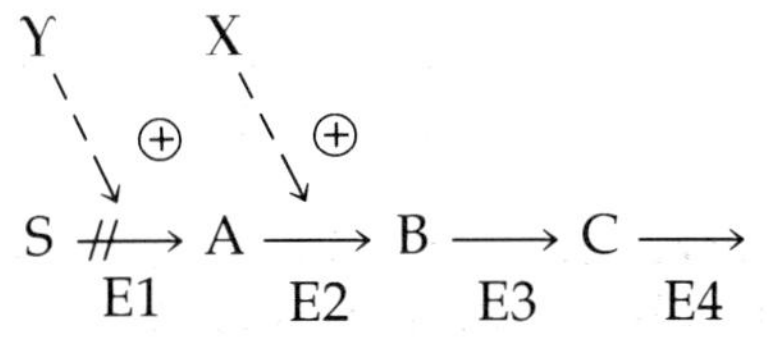

(*J*)

In contrast to X, regulator Y is able to change the flux through the pathway. The important difference is that E1 is the flux-generating step for this flux; in response to a change in Y, any change in the concentration of S will not significantly modify the activity of E1. The activity of E1 is 'buffered' against changes in its substrate concentration by its flux-generating nature. And E1 communicates with all the reactions that transmit the flux, via the concentrations of the pathway-substrates (A, B, and C) so that it is regulatory for the flux. Note that E2 can communicate with E3 via B but is unable to communicate with the flux-generating step E1. It is not regulatory for the flux. However, if the system is modified to include an inhibitory effect of A on enzyme E1, as follows:

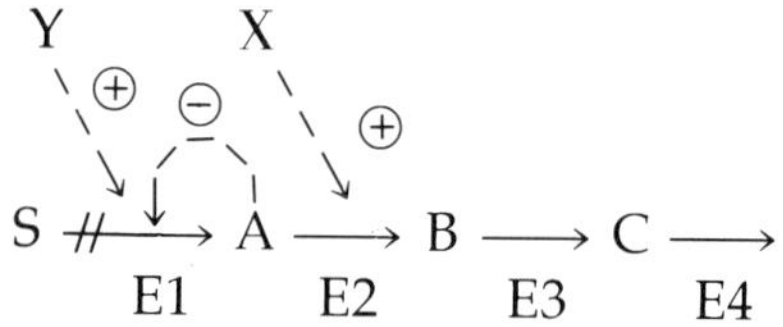

(*J*)

a feedback communication from E2 to E1 is established so that both E1 and E2 communicate with all the reactions that constitute the flux. Reactions that communicate with the flux are defined as regulatory for the flux. (Although this may appear to be very straightforward, some care is needed when defining the flux that is being modified, especially when the pathway is branched, see below.)

It has been emphasized that, in the absence of the feedback communication between E2 and E1, compound X cannot communicate with the flux, and therefore cannot regulate the flux through the reaction. However, such an effect may not be redundant *in vivo*. An increase in the rate of E1, produced by a change in the concentration of Y, is communicated to E2 via the increase in concentration of A. However, if at the same time that the concentration of Y is increased, that of X is also increased, the change in the concentration of A needed to communicate between E1 and E2 will be less. This would be important, for example, if A was utilized in some other pathway (i.e. if the pathway was branched), if A was a labile or a chemically reactive intermediate so that a high concentration could lead to unwanted side-reactions, or if the change in flux had to be very large and the concentration change in A would not be able to produce the necessary increase in the rate of E2. Such considerations may explain why many enzymes in the physiological pathway, although they do not catalyse flux-generating steps, are subject to regulation by factors other than the pathway-substrate (i.e. other than internal regulation). Effectors other than the pathway-substrate are known as external effectors, and they can include allosteric effectors or cofactors. Hormones, or the second messengers of hormones, in this context, classify as allosteric regulators, although the effects they have on the flux may be achieved by changing the concentration of a co-substrate for a reaction.

From the above discussions it might seem that the flux-generating step in a physiological pathway must be the 'rate-limiting' or 'pacemaker' step for the flux. Although this may be the case, if there are several regulatory reactions (and this is usually the case), none of these can be regarded as 'rate limiting'. Their significance in control must be based on knowledge of communication within the pathway. The interpretation is also more difficult if the pathway is branched.

5.2.4 Control of flux in branched pathways

In practice, metabolic pathways are not always as straightforward as described above. They are usually branched, so that one flux divides into two or more fluxes which may also be further subdivided.

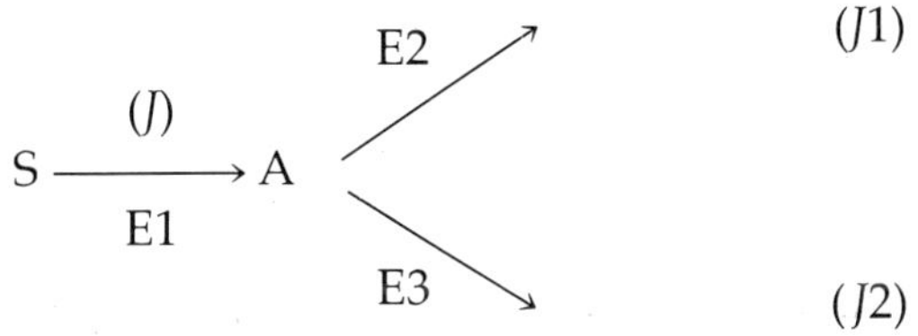

In the above system, the flux generated at E1 divides at A into J_1 and J_2 such that $J = J_1 + J_2$. This complexity influences the factors that affect the concentration of A and the relationship of A to the various fluxes. The concentration of A is now determined by the total flux, J, plus the kinetics of both enzymes E2 and E3; hence, the concentration of A is only partially determined by either flux J_1 or J_2. Consequently, although A is a simple internal effector for flux J, it is a partial internal (or partial external) effector for fluxes J_1 and J_2. (This shows that the terms 'internal' and 'external', when applied to effectors, must relate to a given flux and cannot be used in an absolute sense.)

In this particular branched pathway, enzyme E1 is regulatory for all three fluxes, J, J_1, and J_2, whereas, for example, enzyme E2 is regulatory for J_1 and J_2 but not for J. Consequently, an effector of E2 could change the fluxes J_1 and J_2 (in opposite directions), but J would not be affected. This re-emphasizes the important point that to understand the control of flux, the precise flux under consideration must be defined. The important point in relation to the effects of hormones is that by influencing the activity of an enzyme at a branch point in metabolism, there may be changes of flux in the two arms of the pathway without affecting the flux-generating step. If, however, one of the enzymes at the branch point catalyses a flux-generating step, affecting the activity of the other enzyme at the branch (e.g. by a hormone) will not change the flux in either arm of the branched pathway. This shows the importance of knowledge of the kinetic structure of the pathway before the effects of hormones on enzymes, usually observed in the test-tube, are converted into a biochemical hypothesis of control. (Note that this is not the case for the branch point in glycolysis at the level of glucose 6-phosphate; glycolysis via phosphofructokinase has no flux-generating step, whereas the glycogen synthesis pathway does: this latter pathway is influenced directly by insulin, whereas glycolysis via phosphofructokinase is not (see below).

5.3 Effect of insulin on glucose utilization and glycogen synthesis

5.3.1 Glucose transport across the cell membrane

The first reaction of the glycolytic pathway in any tissue is considered to be the transport of glucose across the cell membrane. Transport involves the transient combination of a glucose molecule with the protein carrier at the outer surface of the membrane and the subsequent release of the glucose at the inner surface. In most tissues, this process of glucose transport is down a concentration gradient and is therefore a passive process. (The active transport of glucose is seen in the epithelial cells of the intestine, in the epithelial cells of the kidney tubule, and possibly into the choroid plexus (where the cerebrospinal fluid is formed).) Since it has been known since 1950 that insulin increases the rate of transport of glucose across the cell membrane in both muscle and adipose tissue, this process has, for many years, been a focus of attention. Thus it is now known that insulin increases the number of glucose transporters (GLUT-4) in the plasma membrane: and insulin increases the rate of translocation of the GLUT-4 transporter from an intracellular membrane store to the cell membrane in skeletal muscle (Klip *et al.* 1987).

However, there are a number of important questions and points of discussion that are raised by this effect, which influence our understanding of the physiological action of insulin. An important point is that, in muscle and adipose tissue, the glucose transport process is non-equilibrium so that it can respond to the influence of insulin, or rather its intracellular second messenger (an allosteric regulator): in liver, the process is near-equilibrium, so there would be no point for this process to be stimulated by insulin. The control of the rate and direction of glucose flux in the liver therefore shifts to the enzymes glucokinase and glucose 6-phosphatase (see Newsholme and Leech 1983).

The transport process appears also to be near-equilibrium in many cells undergoing proliferation; this means that transport is no longer a regulatory step; this is transferred to the hexokinase reaction, which then becomes the flux-generating step for glucose utilization. This means that the rate of glycolysis in these cells can be maintained constant despite marked hypoglycaemia, since the rate of glycolysis now depends upon the K_m of hexokinase (<0.1 mM glucose) rather than that of the glucose transporter (5 mM glucose). The importance of a high, yet constant, rate of glycolysis in rapidly dividing cells has been given elsewhere (Newsholme *et al.* 1985). In view of the role of glycolysis in such cells, it is not surprising that insulin is without effect on them.

5.3.2 Glucose transport and glycolysis

Insulin increases the rate of glucose transport into muscle cells and also increases the rate of glycolysis, that is, glucose conversion to lactate. How do the glycolytic enzymes respond to insulin? Is this stimulation of the glycolytic enzymes due solely to the increase in the rate of glucose transport? In this case, this would cause an increase in the rate of all the glycolytic enzymes by internal regulation—that is, by increasing the concentration of each intermediate of glycolysis. Alternatively, insulin might, in addition to its effect on transport, lead to a stimulation of one or more of other glycolytic enzymes. Such effects could not be considered to control glycolysis but would facilitate the control of glycolysis caused by the effect of insulin on glucose transport. As yet, however, there is no firm evidence that insulin stimulates the activities of glycolytic enzymes. Two candidates are hexokinase and 6-phosphofructokinase. In order to provide evidence that insulin affects the activity of hexokinase, it must be shown that the level of intracellular glucose decreases as the flux through hexokinase, in response to insulin, increases: this is difficult because of the problem of the measurement of the precise concentration of intracellular glucose in muscle. Experiments with 2-deoxyglucose, following the formation of 2-deoxyglucose 6-phosphate, could be fruitful.

Insulin could increase the activity of phosphofructokinase by changing the concentration of one of its many regulators, such as fructose 2,6-bisphosphate. So far, however, consistent changes in the concentration of any one of the regulators, except for fructose 6-phosphate, in muscle in response to insulin have not been reported.

5.3.3 Glucose transport and the flux-generating step for glycolysis in muscle

The process of glycolysis-from-glucose, in contrast to glycolysis-from-glycogen, is not a physiological pathway as defined above. The non-equilibrium reactions in glycolysis are catalysed by the glucose transport system, hexokinase, 6-phosphofructokinase, and pyruvate kinase, and none of these reactions is saturated with substrate (Newsholme and Leech 1983). The flux-generating step for this process is either the absorption of glucose from the intestine or, when glucose arises endogenously, the breakdown of glycogen in the liver (Fig. 1). Thus, in these cases, a metabolic pathway spans more than one tissue: it includes the intestine or liver, the blood, and the target tissue, for example muscle. The metabolic importance of this interpretation in the present context is related to the control of the rate of glucose utilization by muscle. When a high-carbohydrate-containing meal has been digested, the rate of absorption of glucose by the intestine will be high, and, in order to prevent a very large increase in the blood glucose concentration, the rate of glucose utilization by muscle is increased. (The amount of carbohydrate ingested in the average meal is approximately 20–30 times greater than that normally present in the blood, yet the normal increase in the blood glucose level is <50 per cent.) But this cannot be due to increased demand for energy by muscle since it occurs even when muscles are resting after a meal. (The response of muscle to insulin can therefore be considered to be physiologically abnormal: muscle normally takes up glucose in response to a demand for energy, when the glucose will be

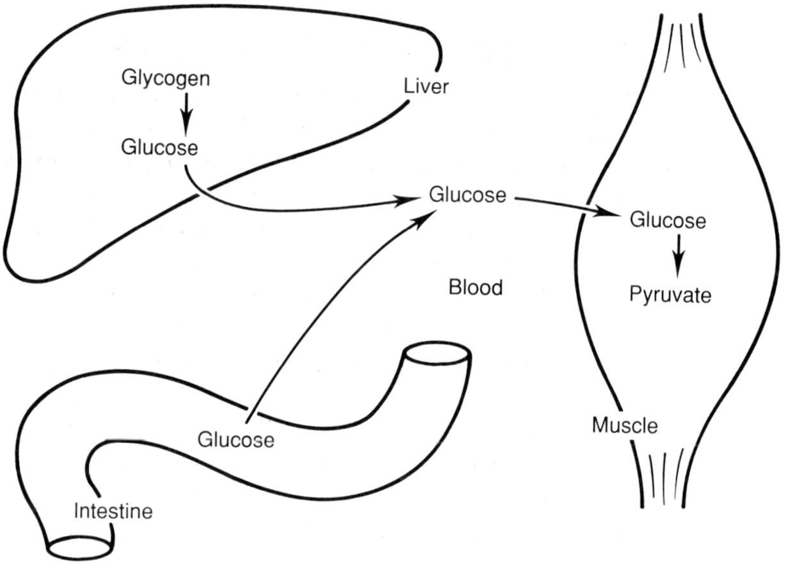

Fig. 1 Source of glucose for glycolysis in muscle. Other glucose-utilizing tissues (e.g. brain, kidney) receive glucose from the same sources. Release of glucose from the intestine is probably important as a flux-generating step in the absorptive state, whereas the degradation of liver glycogen is important in the post-absorptive state. (This is taken from *Biochemistry for the medical sciences* by E. A. Newsholme and A. R. Leech, with permission of the publishers, John Wiley and Sons Ltd)

oxidized to provide ATP.) Not surprisingly, therefore, not all of the glucose taken up by muscle at the behest of insulin is oxidized to CO_2: some, and possibly a high proportion, appears as lactate (this is best seen in isolated incubated muscle preparation—see Challiss *et al.* 1987).

To understand the physiology behind this process, the complete metabolic pathway must be considered under the conditions of feeding. Thus, the flux-generating step is the transport of glucose across the intestine into the bloodstream. Since insulin obviously does not affect this step—it affects glucose transport into muscle—it can be seen to be stimulating a step equivalent to E2 in the hypothetical pathway given above. This does not control glucose flux through the whole pathway, which is regulated by the intestine, but it controls the branch that is present in muscle and, in particular, it controls the concentration of an intermediate in that pathway—an important intermediate, blood glucose! Insulin can be viewed as a feed-forward effector of the glycolytic process in muscle, which is dependent upon the rate of entry of glucose into the bloodstream, and the main physiological role of this is to prevent a massive increase in the blood glucose level: the latter would result in loss of glucose through the kidneys and unwanted and dangerous glycosylation of proteins as a side-reaction. Therefore, secretion of insulin must be strictly related to the entry of glucose into the bloodstream from the intestine, which is known to be the case.

It follows, therefore, that a rise in the level of insulin at other times would be highly dangerous: injection of insulin between meals can result in dangerous hypoglycaemia. Insulin is probably the only hormone that has been used as a murder weapon (Birkinshaw *et al.* 1958).

5.3.4 Insulin and glycolysis in muscle

If the insulin-stimulated increase in the rates of glucose transport and glycolysis in muscle is considered to be physiologically abnormal, what is the fate of the end-product of glucose metabolism? An increase in the rate of glycolysis will provide more lactate, which is released from the muscle. What is the fate of this end-product? There is now considerable evidence that some/most of the glycogen that is synthesized in the liver after a meal is derived not so much from glucose but from lactate, alanine, and other gluconeogenic processors (known as the indirect pathway, see Newsholme and Leech 1983 for an historical account of this topic). The conversion of glucose to lactate in muscle and the conversion of lactate to glycogen in liver may be part of an inter-organ communication system designed to provide precision in regulation. The conversion in muscle controls the blood level of glucose, an excess of which could be dangerous. The increase in the blood lactate concentration, together with an increased hepatic blood flow, will facilitate the conversion of lactate to glycogen in the liver. Thus it is interesting to consider that the stimulation by insulin of the rate of glucose transport in muscle, via a feed-forward regulation from digestion and absorption of carbohydrate in the intestine, may lead directly, but via a long and complex inter-tissue metabolic pathway, to increased levels of liver glycogen (Fig. 2). The possible physiological significance of this is discussed in detail below.

5.3.5 The physiological pathway of glycogen synthesis in muscle and the effects of insulin

It is known that, of the reactions involved in the process of glycogen synthesis (Fig. 3), hexokinase, glycogen synthase, and branching enzyme catalyse non-equilibrium reactions, whereas phosphoglucomutase and uridylyltransferase catalyse near-equilibrium reactions. Of the non-equilibrium reactions, only glycogen synthase approaches saturation with pathway-substrate (UDP-glucose) (Newsholme and Leech 1983). This therefore is the flux-generating reaction, so that the physiological pathway for glycogen synthesis starts at UDP-glucose and comprises only two reactions, those catalysed by the synthase and the branching enzyme. However, since the concentration of UDP-glucose is low in comparison to the rate of glycogen synthesis, its concentration must be maintained whenever glycogen is being synthesized. This is achieved by the reactions catalysed by uridylyltransferase and phosphoglucomutase, both near-equilibrium reactions, utilizing glucose 6-phosphate produced via the hexokinase reaction. Of all of the reactions in the pathway, metabolic control logic predicts that glycogen synthase would be regulated by insulin; this is, of course, a well established effect of this hormone!

Another important principle that emerges from these considerations is that, in relation to the effects of insulin, glucose transport and glycogen synthase are

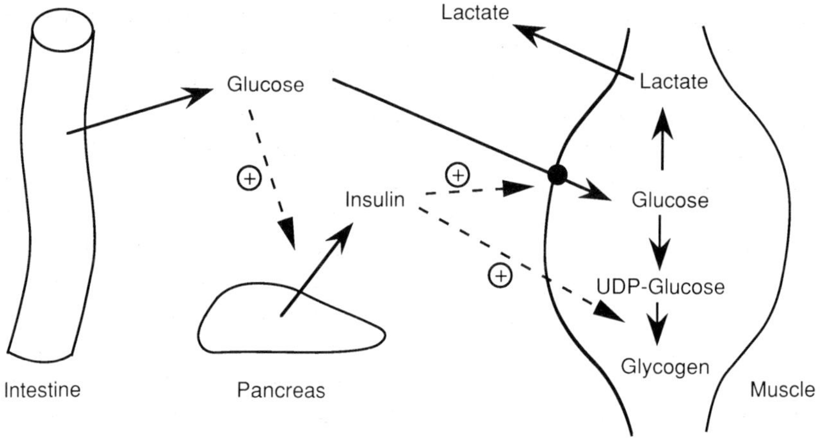

Fig. 2 Insulin effects on carbohydrate metabolism in muscle as part of a regulatory loop in the physiological pathway of glycolysis in the body. Glucose, together with hormones from the intestine, stimulates insulin release from the pancreas, and insulin facilitates the uptake of glucose by muscle and the synthesis of glycogen in muscle. It may also, via the effect on glucose transport in muscle, stimulate the synthesis of glycogen in the liver (see Fig 8)

independent reactions. Insulin stimulation of glucose transport does not necessarily result in a higher rate of glycogen synthesis (all of the increased glucose entering the muscle could be converted to lactate): similarly, insulin could increase markedly the rate of glycogen synthesis without affecting transport, since it would lead to more glucose residues being converted to glycogen and hence fewer to lactate—this illustrates the importance of control at a branch point (see above). (Of course, it is necessary that some glucose enters the muscle in order to allow glycogen synthesis to proceed.)

In a normal, physically active animal that eats only sufficient food for maintenance, insulin may control the rate of glucose utilization primarily through its effects on glucose transport and glycogen synthesis in muscle. In this way, glycogen that had been used by the muscle during exercise would be repleted after the

Fig 3 Glycogen synthesis in muscle and steps controlled by insulin. Insulin binds to its receptor, which results in the production or activation of the unknown second messenger. This leads to activation of glucose transport and glycogen synthesis

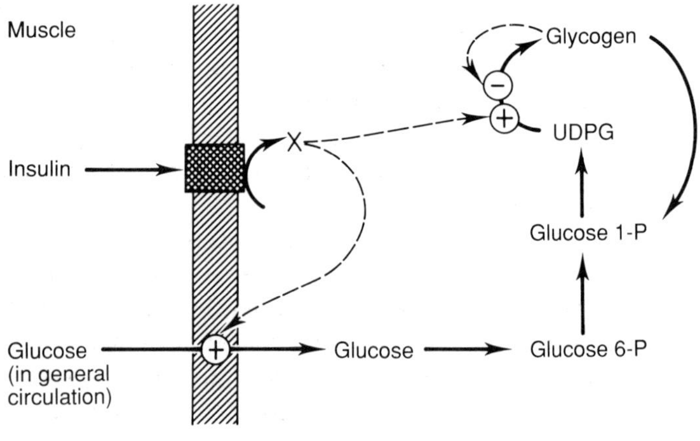

carbohydrate meal, and this would occur before glycogen was synthesized in the liver, since this latter process depends upon control of the direction of glucose metabolism in muscle (see above). This may be an advantage of the indirect pathway for glycogen synthesis in liver. It is well established that, after prolonged exercise, glycogen repletion occurs in the muscle before it occurs in the liver (Hultman 1978). Indeed, glycogen is synthesized in muscle after exercise even during a period of starvation, when liver glycogen would be broken down (Maehlum and Hermansen 1978). However, when the amount of food ingested is above that required for repletion of muscle glycogen, it is suggested that insulin controls the blood glucose level by increasing the rate of glycolysis in muscle via internal control, and hence the formation of lactate. This then, via glyconeogenesis, will cause repletion of the glycogen store in the liver. And this physiological result is dependent simply upon the kinetic structure of the branch point at the level of glucose-6-phosphate in muscle.

5.4 Insulin and the importance of its anti-lipolytic effect

The major store of triacylglycerol in the body is present in adipose tissue but it is mobilized in the form of long chain fatty acids, which are carried to other tissues via the bloodstream. Hence muscle oxidizes fatty acids derived from adipose tissue.

The pathway for fatty acid oxidation in muscle involves β-oxidation and the Krebs cycle. But the flux-generating step for β-oxidation of fatty acids is triacylglycerol lipase in adipose tissue, since increasing the blood concentration of fatty acids increases the rate of fatty acid oxidation by muscle (and other tissues): in other words, the plasma fatty acid concentration can be considered to be an internal regulator for the rate of β-oxidation in muscles (Fig. 4). (It should be noted that increasing the amount of sustained exercise performed by the muscle can increase the rate of oxidation of fatty acids without an increase in their extracellular concentration. In other words, the pathway in muscle can also be controlled by the muscle—see Newsholme and Leech 1983.) The significance of this is that insulin causes inhibition of adipose tissue lipase and so decreases the rate of lipolysis in adipose tissue. This will result in a decrease in the plasma level of fatty acid and therefore less oxidation of fatty acids by muscle. Thus insulin affects the rate of fatty acid oxidation in muscle by a direct regulatory effect on a flux-generating step in adipose tissue.

The fuel reserves in the average human subject are given in Table 1. These data provide part of the answer to the question as to why carbohydrate and lipid metabolism must be integrated. The amount of carbohydrate stored is very small; it represents only about 2 per cent of that of lipid. Liver contains the only store of glycogen that can be broken down into glucose and released into the bloodstream for use by other tissues (muscle glycogen is used solely within the muscle). How-

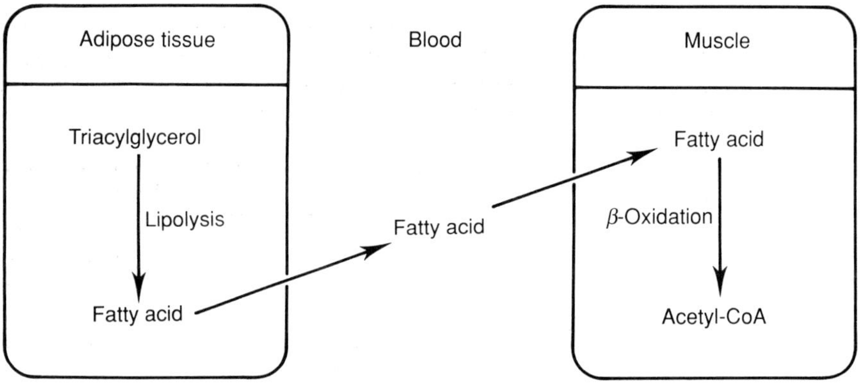

Fig 4 The physiological pathway of lipid oxidation in muscle. In addition to muscle, other tissues such as kidney will also oxidize fatty acids. The flux-generating step is catalysed by triglyceride lipase within the adipose tissue, which is controlled, in part, by insulin. Control at this step is amplified by the existence of a substrate cycle, the triglyceride/fatty acid cycle, which is stimulated by catecholamines (Newsholme and Craftree 1976, Newsholme *et al.* 1986), and this stimulation may be decreased in obese and diabetic patients

ever, the quantity of stored carbohydrate in the liver (approx. 100 g) is very small in relation to the glucose requirement of the tissues. At rest, the total glucose requirement of the major carbohydrate-utilizing tissues of the body (brain, kidney, heart, muscle) is over 300 g per day and this is normally met from the dietary intake of carbohydrate. But in starvation, liver glycogen is broken down to provide glucose for the tissues; measurement of the glycogen content in small samples of human liver, removed by biopsy needle, have shown that this store of glycogen is largely depleted after 24 h starvation (Hultman 1978). Despite this, the blood glucose level remains remarkably constant, as shown by the now classic work of Cahill on Harvard divinity students (Cahill 1970). One answer to the limited store of glycogen is for tissues to use an alternative fuel—fatty acids. Mobilization of fatty acids from the adipose tissue triacylglycerol store will probably begin during the overnight fast and will increase particularly if breakfast is missed. It is well established, in both experimental animals and man, that even short periods of starvation raise the plasma fatty acid concentration. Calculations based on the increase in the concentration of fatty acids in starvation and their turnover rate indicate that fatty acid oxidation can account for most of the energy requirements of the tissues after 24 h starvation (Newsholme and Leech 1983). But this has an additional role—the conservation of blood glucose and maintenance of its level.

In the early stages of both starvation and exercise, skeletal muscle will use primarily glycogen and blood glucose to satisfy its energy demands. As the carbohydrate stores become depleted, fatty acids are mobilized and their rate of oxidation increases (Fig. 5). This occurs despite the fact that the blood glucose concentration falls very little (by no more than about 25 per cent of the normal) and while the absolute blood glucose concentration remains considerably higher than that of

Table 1 Approximate fuel stores* in the average man (from *Biochemistry for the medical sciences* by E.A. Newsholme and A.R. Leech, with kind permission of the publishers, John Wiley and Sons)

Tissue fuel store	Approximate total fuel reserve		Estimated period for which fuel store would provide energy		
	g	kJ	Days of starvation[a]	Days of walking[b]	Minutes of marathon running[c]
Adipose tissue triacyglycerol	9000	337 000	34	10.8	4018
Liver glycogen	90	1500	0.15	0.05	18
Muscle glycogen	350	6000	0.6	0.20	71
Blood and extracellular glucose	20	320	0.03	0.01	4
Body protein	8800	150 000	15	4.8	1800

* Normal man possesses 12 per cent of the body weight as triacylglycerol and normal woman about 26 per cent. Periods for which the fuel will last are calculated as below.
[a] Assuming that energy expenditure during starvation is 10 050 kJ day^{-1} (i.e. normal energy expenditure of 13 400 kJ day^{-1} is reduced by 25 per cent on starvation.
[b] Assuming that energy expenditure during walking (4 miles h^{-1}) for a 65 kg man is 31 248 kJ day^{-1}.
[c] Assuming energy expenditure of 84 kJ min^{-1}.
The data illustrate the time for which the fuel stores would provide energy, provided this was the only fuel utilized by the body.

fatty acid (see Table 2). Since both fuels are available in the bloodstream at the same time, the question arises why does muscle utilize fatty acids in preference to glucose? The answer to this question is, in principle, very simple. The elevated concentration of fatty acid in the bloodstream increases the rate of fatty acid oxidation in the muscle and this specifically decreases the rate of glucose utilization and oxidation (Fig. 6). This control mechanism is known as the 'glucose/fatty acid cycle'.

5.4.1 The glucose/fatty acid cycle

The concept of the glucose/fatty acid cycle, put forward by Randle *et al.* (1963), provides a mechanism by which fatty acid oxidation decreases the rate of glucose utilization and particularly glucose oxidation by muscle. There is now considerable evidence to support the important proposal that, under conditions of 'carbohydrate stress' (defined here as when the glycogen store in the liver is reduced), fatty acids are mobilized from adipose tissue so that their rate of oxidation by muscle increases and this, in turn, decreases the rate of glucose utilization and oxidation. Conversely, when the carbohydrate stress is removed (e.g. by refeeding a starved subject) the rate of fatty acid release by adipose tissue is reduced. Consequently, the plasma level of fatty acids and hence their rate of oxidation is

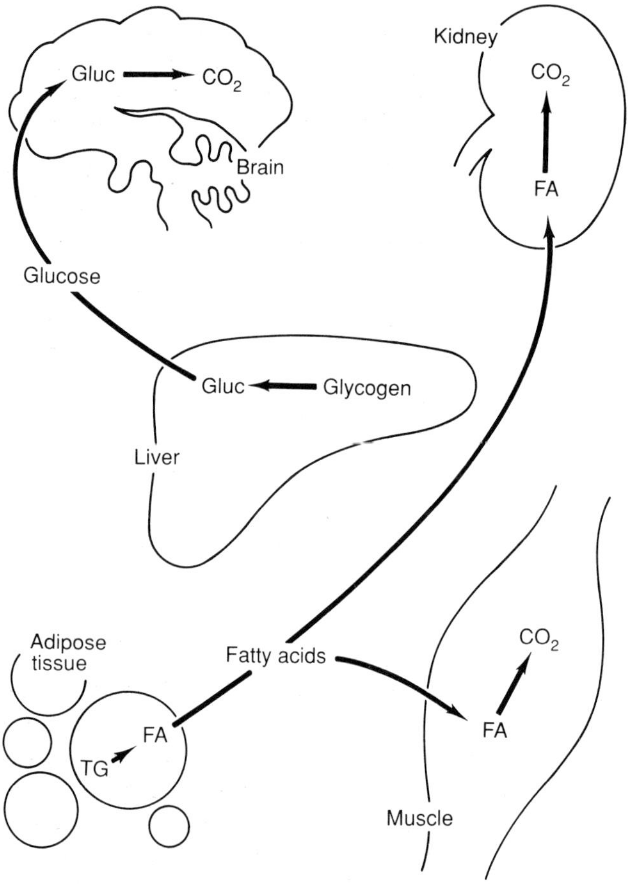

Fig. 5 Pattern of fuel utilization during the early period of starvation. The use of fatty acid by muscle 'protects' glucose for use by the brain and other tissues (e.g. red blood cells) taken from *Biochemistry for the medical sciences* by E. A. Newsholme and A. R. Leech, with permission of the publishers, John Wiley and Sons Ltd)

decreased, so that the rate of glucose utilization by the muscle increases. And this is controlled by the increase in the concentration of insulin after feeding, since insulin inhibits the lipase in adipose tissue; indeed it is one of the few agents that inhibit lipolysis (Newsholme and Leech 1983). These responses stabilize the blood glucose concentration and conserve glucose. An important point is that insulin does not directly affect the rate of fatty acid oxidation by muscle: it occurs indirectly.

This interpretation of the role of insulin is based on the application of metabolic control logic: the pathway for fatty acid oxidation in muscle starts with the flux-generating step, lipolysis in adipose tissue, so that *this* is the key process to be controlled by insulin (Fig. 6).

The regulatory effect of fatty acid on glucose utilization can be seen as a logical necessity when the small reserves of carbohydrate are taken into account together with the fact that some tissues have an obligatory requirement for glucose: these tissues include the brain and the red blood cells. The biochemical mechanisms by which fatty acid oxidation decreases glucose utilization and oxidation in muscle are given elsewhere (Newsholme and Leech 1983).

Table 2 Concentrations of blood glucose, fatty acids, ketone bodies, and insulin during starvation in man and rat (from *Biochemistry for the medical sciences* by E.A. Newsholme and A.R. Leech, with kind permission of the publishers, John Wiley and Sons)

Animal	Fuel or hormone	Fed	Days of starvation 1	2	3	4	5	6	8
Man	Glucose	5.5	4.7	4.1	3.8	3.6	3.6	3.5	3.5
	Fatty acids	0.30	0.42	0.82	1.04	1.15	1.27	1.18	1.88
	Ketone bodies	0.01	0.03	0.55	2.15	2.89	3.64	3.98	5.34
	Insulin*	>40	15.2	9.2	8.0	7.7	8.6	7.7	8.3
Rat	Glucose	6.3	–	4.8	4.4	4.3	–	–	–
	Fatty acids	.66	–	1.3	–	–	–	–	–
	Ketone bodies	0.22	–	2.8	3.0	3.3	–	–	–
	Insulin	28.7	–	4.2	–	–	–	–	–

*The units of insulin are $\mu U\ ml^{-1}$.

5.5 Insulin and gluconeogenesis

The major precursors for gluconeogenesis in the liver are lactate, glycerol, and amino acids, of which alanine is quantitatively important. The gluconeogenic pathway can, in fact, be considered to start with the generation of alanine from muscle and the intestine, lactate from muscle (and other tissues), and glycerol from adipose (and other) tissues. Gluconeogenesis is thus seen as a complex, branched pathway that spans more than one tissue (Fig. 7).

The normal blood lactate concentration (about 1 mM) represents a steady-state concentration that reflects the balance between the rates of production and utilization. Although the principal tissue for lactate utilization is liver, other tissues, such as heart and red muscles (type I and IIA fibres) can remove lactate from the bloodstream and oxidize it for energy production. Most of the lactate that is removed by the liver is converted to glucose or glycogen, via gluconeogenesis, although a small proportion may be oxidized or converted to triacylglycerol under lipogenic conditions. The conversion of lactate to glucose in the liver and the continuous formation of lactate from glucose in the other tissues of the body represents a cyclical flow of carbon that has been termed the Cori cycle (after Carl Cori, who originally put forward the idea) (Fig. 8). It may have greater physiological significance than just that of a carbon link between peripheral tissues and the liver. By maintaining a high flux through the Cori cycle, a dynamic buffer of key metabolic intermediates is provided, both in the tissues and in the bloodstream, so that they can be used by tissues when required. If the rate of utilization of the intermediates is low, compared to the flux through the Cori cycle, this may provide for precision in regulation as described and defined elsewhere (Newsholme *et al.* 1985). This may be one reason for increased Cori cycle flux during injury, sepsis,

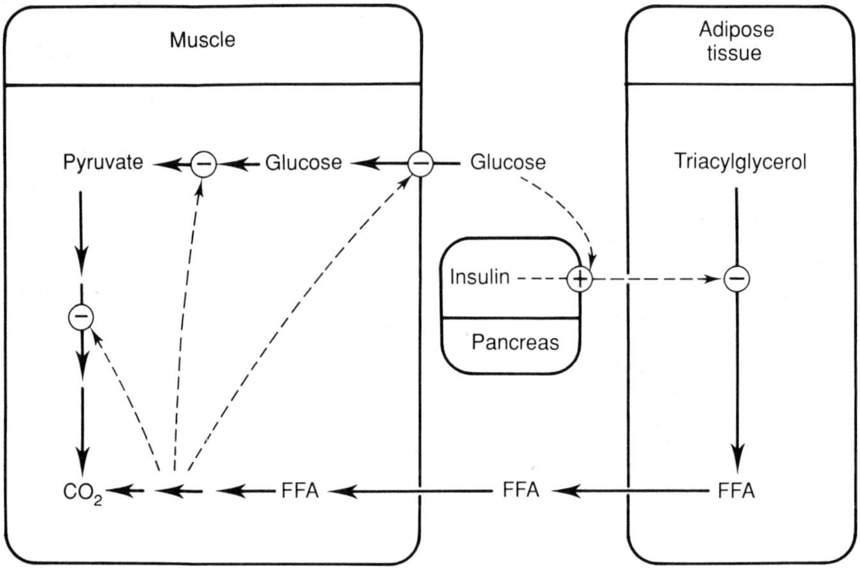

Fig. 6 The glucose/fatty acid cycle. Note that a change in the peripheral blood glucose concentration is less important than changes in the glucose concentration in hepatic portal blood in eliciting insulin release from the pancreas, since glucose absorption from the intestine is accompanied by secretion of duodenal hormones which increase the sensitivity of the β-cells to the stimulatory effect of glucose (taken from *Biochemistry for the medical sciences* by E. A. Newsholme and A. R. Leech, with permission of the publishers, John Wiley and Sons Ltd)

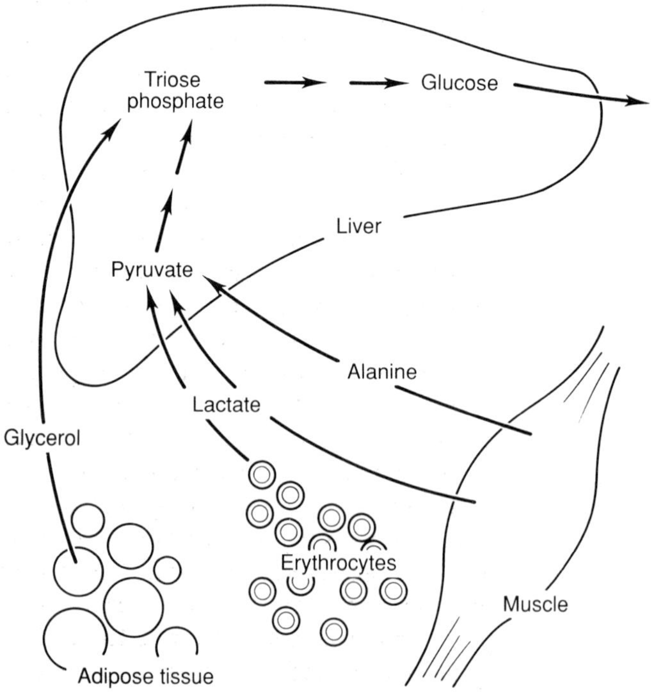

Fig. 7 The tissues and the pathway involved in gluconeogenesis. This pathway is also used in the synthesis of liver glycogen. The rate of gluconeogenesis in the kidney is quantitatively much less important than that in the liver, except in prolonged starvation (taken from *Biochemistry for the Medical sciences* by E. A. Newsholme and A. R. Leech, with permission of the publishers, John Wiley and Sons Ltd)

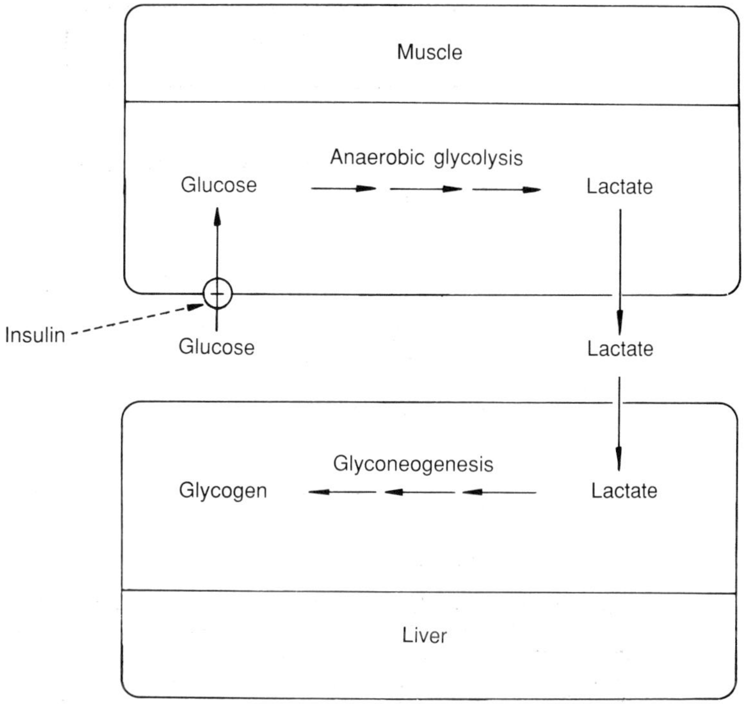

Fig. 8 A modified Cori cycle. Insulin stimulates the conversion of glucose to lactate in muscle; the lactate is release by muscle and the blood concentration is increased (slightly): this stimulates the rate of conversion of lactate to glycogen in the liver (glyconeogenesis). In this way, insulin indirectly stimulates the rate of glycogen synthesis in the liver

post-surgery and burns, when, although there is generalized resistance to insulin, muscle may have a higher rate of conversion of glucose to lactate (Leighton *et al.* 1989). The possible physiological significance of the insulin effect on glucose transport in muscle, which indirectly stimulates the rate of glycogen synthesis in the liver, is discussed above.

Glycerol is released into the bloodstream as a result of triacylglycerol hydrolysis from a number of tissues, the most important being adipose tissue.

Experiments in the 1930s demonstrated that amino acids could be divided into three classes; glucogenic, ketogenic, and glucogenic-plus-ketogenic. A glucogenic amino acid was one that, upon administration to a starving animal, increased the blood concentration of glucose; a ketogenic amino acid raised the blood concentration of ketone bodies; and glucogenic-plus-ketogenic amino acids increased the bood concentrations of both. Knowledge of the metabolism of the various amino acids provides an explanation for these observations and this classification. Most amino acids are catabolized to pyruvate or intermediates of the TCA cycle, all of which can be converted to glucose; these amino acids are glucogenic. On the other hand, there are just two amino acids (leucine and lysine) that give rise solely to acetyl-CoA, which cannot be converted to glucose but which can, particularly in the starving animal, be converted to ketone bodies. Obviously, amino acids (e.g. phenylalanine) that give rise to both acetyl-CoA and an intermediate of the TCA cycle will be both ketogenic and glucogenic.

Quantitatively, the most important amino acid precursor for gluconeogenesis is alanine, because it is released from muscle due to metabolism of other amino acids, from the intestine due to metabolism of glutamine and other amino acids, and perhaps from other tissues (e.g. immune system, lung). Whether insulin can influence the rate of alanine release from the intestine, lung or immune cells has not been studied.

5.5.1 Control of gluconeogenesis

Since the gluconeogenic pathway has several substrates and spans several tissues, its control will be complex. Here two factors are considered because of their importance in relation to insulin. First, control can be achieved by variations in concentration of the precursors, glycerol, lactate, and amino acids, and of the end-product, glucose. Raising the concentration of these precursors will increase the rate of gluconeogenesis by the liver, whereas raising the concentration of glucose will decrease it. Secondly, glucagon and glucocorticoids increase the rate of gluconeogenesis, whereas insulin counteracts the effect of these hormones and decreases it. However, it is possible to argue that, for normal, short-term diurnal control, hormonal effects on the liver are much less important than the change in the blood concentrations of precursors and product. Therefore, although insulin is considered to inhibit gluconeogenesis by antagonizing the effect of glucagon, it may stimulate, after a carbohydrate-rich meal, the rate of gluconeogenesis in the liver by increasing the concentration of plasma lactate via its effect on glucose transport in muscle! This may also be facilitated by an increase in hepatic blood flow after feeding.

5.6 How does insulin regulate the blood glucose level?: I

The success of insulin in decreasing the blood glucose concentration in the early work on diabetic animals stressed the important role of insulin in facilitating glucose utilization. Consequently, when, in the 1950s, it was discovered that insulin increased the rate of transport of glucose into the muscle cell (Levine and Goldstein 1955), it was obviously considered that this was the major means by which insulin regulated the blood glucose concentration *in vivo*.

The advent of the concept of the glucose/fatty acid regulatory cycle in the 1960s enabled a different interpretation of how insulin regulates the blood glucose level to be put forward. Since insulin is a potent anti-lipolytic hormone, an increased concentration of this hormone would decrease the rate of adipose tissue lipolysis, which would result in a decrease in the blood fatty acid concentration and this would then stimulate the rate of glucose utilization by muscle. This is insulin influencing the rate of glucose utilization in muscle via the glucose–fatty-acid cycle.

The process of gluconeogenesis can play an important role in regulating the blood glucose concentration. The rate of this process is regulated, in part, by the balance of the two hormones, glucagon and insulin. Both these hormones are secreted by the cells of the islets of Langerhans, glucagon from the α-cells and insulin from the β-cells. The endocrine secretions of the pancreas enter the hepatic portal vein, so that the liver is the first tissue to be exposed to these hormones and here the balance of the two hormones could play an important role in regulation of the blood glucose level—insulin decreasing and glucagon increasing the rate of release.

All three mechanisms described above, namely stimulation of glucose transport into muscle (and adipose tissue), increased rates of glucose utilization and oxidation in many tissues, dependent upon a decreased level of plasma fatty acid (inhibition of adipose tissue lipase by insulin), and decreased rates of glucose production and release by the liver (antagonism of the glucagon effect by insulin), could all play a role in increasing the rate, or the apparent rate, of glucose utilization when the plasma insulin concentration is increased (summarized in Fig. 9). The quantitative importance of each process will depend on the magnitude of

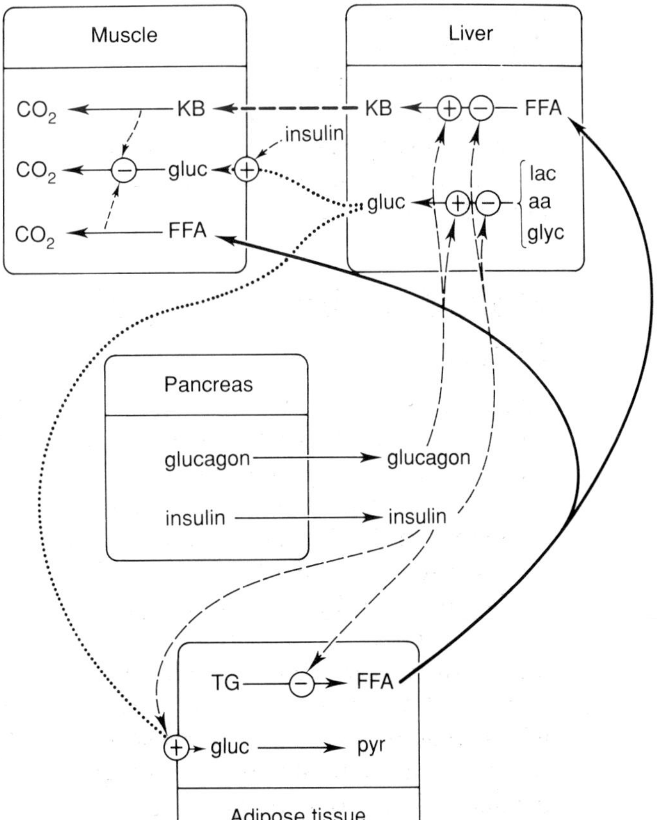

Fig. 9 Control steps of possible importance in how insulin regulates the blood glucose level. Four effects of insulin are depicted in the diagram: stimulation of glucose uptake by muscle and adipose; inhibition of lipolysis in adipose tissue; inhibition of gluconeogenesis in the liver; inhibition of ketosis in the liver (this effect may be of importance only after several days of starvation or after total depletion of liver glycogen) (see Newsholme and Leech 1983). Abbreviations: KB, keton bodies; gluc, glucose; lac, lactate; aa, amino acids; glyc, glycerol; pyr, pyruvate; TG, triacylglycerol. (Taken from *Biochemistry for the medical sciences* by E. A. Newsholme and A. R. Leech, with permission of the publishers, John Wiley and Sons Ltd)

the response of each process in each tissue to the change in insulin concentration *in vivo* at that particular time. Thus it is not possible at present to provide quantitative information as to how changes in the insulin concentration, after a meal, control the blood glucose level.

Furthermore, it is now known that the sensitivity of these processes, at least in muscle and adipose tissue, to insulin can be changed markedly by the presence of various local hormones, by small molecules, and by various other conditions, so that the above difficulties are compounded by the possibility that different muscles and/or different adipose tissue depots may respond quantitatively differently to a change in the concentration of insulin: indeed sensitivity may change rapidly from condition to condition.

5.6.1 Insulin sensitivity: its possible importance in physiology

The ability of the blood glucose level to respond to a given dose of insulin is impaired in a number of relatively common conditions (e.g. diabetes mellitus, ageing, obesity, gestational diabetes). It is described as glucose intolerance and is considered to be caused by resistance of the tissues to the effect of insulin. The frequency with which these glucose-intolerant states can occur, and the severity of the symptoms which may develop from them, has engendered a massive medical interest (and a correspondingly large literature) on insulin resistance as a pathological entity. The possibility that changes in insulin sensitivity, including resistance, may also be physiological appears to have been neglected. To attempt to redress the balance, insulin sensitivity is discussed below.

Definition of sensitivity

Sensitivity has been defined as the concentration of insulin producing 50 per cent of the maximal response. This quantity is termed the EC_{50}. A change in EC_{50} alters the metabolic response produced by any given level of insulin (except for the basal and maximal responses) (Fig. 10). The advantage of EC_{50} is that it is quantifiable and can be measured *in vitro* or *in vivo*.

It is particularly important to recognize what factors can alter the sensitivity and whether they have physiological significance. At present, this may be more important than their molecular mechanism. Some factors that are known to influence insulin sensitivity are introduced briefly below.

Adenosine

Adenosine is produced continuously by most tissues: some, or all, of it is released and may produce extracellular effects. Adenosine increases the sensitivity to insulin of the rate of glucose utilization and lipolysis in adipose tissue but it decreases the sensitivity of glucose utilization to insulin in skeletal muscle. The effects on muscle have been studied mainly with the isolated, incubated soleus muscle of the

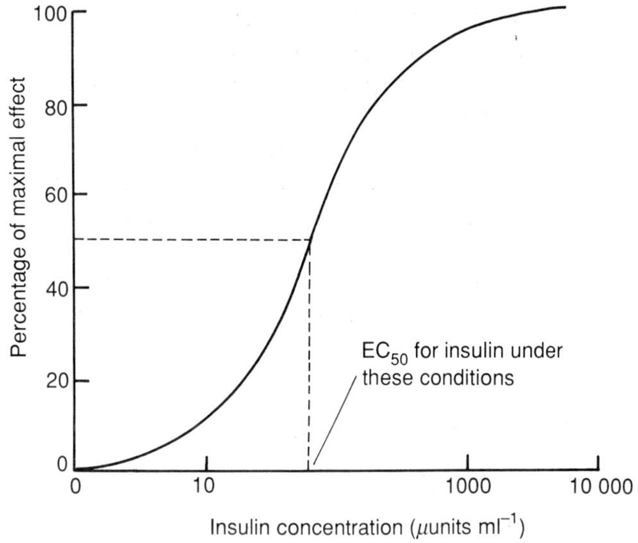

Fig. 10 The concentration of insulin that produces the half-maximal response (EC_{50})

rat. Addition of adenosine deaminase to the incubation medium, which is known to lower the concentration of adenosine, dramatically increases the sensitivity of glucose utilization to insulin. Addition of an adenosine-receptor agonist (e.g. 2-chloroadenosine) to the incubation medium decreases about tenfold the sensitivity of this process to insulin, whereas addition to an adenosine-receptor antagonist (e.g. 8-phenyl-theophylline) increases sensitivity by about tenfold. This receptor antagonist not only influences the insulin sensitivity of soleus muscle from normal animals but also removes the resistance of glycolysis to insulin in isolated soleus muscle obtained from either genetically obese rats or rats fed on a high sucrose diet. In soleus muscles obtained from rats subjected to cold-exposure for 2 days, addition of a receptor agonist totally removes the increased insulin sensitivity of glycolysis (reviewed by Challiss *et al.* 1987).

These data suggest that a specific adenosine receptor is involved in changing the sensitivity to insulin of both muscle and adipose tissue. However, the *physiological* significance of these effects is unclear for at least two reasons. First, the identity of the type of adenosine receptor(s) that is present in skeletal muscle is not known; if a unique adenosine receptor is present, this might provide an opportunity for medicinal chemists to produce specific and novel antagonists that could improve the sensitivity of glucose utilization in muscle to insulin. Secondly, the effects of adenosine receptor agonists and antagonists on the sensitivity of muscle and adipose tissue to insulin needs to be investigated *in vivo*.

Prostaglandins

It has been found that prostaglandins of the E series (PGE_1 and PGE_2) can improve insulin sensitivity in the isolated incubated soleus muscle. And indomethacin, which inhibits prostaglandin formation, markedly decreases the sensitivity of glycolysis to insulin (Leighton *et al.* 1985). Since PGE_2 is produced from arachidonic acid, these findings suggest that diets that are high in polyunsaturated fatty acids could improve insulin sensitivity. Of interest is the idea that conversion of linoleic acid, a common essential fatty acid in the diet, to arachidonic acid is severely limited in man by the activity of Δ-6-desaturase; but this limitation can be overcome by supplementation of the diet with γ-linolenic acid (GLA) (obtained from oil of the evening primrose) which bypasses this limiting step. This, therefore, should allow, if the hypothesis is correct, a higher rate of synthesis of arachidonic acid, a higher level of this fatty acid in the cell membrane, and hence more substrate for PGE_2 production (provided the cyclo-oxygenase does not catalyse a flux-generating step—a process that has not been subjected to metabolic control-logic analysis). It is of importance to note that dietary supplementation with γ-linolenic acid improves some of the complications of diabetes mellitus (Jamal 1990).

Catecholamines

The effect of an acute injection of catecholamines is to decrease the sensitivity of glucose utilization to insulin. This is probably due to enhanced rates of glycogen breakdown and increased rates of mobilization of fatty acids. But what about chronic effects? Several physiological conditions in which the catecholamine level is raised are associated with an *increase* in the insulin sensitivity of glucose metabolism (e.g. cold-exposure, exercise-training). It has been shown that the soleus muscles removed from animals subjected to the above conditions and incubated *in vitro* exhibit an increased sensitivity of glucose utilization to insulin. And it has been shown that chronic treatment *in vivo* with adrenalin or β-adrenoceptor agonists increases the sensitivity of glucose utilization by muscle incubated *in vitro* to insulin (Challiss *et al.* 1985; Budohoski *et al.* 1987). Does this explain the effects of exercise and exercise-training on sensitivity?

Exercise

Endurance exercise increases the rate of utilization of all metabolic fuels. Hence it decreases the glycogen content of both the liver and muscle. The latter may result in an increase in the activity of glycogen synthase and consequently may decrease the requirement for insulin in the stimulation of glycogen synthesis after a carbohydrate meal. In addition, it has been shown that, in normal subjects, endurance-training markedly increases insulin sensitivity, so that very much lower concentrations of insulin are required to control the blood glucose concentration after an oral glucose load (Bjorntorp *et al.* 1977). And it has been shown that, in diabetic subjects who were exercised to deplete muscle glycogen stores (by 80 per cent), the rate of glycogen synthesis in the muscles for 4 h post-exercise, after a carbohydrate-rich meal, was the same whether the subjects took their normal insulin or were deprived of their insulin (Maehlum *et al.* 1978).

In rats, exercise-training increases the sensitivity of glucose utilization to insulin by muscle isolated from these animals for more than 48 h (Langfort *et al.* 1988). The mechanism underlying this effect may be very important. It would be of interest if it was related to the chronic effect of catecholamines!

5.7 How does insulin regulate the blood glucose level?: II

If, in a given condition, an increase in insulin sensitivity occurred in one or more skeletal muscles, the increased entry of glucose into the body after a meal could be dealt with in the absence of a marked (or without any) change in the blood concentration of insulin; the blood glucose level would be controlled by changes in insulin *sensitivity* at the tissue level rather than by a change in the secretion of insulin from the pancreas and hence in the plasma insulin concentration (see Fig. 11). In this case, it is the presence of insulin rather than any change in its concentration which is necessary for regulation of the blood glucose level.

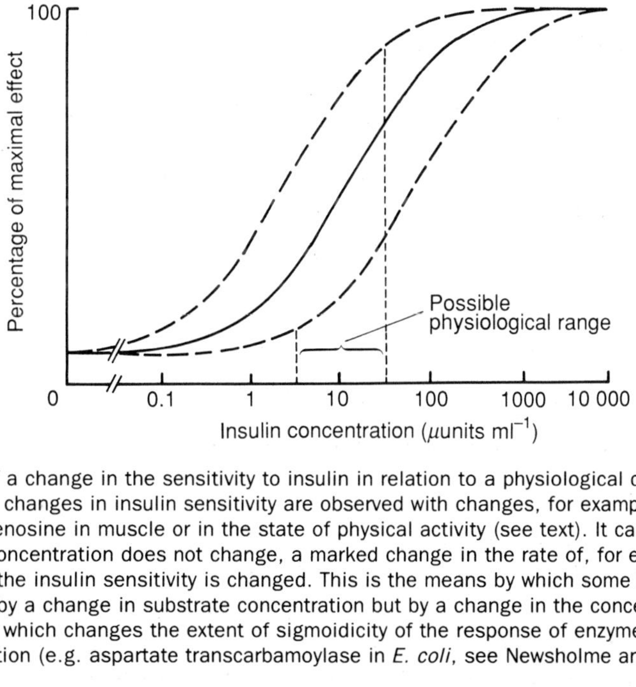

Fig. 11 The effect of a change in the sensitivity to insulin in relation to a physiological change in insulin concentration. Such changes in insulin sensitivity are observed with changes, for example, in the concentration of adenosine in muscle or in the state of physical activity (see text). It can be seen that even if the insulin concentration does not change, a marked change in the rate of, for example, glucose uptake can occur if the insulin sensitivity is changed. This is the means by which some bacterial enzymes are controlled—not by a change in substrate concentration but by a change in the concentration of allosteric regulators which changes the extent of sigmoidicity of the response of enzyme activity to substrate concentration (e.g. aspartate transcarbamoylase in *E. coli*, see Newsholme and Start 1973)

5.7.1 A common mechanism for insulin resistance and decreased thermogenesis in type II diabetes mellitus and obesity

Whatever the mechanism for the chronic effect of catecholamines in increasing insulin sensitivity, the finding leads to an hypothesis for a common mechanism to explain both insulin resistance and the decreased rate of thermogenesis (not in the basal state, but in response to thermogenic stimuli). A decreased rate of thermogenesis in response to thermogenic stimuli is common in obesity and type II diabetes.

There is considerable evidence to support the view that catecholamines can stimulate the rate of substrate cycles (see Newsholme *et al.* 1984, 1988). Increased rates of cycling not only increase sensitivity in metabolic control but also increase the rate of thermogenesis (Newsholme and Crabtree 1978). It is suggested, therefore, that the increased rate of thermogenesis under conditions when catecholamines are elevated is due, in part, to increased rates of substrate cycling. It is also known that catecholamines chronically increase the sensitivity of glucose utilization in muscle to insulin (see above). If these two quite separate effects of catecholamine are considered together, they could explain why, at least in some cases of obesity and diabetes, insulin resistance coexists with a decreased rate of thermogenesis in response to thermogenic stimuli (e.g. the rate of thermogenesis is decreased in response to glucose and/or in response to exercise (see Ravassin *et al.* 1985; Katyeff *et al.* 1986)). Hence, if the muscles of these animals/subjects were insensitive to catecholamines, the rates of cycling would be lower, resulting in a decreased rate of thermogenesis, together with a decreased insulin sensitivity of muscle, which would result in insulin resistance *in vivo*. A similar condition would result from a lower activity of the sympathetic nervous system, which has been reported in some obese subjects (James 1981), or by pharmacological blockade of β-adrenoreceptors (Leibel and Hirsch 1985).

The logic behind the control mechanism that underpins this hypothesis may have had some influence on the interest of some pharmaceutical firms in the development of novel β-adrenoreceptor agonists (β_3-receptor agonists) as both anti-obesity *and* hypoglycaemic drugs (Mitchel *et al.* 1989; Smith *et al.* 1990).

5.7.2 Insulin and protein synthesis

It has been known for many years that insulin increases the rate of protein synthesis in many tissues and that this is a rapid effect (observed within 5 minutes). But which steps are influenced by insulin? There is no doubt that insulin increases the rate of transport of some amino acids into cells and that it also increases the rate of initiation of protein synthesis (reviewed by Sugden and Fuller 1991). But does this agree with a logical approach to the problem?

The first step in the logical approach is to identify the equilibrium nature of the reactions. With the available information, this has been done (Table 3). Flux-generating steps can be identified by comparison of the substrate concentration with the K_m of the enzyme for that substrate. The available evidence suggests that some, and perhaps all, of the activating enzymes (aminoacyl tRNA synthetases) catalyse flux-generating steps (Table 4): if it is assumed that this is the case for the important initial enzyme, methionyl-tRNA synthetase, then it can be considered to catalyse a flux-generating step for the overall process. Since the formation of the ternary complex is near-equilibrium, it cannot be flux-generating, although it may well be controlled, despite the fact that the concentration of the important initial pathway-substrate, methionyl-tRNA, appears to remain constant; more than 90 per cent of methionyl-tRNA is always covalently linked to methionine, i.e. charged.

For control of protein synthesis the following points of the hypothesis are

Table 3 Equilibrium nature of some key reactions in protein synthesis in muscle

Process		ΔG kJ mol^{-1}	Suggested equilibrium nature
1	Methionine transport	−8.6	near-equilibrium[a]
2	Methionine activation	−13.2	non-equilibrium[b]
3	Ternary complex formation		near-equilibrium[c]
4	40S addition		near-equilibrium[d]
5	Addition of mRNA		non-equilibrium[e]
6	60S addition		non-equilibrium[f]
7	Methionine activation	−20	non-equilibrium[g]

[a] Amino-acid uptake: it is known that there are at least seven amino-acid uptake mechanisms, each transporting a variety of amino acids. Only two have been reported to be stimulated by insulin. The ΔG value given is for transport of methionine, assuming a 1:1 symport with a sodium ion. The nearness to equilibrium agrees with the transport of methionine not being stimulated by insulin, and allows regulation by product concentration, i.e. the concentration of intracellular methionine. This value of ΔG suggests that the transport process is not very sensitive to internal amino-acid concentrations. This is in agreement with the substantial fall in internal amino-acid concentrations reported on stimulation of protein synthesis by insulin.

[b] Amino-acid activation by the aminoacyl tRNA (aatRNA) synthetase: the ΔG value was determined by comparison of the mass-action ratio with the equilibrium constant. The non-equilibrium nature accords with hydrolysis of ATP at this step.

[c] The near-equilibrium nature is indicated by the rapid exchange of added labelled aatRNA with aatRNA bound to eIF2.

[d] Formation of 40S initiation complex: this step seems to be near-equilibrium as binding of Met-tRNA to 40S subunits is reversible until the complex binds to mRNA.

[e] Formation of 48S initiation complex: this step is non-equilibrium, as Met-tRNA bound to 40S subunits can no longer exchange with added, labelled, Met-tRNA once mRNA is bound. This agrees with the hydrolysis of ATP occurring at this step.

[f] Addition of 60S subunit: this step has not been studied in detail, but as eIF2.GDP is released at this step it may well be non-equilibrium. It is not yet known at which step the GTP is hydrolysed.

[g] Using [ATP] and [AMP] appropriate to muscle.

important (Fig. 12). The flux-generating step is formation of methionyl-tRNA by the synthetase enzyme. The activity of this enzyme must therefore be increased by insulin, but the mechanism is not known. Formation of ternary complex is near-equilibrium, but does not appear to be regulated by changes in the concentration of its pathway-substrate. Insulin probably regulates this step by increasing the availability of the co-substrate, eukaryotic initiation factor 2 (eIF2). This is achieved by decreasing the phosphorylation of the subunit of eIF2. It will be of importance to understand how the change in activity of this initiation factor, due to insulin, is co-ordinated, or communicates, with the flux-generating step, the methionyl-tRNA synthetases. Such a communication or co-ordination is suggested by the maintenance of the constant level of tRNA charging. A simple co-ordinating mechanism would be the availability of tRNA—as initiation is stimulated, more methionyl-tRNA is converted to its tRNA uncharged form, which stimulates, via co-substrate concentration, the activity of the synthetase. This would, however, require changes in the concentration of both methionyl tRNA and the uncharged tRNA.

Table 4 Concentration of some amino acids in the liver, and K_m values of aminoacyl tRNA synthetases for these amino acids

Amino acid	K_m of synthetase (μM)	Intracellular concentration (μM)
Leucine	6	2300
Glycine	4	7000
Threonine	4	1000
Glutamine	18	19 000
Serine	500	2000
Arginine	2	200

The subsequent reactions may be controlled by internal regulation: external regulation by insulin might be redundant. An interesting question remains as to the significance of the increased phosphorylation of ribosomal protein S6 elicited by insulin. Is it merely a silent phosphorylation?

The presence of the flux-generating step for protein synthesis at the aminoacyl tRNA reactions is equivalent to the status of the glycogen synthase reaction in relation to the physiological effects of insulin on glucose and amino-acid metabolism. Thus, just as this flux-generating nature allows changes in the rate of glycogen synthesis to occur independently of any effect on glucose transport, so changes in the rate of protein synthesis can occur independently of rates of amino-acid transport or metabolism: changes in the rate of utilization of amino acids (transamination, oxidation) will depend, in part, upon concentration changes in the intracellular concentrations of amino acids, that is, internal regulation. The key process of protein synthesis, just as the process of glycogen synthesis, is protected against the vicissitudes of internal control.

5.7.3 Insulin and the insulin-like growth factors in the control of growth and glucose utilization

The consequence of growth is an increase in size. To accommodate any increase in the size of the organs, the structural elements of the body—the connective tissues—must enlarge. All connective tissues consist of protein fibres (collagen or elastin) embedded in a matrix of proteoglycan. In all these tissues, both protein fibrils and proteoglycan matrix are secreted and metabolized by cells that remain in the connective tissue. These cells are known by different names according to the connective tissue that they form. Thus, fibroblasts form connective tissue, chondroblasts form cartilage, and osteoblasts form bone. A second type of cell, the osteoclast, is also found in bone. Its function is to resorb bone, that is, to remove the minerals and hydrolyse some of the matrix, so that the growth of bone depends on a balance between osteoblast and osteoclast activities.

Hormones have long been known to play a major role in the co-ordination of the

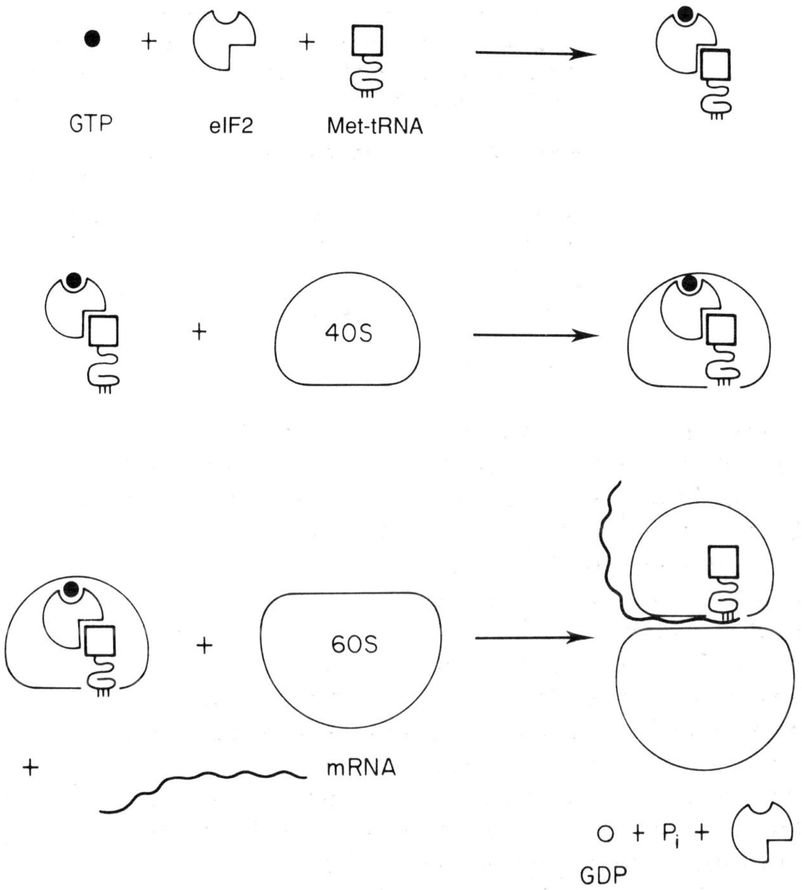

Fig. 12 Initiation of protein synthesis. Initiation of the process of translation is defined as the positioning of the first amino acid into the correct site on the ribosome. The first amino acid is always methionine, for which the codon is AUG. Although the same codon is used to place methionine in an internal position in the polypeptide, a different transfer RNA molecule is involved. The sequence of events in eukaryotes is as follows. Methionyl-tRNA interacts with GTP and an initiation factor known as eIF2 to produce a complex that associates with the smaller ribosomal subunit (the 40S subunit) to form a larger complex. At this stage, a molecule of mRNA is bound to the ribosome subunit. This latter reaction involves other initiation factors and results in the hydrolysis of a molecule of ATP. Finally, in the initiation process, a 60S subunit joins the complex to produce the 80S ribosomal complex that is involved in translation; the original GTP molecule is hydrolysed to GDP. The available evidence suggests that the first reaction in this sequence is regulatory and is controlled by the activity of the initiation factor eIF2, which is inflienced by insulin

anabolic processes involved in growth, although many of the mechanisms involved have remained obscure. It would be expected that the cells referred to above would represent target sites for the effect of growth-promoting hormones including, perhaps, insulin. Although a number of growth factors have been identified in recent years, considerable interest is currently being focused upon the growth factors known as insulin-like growth factors. Their role, as far as it is understood at the present, and in relation to that of insulin, is described below.

The biochemical regulation of growth by extracts of endocrine glands was examined in the 1930s. Since growth involves proliferation of cartilage, the effects were measured by following the incorporation of $^{35}SO_4$ into chondroitin sulphate. The incorporation into chondroitin sulphate in isolated pieces of cartilage was unaffected by pituitary extracts or by serum from hypophysectomized rats. However, incorporation was stimulated when cartilage was incubated with serum from hypophysectomized rats that had been treated with pituitary extracts. An hypothesis was proposed which suggested that stimulation of growth by growth hormone was not mediated directly but required the production of a factor that could be detected in serum, and which was initially named 'sulfation factor' (Salmon and Daughaday 1957). Further, 'sulfation factor' had wide effects on growth and anabolism in different tissues and it was considered to represent a family of serum peptides. These were first called somatomedins, which were thought to be a family of peptides including somatomedins A, B, and C, insulin-like growth factors (IGF-I and IGF-II), and multiplication stimulating activity (MSA). Uncertainty surrounded the identity of these peptides because of the difficulty of purification, since they are present at very low free concentrations in serum and are attached to much larger transport proteins from which they must be separated before purification.

Insulin-like growth factors

Recent work has now established that there are only two somatomedins, termed insulin-like growth factors I and II (IGF-I and IGF-II). These growth factors are single-chain peptides with molecular weights of 7649 and 7471, containing 70 and 67 amino-acid residues, respectively, with three interchain disulphide bridges. They have a structural homology to proinsulin which has 86 amino acids. The amino-acid sequence identity between the two growth factors is 62 per cent (for a review of the field, see Sara and Hall 1990).

IGF-I and IGF-II are carried in plasma and extracellular fluids bound to specific carrier proteins. There are several types of binding proteins: e.g. a high (150 kDa) and several low molecular weight proteins. Although it is suggested that IGFs in blood are entirely bound to these proteins, there should also be a small amount of the IGFs in the free state in the plasma, comprising, in part, peptides newly synthesized and released from their sites of production which have not yet been bound. In addition, there should be a given concentration of free peptide which is in equilibrium with the complex according to the binding constant:

$$\text{BP-IGF} \rightleftarrows \text{BP} + \text{IGF}$$

The high molecular weight protein binds both IGF-I and IGF-II with a high affinity. It is secreted by the liver and its rate of secretion is increased by growth hormone: its concentration is not detectable in the serum from hypophysectomized rats but is restored by growth hormone administration. It is considered to act as a carrier and buffer for the IGFs in the blood; it does not appear to cross the normal capillary barrier.

Low molecular weight binding proteins are synthesized and secreted by several tissues, including the liver, vascular endothelial cells, fibroblasts, pituitary cells, and endometrium. Plasma levels of the low molecular weight binding proteins may be regulated by the levels of growth hormone, insulin, IGFs themselves, and the plasma level of glucose. Thus, the binding protein level is decreased by an oral glucose load or a mixed meal and it is increased by hypoglycaemia (induced by insulin administration). The level is increased during the overnight fast. It is suggested that these acute changes in level of IGF-binding proteins are important in controlling the levels of free IGF-I and II: in this way the binding proteins may influence, quite markedly, the biological effectiveness of the IGFs.

Insulin and IGF receptors

Specific receptors for IGF-I and II are present on membranes of cells from a large number of tissues. There is substantial cross-reactivity between these receptors and also that of insulin. In particular, the receptors for insulin and IGF-I have a high degree of structural homology: they are both tetramers with α and β-subunits linked by sulphydryl bonds. The α-subunits are completely extracellular and contain the peptide-binding site. The IGF-I receptor binds IGF-II and insulin, although there is a difference in the binding affinities. Similarly, the insulin receptor binds insulin and IGF-I and II. This cross-reactivity makes it difficult to differentiate between the effects and effectiveness of these three hormones. If an effect of insulin upon a cell is observed, is this due to an interaction with the insulin receptor or an interaction with an IGF receptor? It is always important to carry out concentration dependence studies on the effects of hormones, particularly insulin, to assess whether the effect being observed occurs at an approximately physiological concentration of the hormone (i.e. what is the value of the EC_{50} and how does this compare with the plasma level of the hormone?).

Growth promoting effects of IGFs

The IGFs have a number of effects that support the view that they play a major role in promoting growth. They increase the rate of cartilage formation by stimulating the following processes in chondrocytes: transport of amino acids across the cell membrane, the rate of protein synthesis, the rates of synthesis of DNA and RNA, incorporation of sulphate into proteoglycans and of proline into collagen; they also decrease the rate of protein degradation. Cartilage from young animals is considerably more sensitive to these effects than is cartilage from older animals.

It is possible that some of these effects stem from a more fundamental effect on cell division, since the rate of mitosis of both fibroblasts in culture and chondrocytes in cartilage is stimulated by the IGFs. The effect of IGFs may account, in part, for the requirement for serum for proliferation of fibroblasts in culture. However, the reactions that these growth factors affect and the metabolic control logic of the effects have not been analysed. A clear unambiguous understanding of their role in growth promotion awaits such an analysis.

An important question, which is still not resolved, is whether the stimulation of growth by growth hormone is caused solely by its ability to enhance the synthesis and secretion of the IGFs from liver and also locally in the tissues. If the major growth-producing role of growth hormone is its ability to simulate the release of the IGFs, growth hormone should be considered as a trophic hormone. Consistent with this suggestion, is the observation that IGFs can inhibit growth hormone secretion. Such a feedback inhibitory mechanism is typical of the trophic hormones.

However, infusion of growth hormone into normal or hypophysectomized rats for 6 days resulted in a greater rate of growth than infusion of IGF-I. It may be that growth hormone has a more co-ordinating effect on growth than can be achieved simply by changing the concentration of IGF-I.

There is some evidence to suggest that insulin may be as important as growth hormone in promoting growth. This could be achieved either by the hormone stimulating the synthesis or release of IGFs from the tissues, mimicking the effects of the IGFs, or changing the concentration of the low molecular weight binding proteins to make the IGFs more effective on tissues.

Effects of IGFs on carbohydrate metabolism

Intravenous bolus injections of IGF-I or IGF-II in normal or hypophysectomized rats have acute effects on glucose utilization: they lower the blood glucose concentration and increase the rate of glycogen synthesis in muscle. Indeed, IGF-I was more effective than insulin in increasing the rate of glycogen synthesis. There was no evidence of an effect of the IGFs on glucose release by the liver. The IGFs had no effect on the plasma FFA levels and did not alter the rate of lipid synthesis from glucose. It appears that the IGFs cause hypoglycaemia by increasing the rate of glucose utilization by peripheral tissues, probably skeletal muscle.

Although skeletal muscle is considered to be the most important tissue for disposal of glucose in response to insulin after a meal (De Fronzo and Ferrannini 1987), there has been no information concerning the effects of IGF-I or IGF-II and their interactions with insulin on glucose uptake and utilization in this tissue. For this reason, studies have recently been undertaken to examine the effects of IGF-I on the rates of glucose transport, glucose phosphorylation, glycogen synthesis, and glycolysis at various concentrations of insulin in the isolated rat soleus muscle preparation (Dimitriadis *et al.* 1991). Effects of IGF-II on the rates of glucose utilization, glycogen synthesis, and lactate formation have also been reported (Bevan *et al.* 1991).

In isolated incubated rat soleus muscle, IGF-I increased the rates of glucose transport and glucose conversion to lactate (glycolysis) at all insulin concentrations except the maximal. The maximal increase in the rate of lactate formation caused by IGF-I was apparent even at the lowest concentration of insulin, which normally has no effect on rates of glucose transport, glycolysis, or glycogen synthesis (Fig. 13).

This indicates that IGF-I can stimulate the rate of glycolysis independently of insulin. And perhaps of importance, it shows that, at least *in vitro*, a physiological level of IGF-I can produce an increase in the rate of glycolysis that is only seen with maximal and unphysiological levels of insulin. These results suggest that IGF-I increases the rate of glycolysis by a mechanism identical to that of insulin, that is by stimulating the rate of glucose transport, thus leading to increased rates of glucose phosphorylation and glycolysis. The fact that the increase in the rate of glucose transport by IGF-I was similar to that produced by maximal concentrations of insulin suggests that IGF-I induced a redistribution of all available glucose transporters from the intracellular pool into the plasma membrane. In this case, insulin would not be expected to have any additive effects with IGF-I on the rate of glucose transport. Indeed, maximal rates of glucose transport and glucose phosphorylation were the same either in the presence of insulin alone or insulin plus IGF-I (Fig. 13).

Not only did IGF-I stimulate the rate of glucose transport and glycolysis, it also increased the rate of glycogen synthesis to a value similar to that observed in the presence of maximal concentrations of insulin (Fig. 13). However, the stimulation of the rate of glucose transport cannot explain the effect of IGF-I on glycogen synthesis, since glycogen synthase is considered to catalyse a flux-generating step.

Despite the similarities of IGF-I and insulin on glucose and glycogen metabolism, the effects of IGF-I are considered to occur via the IGF-I receptor and not the insulin receptor: the level of IGF-I present in the incubation medium was physiological and it is considered that such low levels of IGF-I would not interact significantly with the insulin receptor. However, because of the similarity of effects, it is tempt-

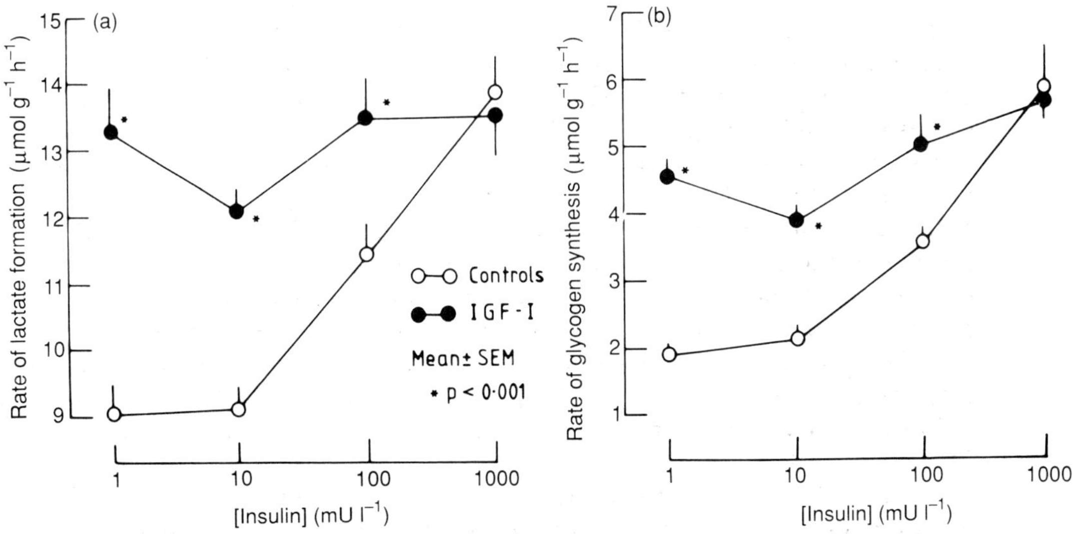

Fig. 13 Effect of IGF-I on the rates of (a) lactate formation and (b) glycogen synthesis in the isolated, incubated, stripped soleus muscle of the rat at various concentrations of insulin. Each point is mean of at least 10 separate incubations; ○—○, control; ●—●, IGF-I * indicates that the difference from control is statistically significant (see Dimitriadis *et al.* 1991 for full details)

ing to speculate that IGF-I and insulin may increase the rate of glucose transport and the activity of glycogen synthase via the same second messenger.

Somewhat in contrast to these effects of IGF-I, the other insulin-like growth factor, IGF-II, had no effect in the absence of insulin. But at low physiological levels of insulin it increased the rates of both glucose uptake and glycogen synthesis by the isolated soleus muscle. This evidence supports the view that IGF-II increases the sensitivity of glucose transport and glycogen synthase to insulin (Bevan *et al.* 1991). Since these effects were observed at physiological levels of the peptide, it is assumed to be acting via the IGF-II receptor on skeletal muscle.

5.8 How does insulin regulate the blood glucose level?: III

It has been shown that IGF-I does not increase insulin sensitivity but is able to mimic the effect of insulin (see above). Therefore, if insulin could influence the local level of IGF-I in tissues or influence the free level via a change in the concentration of the binding protein (low molecular weight), the resultant change in the IGF-I concentration might influence glucose utilization by muscle to a greater extent than could be produced by normal physiological changes in the insulin (see Fig. 13). Could this be the means by which insulin normally regulates the blood glucose level?

Finally, it has recently been shown that IGF-II, at physiological concentration, increases the sensitivity of glucose utilization and glycogen synthesis to insulin in the isolated soleus muscle (see above). This result suggests that the list of factors presented above, under insulin sensitivity, should be increased by inclusion of IGF-II. It raises the intriguing question as to whether changes in sensitivity caused by exercise, chronic treatment with catecholamines or even adenosine, could be caused by local changes in the concentration of IGF-II? Such a change could influence the rate of glucose uptake by muscle without any need for a change in the plasma level of insulin.

It is interesting to note that with the development of knowledge of biochemical physiology, rather than molecular biology, our appreciation of how the blood glucose could be controlled has dramatically expanded. Is control due to insulin, to fatty acids, to local hormones, to changes in sensitivity at specific sites in the body, or to insulin-like growth factors? With this complexity, the definition of the basic cause of diabetes mellitus provided by Renold *et al.* (1978) as 'inappropriate hyperglycaemia' showed considerable foresight by these earlier workers!

References

Bevan, S. J., Parry-Billings, M., Opara, E., Dungar, D., and Newsholme, E. A. (1992). The effect of insulin-like growth factor 2 (IGFII) on glucose utilisation in soleus muscle of the rat *in vitro*, *Diabetes* **40** (Suppl 1) 172A.

Birkinshaw, V. P., Gurd, M. R., Randall, S. S., Curry, A. S., Price, D. E., and Wright, P. H. (1958). *Br. Med. J.* **2,** 463–8.

Bjorntorp, P., Fahten, M., and Grimb, G. (1977). Carbohydrate and lipid metabolism in middle-aged physically well trained men. *Metabolism* **21,** 1037–44.

Budohoski, L., Challiss, R. A. J., Dubaniewicz, A., Kaciuba-Uscilko, H., Leighton, B., Lozeman, F. J., Nazar, K., Newsholme, E. A., and Porta, S. (1987). Effect of a prolonged elevation of plasma adrenaline concentration *in vivo* on insulin sensitivity in soleus muscle of the rat. *Biochem. J.* **244,** 655–60.

Cahill, G. F. (1970). Starvation in man. *New Engl. J. Med.* **282,** 668–75.

Challiss, R. A. J., Budohoski, L., Newsholme, E. A., Sennitt, M. V., and Cawthorne, M. A. (1985). Effect of a novel thermogenic β-adrenoceptor agonist (BRL 26830) on insulin resistance in soleus muscle from obese Zucker rats. *Biochem. Biophys. Res. Comm.* **128,** 928–35.

Challiss, R. A. J., Leighton, B., Lozeman, F. J., and Newsholme, E. A. (1987). The hormone-modulatory effects of adenosine in skeletal muscle. In *Topics and perspectives in adenosine research* (ed. E. Gerlock and B. F. Becker), pp. 275–85. Springer-Verlag, Berlin.

De Fronzo, R. A. and Ferrannini, E. (1987). Regulation of hepatic glucose metabolism in hormones. *Diabetes/Metabolism Rev.* **3,** 415–59.

Dimitriadis, G., Parry-Billings, M., Piva, T., Dunger, D., Wegener, G., Krause, U., and Newsholme, E. (1990). Effects of IGF-I on the rates of glucose transport and utilization in rat skeletal muscle in vitro. *Diabetologia* **33,** A71.

Hultman, E. (1978). *Regulation of carbohydrate metabolism in the liver during rest and exercise with special reference to diet,* 3rd International Symposium on Biochemistry of Exercise (ed. P. Landry and W. A. R. Orban), pp. 99–126. Symposia Specialist Inc, Miami, Florida.

Jamal, G. A. (1990). Prevention and treatment of diabetic distal polyneuropathy by the supplementation of gamma-linolenic acid. In *Omega-6-essential fatty acids* (ed. D. F. Horrobin), pp. 485–504. Wiley-Liss, New York.

James, W. P. T. (1981). Adrenergic mechanisms in obesity. *Clin. Physiol. (Suppl. 1)* **1,** 60–76.

Katzeff, H. L., O'Connell, M., and Horton, E. S. (1986). Metabolic studies in human obesity during undernutrition and overnutrition: thermogenic and hormonal responses to norepinephrine. *Metabolism* **35,** 166.

Klip, A., Ramlal, T., Young, D. A., and Holloszy, J. O. (1987). Insulin-induced translocation of glucose transporters in rat hindlimb muscles. *FEBS Lett.* **224,** 224–30.

Langfort, J., Budohoski, L., and Newsholme, E. A. (1988). Effect of various types of acute exercise and exercise training on the insulin sensitivity of rat soleus muscle measured *in vitro. Pflügers Arch.* **412,** 101–5.

Leibel, R. L. and Hirsch, J. (1985). A radioisotopic technique for analysis of free fatty acid re-esterification in human adipose tissue. *Am. J. Physiol.* **248,** E140–E146.

Leighton, B., Budohoski, L., Lozeman, F. J., Challiss, R. A. J., and Newsholme, E. A. (1985). The effect of prostaglandins E1, E2 and F2 and indomethacin on the sensitivity of glycolysis and glycogen synthesis to insulin in stripped soleus muscles of the rat. *Biochem. J.* **227,** 337–40.

Leighton, B., Dimitriadis, G. D., Parry-Billings, M., Bond, J., de Vasconcelos, P. R. L., and Newsholme, E. A. (1989). Effects of insulin on glucose metabolism in skeletal muscle from peptic and endotoxaemic rats. *Clin. Science* **77,** 61–7.

Levine, R. and Goldstein, M. S. (1955). On the mechanism of action of insulin. *Rec. Progr. Hormone Res.* **11,** 343–80.

Maehlum, S. and Hermansen, L. (1978). Muscle glycogen concentration during recovery after prolonged severe exercise in fasting subjects. *Scand. J. Clin. Lab. Invest.* **38,** 557–60.

Maehlum, S., Hostmark, A. T., and Hermansen, L. (1978). Synthesis of muscle glycogen during recovery after prolonged severe exercise in diabetic subjects. Effect of insulin deprivation. *Scand. J. Clin. Lab. Invest.* **38,** 35–9.

Mitchell, T. H., Ellis, R. D. M., Smith, S. A., Robb, G., and Cawthorne, M. A. (1989). Effects of BRL 35135, a β-adrenoceptor agonist with novel selectivity, on glucose tolerance and insulin sensitivity in obese subjects. *Int. J. Obesity* **13,** 757–66.

Newsholme, E. A. and Crabtree, B. (1976). Substrate cycles in metabolic regulation and heat generation. *Biochem. Soc. Symp.* **41,** 61–110.

Newsholme, E. A. and Leech, A. R. (1983). *Biochemistry for the medical sciences.* John Wiley and Sons, Chichester, England.

Newsholme, E. A. and Start, C. (1973). *Regulation in metabolism.* John Wiley and Sons Ltd, Chichester, England.

Newsholme, E. A., Challiss, R. A. J., and Crabtree, B. (1984). Substrates cycles; their role in improving sensitivity in metabolic control. *TIBS* **9,** 27–32.

Newsholme, E. A., Crabtree, B., and Ardawi, M. S. M. (1985). Glutamine metabolism in lymphocytes: its biochemical, physiological and clinical importance. *Q. J. Exp. Physiol.* **70,** 473–95.

Newsholme, E. A., Challiss, R. A., Leighton, B., Lozeman, F. J., and Budohoski, L. A. (1988). A possible common mechanism for defective thermogenesis and insulin resistance in obesity. *Nutrition* **3,** 195–200.

Randle, P. J., Garland, P. B., Hales, C. N., and Newsholme, E. A. (1963). The glucose fatty acid cycle. Its role in insulin sensitivity and the metabolic disturbances of diabetes mellitus. *Lancet* **i,** 785–9.

Ravussin, E., Cheson, K. J., and Vernet, O. (1985). Evidence that insulin resistance is responsible for the decreased thermic effect of glucose in human obesity. *J. Clin. Invest.* **76,** 1268.

Renold, A. E., Muller, W. A., and Mintz, P. H. (1978). Diabetes mellitus. In *The metabolic basis of inherited disease* (ed. J. B. Stanbury, J. B. Wyngaarden, and D. S. Fredrickson), pp. 80–109. McGraw-Hill, New York.

Salmon, W. D. and Daughaday, W. H. (1957). A hormonally controlled serum factor which stimulates sulphate incorporation by cartilage *in vitro. J. Lab. Clin. Med.* **49,** 825–36.

Sara, V. R. and Hall, K. (1990). Insulin-like growth factors and their binding proteins. *Physiol. Rev.* **70,** 591–614.

Smith, S. A., Sennitt, M. V., and Cawthorne, M. A. (1990). BRL 35135: an orally active antihyperglycaemic agent with weight reducing effects. In *New antidiabetic drugs* (ed. C. J. Bailey and P. R. Flat), pp. 177–89. Smith-Gordon.

Sugden, P. H. and Fuller, S. J. (1991). Regulation of protein turnover in skeletal and cardiac muscle. *Biochem. J.* **273,** 21–37.

6 The insulin receptor

KENNETH SIDDLE

It has been widely accepted for many years that the metabolic effects of insulin are initiated by the binding of insulin to cell-surface receptors. It has become clear that there is only one type of insulin receptor, encoded by a single gene. The structure and activity of this receptor is understood in considerable detail, although many important aspects remain to be eludicated, from the mechanism of receptor activation at the molecular level to its involvement in pathophysiological processes *in vivo*.

In common with all surface receptors, the insulin receptor must display two fundamental properties: first, the reversible binding of ligand with high affinity and specificity at the extracellular face of the plasma membrane; secondly, the generation of specific intracellular regulatory signals as a consequence of ligand binding. Both these aspects of receptor function have important clinical implications, apart from their intrinsic scientific interest in terms of structure–function relationships. Detailed knowledge of the binding interaction may facilitate the design of novel insulin analogues or even non-peptide alternatives for therapeutic use in states of insulin deficiency. A full understanding of signalling mechanisms and their regulation may permit new approaches to the treatment of insulin resistance.

Studies of insulin receptor structure and signalling have been the subject of numerous reviews in recent years (Czech 1985; Goldfine 1987; Rosen 1987; Kahn and White 1988; Houslay and Siddle 1989; Zick 1989; Becker and Roth 1990; Olefsky 1990). This chapter will attempt to trace the development of ideas about receptor function, and to highlight areas of current interest.

6.1 Structural overview

The initial phase of receptor characterization in the 1970s was dominated by studies of insulin binding, using [^{125}I]insulin as a tracer. These studies indicated that receptors for insulin were widely distributed in mammalian tissues, varying in number from approximately 10^5 on hepatocytes and adipocytes to 10^3 on fibroblasts. Insulin binding to membranes or detergent-solubilized receptors was readily reversible, with a dissociation constant of the order of 1 nM. However, the binding kinetics were complex and suggested site heterogeneity and/or negative co-operativity (Gammeltoft 1984). It became apparent that in intact cells at physiological temperatures the insulin–receptor complex was rapidly internalized, and much of the insulin was degraded intracellularly (Sonne 1988).

6.1.1 Biochemical characterization

Structural characterization of the receptor developed rapidly in the late 1970s and early 1980s (Czech 1985). Several techniques combined to permit these advances: covalent affinity labelling with radioactive ligands; metabolic labelling with [^{35}S]methionine, [^{3}H]sugars, or [^{32}P]phosphate; use of specific antibodies for immunoprecipitation; purification of receptors by affinity chromatography. The receptor contains two distinct subunits, termed α and β, which are derived from a common polypeptide precursor by proteolysis, and are therefore the products of a single gene. Both subunits are glycosylated, with apparent M_r on SDS-polyacrylamide gels of approximately 135 000 and 95 000, respectively. In the brain, the size of the subunits is a little smaller than in other tissues, apparently reflecting a tissue-specific pattern of glycosylation.

The α-subunit is wholly extracellular and contains the insulin-binding site (Fig. 1). The β-subunit is a transmembrane protein, the intracellular portion being a tyrosine-specific protein kinase which is activated by insulin. This kinase catalyses both an autophosphorylation reaction and the phosphorylation of intracellular substrates. The native receptor is a disulphide-linked β–α–α–β heterotetramer, and this structure is required for insulin-stimulated tyrosine kinase activation (Boni-Schnetzler *et al.* 1988). Although other structures have been reported, including αβ heterodimers and $\alpha_2\beta$ heterotrimers, these may arise in part as an artefact of experimental procedures (Boyle *et al.* 1985) and they do not appear to be functionally significant (Le Marchand-Brustel *et al.* 1989; Treadway *et al.* 1990). There is no evidence that other proteins are stoichiometrically or stably associated with the

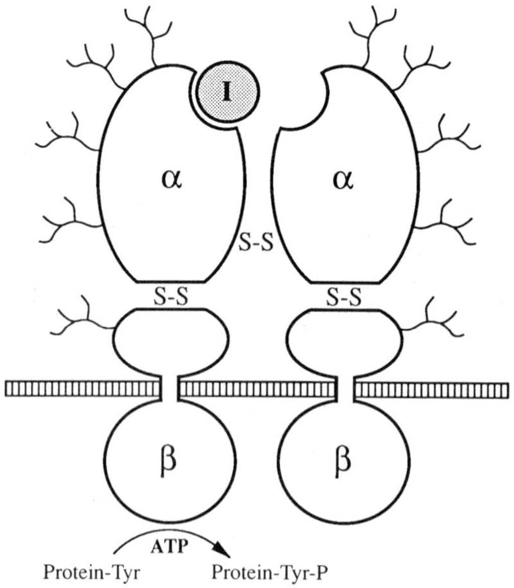

Fig. 1 Basic structure of the insulin receptor. The receptor is shown as a disulphide-linked, heterotetrameric, transmembrane glycoprotein. Insulin (I) binds to the α-subunit. The intracellular portion of the β-subunit is a tyrosine-specific protein kinase

receptor as a functional complex, although transient associations may occur as part of the signalling mechanism. Thus, under some conditions serine-specific protein kinase activity (Smith *et al.* 1988; Lewis *et al.* 1990*b*) and phosphatidylinositol kinase activity (Ruderman *et al.* 1990) are present in purified receptor preparations or specific immunoprecipitates. In some tissues a significant fraction of insulin receptors appears to be associated with MHC class I antigens (Fehlmann *et al.* 1985; Due *et al.* 1986) but the functional significance of this interaction is unclear.

6.1.2 Molecular biology

An achievement of major significance was the cloning of cDNA encoding the human insulin receptor precursor (Ebina *et al.* 1985; Ullrich *et al.* 1985). This allowed the complete amino-acid sequence to be determined, providing structural detail and a framework for the analysis of functional domains. The availability of cDNA clones also opened up new experimental approaches in the study of structure–function relationships, including expression of laboratory-engineered mutant and chimeric receptors by transfection of cultured cells. Subsequently, the mouse and rat receptors were also cloned and shown to be very similar in sequence to the human receptor (Flores-Riveros *et al.* 1989; Goldstein and Dudley 1990). In addition, the exon–intron organization and promoter region of the insulin receptor gene have been characterized (Seino *et al.* 1989, 1990; Sibley *et al.* 1989).

The receptor cDNA encodes an open reading frame of 1382 (1370) amino acids (Fig. 2). The published sequences differ by the presence or absence of a block of 12 amino acids near the C-terminus of the α-subunit. (In this article amino acids will be designated by the numbering system of Ebina *et al.* 1985.) The polymorphism is the result of alternative splicing at the level of mRNA processing, involving a

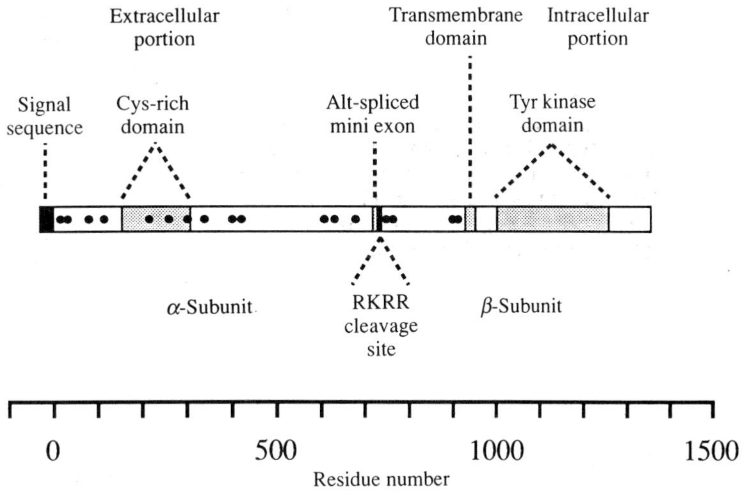

Fig. 2 The insulin receptor primary sequence. The principal features of the amino-acid sequence are indicated, as deduced from the cDNA sequence (Ebina *et al.* 1985; Ullrich *et al.* 1985). Potential N-linked glycosylation sites are also shown (●). The cleavage site, RKRR, signifies Arg–Lys–Arg–Arg

discrete mini-exon of 36 bp (Moller *et al.* 1989*a*; Seino and Bell 1989). In other respects the published sequences are almost identical. There is a typical N-terminal signal sequence of 27 amino acids, which is not present in the mature protein. A tetrabasic sequence Arg–Lys–Arg–Arg marks the point of cleavage to generate an α-subunit of 732 (720) amino acids and β-subunit of 620 amino acids (with predicted peptide M_r of approximately 84 000 and 70 000, respectively). The α-subunit has 13 and the β-subunit four potential N-linked (asparagine) glycosylation sites, most of which appear to be utilized. Hydropathy analysis predicts a single membrane-spanning sequence (an α-helix of 23 residues) in the β-subunit. The intracellular portion of the β-subunit shows clear sequence similarity to several other receptors and products of oncogenes with tyrosine-specific protein kinase activity (Yarden and Ullrich 1988).

Almost nothing is known about the three-dimensional structure of the receptor. Visualization of solubilized receptor by electron microscopy with negative staining shows a T-shaped structure (Christiansen *et al.* 1991). Detailed information on conformation will require X-ray crystallographic or NMR analysis, and a protein of the size and complexity of the insulin receptor presents a considerable challenge for the application of these techniques. The most fruitful approach in the first instance may be the separate analysis of the extracellular and intracellular portions of the receptor. It is possible to generate these, or smaller subdomains, by expression of appropriately truncated DNA constructs in mammalian or insect cells (Ellis *et al.* 1988; Whittaker and Okamoto 1988; Sissom and Ellis 1989; Villalba *et al.* 1989; Kallen *et al.* 1990; Paul *et al.* 1990; Schaefer *et al.* 1990). For the present, only tentative theoretical models based on sequence comparisons and secondary structure prediction are available (Bajaj *et al.* 1987).

6.1.3 Related receptors and receptor heterogeneity

The family of receptors with tyrosine kinase activity can be divided into three subgroups (Fig. 3), typified by the receptors for insulin, epidermal growth factor (EGF), and platelet derived growth factor (PDGF) (Yarden and Ullrich 1988). The EGF and PDGF receptors each consist of only a single polypeptide, although these almost certainly form non-covalent dimers during ligand activation. The PDGF receptor is further distinguished by the presence of an inserted sequence, relative to other receptor subgroups, within the tyrosine kinase domain. The insulin receptor subgroup consists of three members which are very closely related in structure, having 50–60 per cent sequence identity generally, but 80 per cent within the kinase domain (Fig. 3). These are the insulin receptor itself (IR), the type I IGF receptor (IGFR) (Ullrich *et al.* 1986), and a so-called insulin receptor-related receptor (IRR) (Shier and Watt 1989). The type I IGF receptor binds both IGF-I and IGF-II with high affinity, and is probably responsible for most, if not all, of the biological effects of these ligands (Czech 1989; Humbel 1990). The IRR has so far been identified only by gene cloning, and its function is obscure. The gene is apparently

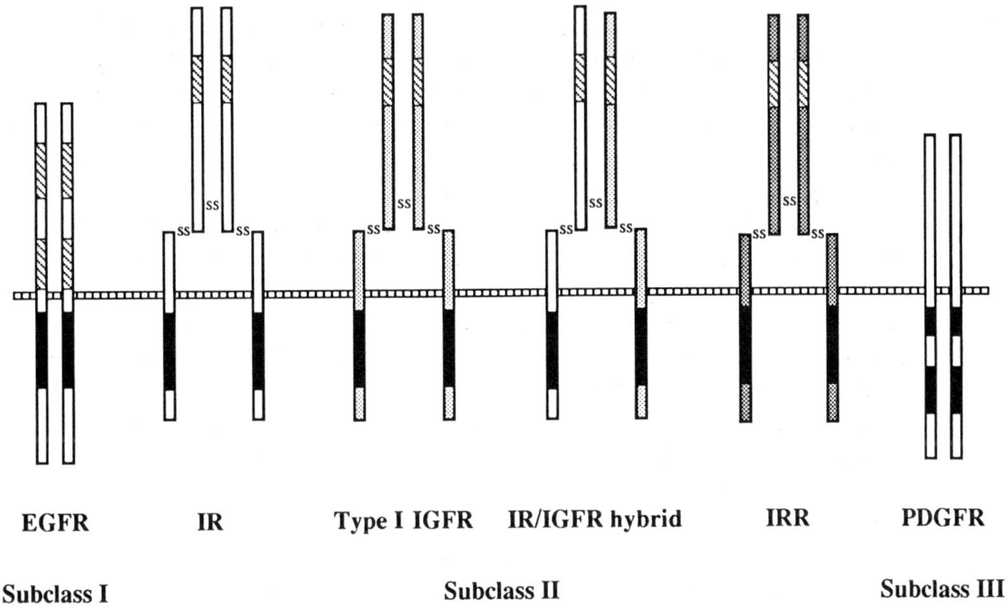

Fig. 3 The family of receptor tyrosine kinases. Three subtypes of receptors with tyrosine kinase activity are recognized (Yarden and Ullrich 1988), typified by receptors for EGF (subclass I), insulin (subclass II), and PDGF (subclass III). Other members of subclass II are the IGF-I receptor, the insulin receptor-related receptor, and hybrid receptors. The EGF and PDGF receptors are shown as non-covalently associated dimers, which probably represent the active forms, while the insulin receptor subunits are disulphide-linked. Tyrosine kinase domains are indicated by solid areas, and cysteine-rich domains by hatched areas

expressed in stomach and kidney but the ligand for this receptor is unknown and could even be a previously unrecognized peptide.

A novel form of receptor, which has been described recently, is a hybrid containing one half of insulin receptor and the other of IGF receptor, in a β–α–α^*–β^* structure (Soos and Siddle 1989; Soos *et al.* 1990). These hybrids bind IGF with high affinity, and are functional in terms of IGF-stimulated autophosphorylation (Moxham *et al.* 1989). However, the affinity for insulin appears to be relatively low, so that the properties of hybrids are not simply the sum of their component halves. The reassembly of isolated IR and IGFR $\alpha\beta$ heterodimers into heterotetramers *in vitro* appears to occur randomly (Treadway *et al.* 1989), implying a high degree of structural conservation at the interface between β-subunits. The physiological significance of hybrid receptors remains to be investigated, but it is an attractive possibility that the regulated assembly of hybrids *in vivo* might provide a mechanism to modulate ligand binding or signalling potential.

As indicated previously, heterogeneity of insulin receptors can also result from at least two other mechanisms, namely tissue-specific glycosylation and alternative splicing. No functional consequences of differential glycosylation are known. In the case of the splice variants, an affinity difference between the two forms has been demonstrated (Mosthaf *et al.* 1990). When receptors are expressed in cultured fibroblasts by cDNA transfection, the A-isoform (which lacks exon 11 and is the predominant form in muscle, fat, and brain) has approximately twofold higher affinity than the B-isoform (which includes exon 11 and is the predominant form in liver). Thus the affinity difference may provide a mechanism to compensate for the higher concentration of insulin in portal blood than in the peripheral circulation. However, the predicted difference in affinity has yet to be confirmed with receptors isolated from liver and peripheral tissues (see, for example, Gammeltoft 1984).

6.2 The extracellular portion: insulin binding

The extracellular portion of the insulin receptor consists of the whole 732 residues of the α-subunit and 194 residues of the β-subunit (Ebina *et al.* 1984; Ullrich *et al.* 1985). A conspicuous feature in the sequence of the α-subunit is a predominantly hydrophilic and cysteine-rich region (residues 155–312) containing 26 cysteine residues which presumably exist as disulphide bridges. The EGF receptor contains two analogous cysteine-rich regions (Yarden and Ullrich 1988). Either side of the insulin-receptor cysteine-rich region are sequences of approximately 120 amino acids (designated the L1 and L2 domans by Bajaj *et al.* 1987) with some similarity to each other and to sequences in the EGF receptor. The C-terminal portion of the α-subunit (440–732) and extracellular portion of the β-subunit (736–929) have no obvious structural features but contain seven and four cysteines, respectively, some of which may be responsible for the α–α (class I) and α–β (class II) disulphide bridges. However, at present the precise location of inter-subunit disulphides is unclear, and the cysteine-rich domain may also be involved in α–α cross-linking (Finn *et al.* 1990; Xu *et al.* 1990).

6.2.1 Insulin-binding site

The receptor-binding region of the insulin molecule has been defined by comparing the binding properties of a wide range of sequence variants from different species or chemically synthesized insulin analogues. It appears that a relatively large area containing both A- and B-chain residues is involved in receptor binding (Gammeltoft 1984). Analogues with binding affinities considerably greater than those of naturally occurring insulins have been synthesized, as well as many with decreased affinity. In almost all cases, the biological potency closely parallels binding affinity so that it has not been possible to identify separate regions of the insulin molecule involved in receptor recognition and activation. However, a covalently dimerized insulin derivative has been synthesized which has high binding affinity but very low biological potency (Weiland *et al.* 1990). This derivative therefore acts as a competitive antagonist, and indicates that binding of insulin alone is not sufficient to activate the receptor. Presumably, further interactions or conformational changes within the insulin molecule are required for receptor activation to occur.

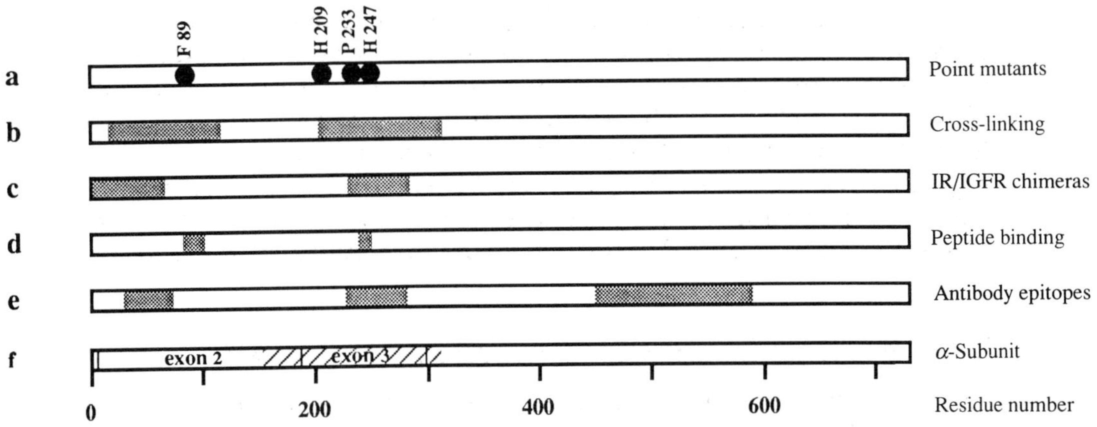

Fig. 4 The insulin-binding site. Regions of the α-subunit which may contribute to the insulin-binding site are indicated, as deduced from various experimental approaches: (a) point mutations that affect insulin binding (Rafaeloff *et al.* 1989; De Meyts *et al.* 1990; Taylor *et al.* 1990); (b) proteolytic fragments labelled after affinity cross-linking (Yip *et al.* 1988; Wedekind *et al.* 1989); (c) regions conferring binding specificity in chimeric receptors (Gustafson and Rutter 1990; Kjeldsen *et al.* 1991); (d) synthetic peptides that bind to insulin (Yip *et al.* 1988; De Meyts *et al.* 1990); (e) epitopes for antibodies that inhibit ligand binding (Toyoshige *et al.* 1989; Gustafson and Rutter 1990; Prigent *et al.* 1990); (f) relative contributions of exons 2 and 3, and the cysteine-rich domain (cross-hatched) to the α-subunit

Identification of residues on the receptor which interact with insulin is a much more difficult proposition (Fig. 4). Experiments involving affinity cross-linking initially showed that the α-subunit contributes most, if not all, of the insulin-binding site. More detailed studies of this type have now implicated residues in the N-terminal (20–120) and cysteine-rich (205–316) regions in the interaction with insulin (Yip *et al.* 1988; Waugh *et al.* 1989; Wedekind *et al.* 1989). The effects of site-directed mutagenesis, and binding studies with synthetic peptides, have also been interpreted as evidence for a role of specific residues in both these regions (Rafaeloff *et al.* 1989; De Meyts *et al.* 1990). Analyses of chimeric receptors, engineered by replacing segments of insulin receptor with corresponding sequences of IGF receptor or vice versa, have confirmed a major role for the cysteine-rich domain in determining binding specificity (Andersen *et al.* 1990; Gustafson and Rutter 1990; Kjeldsen *et al.* 1991). Some of these experiments have suggested the involvement of the N-terminal sequence (1–68) in the binding of insulin (Kjeldsen *et al.* 1991). Other parts of the receptor may also contribute to the binding site or be in close proximity to it. It has been shown, for instance, that insulin binding is inhibited by several antibodies which recognize epitopes within the region 450–590 of the α-subunit (Gustafson and Rutter 1990; Prigent *et al.* 1990). It is, of course, probable that the complete binding site will involve groups of amino acids which are quite distant in the primary sequence but which are brought together in the native conformation. Portions of the receptor that do not contribute directly to the binding site may be crucial for overall conformation and hence profoundly influence binding. This is certainly a complicating factor in interpreting binding studies with mutant or chimeric receptors. For instance, the $(\alpha\beta')_2$ ectodomain produced as a soluble protein has a significantly lower affinity for insulin than the intact solubilized receptor, while the free α-subunit binds insulin poorly, and truncated α-subunits not at all (Schaefer *et al.* 1990). The more subtle influence of the alternatively spliced sequence at the C-terminus of the α-subunit was discussed previously.

6.2.2 Conformational changes

A major challenge remaining after identification of the binding site itself will be to understand the conformational changes induced by ligand binding, which not only activate the receptor kinase of the β-subunit but also co-operatively affect insulin binding to the adjacent α-subunit. Little is known as yet about regions of the receptor that are critically involved in these conformational transitions. Effects of insulin on receptor susceptibility to chemical modification (Wilden and Pessin 1987; Waugh and Pilch 1989), sensitivity to proteolysis (Donner and Yonkers 1983; Lipson *et al.* 1986), and sedimentation properties (Florke *et al.* 1990) suggest that the conformational change may be quite substantial. Further insights can be expected from the engineering of mutant receptors that display normal binding but impaired activation or co-operativity.

Distinct conformation states of the receptor are also indicated by binding studies (Gammeltoft 1984; Gu *et al.* 1988). Although the native molecule contains two apparently equivalent α-subunits, only a single molecule of insulin binds with high affinity (Pang and Shafer 1984; Sweet *et al.* 1987). Binding of insulin to one subunit dramatically decreases the affinity of the unoccupied site on the other subunit, a process of negative co-operativity. This is reflected in curvilinear Scatchard plots of equilibrium binding, and in the accelerated dissociation of receptor-bound $[^{125}I]$insulin by addition of excess unlabelled insulin (Gammeltoft 1984). Isolated αβ heterodimers, obtained by mild reduction of native receptors, display linear Scatchard plots with a binding affinity intermediate between the high and low affinity states of $(\alpha\beta)_2$ heterotetramers (Boni-Schnetzler *et al.* 1987; Sweet *et al.* 1987). Evidence for a role of residue lysine 460 in co-operative interactions between α-subunits has been provided by the analysis of mutant receptors (Kadowaki *et al.* 1990*c*).

The conformational changes responsible for negative co-operativity and receptor activation apparently can occur independently of each other. Insulin analogues have been produced that bind and activate the receptor with high potency but do not induce negative co-operativity (Gammeltoft 1984). Monoclonal antibodies are also able to activate the receptor without inducing negative co-operativity, or vice versa, depending on the epitope recognized (Siddle *et al.* 1987; Prigent *et al.* 1990). Binding of insulin, as the natural ligand, must therefore trigger a complex cascade of conformational changes which are propagated both across the plasma membrane and to the 'trans' α-subunit by divergent mechanisms.

6.3 The intracellular portion: tyrosine kinase

The cytoplasmic portion of the insulin receptor has been a focus of great interest, as the presumed effector domain from which the signals regulating intracellular metabolic activity must arise. The initial indication that the insulin receptor possessed tyrosine kinase activity was the observation that insulin induced rapid tyrosine phosphorylation of the receptor itself (Kasuga *et al.* 1982). Several lines of evidence subsequently established that this activity was intrinsic to the receptor, including affinity labelling of receptor with ATP analogues (Shia and Pilch 1983) and co-purification of insulin binding and tyrosine kinase activities (Petruzzelli *et al.* 1984). Finally and conclusively, the deduced amino acid sequence contains a region (1002–1257) with obvious similarity to other tyrosine kinases (Ebina *et al.* 1985; Ullrich *et al.* 1985). There is a consensus ATP-binding site, Gly–X-Gly–X–X–Gly, located approximately 50 residues from the membrane-spanning region, and, a further 25 residues downstream, a conserved lysine (1030) which is essential for kinase activity (Fig. 5). Within the cytoplasmic portion of 402 amino acids, the kinase domain proper is flanked by the so-called juxtamembrane and C-terminal regions (953–1002 and 1258–1355 respectively), which show sequence similarity only to other members of the insulin receptor family (IGFR and IRR).

6.3.1 Properties of the tyrosine kinase

The stimulation of autophosphorylation and receptor kinase is the earliest measurable consequence of insulin binding in intact cells, occurring within seconds. A large body of evidence now supports the idea that the tyrosine kinase plays an essential role in signalling (Rosen 1987; Kahn and White 1988; Zick 1989; Olefsky 1990). A variety of insulin-mimetic agents including trypsin (Leef and Larner 1987; Shoelson *et al.* 1988), hydrogen peroxide (Koshio *et al.* 1988; Heffetz *et al.* 1990) and receptor antibodies (Brindle *et al.* 1990; Steele-Perkins and Roth 1990) stimulate kinase activity. Insulin action is blocked by microinjection of kinase-inhibitory antibodies into otherwise responsive cells (Morgan and Roth 1987). Most conclusively, mutant receptors lacking kinase activity (produced by substitution of the critical lysine residue 1030) fail to transmit an insulin signal when expressed in cultured cells (Chou *et al.* 1987; Ebina *et al.* 1987). Further, naturally occurring mutations of the insulin receptor which impair tyrosine kinase activity result in extreme insulin resistance in patients (Taylor *et al.* 1990). However, it is difficult completely to exclude the possibility that certain effects of insulin could be triggered by kinase-independent mechanisms, perhaps involving receptor aggregation or conformational change (Sung *et al.* 1989).

The properties of the insulin receptor kinase have been studied in considerable detail (Gammeltoft and Van Obberghen 1986; Avruch 1989; Zick 1989; O'Hare and Pilch 1990). Protein tyrosines in an acidic environment, with nearby glutamic or aspartic acid, are preferred substrates. Reduced carboxymethylated lysozyme, or

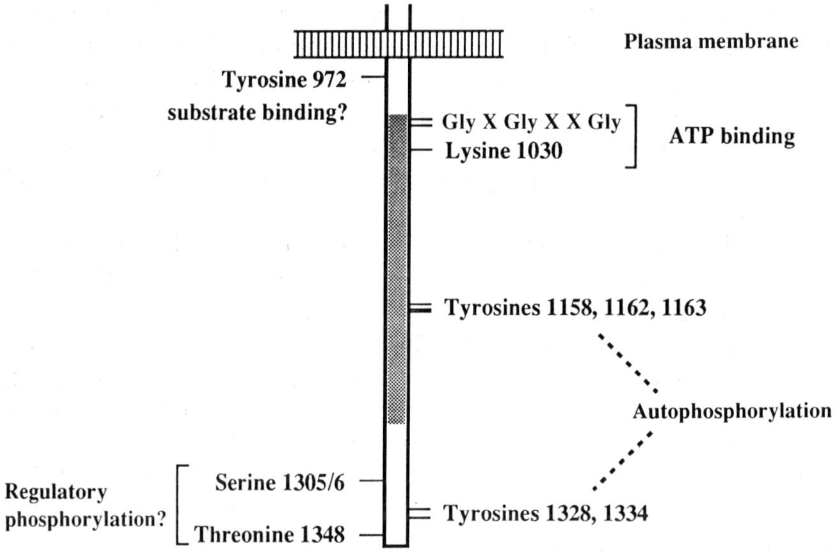

Fig. 5 Intracellular portion of the receptor. Functionally significant residues are indicated, including phosphorylation sites, the ATP-binding site, and a putative substrate-binding site. The conserved kinase domain is indicated by shading

synthetic peptides corresponding to sites of receptor autophosphorylation, are commonly used artificial substrates for receptor kinase assays. Cellular substrates of potential physiological significance are discussed below. The enzyme has an absolute requirement for ATP as phosphate donor, and for bivalent metal ions (Mn or Mg) as cofactors. Insulin increases the V_{max} of the enzyme without affecting the K_m for ATP (approximately 100 μM) or peptide (μM to mM, depending on substrates). The reported extent of insulin stimulation is very variable, depending on the quality of the receptor preparation and the assay system used, but can be up to 80-fold (Zhang *et al.* 1991).

6.3.2 Autophosphorylation

The most conspicuous substrate for the receptor kinase is the receptor itself. Autophosphorylation in fact plays a key role in regulating kinase activity and is essential for maximum activation. Stimulation by insulin is greatly diminished under conditions where autophosphorylation is prevented (such as high concentrations of peptide substrate), while the autophosphorylated enzyme remains fully active even if insulin is removed (Avruch 1989; Zick 1989). The sites of autophosphorylation have been identified by analysis of tryptic phosphopeptides (Tavare *et al.* 1988; Tornqvist *et al.* 1988; White *et al.* 1988*b*). The sites that become phosphorylated in intact cells are in two clusters, one within the kinase domain itself (tyrosines 1158, 1162, 1163) and the other in the C-terminal region (tyrosines 1328, 1334) (Fig. 5). Phosphorylation of the 1158/1162/1163 cluster in response to insulin is ubiquitous, both in intact cells and solubilized receptor. However, although residues 1328/1334 are conspicuously phosphorylated *in vitro* and to a lesser extent in transfected cells, there is apparently little or no phosphorylation of these sites in response to insulin in hepatocytes (Issad *et al.* 1991). Moreover, the kinase activity of C-terminally truncated receptors is stimulated normally by insulin (Maegawa *et al.* 1988; Myers *et al.* 1991), indicating that phosphorylation of the 1328/1334 sites is not required for the activation process.

Attention has therefore focused on the 1158/1162/1163 cluster as the critical sites for kinase activation. These residues are conserved in the IR, IGFR, and IRR, but are not present in other tyrosine kinase receptors. From a regulatory point of view, a key question is whether individual sites have specific roles in the activation process. This has been examined by studying the relationship between activity and phosphorylation state in both wild-type receptor and in mutants with phenylalanine substituted for one or more of the tyrosines. In intact cells, stimulated with insulin, a mixture of bis- and tris-phosphorylated forms of the 1158/1162/1163 cluster is observed. Some data suggest that tris-phosphorylation is necessary for full kinase activation (Tornqvist and Avruch 1988; White *et al.* 1988*b*; Flores-Riveros *et al.* 1989). Stimulation of kinase activity by insulin is clearly impaired in the F1162 single and F1162/1163 double mutants (Ellis *et al.* 1986; Zhang *et al.* 1991). However, the F1158 single mutant, which showed greatly diminished insulin stimulation in one study (Wilden *et al.* 1990), appeared normal in another (Zhang *et al.* 1991). Further work is needed to clarify the importance of the overall level of autophosphorylation and the significance of individual sites for kinase activity.

It is certainly a remarkable enzyme that can potentially autophosphorylate five different tyrosines in two clusters, given the likely constraints on the presentation of the substrate sequences to the catalytic site. The rapidity and apparently concerted nature of the phosphorylation cascade (Avruch 1989) makes it difficult to establish a precise sequence of reactions, although phosphorylation of the kinase domain may precede that of the C-terminus (Tavare *et al.* 1988; Tavare and Dickens 1991). Within the 1158/1162/1163 cluster there does not appear to be an obligatory order of reaction, as when sites are deleted in mutant receptors autophosphorylation of the remaining sites still occurs (Tavare and Dickens 1991; Zhang *et al.* 1991).

6.3.3 Activation mechanism

The mechanism whereby insulin binding to the α-subunit triggers the autophosphorylation reaction, and thus the activation of the tyrosine kinase towards other substrates, is not completely clear. Conformational changes in the extracellular domains, transmitted via lateral, vectorial, or rotational movement of the transmembrane α-helices, must induce changes in conformation or relative positioning of the intracellular domains. The unoccupied α-subunit exerts an inhibitory constraint on autophosphorylation within the native $(\alpha\beta)_2$ heterotetramer. This inhibition can be relieved by binding of insulin or by limited proteolysis (Shoelson *et al.* 1988). Insulin-stimulated autophosphorylation is an intramolecular reaction (White *et al.* 1984) which is dependent on assembly of $(\alpha\beta)_2$ heterotetramers. Insulin does not stimulate autophosphorylation or kinase activity of isolated αβ heterodimers (Boni-Schnetzler *et al.* 1986). Recent evidence suggests that activation occurs by transphosphorylation between the two β-subunits of a heterotetramer, one kinase domain being responsible for phosphorylating the other, rather than each phosphorylating itself (Treadway *et al.* 1991) (Fig. 6). In agreement with this, autophosphorylation of truncated soluble kinase domains is a concentration-dependent, intermolecular reaction (Cobb *et al.* 1989). In insulin/IGF-I hybrid receptors, IGF-I

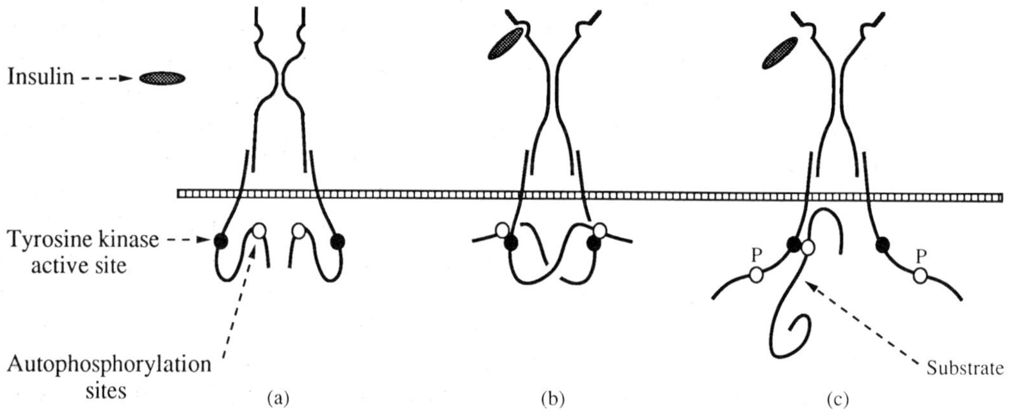

Fig. 6 A model for receptor activation. Activation is considered as a two-step process according to this hypothetical scheme. (a) In the basal state, autophosphorylation is prevented, the kinase active site is inaccessible, and kinase activity is consequently very low. (b) Binding of insulin induces conformational changes in both extracellular and intracellular portions of the receptor. As a result, intramolecular autophosphorylation becomes possible by a 'trans' reaction between the two β-subunits. (c) autophosphorylation triggers further conformational changes, activating the kinase and permitting access of other substrates. These may interact with the juxtamembrane region of the receptor while undergoing tyrosine phosphorylation

induces autophosphorylation of both IR and IGFR β-subunits (Moxham *et al.* 1989). This implies that binding of a single ligand molecule causes reciprocal transphosphorylation and activation of both β-subunits within a heterotetramer. Such a mechanism provides a rationalization for the existence of negative co-operativity in insulin receptors, as a means of preventing the unproductive binding of a second molecule of ligand to receptors which are already fully activated.

Some experiments suggest that intermolecular phosphorylation can also occur between heterotetramers in intact cells (Ballotti *et al.* 1989; Lammers *et al.* 1990). This could, in principle, provide an amplification mechanism whereby binding of insulin to one receptor molecule secondarily activates other receptors. In practice, such amplification does not appear to occur to a significant extent, as stimulation of autophosphorylation by insulin closely parallels receptor occupancy, with no evidence for a 'spare receptor' phenomenon at this level.

6.3.4 Modulation of tyrosine kinase activity

Many examples are known of the regulation of enzyme activity by multisite phosphorylation. In the case of the insulin receptor there is evidence that phosphorylation on serine and threonine residues has an inhibitory effect on tyrosine autophosphorylation and, consequently, on tyrosine kinase activity towards other substrates. This possibility is of obvious interest as a potential mechanism for modulating insulin sensitivity of tissues *in vivo*.

In intact cells, the receptor is phosphorylated to some extent on serine/threonine under basal conditions, and the amount of phosphorylation increases in response to insulin itself, or the addition of phorbol esters, and possibly also cyclic AMP analogues (Rosen 1987; Zick 1989). Analysis of tryptic phosphopeptides indicates a single site of threonine phosphorylation, which has been identified as Thr1348, very close to the C-terminus (Lewis *et al.* 1990*a*) (Fig. 5). The pattern of serine phosphorylation is more complex, with at least three distinct tryptic phosphopeptides (Duronio and Jacobs 1990; Lewis *et al.* 1990*b*; Issad *et al.* 1991). One of these sites, though possibly not the major one in all cell types, is Ser1305 and/or 1306, again in the C-terminal portion of the molecule (Lewis *et al.* 1990*b*). These findings have given rise to the idea that the C-terminal domain, which also contains two tyrosine autophosphorylation sites (1328 and 1334), may have a regulatory function, modulating overall kinase activity or the interaction with specific substrates. However, direct evidence for such a role of the specific C-terminal phosphorylation sites identified so far is still lacking. Further, the possibility still remains that there are important sites of serine phosphorylation within the kinase domain proper.

There is evidence that serine/threonine phosphorylation in general antagonizes kinase activation. In intact cells, insulin-stimulated tyrosine autophosphorylation occurs predominantly in receptors that do not contain phosphoserine (Pang *et al.* 1985; Ballotti *et al.* 1987). Pretreatment of cells with phorbol esters or cyclic AMP analogues inhibits insulin-stimulated autophosphorylation *in situ* and tyrosine kinase activity assayed *in vitro* (Haring *et al.* 1986; Stadtmauer and Rosen 1986; Takayama *et al.* 1988*b*). Direct phosphorylation of insulin receptors with protein kinase C *in vitro* also inhibits tyrosine kinase activity (Bollag *et al.* 1986), but effects of cyclic AMP-dependent protein kinase *in vitro* have not been consistent (Tanti *et al.* 1987). The significance of all these observations must remain in some doubt until there is a clear demonstration of tyrosine kinase inhibition as a result of increased serine/threonine phosphorylation in response to a physiologically important stimulus.

The kinases that phosphorylate the insulin receptor *in situ* have not been identified. Phorbol esters and cyclic AMP analogues must be presumed to activate protein kinase C and cyclic AMP-dependent kinase, respectively, but that does not rule out the participation of intermediate kinases acting on the receptor itself. Indeed, the receptor lacks obvious consensus sites for cyclic AMP-dependent kinase, and phosphorylation of receptor by cyclic AMP-dependent protein kinase *in vitro* has been difficult to demonstrate (Tanti *et al.* 1987). Similarly, the sequence flanking Thr1348 suggests this would only be a relatively poor substrate for direct phosphorylation by protein kinase C.

Particular interest attaches to the insulin-stimulated receptor serine kinase (IRSK) (Czech *et al.* 1988). It is possible that this might participate more generally in signalling mechanisms, as well as exerting feedback control on receptor tyrosine kinase. The pattern of receptor serine/threonine phosphorylations induced by insulin and phorbol esters shows some overlap (Duronio and Jacobs 1990; Lewis *et al.* 1990*a,b*). However, there are significant differences, notably the greater prominence of phosphothreonine compared to phosphoserine in the phorbol ester response, suggesting that distinct kinases are involved. Further, inhibition or down-regulation of protein kinase C does not affect the ability of insulin to stimulate serine phosphorylation of the receptor (Duronio and Jacobs 1990). Evidence has also been presented that different insulin-stimulated kinases may be involved in the phosphorylation of receptor serine and threonine residues (Tavare and Dickens 1991). The insulin-stimulated receptor serine kinase is present as a contaminant in purified insulin receptors (Smith *et al.* 1988; Lewis *et al.* 1990*b*). It has so far proved difficult to isolate and characterize the kinase itself, although it appears to be distinct in its properties from known kinases.

There have been suggestions that receptor tyrosine kinase activity and/or specificity may be regulated by other quite different mechanisms. These include plasma membrane phospholipids (Lewis and Czech 1987) and MHC class I antigens (Hansen *et al.* 1989), soluble proteins, especially polyamines (Fujita-Yamaguchi *et al.* 1989; Kohanski 1989; Yonezawa and Roth 1990), and specific secreted proteins (Auberger *et al.* 1989). The physiological significance of these phenomena is presently unclear.

6.3.5 Tyrosine-specific protein phosphatases

With phosphorylation as with gravity, what goes up must come down. There have been rapid advances very recently in the characterization of tyrosine-specific protein phosphatases (Hunter 1989; Lau *et al.* 1989; Tonks and Charbonneau 1989; Krueger *et al.* 1990). One or more such enzymes must be responsible for dephosphorylating and hence deactivating the insulin receptor when insulin is no longer bound. Dissociation of insulin will occur following internalization of the receptor and transfer to acidified endosomal compartments. It is not clear what fraction of the signalling potential of an activated receptor is expressed after internalization, as opposed to while it is still on the cell surface (Khan *et al.* 1989).

Multiple tyrosine-specific protein phosphatases have already been identified. This multiplicity suggests that these enzymes might exhibit specificity for particular substrates, or different forms of regulation. The insulin receptor is, of course, just one of the many potential phosphotyrosyl-protein substrates which are the products of diverse tyrosine kinases. However, to date only limited specificity has been demonstrated experimentally. Phosphatases from liver and placenta dephosphorylate insulin receptor *in vitro* (Roome *et al.* 1988; Tonks *et al.* 1988; Sale 1990). A placental phosphatase blocks receptor autophosphorylation and insulin action when microinjected into intact cells (Cicirelli *et al.* 1990). The various phosphorylation sites on the insulin receptor differ in their susceptibility to dephosphorylation by protein phosphatases (King and Sale 1990). It is certainly a possibility that phosphatases may have a more complex role in receptor function than the passive termination of activity. Acute or chronic regulation of phosphatases, physiologically or pharmacologically, may provide an important mechanism for controlling the steady-state phosphorylation and activation of insulin receptors.

6.4 Receptor signalling mechanisms

Assuming that activation of the receptor tyrosine kinase is an essential first step in mediating most, if not all, effects of insulin, the obvious question is, what then? Surprisingly, there is still a great deal of uncertainty as to how the tyrosine kinase participates in signalling, though in principle this might occur in one of two ways. The activated tyrosine kinase could directly phosphorylate and regulate cellular substrates, whatever they may be. Alternatively, autophosphorylation might be the critical reaction, generating a conformational change or providing new binding sites in the receptor itself, for a non-covalent interaction with other proteins. Because of the reciprocal interdependence of tyrosine kinase activity and autophosphorylation, it is not a straightforward matter to decide between these possibilities, although the former is undoubtedly the more popular hypothesis. Several different strategies have been adopted in the attempt to identify reactions that immediately follow activation of the receptor itself. First, links have been sought with distal steps in signalling pathways that have already been identified. Secondly, efforts have been made to characterize cellular substrates of the receptor kinase. Thirdly, interactions of other proteins with the activated receptor have been investigated. Finally, signalling properties of mutant receptors have been studied.

6.4.1 Links to known regulatory mechanisms

Many effects of insulin ultimately result from changes in the serine/threonine phosphorylation state of key regulatory enzymes (see Chapter 7). Others involve translocation of membrane vesicles or control of gene transcription, processes which themselves might be regulated by phosphorylation. Perhaps the most straightforward and attractive hypothesis is that the insulin receptor kinase initiates a cascade of phosphorylation, at some point in which 'switch' kinases participate in changing the 'currency' of phosphorylation from tyrosine to serine (Czech *et al.* 1988; Kahn and White 1988) (Fig. 7). To date, serine kinases that are directly phosphorylated and regulated by the insulin receptor tyrosine kinase have proved difficult to identify. The best candidate is a so-called MAP kinase (originally microtubule-associated protein kinase, otherwise mitogen activated protein kinase) which may be involved in cell cycle control. This enzyme is activated by insulin

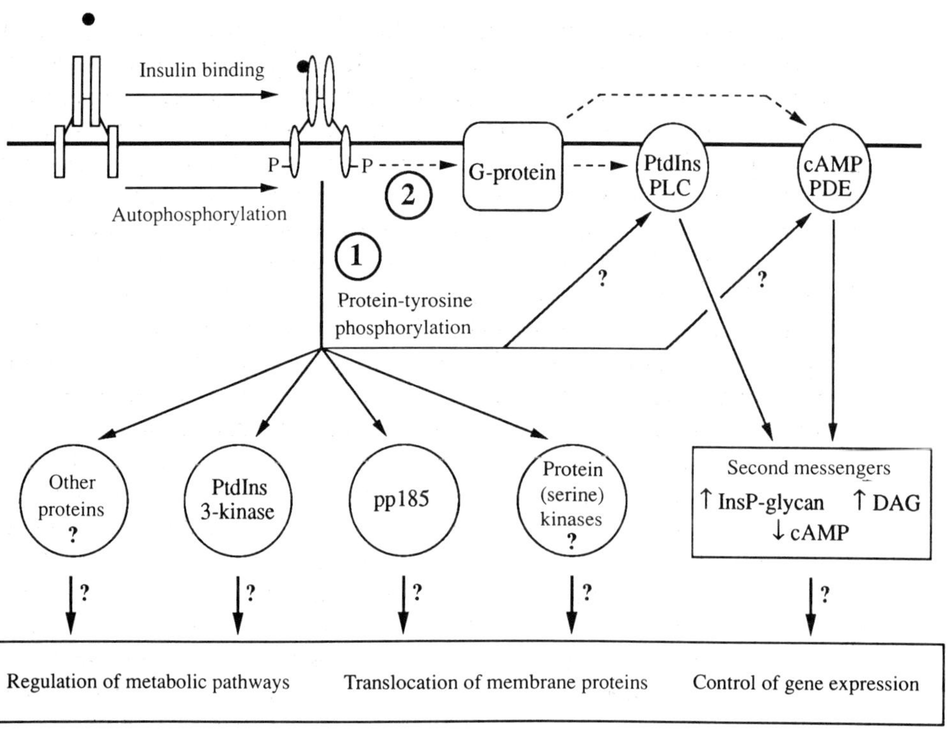

Fig. 7 Signalling pathways from the activated receptor. Receptor autophosphorylation, triggered by insulin binding, may initiate signalling pathways in one of two ways: (1) tyrosine phosphorylation of intracellular substrates by the activated kinase or; (2) conformationally dependent interactions with plasma membrane proteins. The steps by which initial events in signalling pathways are coupled to eventual metabolic effects are unclear (but see Chapter 7). Abbreviations used are: PLC, phospholipase C; DAG, diacylglycerol; PDE, phosphodiesterase; InsP-glycan, inositol phosphate glycan

and other mitogens, accompanied by increased phosphorylation on both tyrosine and threonine residues (Ray and Sturgill 1988; Rossomando *et al.* 1989). Treatment with tyrosine-specific and serine/threonine-specific phosphatases *in vitro* deactivates the enzyme (Anderson *et al.* 1990). Thus there appears to be a direct relationship between the tyrosine phosphorylation and activity of MAP kinase. However, the enzyme has not been shown to be a substrate for tyrosine phosphorylation by the insulin receptor *in vitro*. Another serine/threonine kinase, Raf-1, is also tyrosine-phosphorylated under some conditions (Li *et al.* 1991), but its stimulation by insulin appears to reflect phosphorylation on serine and not tyrosine residues (Blackshear *et al.* 1990; Kovacina *et al.* 1990). The search for serine-specific protein kinases which are activated by tyrosine phosphorylation under physiological conditions therefore continues.

An alternative hypothesis has been that signalling is achieved by generation of diffusible low molecular weight second messengers or 'insulin mediators', which allosterically regulate the activity of kinases and other target enzymes (Fig. 7). The main candidates for such a role are phospho-oligosaccharides (inositol phosphate glycans), released from membrane glycolipids by the action of a specific phospholipase C (Low and Saltiel 1988; Saltiel 1990). Diacylglycerol, a known activator of protein kinase C, has also been proposed as an insulin second messenger (Farese and Cooper 1989), and increases in diacylglycerol might again result in part from the activation of phospholipases. Conceivably, the insulin receptor could regulate one or more phospholipases directly by tyrosine phosphorylation, although the PtdIns-specific phospholipase C-γ which is activated by EGF is not a substrate for the insulin receptor kinase (Nishibe *et al.* 1990). Alternatively, activation might require the participation of a GTP-binding protein, as is the case with many other receptors (Birnbaumer 1990; Bourne *et al.* 1990), in which case receptor conformation could be the important trigger, as discussed below. However, evidence supporting a messenger role for phospho-oligosaccharides or diacylglycerol has not been consistently obtained (Houslay and Siddle 1989; Zick 1989), and the significance of these compounds for insulin action remains in considerable doubt.

A lowering of the concentration of cyclic AMP contributes to some of the effects of insulin, at least in liver and adipose tissue (Houslay 1986). In hepatocytes, insulin induces tyrosine phosphorylation and activation of a plasma membrane cyclic AMP phosphodiesterase (Pyne *et al.* 1989*a*). It remains to be shown that the tyrosine phosphorylation is directly responsible for enzyme activation, and that this activation contributes significantly to lowering of cyclic AMP concentration. Other phosphodiesterases are also activated by insulin, apparently by different mechanisms, and inhibition of adenylate cyclase may also be involved in the overall effect.

6.4.2 Substrates for the receptor tyrosine kinase

Two principal approaches have been taken in the search for physiologically important kinase substrates, both of which make use of anti-phosphotyrosine antibodies. First, such antibodies may be used to identify phosphotyrosyl proteins in insulin-stimulated cells, either by immunoprecipitation from extracts of ^{32}P-labelled cells or by immunoblotting. Secondly, immunoprecipitates obtained with anti-phosphotyrosine antibodies may be tested for known enzyme activities. The first approach has revealed insulin-stimulated tyrosine phosphorylation of a variety of proteins, especially in the presence of inhibitors of tyrosine-specific phosphatases (White and Kahn 1988; Zick 1989). The most widespread and prominent of these intracellular substrates is a cytosolic protein, or family of proteins, of M_r 160 000–185 000 originally referred to as pp185. The cDNA for this protein has recently been cloned and its amino-acid sequence deduced (Sun *et al.* 1991). The sequence contains multiple potential sites for tyrosine phosphorylation, including several in a YMXM motif, but otherwise offers few clues to the function of the protein. A mutant receptor in which tyrosine 972 is replaced by phenylalanine shows greatly diminished capacity to phosphorylate pp185, as well as impaired metabolic signalling, when expressed in cultured fibroblasts (White *et al.* 1988*a*). Receptor autophosphorylation and kinase activity towards artificial peptide substrates are both normal in this mutant (tyrosine 972 is not an autophosphorylation site). These observations therefore provide evidence that phosphorylation of pp185 is necessary for normal signalling, and is dependent on the integrity of the juxtamembrane region of the receptor, as well as its kinase activity. The juxtamembrane region may therefore participate actively in the binding of physiologically important substrates.

Studies of other growth factor receptors and oncoproteins have identified various substrates for tyrosine phosphorylation which may be involved in signal transduction pathways. These include PtdIns-specific phospholipase C-γ, PtdIns3-kinase, and ras GAP, which are phosphorylated and activated *in situ* by EGF and PDGF receptors (Ullrich and Schlessinger 1990; Cantley *et al.* 1991). Of these proteins only the PtdIns3-kinase is phosphorylated in response to insulin, as determined by measuring enzyme activity in anti-phosphotyrosine immunoprecipitates from insulin-treated cells (Endemann *et al.* 1990; Ruderman *et al.* 1990). Unlike the situation with PDGF receptors, there is little, if any, stable association of the activated PtdIns3-kinase with insulin receptors. The physiological role of PtdIns3-P and its further metabolites is unknown, so the significance of the activation of PtdIns3-kinase is unclear. However, tyrosine phosphorylation of the enzyme is associated under various conditions with mitogenesis or transformation, suggesting a key role in signalling in relation to cell growth (Cantley *et al.* 1991). Whether it might also be involved in acute metabolic effects of insulin is much less certain.

Various proteins are phosphorylated by purified insulin receptors *in vitro* (Zick 1989), but the physiological significance of such reactions is difficult to assess.

6.4.3 Autophosphorylation and conformational change

Studies with other growth factor receptors also lend credence to the idea that autophosphorylation itself could play a direct role in signalling. It has been shown that substrates for the PDGF and EGF receptor kinases remain associated to some extent with the activated receptors in a complex, or putative signal transfer particle (Ullrich and Schlessinger 1990; Cantley *et al.* 1991). In part, at least, this involves the interaction of conserved sequences (SH2 domains) of one protein with phosphotyrosines of another. In the case of PDGF receptors, the autophosphorylated tyrosine 751, within the kinase-insert domain, is a binding site for PtdIns3-kinase. In principle, one or more of the autophosphorylation sites of the insulin receptor could similarly function in binding specific proteins. However, at present there is no evidence for a stable stoichiometric association of other proteins, and certainly not phosphoproteins, with the activated insulin receptor. A truncated receptor lacking C-terminal autophosphorylation sites functions relatively normally, ruling out an obligatory role of these sites in signalling at least some metabolic effects (Maegawa *et al.* 1988; Myers *et al.* 1991). The autophosphorylation sites within the insulin receptor kinase domain do not satisfy the apparent consensus sequence

Y(M)XM for interaction with SH2 domains. It is an intriguing possibility that the pp185 substrate for the insulin receptor kinase, rather than the receptor itself, provides the basis for assembly of signal transfer particles involved in the insulin signalling pathway (Sun *et al.* 1991).

A further possibility is that interaction of insulin receptors with potential signalling molecules is dependent on autophosphorylation-induced conformational change, rather than specific phosphotyrosine-binding sites. Conformational changes affecting both the kinase domain and C-terminal region can be detected using cross-linking agents (Schenker and Kohanski 1988) and anti-peptide antibodies (Perlman *et al.* 1989; Baron *et al.* 1990). Such conformational change could either facilitate association of proteins with the activated receptor, or cause activation and/or release of proteins which were associated with the receptor in the basal state. The insulin receptor serine kinase (Smith *et al.* 1988; Lewis *et al.* 1990*b*) is one candidate for an interaction of this type. Alternatively, interaction might occur transiently by collision coupling rather than formation of stable complexes, in which case it must be inferred from indirect experiments.

It has been proposed that insulin receptors interact with one or more G-proteins as signal transducers (Houslay *et al.* 1989; Zick 1989). Such G-proteins might regulate the activity of membrane-bound enzymes such as adenylate cyclase or phospholipases, and thereby control the production of potential second messengers or mediators, such as cyclic AMP, phospho-oligosaccharides, and diacylglycerol. Involvement of G-proteins has been inferred from the attenuation of insulin action by toxins that catalyse the ADP-ribosylation and inactivation of G-proteins (Luttrell *et al.* 1988; Ciaraldi and Maisel 1989; Moises and Heidenreich 1990). Conversely, insulin modulates toxin-induced ADP-ribosylation reactions (Irvine and Houslay 1988; Rothenberg and Kahn 1988; Pyne *et al.* 1989*b*). There is no evidence for tyrosine phosphorylation of G-proteins *in situ* (Joost *et al.* 1989; Pyne *et al.* 1989*b*; Luttrell *et al.* 1990). In other systems, interaction of G-proteins with plasma membrane receptors is by collision coupling (Gilman 1987; Birnbaumer 1990; Bourne *et al.* 1990). However, these receptors are of a very different type to the insulin receptor, and uniformly possess seven membrane-spanning domains. The mechanism and importance of interaction between the insulin receptor and G-proteins remains unclear, and any cross-talk that does occur may be indirect and secondary rather than of primary significance in signalling.

6.4.4 Signalling properties of mutant receptors

Many different possibilities are therefore still under active consideration in relation to the reactions that immediately follow receptor autophosphorylation and kinase activation (Fig. 7). These are by no means mutually exclusive, and indeed the diversity of metabolic effects elicited by insulin lends encouragement to the idea that more than one signal could be produced by the activated receptor. This could be a simple matter of the kinase phosphorylating more than one substrate, or a more fundamental divergence of signals of different types. Evidence for the former possibility is provided by tyrosine kinase inhibitors which differentially antagonize the lipogenic and anti-lipolytic effects of insulin in adipocytes (Schechter *et al.* 1989). Particular attention has focused on the question of whether there may be different signals for acute metabolic effects (such as stimulation of glucose transport and glycogen synthesis) and the longer-term growth-promoting actions (for which stimulation of thymidine incorporation into DNA provides an index). This issue has been approached by studying the properties of mutant receptors (either engineered in the laboratory or identified in insulin-resistant patients). Receptors have been expressed at high levels in a variety of fibroblast cell lines, by cDNA transfection, in order to investigate their signalling activity.

Differential effects on 'metabolic' and 'mitogenic' signalling have been reported for several mutants. A truncated receptor lacking 43 amino acids at the C-terminus of the β-subunit (ΔCT) showed impaired 'metabolic' signalling but enhanced 'mitogenic' potential when expressed in rat-1 fibroblasts (Thies *et al.* 1989). This led to the suggestion that the C-terminal region has a regulatory role, possibly via substrate selection, which influences the relative efficiency of coupling of receptor to distinct signalling pathways. However the same ΔCT mutant expressed in CHO cells was almost indistinguishable from wild type in its activity, with no evidence for differential effects on 'metabolic' and 'mitogenic' responses (Myers *et al.* 1991). Similar controversy surrounds the properties of other mutants, involving autophosphorylation sites. A double mutant with tyrosines 1162/1163 replaced by phenylalanines reportedly was unable to mediate stimulation of glucose uptake, but supported a normal mitogenic response (Ellis *et al.* 1986; Debant *et al.* 1988). Conversely, replacement of tyrosine 1158 with phenylalanine apparently prevented mitogenic signalling, but metabolic responses were normal (Wilden *et al.* 1990). However, in other work the F1162/1163 mutant appeared essentially inactive, while the F1158 mutant was fully active (Ellis *et al.* 1991; Zhang *et al.* 1991). These discrepancies make it impossible to draw firm conclusions either on the divergence or otherwise of signalling pathways, or on the role of specific residues in signalling. Cultured cells display only a limited range of rather small responses to insulin, compared to physiologically important target cells, which makes it difficult to analyse potentially subtle changes in receptor activity. Clonal variability of cell lines and factors specific to different cell backgrounds may also be significant. These difficulties will need to be overcome before analysis of mutant receptors can fulfil its potential as a means of relating structure to function.

6.5 Life history of the insulin receptor

The level of expression of insulin receptors at the cell surface is a major determinant of insulin sensitivity in different tissues. Although the main steps in the synthesis, transport, and turnover of receptors are known, the factors regulating these processes are not yet well understood.

6.5.1 The receptor gene

The human insulin receptor gene spans a region of over 130 kbp on chromosome 19, and is made up of 22 exons (Seino *et al.* 1989, 1990). Some of the exons correspond approximately to structural units, such as the signal peptide (exon 1), cysteine-rich domain (exon 3), transmembrane domain (exon 15), or C-terminal tail (exon 22) (Fig. 8). Multiple mRNA transcripts are detected in most cell types, largely reflecting different lengths of 3′ untranslated sequence. Reference has already been made to the generation of isoforms by alternative splicing, which results in the retention or deletion of the sequence encoded by exon 11. This occurs in a tissue-specific manner which is conserved between species (Goldstein and Dudley 1990).

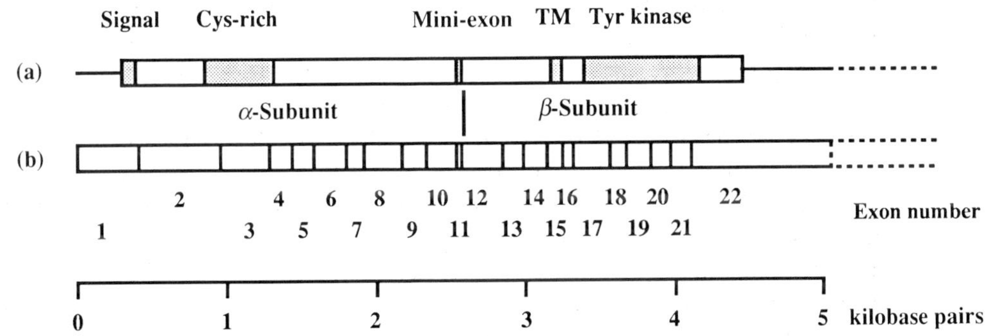

Fig. 8 Exons of the insulin receptor gene. The contribution of different exons (b) to the receptor mRNA (a) is shown. Untranslated sequences are indicated by a single line. Specific structural domains are shown by shading (TM, transmembrane domain)

The promoter region of the gene has been characterized, and has features in common with other constitutively expressed 'housekeeping' genes (Araki *et al.* 1987; Mamula *et al.* 1988; Seino *et al.* 1989; Sibley *et al.* 1989). It contains neither TATA nor CAAT boxes, but possesses several potential binding sites for the transcription factor Sp1. Surprisingly little information is available on the regulation of receptor gene expression (Mamula *et al.* 1990). Dramatic changes of insulin receptor expression during differentiation of cultured BC3H-1 myocytes and 3T3-L1 adipocytes are a consequence of control at the transcriptional level. Receptor biosynthesis is enhanced by glucocorticoids in several cell types, reflecting, at least in part, a stimulation of mRNA transcription, while insulin itself decreases receptor mRNA levels in some cells.

6.5.2 Receptor biosynthesis

The receptor is synthesized initially as a single polypeptide which is co-translationally glycosylated in the endoplasmic reticulum, giving a high-mannose proreceptor of approximately 190 kDa (Hedo and Simpson 1985; Forsayeth *et al.* 1986). Formation of disulphide-linked proreceptor dimers probably also takes place in the endoplasmic reticulum (Olson *et al.* 1988). After transfer to the Golgi apparatus the receptor is proteolytically cleaved to generate the α- and β-subunits, and further processing of the oligosaccharide takes place. In at least some cells the receptor undergoes an additional covalent modification at an early stage of biosynthesis, by fatty acylation of the β-subunit on a cysteine residue (Magee and Siddle 1988). The functional significance of this modification is unknown. The mature receptor is finally transferred from the Golgi to the plasma membrane, and, in polarized cells such as hepatocytes, to specific regions of the plasma membrane (Fig. 9). The targeting mechanisms which direct this traffic are unknown. A small fraction of

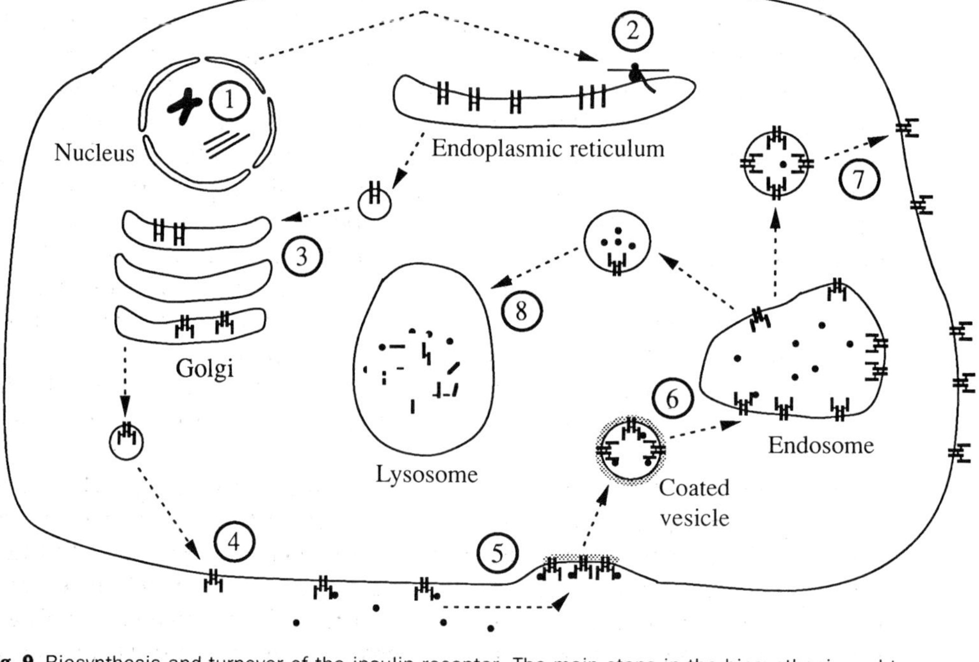

Fig. 9 Biosynthesis and turnover of the insulin receptor. The main steps in the biosynthesis and turnover of insulin receptors are shown diagrammatically (see text for references). (1) Within the nucleus, the receptor gene is transcribed and the RNA is spliced and processed to give mRNAs of various sizes. (2) Receptor mRNA is translated on membrane-bound ribosomes, and nascent polypeptide is co-translationally glycosylated within the endoplasmic recticulum. Protein folding and disulphide bond formation also occur in the endoplasmic reticulum to generate proreceptor dimers. (3) Transport vesicles take the proreceptor to the Golgi, where further glycosylation and proteolysis produce mature receptor. (4) The mature receptor is transported to the plasma membrane, again by a specific class of vesicles. (5) Binding of insulin induces clustering of occupied receptors in clathrin-coated pits, and endocytosis of the insulin/receptor complexes in coated vesicles. (6) Internalized receptors enter the endosome system, where acid pH causes dissociation of insulin and segregation of receptor and ligand occurs. (7) The majority of receptors are recycled from endosomes to the plasma membrane by way of transport vesicles. Limited retro-endocytosis of insulin may also occur. (8) Most of the internalized insulin is degraded within the endosome system and after delivery to lysosomes. A proportion of the internalized receptor is also degraded by this route.

receptors may escape proteolysis to appear on the cell surface as a fully glycosylated 210 kDa polypeptide (Hedo and Gorden 1985). However, the uncleaved receptor has a low affinity for insulin and defective insulin-induced autophosphorylation, and is probably non-functional at physiological insulin concentrations (Williams *et al.* 1990).

6.5.3 Endocytosis and turnover

Mature insulin receptors are located predominantly in the plasma membrane under basal conditions, and have a half-life of the order of 12 h. However, binding of insulin induces rapid endocytosis of the hormone/receptor complex, influencing both the distribution and degradation of receptors (Heidenreich and Olefsky 1985; Sonne 1988). Receptor-mediated endocytosis also serves as an important mechanism for degradation of insulin, particularly in the liver (Duckworth 1988; Sonne 1988). The extent to which endocytosis is important in signalling is unclear. As discussed earlier, the internalized receptor will presumably remain active as long as it is autophosphorylated, and endocytosis could therefore serve to bring the receptor kinase into greater proximity with potential substrates (Backer *et al.* 1989; Khan *et al.* 1989). The possibility that endocytosis might serve to deliver insulin itself to intracellular sites of action has also been considered. Nuclear localization of insulin has been demonstrated (Soler *et al.* 1989; Smith and Jarett 1990), but the mechanism by which insulin could get from endocytic vesicles into the nucleoplasm is unknown, and it remains to be shown whether the process has any physiological significance. The weight of evidence supports the view that insulin does not itself participate in signalling once it has activated the receptor kinase, but it has to be said that the mechanism of insulin action on gene transcription in particular remains obscure.

The process of receptor internalization occurs in two stages (Carpentier 1989). Unoccupied receptors are generally distributed in the plasma membrane, or concentrated in microvilli, but move into clathrin-coated pits after binding insulin. Endocytosis then takes place via coated vesicles. The mechanism that triggers these events is unknown, but there is substantial evidence supporting a requirement for autophosphorylation and/or tyrosine kinase activity (McClain *et al.* 1987; Russell *et al.* 1987). The juxtamembrane region of the receptor, and specifically an NPXY sequence motif in this domain, appears to play an important role (Backer *et al.* 1990; Thies *et al.* 1990*b*). This may provide a recognition site for proteins involved in receptor redistribution or internalization *per se*, which becomes accessible only following autophosphorylation. A similar motif has been implicated in the internalization of other cell-surface receptors, although it is not universally present in proteins undergoing endocytosis via coated pits (Chen *et al.* 1990).

Once internalized, the insulin/receptor complex is rapidly transferred to an endosomal compartment, where acidification results in the dissociation of insulin. Thereafter the ligand and receptor follow separate paths (Fig. 9). The insulin is largely degraded, both within the endosomal system itself and after transfer to lysosomes. The receptor is largely recycled to the plasma membrane in most cells (Carpentier 1989). However, the increased rate of internalization may result in an increase in the size of the intracellular pool of receptors and decreased expression at the cell surface. This redistribution is rapidly reversible on removal of hormone, but if maintained may, in turn, result in increased degradation and net loss of receptors, which is then only slowly reversed by *de novo* synthesis. It is unclear whether degradation of the internalized receptor is itself is a regulated step. In any case, prolonged exposure of cells to insulin concentrations sufficient to occupy a significant fraction of receptors results in down-regulation of receptor numbers (Sonne 1988). The internalization of insulin/receptor complexes may thus provide a mechanism for regulation of receptor concentration in response to ambient hormone concentrations. A reciprocal relationship between insulin concentration and receptor number has sometimes been observed *in vivo*, receptors increasing during fasting and decreasing in hyperinsulinaemic states.

In endothelial cells, internalization of insulin/receptor complexes serves a different role, as part of a process of transcytosis which may be important for delivery of insulin to target tissues. Receptor phosphorylation on serine residues may be involved in regulating this traffic (Bottaro *et al.* 1989).

The insulin receptor is only one of many proteins that follow complex intracellular itineraries. It is likely that further insights into the mechanisms that regulate endocytosis and determine the fate of the internalized receptor will come from the study of other receptors as much as the insulin receptor itself.

6.6 Insulin receptors and insulin resistance

Insulin resistance, loosely defined, has been demonstrated in a wide variety of physiological and pathological states (Reaven 1988). *In vivo*, the presence of insulin resistance may be deduced if the concentration of insulin is inappropriately high for a given level of glycaemia, or more rigorously demonstrated by measurements of glucose utilization as a function of insulin concentration under steady-state (glucose clamp) conditions. Glucose utilization in skeletal muscle, and hepatic glucose output, are the major insulin-regulated processes that determine blood glucose level, and in which defects could give rise to insulin resistance *in vivo*. *In vitro*, it is possible to investigate the insulin sensitivity of a wider range of tissues and metabolic processes, and thus to examine whether insulin resistance occurs selectively in certain cell types or pathways. It should be recognized that there must be many different steps in intracellular signalling and metabolic pathways at which a defect could give rise to insulin resistance. However, studies of the molecular basis of insulin resistance have to date concentrated very much on just two components, the insulin receptor and the glucose transporter, in part because these are relatively well characterized and amenable to investigation.

Defects in insulin receptor function have, in fact, been implicated in various states of insulin resistance (Roth and Taylor 1982; Reddy and Kahn 1988; Caro *et al.*

1989; Haring and Obermaier-Kusser 1989). The most detailed characterization of receptor involvement has been in rare syndromes of extreme insulin resistance, which can result from auto-immune production of receptor antibodies or from mutations in the insulin receptor gene. Some degree of insulin resistance, which may involve insulin receptors, is also a feature of obesity and non-insulin-dependent diabetes, and of diverse physiological conditions including starvation, ageing, and pregnancy.

6.6.1 Auto-antibodies to the insulin receptor

Syndromes of extreme insulin resistance were initially classified as types A or B, depending on whether or not insulin resistance persisted in cultured fibroblasts taken from affected patients. In type B syndrome, cultured fibroblasts showed normal insulin binding and receptor activity. Insulin resistance *in vivo* or in freshly isolated cells resulted from the effects of circulating auto-antibodies to the insulin receptor (Roth and Taylor 1982). In fact, receptor antibodies obtained from patient serum proved very valuable reagents in the study of receptor structure and function, and their biological effects were well characterized (Kahn *et al.* 1981). This characterization was facilitated by the fact that the antibodies cross-reacted with rat receptors, and so could be used in experiments with rat cells. The detailed properties of the antibodies varied to some extent between patients. Many sera contained antibodies which inhibited insulin binding, but mimicked most, or all, aspects of insulin action when added acutely to cells *in vitro*. Most probably, the inhibition of insulin action by antibodies acting chronically *in vivo* reflects in part their ability to desensitize cells by down-regulation of receptors or other mechanisms. Interestingly, some patients with auto-immune insulin resistance also display episodes of hypoglycaemia, which could result from changing patterns of antibody secretion and bioeffects.

Studies with monoclonal antibodies for the insulin receptor have confirmed the diversity of bioeffects that can be elicited by antibodies recognizing different epitopes (Siddle *et al.* 1987, 1988). Antibodies generated in mice have generally been very species specific for human receptors, and failed to react with rat and mouse receptors. Such antibodies, therefore, may not be completely representative of the antibody subpopulations in auto-immune sera. Nevertheless, individual monoclonal antibodies exhibit many of the properties associated with polyclonal sera. Some antibodies inhibit insulin binding, but others even stimulate binding. Many antibodies, recognizing epitopes on both α- and β-subunits, elicit insulin-like metabolic effects on human cells, and also induce the down-regulation of receptors by stimulation of internalization and degradation. Antibodies have thus proved to be valuable reagents in the analysis of structure–function relationships.

Receptor antibodies have also been detected in subjects with newly diagnosed IDDM (Maron *et al.* 1983; Rochet *et al.* 1990), and in NIDDM (Batarseh *et al.* 1989; Shimoyama *et al.* 1989). It has been suggested that these antibodies may arise in part as anti-idiotypes of anti-insulin antibodies (Schechter *et al.* 1984; Ludwig *et al.*

1987). Both in IDDM and NIDDM the proportion of patients with antibodies is small, and the antibody titres are generally low. Thus, it does not appear that receptor antibodies contribute significantly to the pathology of diabetes.

6.6.2 Mutations in the insulin receptor gene

The presence of a primary, heritable defect in the insulin receptor was indicated by impaired insulin binding and/or tyrosine kinase activity in cultured cells from patients with several syndromes of extreme insulin resistance, namely the type A syndrome, leprechaunism, lipoatrophic diabetes, and Rabson–Mendenhall syndrome (Reddy and Kahn 1988). Mutations in the insulin receptor gene have now been identified in several such patients (Kadowaki *et al.* 1990*a*; Taylor *et al.* 1990). It is unclear whether the diverse clinical presentations reflect the nature or severity of the receptor defect, the involvement of additional lesions at other loci, or more subtle influences of genetic background. Although extreme resistance to the actions of insulin is a common feature of all the syndromes, many patients are not diabetic. Normal glucose tolerance is frequently maintained with levels of insulin that may be 10–100-fold above the normal range.

The characterization of mutant insulin receptors is a considerable undertaking, given the size and complexity of the gene. In some cases, mutations were identified by cloning and sequencing cDNA or genomic DNA. More recently, the application of the polymerase chain reaction to amplify regions of cDNA or individual exons of genomic DNA for direct sequencing has reduced the workload significantly and accelerated the rate of analysis (Kadowaki *et al.* 1990*b*; Moller *et al.* 1990*b*). Functional characterizatioin of mutant receptors may be undertaken in patient fibroblasts or Epstein–Barr-virus-transformed lymphocytes (Reddy and Kahn 1988), or more comprehensively by transfection of cultured fibroblasts with mutant receptor cDNA (Moller *et al.* 1990*a*,*b*; Taylor *et al.* 1990; Yamamoto-Honda *et al.* 1990).

Mutations identified to date have been widely scattered throughout the coding sequence and very diverse in their effects on receptor function (Kadowaki *et al.* 1990*a*; Taylor *et al.* 1990). Each family studied has revealed different mutations. In two patients a deletion of the tyrosine kinase domain was demonstrated, but point mutations have predominantly been found. Some of these are nonsense mutations, which result in decreased mRNA levels and/or production of a truncated protein. Most commonly, missense mutations produce single amino-acid subtitutions within an otherwise normal receptor (Fig. 10). In principle, it is to be expected that mutations in non-coding regions, particularly regulatory sequences or intron/exon boundaries, might affect transcription and processing of mRNA, and thus the level of receptor expression. However, no such mutations have yet been identified.

Functionally, mutations manifest themselves as decreases in receptor number, binding affinity, or kinase activity (Taylor *et al.* 1990). A decreased number of receptors may reflect a decrease in the level of mRNA, or more subtle influences of point mutations on biosynthesis and turnover. In some cases (Asn15 $\rightarrow$ Lys,

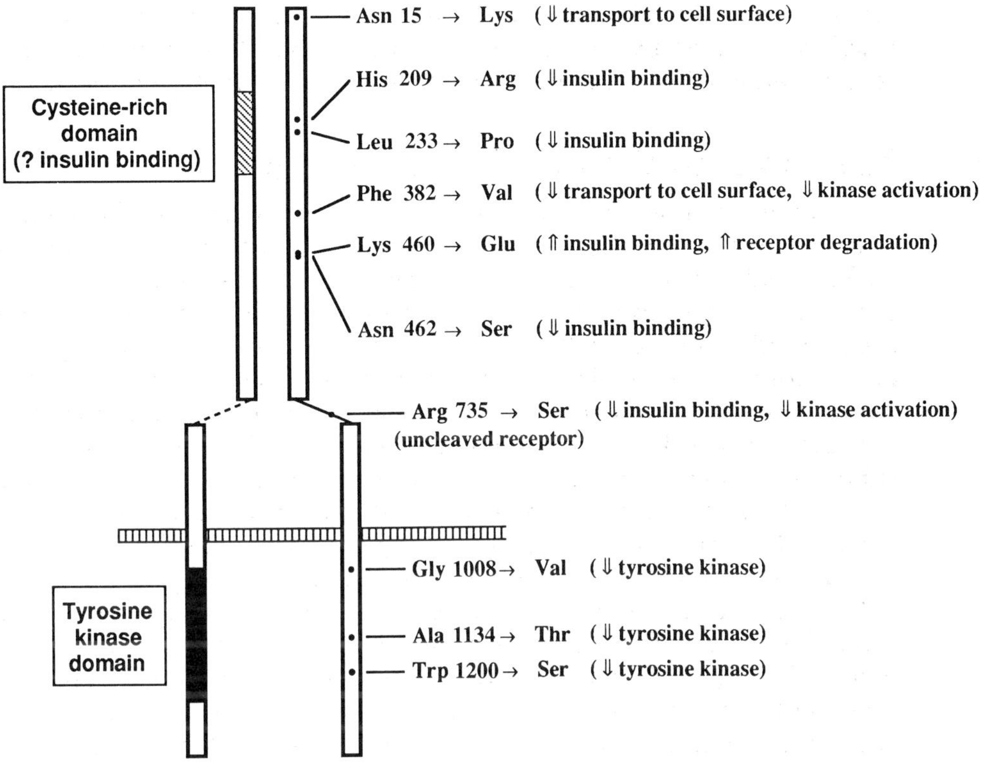

Fig. 10 Mutations of the insulin receptor gene identified in insulin-resistant patients. Point (missense) mutations resulting in single amino-acid substitutions are indicated, together with the known effects on receptor function (see text for references)

Phe382 → Val) it seems that transport to the cell surface may be impaired, probably reflecting the retention and degradation of malfolded protein within the endoplasmic reticulum. In another mutant (Lys460 → Glu) retardation of the dissociation of insulin, especially at the acid pH likely to be encountered in endosomes, apparently leads to accelerated receptor degradation (Kadowaki *et al.* 1990*c*).

Several mutations result in a marked decrease in binding affinity for insulin. Some of these (His209 → Arg, Leu233 → Pro) are within the cysteine-rich domain which has been shown to confer binding specificity. In another instance (Asn462 → Ser) the binding site may be influenced secondarily to conformational effects in other parts of the molecule. One patient has been identified with a mutation (Arg735 → Ser) within the tetrabasic sequence, which is the site for proteolytic cleavage of the proreceptor. This mutation prevents cleavage and results in expression of unprocessed proreceptors with low binding affinity.

The mutations that inhibit kinase activity (so far identified) are mostly in the kinase domain itself. One mutation (Gly1008 → Val) is within the ATP-binding site (Yamamoto-Honda *et al.* 1990). Others (Ala1134 → Thr, Trp1200 → Ser) affect residues that are conserved in different tyrosine kinases but as yet have no known specific role (Moller *et al.* 1990*a,b*). In principle, substitutions in other parts of the receptor might impair the ability of insulin to activate the kinase, and one mutation has been identified (Phe382 → Val) which apparently has this effect (Accili *et al.* 1991). Two patients have been described with deletions of the tyrosine kinase domain (Shimada *et al.* 1990).

Not the least important aspect of these 'experiments of nature' is the potential they provide for detailed studies of structure–function relationships. The mutations identified in patients complement those engineered in the laboratory. It will be of particular interest to see whether the heterogeneous clinical presentation of patients with severe insulin resistance is, in some cases, a reflection of mutations that selectively affect different actions of insulin, and thus provide evidence for divergent signalling pathways. However, to date it appears that it is the severity of insulin resistance that determines the clinical manifestations, with leprechaunism representing the most extreme situation.

Analysis of the mode of inheritance of extreme insulin resistance produced some surprises. In the majority of cases affected individuals were either homozygous offspring of a consanguineous marriage, or compound heterozygotes. However, in some patients one allele appears normal and the mutant allele behaves in a dominant negative fashion (Moller *et al.* 1990*a,b*). This dominance probably reflects the fact that hybrid receptors assembled from one normal and one mutant proreceptor polypeptide are inactive, at least with mutations that impair tyrosine kinase activity (Whittaker *et al.* 1990). It has been shown, by assembly of hybrids *in vitro*, that both kinase domains within an $(\alpha\beta)_2$ heterotetramer must be functional for activation to occur by transphosphorylation (Treadway *et al.* 1991). On a purely statistical basis, random assembly of normal and mutant proreceptors, synthesized in equal amounts, would be expected to produce 50 per cent hybrids as well as 25 per cent of each of the mutant and wild-type homodimers. The fraction of plasma membrane receptors which are wild-type homodimers might be further decreased because of the preferential down-regulation of active receptors in the face of persistent hyperinsulinaemia.

Family studies involving severely insulin-resistant patients have also been of considerable interest. Previously unrecognized insulin resistance has been found in relatives who are heterozygous carriers of mutations affecting insulin binding (Lekanne-Deprez *et al.* 1989; Taylor *et al.* 1990). In these cases, hyperinsulinaemia occurs without hyperglycaemia or obvious symptoms.

The fact that a single abnormal allele of the insulin receptor gene can give rise to insulin resistance suggests that receptor mutations may be a more widespread cause of insulin resistance than had been recognized previously. The question then arises as to whether receptor mutations could be responsible for insulin resistance in the relatively common condition of NIDDM. Present calculations of the likely frequency of mutant alleles in the population suggest that they would occur in only a small fraction of NIDDM patients (Taylor *et al.* 1990). Further, heterozygotes with mild insulin resistance are not diabetic, implying that additional defects would be required before overt diabetes could develop (Lekanne-Deprez *et al.* 1989). The insulin receptor gene is normal in a small number of subjects studied from the insulin-resistant Pima Indian population (Moller *et al.* 1989*b*; Cama *et al.* 1990).

Interest in the incidence of insulin-receptor mutations and their possible pathological consequences remains high. However, it is a considerable undertaking to identify such polymorphisms in populations, as opposed to individuals with clear evidence of receptor defects. Screening methods to detect DNA polymorphisms, as a prelude to sequencing, are still being developed to a stage where they can be efficiently applied to large numbers of samples in order to detect relatively rare events. At the present time, denaturing gradient gel electrophoresis and single-stranded conformation polymorphism analysis appear to offer the best prospects as screening techniques (Rossiter and Caskey 1990). Such methods rely on amplification of genomic DNA by the polymerase chain reaction. In the case of the insulin receptor, the division of the gene into 22 separate exons prohibits comprehensive analysis. Instead, effort may be focused initially on regions of the receptor involved in insulin binding (exon 3) or tyrosine kinase (exons 17–20). This approach has led to the identification of two relatively common polymorphisms of the insulin receptor gene (Val985 → Met, Lys1069 → Glu) in normal and diabetic subjects (O'Rahilly *et al.* 1991). It remains to be seen whether either of these affects receptor function, and whether other polymorphisms are detected when larger numbers of subjects are screened.

6.6.3 Non-insulin-dependent diabetes

The extent to which insulin resistance is a causative factor in the development of NIDDM is still debated (DeFronzo 1988; Taylor 1989; Chapter 11, this volume). Insulin resistance could, in part, arise secondarily to insulin deficiency or hyperglycaemia, both of which may affect the capacity or acute responsiveness of insulin-sensitive metabolic pathways. Considerable attention has been devoted to the functioning of the insulin receptor in diabetes (Caro *et al.* 1989; Haring and Obermaier- Kusser 1989). One difficulty in assessing experimental data in this area is that relatively small changes in insulin binding or receptor kinase activity could be of considerable physiological significance. Such changes are not necessarily easy to demonstrate in accessible human tissues with available methods by comparison of mean responses in diabetic or normal populations. In addition, it is important to separate changes due to obesity from those associated with diabetes *per se.*

As discussed previously, receptor antibodies or mutations are unlikely to be important causes of insulin resistance in more than minor subsets of NIDDM, if at all. Major changes in insulin binding are not seen, although both in humans and in animal models the receptor number on target cells may be decreased in obesity (Roth and Taylor 1982; Sonne 1988). This appears to be a secondary defect resulting from down-regulation induced by hyperinsulinaemia, which is reversible by dietary restriction and weight loss. Interest has therefore turned to the possibility that receptor tyrosine kinase activity is impaired in NIDDM. Decreased kinase activity has been described in liver and adipose tissue of obese diabetics (Caro *et al.* 1986; Friedenberg *et al.* 1987; Sinha *et al.* 1987; Takayama *et al.* 1988*a*; Nyomba *et al.* 1990; Thies *et al.* 1990*a*). Although the defect was not seen in tissues of obese non-diabetics, it was reversed by weight reduction (Friedenberg *et al.* 1988). In skeletal muscle, kinase activity was reportedly impaired in obesity with no additional effect of diabetes (Arner *et al.* 1987; Caro *et al.* 1987).

The mechanism of kinase impairment observed in such studies is unclear. In all cases, results were normalized for expression relative to equal amounts of insulin binding. One study reported an altered pattern of autophosphorylation in skeletal muscle from diabetics, with a decreased proportion of the 1158/1162/1163 tris-phosphorylated state which may be required for maximal kinase activity (Obermaier-Kusser *et al.* 1989). Other work, on adipose tissue, suggested that the decreased kinase activity reflected an increased subpopulation of receptors which bound insulin but did not undergo autophosphorylation (Brillon *et al.* 1989). Mechanisms regulating autophosphorylation might include phosphorylation on serine/threonine residues as discussed previously. Alternatively, limited proteolysis of insulin receptors, perhaps secondary to increased internalization, might be responsible in part for the loss of kinase activity (O'Hare and Pilch 1988). For the time being, no firm conclusions can be reached regarding either the significance or mechanisms of changes in receptor kinase activity as a contribution to insulin resistance in NIDDM.

6.6.4 Other insulin-resistant states

Uncertainty also surrounds receptor function in other insulin-resistant states. In starvation there is evidence for insulin resistance in liver and fat, while muscle displays increased insulin sensitivity. This may be related to decreased kinase activity in liver but not muscle, although receptor number is increased in both tissues (Balage *et al.* 1990; Contreras *et al.* 1990; Karasik *et al.* 1990). Decreased kinase activity of hepatic insulin receptors, associated with insulin resistance, has also been reported as a result of high fat feeding (Boyd *et al.* 1990; Nagy *et al.* 1990) and during late pregnancy (Martinez *et al.* 1989) but does not occur as a result of ageing (Kono *et al.* 1990). The mechanism of inhibition of kinase activity in all cases remains to be investigated, although in starvation the decreased capacity of hepatic insulin receptors for autophosphorylation was associated with an increase in the tissue level of protein kinase C (Karasik *et al.* 1990).

6.7 Conclusions

The insulin receptor is relatively well characterized by the standards of most other membrane glycoproteins. It is a complex multidomain structure in which binding of insulin, involving a cysteine-rich domain of the extracellular α-subunit, activates the tyrosine kinase which makes up most of the intracellular portion of the β-subunit. Other domains within the α-subunit and extracellular portion of the β-subunit may be critical for the conformational changes that mediate both the transmembrane activation and co-operative effects on ligand binding. It is postulated, but so far unproven, that juxtamembrane and C-terminal domains of the intracellular portion of the β-subunit have a role in interaction of the kinase with specific substrates, or in regulation of its overall activity. The receptor kinase is generally considered essential for insulin signalling, but the nature of its involvement is still unclear. Most probably, it is required to phosphorylate other intracellular proteins further down the signalling pathway, although it may be that the receptor sends more than one type of signal into the cell.

In spite of the enormous physiological importance of normal receptor function, defects in the receptor have so far been clearly associated only with certain rare syndromes of insulin resistance. It remains possible, however, that acquired alterations of receptor function contribute to other states of insulin resistance, and to the development of NIDDM. A wide range of powerful techniques of modern molecular biology and biophysics are now being brought to bear on the outstanding issues of receptor structure and function. Detailed structural information can be expected in the future from NMR analysis and X-ray crystallography. Further insights into function and signalling are likely to come from analysis of mutant receptors in cultured cells and in transgenic animals. These are exciting times in which the insulin receptor surely cannot withhold its remaining secrets much longer.

Acknowledgements

Work in my laboratory is supported by the Wellcome Trust, Medical Research Council, British Diabetic Association, and Serono Diagnostics Ltd. I am grateful to many collaborators for stimulating discussions, particularly Maria Soos in Cambridge and my sabbatical colleagues at the Joslin Diabetes Center in Boston. I also thank Jackie Sharpe for her help in preparation of the manuscript.

References

Accili, D., Mosthaf, L., Ullrich, A., and Taylor, S. I. (1991). A mutation in the extracellular domain of the insulin receptor impairs the ability of insulin to stimulate receptor autophosphorylation. *J. Biol. Chem.* **266,** 434–9.

Andersen, A. S., Kjeldsen, T., Wiberg, F. C., Christensen, P. M., Rasmussen, J. S., Norris, K., Moller, K. B., and Moller, N. P. H. (1990). Changing the insulin receptor to possess insulin-like growth factor I ligand specificity. *Biochemistry* **29,** 7363–6.

Anderson, N. G., Maller, J. L., Tonks, N. K., and Sturgill, T. W. (1990). Requirement for integration of signals from two distinct phosphorylation pathways for activation of MAP kinase. *Nature* **343,** 651–3.

Araki, E., Shimada, F., Uzawa, H., Mori, M., and Ebina, Y. (1987). Characterization of the promoter region of the human insulin receptor gene. *J. Biol. Chem.* **262,** 16186–91.

Arner, P., Pollare, T., Lithell, H., and Livingston, J. N. (1987). Defective insulin receptor tyrosine kinase in human skeletal muscle in obesity and type 2 (non-insulin dependent) diabetes mellitus. *Diabetologia* **30,** 347–440.

Auberger, P., Falquerho, L., Contreres, J. O., Pages, G., Le Cam, G., Rossi, B., and Le Cam, A. (1989). Characterization of a natural inhibitor of the insulin receptor tyrosine kinase: cDNA cloning, purification and anti-mitogenic activity. *Cell* **58,** 631–40.

Avruch, J. (1989). The insulin receptor kinase: structure–function relationships. In *Insulin action,* pp. 65–77. Alan Liss, New York.

Backer, J. M., Kahn, C. R., and White, M. F. (1989). Tyrosine phosphorylation of the insulin receptor during insulin-stimulated internalization in rat hepatoma cells. *J. Biol. Chem.* **264,** 1694–701.

Backer, J. M., Kahn, C. R., Cahill, D. A., Ullrich, A., and White, M. F. (1990). Receptor-mediated internalization of insulin requires a 12-amino acid sequence in the juxtamembrane region of the insulin receptor β-subunit. *J. Biol. Chem.* **265,** 16450–4.

Bajaj, M., Waterfield, M. D., Schlessinger, J., Taylor, W. R., and Blundell, T. (1987). On the tertiary structure of the extracellular domains of the epidermal growth factor and insulin receptors. *Biochim. Biophys. Acta* **916,** 220–6.

Balage, M, Grizard, J., Sornet, C., Simon, J., Dardevet, D., and Manin, M. (1990). Insulin binding and receptor tyrosine kinase activity in rat liver and skeletal muscle: effect of starvation. *Metabolism* **39,** 366–73.

Ballotti, R., Kowalski, A., White, M. F., Le Marchand-Brustel, Y., and Van Obberghen, E. (1987). Insulin stimulates tyrosine phosphorylation of its receptor β-subunit in intact rat hepatocytes. *Biochem. J.* **241,** 99–104.

Ballotti, R., Lammers, R., Scimeca, J-C., Dull, T., Schlessinger, Ullrich, A., and Van Obberghen, E. (1989). Intermolecular transphosphorylation between insulin receptors and EGF-insulin receptor chimerae. *EMBO J.* **8,** 3303–9.

Baron, V., Gautier, N., Komoriya, A., Hainaut, P., Scimeca, J-C., Mervic, M., Lavielle, S., Dolais-Kitabgi, J., and Van Obberghen, E. (1990). Insulin binding to its receptor induces a conformational change in the receptor C-terminus. *Biochemistry* **29,** 4634–41.

Batarseh, H., Thompson, R. A., Odugbesan, O., and Barnett, A. H. (1988). Insulin receptor antibodies in diabetes mellitus. *Clin. Exp. Immunol.* **71,** 85–90.

Becker, A. B. and Roth, R. A. (1990). Insulin receptor structure and function in normal and pathological conditions. *Ann. Rev. Med.* **41,** 99–115.

Birnbaumer, L. (1990). Transduction of receptor signal into modulation of effector activity by G proteins. *FASEB J.* **4,** 3068–78.

Blackshear, P. J., Haupt, D. M., App, A., and Rapp, U. R. (1990). Insulin activates the Raf-1 protein kinase. *J. Biol. Chem.* **265,** 12131–4.

Bollag, G. E., Roth, R. A., Beaudoin, J., Mochly-Rosen, D., and Koshland, D. E. (1986). Protein kinase C directly phosphorylates the insulin receptor *in vitro* and reduces its protein-tyrosine kinase activity. *Proc. Natl Acad. Sci. USA* **83,** 5822–4.

Boni-Schnetzler, M., Rubin, J. B., and Pilch, P. F. (1986). Structural requirements for the transmembrane activation of the insulin receptor kinase. *J. Biol. Chem.* **261,** 15281–7.

Boni-Schnetzler, M., Scott, W., Waugh, S. E., DiBella, E., and Pilch, P. F. (1987). The insulin receptor: structural basis for high affinity ligand binding. *J. Biol. Chem.* **262,** 8395–401.

Boni-Schnetzler, M., Kaligian, A., Del Vecchio, R., and Pilch, P. F. (1988). Ligand-dependent intersubunit association within the insulin receptor complex activates its intrinsic kinase activity. *J. Biol. Chem.* **263,** 6822–8.

Bottaro, D. P., Bonner-Weir, S., and King, G. L. (1989). Insulin receptor recycling in vascular endothelial cells: regulation by insulin and phorbol ester. *J. Biol. Chem.* **264,** 5916–23.

Bourne, H. R., Sanders, D. A., and McCormick, F. (1990). The GTPase superfamily: a conserved switch for diverse cell functions. *Nature* **348,** 125–32.

Boyd, J. J., Contreras, I., Kern, M., Tapscott, E. B., Downes, D. L., Frisell, W. R., and Dohm, G. L. (1990). Effect of a high-fat-sucrose diet on *in vivo* insulin receptor kinase activation. *Am. J. Physiol.* **259,** E111–E116.

Boyle, T. R., Campana, J., Sweet, L. J., and Pessin, J. E. (1985). Subunit structure of the purified human placental insulin receptor. *J. Biol. Chem.* **260,** 8593–600.

Brillon, D. J., Friedenberg, G. R., Henry, R. R., and Olefsky, J. M. (1989). Mechanism of defective insulin-receptor kinase activity in NIDDM: evidence for two receptor populations. *Diabetes* **38,** 397–403.

Brindle, N. P. J., Tavare, J. M., Dickens, M., Whittaker, J., and Siddle, K. (1990). Anti-(insulin receptor) monoclonal antibody-stimulated tyrosine phosphorylation in cells transfected with human insulin receptor cDNA. *Biochem. J.* **268,** 615–20.

Cama, A., Patterson, A. P., Kadowaki, T., Kadowaki, H., Siegel, G., D'Ambrosio, D., Lillioja, S., Roth, J., and Taylor, S. I. (1990). The amino acid sequence of the insulin receptor is normal in an insulin-resistant Pima Indian. *J. Clin. Endocrin. Metab.* **70,** 1155–61.

Cantley, L. C., Auger, K. R., Carpenter, C., Duckworth, B., Graziani, A., Kapeller, R., and Soltoff, S. (1991). Oncogenes and signal transduction. *Cell* **64,** 281–302.

Caro, J. F., Ittoop, O., Pories, W. J., Meelheim, D., Flickinger, E. G., Thomas, F., Jenquin, M., Silverman, J. F., Khazanie, P. G., and Sinha, M. K. (1986). Studies on the mechanism of insulin resistance in the liver from humans with non-insulin-dependent diabetes. *J. Clin. Invest.* **78,** 249–58.

Caro, J. F., Sinha, M. K., Raju, S. M., Ittoop, D., Pories, W. J., Flickinger, E. G., Meelheim, D., and Dohm, G. L. (1987). Insulin receptor kinase in human skeletal muscle from obese subjects with and without non-insulin-dependent diabetes. *J. Clin. Invest.* **79,** 1330–7.

Caro, J. F., Dohm, L. G., Pories, W. J., and Sinha, M. K. (1989). Cellular alterations in liver, skeletal muscle, and adipose tissue responsible for insulin resistance in obesity and type II diabetes. *Diabetes/Metabolism Rev.* **5,** 665–89.

Carpentier, J.-L. (1989). The cell biology of the insulin receptor. *Diabetologia* **32,** 627–35.

Chen, W.-J., Goldstein, J. L., and Brown, M. S. (1990). NPXY, a sequence often found in cytoplasmic tails, is required for coated pit-mediated internalization of the low density lipoprotein receptor. *J. Biol. Chem.* **265,** 3116–23.

Chou, C. K., Dull, T. J., Russell, D. S., Gherzi, R., Lebwohl, D., Ullrich, A., and Rosen, O. M. (1987). Human insulin receptors mutated at the ATP-binding site lack protein tyrosine kinase activity and fail to mediate postreceptor effects of insulin. *J. Biol. Chem.* **262,** 1842–7.

Christiansen, K., Tranum-Jensen, J., Carlsen, J., and Vinten, J. (1991). A model for the quaternary structure of human placental insulin receptor deduced from electron microscopy. *Proc. Natl Acad. Sci. USA* **88,** 249–52.

Ciaraldi, T. P. and Maisel, A. (1989). Role of guanine nucleotide regulatory proteins in insulin stimulation of glucose transport in rat adipocytes: influence of bacterial toxins. *Biochem. J.* **264,** 389–96.

Cicirelli, M. F., Tonks, N. K., Diltz, C. D., Weiel, J. E., Fischer, E. H., and Krebs, E. G. (1990). Microinjection of a protein-tyrosine-phosphatase inhibits insulin action in *Xenopus* oocytes. *Proc. Natl Acad. Sci. USA* **87,** 5514–18.

Cobb, M. H., Sang, B.-C., Gonzalez, R., Goldsmith, E., and Ellis, E. (1989). Autophosphorylation activates the soluble cytoplasmic domain of the insulin receptor in an intermolecular reaction. *J. Biol. Chem.* **264,** 18701–6.

Contreras, I., Dohm, G. L., Abdallah, S., Wells, J. A., Mooney, N., Rovira, A., and Caro, J. F. (1990). The effect of fasting on the activation *in vivo* of the insulin receptor kinase. *Biochem. J.* **265,** 887–90.

Czech, M. P. (1985). The nature and regulation of the insulin receptor: structure and function. *Ann. Rev. Physiol.* **47,** 357–81.

Czech, M. P. (1989). Signal transmission by the insulin-like growth factors. *Cell.* **59,** 235–8.

Czech, M. P., Klarlund, J. K., Yagaloff, K. A., Bradford, A. P., and Lewis, R. E. (1988). Insulin receptor signalling: activation of multiple serine kinases. *J. Biol. Chem.* **263,** 11017–20.

Debant, A., Clauser, E., Ponzio, G., Filloux, C., Auzan, C., Contreres, J. O., and Rossi, B. (1988). Replacement of insulin receptor tyrosine residues 1162 and 1163 does not alter the mitogenic effect of the hormone. *Proc. Natl Acad. Sci. USA* **85,** 8032–6.

DeFronzo, R. A. (1988). The triumvirate: β-cell, muscle, liver. A collusion responsible for NIDDM. *Diabetes* **37,** 667–87.

De Meyts, P., Gu, J.-L., Shymko, R. M., Kaplan, B. E., Bell, G. I., and Whittaker, J. (1990). Identification of a ligand-binding region of the human insulin receptor encoded by the second exon of the gene. *Mol. Endocrinol.* **4,** 409–16.

Donner, D. B. and Yonkers, K. (1983). Hormone-induced conformational changes in the hepatic insulin receptor. *J. Biol. Chem.* **258,** 9413–18.

Duckworth, W. C. (1988). Insulin degradation: mechanisms, products and significance. *Endocrine Rev.* **9,** 319–45.

Due, C., Sinonsen, M., and Olsson, L. (1986). The major histocompatibility complex class I heavy chain as a structural subunit of the human cell membrane insulin receptor: implications for the range of biological functions of histocompatibility antigens. *Proc. Natl Acad. Sci. USA* **83,** 6007–11.

Duronio, V. and Jacobs, S. (1990). The effect of protein kinase-C inhibition on insulin receptor phosphorylation. *Endocrinology* **127,** 481–7.

Ebina, Y., Ellis, L., Jarnagin, K., Edery, M., Graf, L., Clauser, E., Ou, J.-H., Masiarz, F., Kan, Y. W., Goldfine, I. D., Roth, R. A., and Rutter, W. J. (1985). The human insulin receptor cDNA: the structural basis for hormone-activated transmembrane signalling. *Cell* **40,** 747–58.

Ebina, Y., Araki, E., Taira, M., Shimada, F., Mori, M., Craik, C. S., Siddle, K., Pierce, S. B., Roth, R. A., and Rutter, W. J. (1987). Replacement of lysine residue 1030 in the putative ATP-binding region of the insulin receptor abolishes insulin- and antibody-stimulated glucose uptake and receptor kinase activity. *Proc. Natl Acad. Sci. USA* **84,** 704–8.

Ellis, L., Clauser, E., Morgan, D. O., Edery, M., Roth, R. A., and Rutter, W. J. (1986). Replacement of insulin receptor tyrosine residues 1162 and 1163 compromises insulin-stimulated kinase activity and uptake of 2-deoxyglucose. *Cell* **45,** 721–32.

Ellis, L., Levitan, A., Cobb, M. H., and Ramos, P. (1988). Efficient expression in insect cells of a soluble, active human insulin receptor protein-tyrosine kinase domain by use of a baculovirus vector. *J. Virol.* **62,** 1634–9.

Ellis, L., Tavare, J. M., and Levine, B. (1991). Insulin receptor tyrosine kinase structure and function. *Biochem. Soc. Trans,* **19,** 426–32.

Endemann, G., Yonezawa, K., and Roth, R. A. (1990). Phosphatidylinositol kinase or an associated protein is a substrate for the insulin receptor tyrosine kinase. *J. Biol. Chem.* **265,** 396–400.

Farese, R. V. and Cooper, D. R. (1989). Potential role of phospholipid-signalling systems in insulin action and states of clinical insulin resistance. *Diabetes/Metabolism Rev.* **5,** 455–74.

Fehlmann, M., Peyron, J. F., Samson, M., Van Obberghen, E., Brandenburg, D. and Brossette, N. (1985). Molecular association between major histocompatibility complex class I antigens and insulin receptors in mouse liver membranes. *Proc. Natl Acad. Sci. USA* **82,** 8634–7.

Finn, F. M., Ridge, K. D., and Hofmann, K. (1990). Labile disulfide bonds in human placental insulin receptor. *Proc. Natl Acad. Sci. USA* **87,** 419–23.

Flores-Riveros, J. R., Sibley, E., Kastelic, T., and Lane, D. M. (1989). Substrate phosphorylation catalysed by the insulin receptor tyrosine kinase: kinetic correlation to autophosphorylation of specific sites in the β subunit. *J. Biol. Chem.* **264,** 21557–72.

Florke, R.-R., Klein, H. W., and Reinauer, H. (1990). Structural requirements for signal transduction of the insulin receptor. *Eur. J. Biochem.* **191,** 473–82.

Forsayeth, J., Maddux, B., and Goldfine, I. D. (1986). Biosynthesis and processing of the human insulin receptor. *Diabetes* **35,** 837–46.

Friedenberg, G. R., Henry, R. R., Klein, H. H., Reichart, D. R., and Olefsky, J. M. (1987). Decreased kinase activity of insulin receptors from adipocytes of non-insulin-dependent diabetic subjects. *J. Clin. Invest.* **79,** 240–50.

Friedenberg, G. R., Reichart, D., Olefsky, J. M., and Henry R. R. (1988). Reversibility of defective adipocyte insulin reecptor kinase activity in non-insulin-dependent diabetes mellitus: effect of weight loss. *J. Clin. Invest.* **82,** 1398–406.

Fujita-Yamaguchi, Y., Kathuria, S., Xu, Q.-Y., McDonald, J. M., Nakano, H., and Kamata, T. (1989). *In vitro* tyrosine phosphorylation studies on RAS proteins and calmodulin suggest that polylysine-like basic peptides or domains may be involved in interactions between insulin receptor kinase and its substrate. *Proc. Natl Acad. Sci. USA* **86,** 7306–10.

Gammeltoft, S. (1984). Insulin receptors: binding kinetics and structure–function relationship of insulin. *Physiol. Rev.* **64,** 1322–78.

Gammeltoft, S. and Van Obberghen, E. (1986). Protein kinase activity of the insulin receptor. *Biochem. J.* **235,** 1–11.

Gilman, A. G. (1987). G proteins: transducers of receptor-generated signals. *Ann. Rev. Biochem.* **56,** 615–49.

Goldfine, I. D. (1987). The insulin receptor: molecular biology and transmembrane signaling. *Endocrine Rev.* **8,** 235–55.

Goldstein, B. J. and Dudley, A. L. (1990). The rat insulin receptor: primary structure and conservation of tissue-specific alternative messenger RNA splicing. *Mol. Endocrinol.* **4,** 235–44.

Gu, J. L., Goldfine, I. D., Forsayeth, J. R., and De Meyts, P. (1988). Reversal of insulin-induced negative cooperativity by monoclonal antibodies that stabilise the slowly dissociating (K super) state of the insulin receptor. *Biochem. Biophys. Res. Comm.* **150,** 694–701.

Gustafson, T. A. and Rutter, W. J. (1990). The cysteine-rich domains of the insulin and insulin-like growth factor I receptors are primary determinants of hormone binding specificity: evidence from receptor chimeras. *J. Biol. Chem.* **265,** 18663–7.

Hansen, T., Stagsted, J., Pedersen, L., Roth, R. A., Goldstein, A., and Olsson, L. (1989). Inhibition of insulin receptor phosphorylation by peptides derived from major histocompatibility complex class I antigens. *Proc. Natl Acad. Sci. USA* **86,** 3123–6.

Haring, H., Kirsch, D., Obermaier, B., Ermel, B., and Machicao, F. (1986). Tumor-promoting phorbol esters increase the K_m of the ATP-binding site of the insulin receptor kinase from rat adipocytes. *J. Biol. Chem.* **261,** 3869–75.

Haring, H. and Obermaier-Kusser, B. (1989). Insulin receptor kinase defects in insulin-resistant tissues and their role in the pathogenesis of NIDDM. *Diabetes/Metabolism Rev.* **5**, 431–41.

Hedo, J. O. and Gorden, P. (1985). Biosynthesis of the insulin receptor. *Horm. Metab. Res.* **17,** 487–90.

Hedo, J. A. and Simpson, I. A. (1985). Biosynthesis of the insulin receptor in rat adipose cells: intracellular processing of the M_r-190000 pro-receptor. *Biochem. J.* **232,** 71–8.

Heffetz, D., Bushkin, I., Dror, R., and Zick, Y. (1990). The insulinomimetic agents H_2O_2 and vanadate stimulate protein tyrosine phosphorylation in intact cells. *J. Biol. Chem.* **265**, 2896–902.

Heidenreich, K. A. and Olefsky, J. M. (1985). The metabolism of insulin receptors: internalization, degradation, and recycling. In *Molecular basis of insulin action* (ed. M. P. Czech), pp. 45–65. Plenum Press, New York.

Houslay, M. D. (1986). Insulin, glucagon and the receptor-mediated control of cyclic AMP concentrations in liver. *Biochem. Soc. Trans.* **14,** 183–92.

Houslay, M. D. and Siddle, K. (1989). Molecular basis of insulin receptor function. *Br. Med. Bull.* **45,** 264–84.

Houslay, M. D., Pyne, N. J., O'Brien, R. M., Siddle, K., Strassheim, D., Palmer, T., Spence, S., Woods, M., Wilson, A., Lavan, B., Murphy, G. J., Saville, M., McGregor, M., Kilgour, E., Anderson, N., Knowles, J. T., Griffiths, S., and Milligan, G. (1989). Guanine nucleotide regulatory proteins in insulin's action and in diabetes. *Biochem. Soc. Trans.* **17,** 627–9.

Humbel, R. E. (1990). Insulin-like growth factors I and II. *Eur. J. Biochem.* **190,** 445–62.

Hunter, T. (1989). Protein-tyrosine phosphatases: the other side of the coin. *Cell* **58,** 1013–16.

Irvine, F. J. and Houslay, M. D. (1988). Insulin and glucagon attenuate the ability of cholera toxin to activate adenylate cyclase in intact hepatocytes. *Biochem. J.* **51,** 447–52.

Issad, T., Tavare, J. M., and Denton, R. M. (1991). Analysis of insulin receptor phosphorylation sites in intact rat liver cells by two-dimensional phosphopeptide mapping: predominance of the tris-phosphorylated form of the kinase domain after insulin stimulation. *Biochem. J.* **275,** 15–21.

Joost, H. G., Schmitz-Salne, C., Hinsch, K. D., Schultz, G., and Rosenthal, W. (1989). Phosphorylation of G-protein α-subunits in intact adipose cells: evidence against a mediating role in insulin-dependent metabolic effects. *Eur. J. Pharmacol.* **172,** 461–9.

Kadowaki, T., Kadowaki, M., Rechler, M. M., Serrano-Rios, M., Roth, J., Gorden, P., and Taylor, S. I. (1990*a*). Five mutant alleles of the insulin receptor gene in patients with genetic forms of insulin resistance. *J. Clin. Invest.* **86,** 254–64.

Kadowaki, T., Kadowaki, H., and Taylor, S. I. (1990*b*). A nonsense mutation causing decreased levels of insulin receptor mRNA: detection by a simplified technique for direct sequencing of genomic DNA amplified by the polymerase chain reaction. *Proc. Natl Acad. Sci. USA* **87,** 658–62.

Kadowaki, T., Kadowaki, H., Cama, A., Marcus-Samuels, B., Rovira, A., Bevins, C. L., and Taylor, S. I. (1990*c*). Mutagenesis of Lysine 460 in the human insulin receptor: effects upon receptor recycling and cooperative interactions among binding sites. *J. Biol. Chem.* **265,** 21285–96.

Kahn, C. R. and White, M. F. (1988). The insulin receptor and the molecular mechanism of insulin action. *J. Clin. Invest.* **82,** 1151–6.

Kahn, C. R., Baird, K. L., Flier, J. S., Grunfeld, C., Harmon, J. T., Harrison, L. C.,

Karlsson, F. A., Kasuga, M., King, G. L., Lang, U. C., Poldskalny, J. M., and Van Obberghen, E. (1981). Insulin receptors, receptor antibodies and the mechanism of insulin action. *Rec. Progr. Hormone Res.* **37,** 477–538.

Kallen, R. G., Smith, J. E., Sheng, Z., and Tung, L. (1990). Expression, purification and characterization of a 41 kDa insulin receptor tyrosine kinase domain. *Biochem. Biophys. Res. Comm.* **168,** 616–24.

Karasik, A., Rothenberg, P. L., Yamada, K., White, M. F., and Kahn, C. R. (1990). Increased protein kinase C activity is linked to reduced insulin receptor autophosphorylation in liver of starved rats. *J. Biol. Chem.* **265,** 10226–31.

Kasuga, M., Karlsson, F. A., and Kahn, C. R. (1982). Insulin stimulates the phosphorylation of the 95000-Dalton subunit of its own receptor. *Science* **215,** 185–6.

Khan, M. N., Baquiran, G., Brule, C., Burgess, J., Foster, B., Bergeron, J. J. M., and Posner, B. I. (1989). Internalization and activation of the rat liver insulin receptor kinase *in vivo*. *J. Biol. Chem.* **264,** 12931–40.

King, M. J. and Sale, G. J. (1990). Dephosphorylation of insulin receptor autophosphorylation sites by particulate and soluble phosphotyrosyl-protein phosphatases. *Biochem. J.* **266,** 251–9.

Kjeldsen, T., Andersen, A. S., Wiberg, F. C., Rasmussen, J. S., Schaffer, L., Balschmidt, P., Moller, K. B., and Moller, N. P. H. (1991). The ligand specificities of the insulin receptor and the IGF-1 receptor reside in different regions of a common binding site. *Proc. Natl Acad. Sci. USA* **88,** 4404–8.

Kohanski, R. A. (1989). Insulin receptor aggregation and autophosphorylation in the presence of cationic polyamino acids. *J. Biol. Chem.* **264,** 20984–91.

Kono, S., Kuzuya, H., Okamoto, M., Nishimura, H., Kosaki, A., Kakehi, T., Okamoto, M., Inoue, G., Maeda, I., and Imura, H. (1990). Changes in insulin receptor kinase with aging in rat skeletal muscle and liver. *Am. J. Physiol.* **259,** E27–E35.

Koshio, O., Akanuma, Y., and Kasuga, M. (1988). Hydrogen peroxide stimulates tyrosine phosphorylation of the insulin receptor and its tyrosine kinase activity in intact cells. *Biochem. J.* **250,** 95–101.

Kovacina, A., Yonezawa, K., Brautigan, D. L., Tonks, N. K., Rapp, U. R., and Roth, R. A. (1990). Insulin activates the kinase activity of the Raf-1 proto-oncogene by increasing its serine phosphorylation. *J. Biol. Chem.* **265,** 12115–18.

Krueger, N. X., Streuli, M., and Saito, H. (1990). Structural diversity and evolution of human receptor-like protein tyrosine phosphatases. *EMBO J.* **9,** 3241–52.

Lammers, R., Van Obberghen, E., Ballotti, R., Schlessinger, J., and Ullrich, A. (1990). Transphosphorylation as a possible mechanism for insulin and epidermial growth factor amplification. *J. Biol. Chem.* **265,** 16886–90.

Lau, K. H. W., Farley, J. R., and Baylink, D. J. (1989). Phosphotyrosyl protein phosphatases. *Biochem. J.* **257,** 23–36.

Leef, J. W. and Larner, J. (1987). Insulin-mimetic effect of trypsin on the insulin receptor tyrosine kinase in intact adipocytes. *J. Biol. Chem.* **262,** 14837–42.

Lekanne-Deprez, R. H., Potter, Loon, B. J. van, Zon, G. C. M. van der, Moller, W., Lindhout, D., Klinkhamer, M. P., Krans, H. M. J., and Maassen, J. A. (1989). Individuals with only one allele for a functional insulin receptor have a tendency to hyperinsulinaemia but not to hyperglycaemia. *Diabetologia* **32,** 740–4.

La Marchand-Brustel, Y., Ballotti, R., Gremeaux, T., Tanti, J.-F., Brandenburg, D., and Van Obberghen, E. (1989). Functional labelling of insulin receptor subunits in live cells: $\alpha_2\beta_2$ species is the major autophosphorylated form. *J. Biol. Chem.* **264,** 21316–21.

Lewis, R. E. and Czech, M. P. (1987). Phospholipid environment alters hormone-sensitivity of the purified insulin receptor kinase. *Biochem. J.* **248,** 829–36.

Lewis, R. E., Cao, L., Perregaux, D., and Czech, M. P. (1990*a*). Threonine 1336 of the human insulin receptor is a major target for phosphorylation by protein kinase C. *Biochemistry* **29,** 1807–13.

Lewis, R. E., Wu, G. P., MacDonald, R. G., and Czech, M. P. (1990*b*). Insulin-sensitive phosphorylation of serine 1293/1294 on the human insulin receptor by a tightly associated serine kinase. *J. Biol. Chem.* **265,** 947–54.

Li, P., Wood, K., Mamon, H., Haser, W., and Roberts, T. (1991). Raf-1: a kinase currently without a cause but not lacking in effects. *Cell* **64,** 479–82.

Lipson, K. E., Yamada, K., Kolhatkar, A. A., and Donner, D. B. (1986). Relationship between the affinity and proteolysis of the insulin receptor. *J. Biol. Chem.* **261,** 10833–8.

Low, M. G. and Saltiel, A. R. (1988). Structural and functional roles of glycosyl-phosphatidylinositol in membranes. *Science* **239,** 268–75.

Ludwig, S. M., Faiman, C., and Dean, H. J. (1987). Insulin and insulin-receptor autoantibodies in children with newly diagnosed IDDM before insulin therapy. *Diabetes* **36,** 420–5.

Luttrell, L. M., Hewlett, E. L., Romero, G., and Rogol, A. D. (1988). Pertussis toxin treatment attenuates some effects of insulin in BC3H-1 myocytes. *J. Biol. Chem.* **263,** 6134–41.

Luttrell, L. M., Kilgour, E., Larner, J., and Romero, G. (1990). A pertussis toxin-sensitive G protein mediates some aspects of insulin action in BC3H-1 murine myocytes. *J. Biol. Chem.* **265,** 16873–9.

McClain, D. A., Maegawa, H., Lee, J., Dull, T. J., Ullrich, A., and Olefsky, J. M. (1987). A mutant insulin receptor with defective tyrosine kinase displays no biologic activity and does not undergo endocytosis. *J. Biol. Chem.* **262,** 14663–71.

Maegawa, H., McClain, D. A., Friedenberg, G., Olefsky, J. M., Napier, M., Lipari, T., Dull, T. J., Lee, J., and Ullrich, A. (1988). Properties of a human insulin receptor with a COOH-terminal truncation: II. truncated receptors have normal kinase activity but are defective in signaling metabolic effects. *J. Biol. Chem.* **263,** 8912–17.

Magee, A. I. and Siddle, K. (1988). Insulin and IGF-1 receptors contain covalently bound palmitic acid. *J. Cell. Biochem.* **37,** 347–57.

Mamula, P. W., Wong, K. Y., Maddux, B. A., McDonald, A. R., and Goldfine, I. D. (1988). Sequence and analysis of promoter region of human insulin-receptor gene. *Diabetes* **37,** 1241–6.

Mamula, P. W., McDonald, A. R., Brunetti, A., Okabayashi, Y., Wong, K. Y., Maddux, B. A., Logsdon, C., and Goldfine, I. D. (1990). Regulating insulin receptor gene expression by differentiation and hormones. *Diabetes Care* **13,** 288–301.

Maron, R., Elias, D., Jongh, B. M. de, Bruining, G. J., Rood, J. J. van, Schechter, Y., and Cohen, I. R. (1983). Autoantibodies to the insulin receptor in juvenile onset insulin-dependent diabetes. *Nature* **303,** 817–18.

Martinez, C., Ruiz, P., Andres, A., Satrustegui, J., and Carrascosa, J. M. (1989). Tyrosine kinase activity of liver insulin receptor is inhibited in rats at term gestation. *Biochem. J.* **268,** 267–72.

Moises, R. S. and Heidenreich, K. (1990). Pertussis toxin catalyzed ADP-ribosylation of a 41 kDa G-protein impairs insulin-stimulated glucose metabolism in BC3H-1 myocytes. *J. Cell. Physiol.* **144,** 538–45.

Moller, D. E., Yokota, A., Caro, J. F., and Flier, J. S. (1989*a*). Tissue-specific expression of two alternatively spliced insulin receptor mRNAs in man. *Mol. Endocrinol.* **3,** 1263–9.

Moller, D. E., Yokota, A., and Flier, J. S. (1989*b*). Normal insulin-receptor cDNA sequence in Pima Indians with NIDDM. *Diabetes* **38,** 1496–500.

Moller, D. E., Yokota, A., Ginsberg-Fellner, F., and Flier, J. S. (1990*a*). Functional properties of a naturally occurring Trp^{1200} mutation of the insulin receptor. *Mol. Endocrinol.* **4,** 1183–91.

Moller, D. E., Yokota, A., White, M. F., Pazianos, A. G., and Flier, J. S. (1990*b*). A naturally occurring mutation of insulin receptor alanine 1134 impairs tyrosine kinase function and is associated with dominantly inherited insulin resistance. *J. Biol. Chem.* **265,** 14979–85.

Morgan, D. O. and Roth, R. A. (1987). Acute insulin action requires insulin receptor kinase activity: introduction of an inhibitory monoclonal antibody into mammalian cells blocks the rapid effects of insulin. *Proc. Natl Acad. Sci. USA* **84,** 41–5.

Mosthaf, L., Grako, K., Dull, T. J., Coussens, L., Ullrich, A., and McClain, D. A. (1990). Functionally distinct insulin receptors generated by tissue-specific alternative splicing. *EMBO J.* **9,** 2409–13.

Moxham, C. P., Duronio, V., and Jacobs, S. (1989). Insulin-like growth factor I receptor β subunit heterogeneity: evidence for hybrid tetramers composed of insulin-like growth factor I and insulin receptor heterodimers. *J. Biol. Chem.* **264,** 13238–44.

Myers, M. G., Backer, J. M., Siddle, K., and White, M. F. (1991). The insulin receptor functions normally in Chinese hamster ovary cells after truncation of the C-terminus. *J. Biol. Chem.* **266,** 10616–23.

Nagy, K., Levy, J., and Grunberger, G. (1990). High-fat feeding induces tissue-specific alteration in proportion of activated insulin receptors in rats. *Acta Endocrin.* **122,** 361–8.

Nishibe, S., Wahl, M. I., Wedegaertner, P. B., Kim, J. J., Rhee, S. G., and Carpenter, G. (1990). Selectivity of phospholipase C phosphorylation by the epidermal growth factor receptor, the insulin receptor, and their cytoplasmic domains. *Proc. Natl Acad. Sci. USA* **87,** 424–8.

Nyomba, B. L., Ossowski, V. M., Bogardus, C., and Mott, D. M. (1990). Insulin-sensitive tyrosine kinase: relationship with *in vivo* insulin action in humans. *Am. J. Physiol.* **258,** E964–E974.

Obermaier-Kusser, B., White, M. F., Pongratz, D. E., Su, Z., Ermel, B., Muhlbacher, C., and Haring, H. U. (1989). A defective intramolecular autoactivation cascade may cause the reduced receptor kinase activity of the skeletal muscle insulin receptor from patients with non-insulin-dependent diabetes mellitus. *J. Biol. Chem.* **264,** 9497–504.

O'Hare, T. and Pilch, P. F. (1988). Separation and characterization of three insulin receptor species that differ in subunit composition. *Biochemistry* **27,** 5693–700.

O'Hare, T. and Pilch, P. F. (1990). Intrinsic kinase activity of the insulin receptor. *Int. J. Biochem.* **22,** 315–24.

Olefsky, J. M. (1990). The insulin receptor: a multifunctional protein. *Diabetes* **39,** 1009–16.

Olson, T. S., Bamberger, M. J., and Lane, M. D. (1988). Post-translational changes in tertiary and quaternary structure of the insulin proreceptor: correlation with acquisition of function. *J. Biol. Chem.* **263,** 7342–51.

O'Rahilly, S., Choi, W. H., Patel, P., Turner, R. C., Flier, J. S., and Moller, D. E. (1991). Detection of mutations in the insulin receptor gene in non-insulin-dependent diabetic patients by analysis of single-stranded conformation polymorphisms. *Diabetes* **40,** 777–82.

Pang, D. T. and Shafer, J. A. (1984). Evidence that insulin receptor from human placenta has a high affinity for only one molecule of insulin. *J. Biol. Chem.* **259,** 8589–96.

Pang, D. T., Sharma, B. R., Shafer, J. A., White, M. F., and Kahn, C. R. (1985). Predominance of tyrosine phosphorylation of insulin receptors during the initial response of intact cells to insulin. *J. Biol. Chem.* **260,** 7131–6.

Paul, J. I., Tavare, J. M., Denton, R. M., and Steiner, D. F. (1990). Baculovirus-directed expression of the human insulin receptor and an insulin-binding ectodomain. *J. Biol. Chem.* **265,** 13074–83.

Perlman, R., Bottaro, D. P., White, M. F., and Kahn, C. R. (1989). Conformational changes in the α- and β-subunits of the insulin receptor identified by anti-peptide antibodies. *J. Biol. Chem.* **264,** 8946–50.

Petruzzelli, L., Herrera, R., and Rosen, O. M. (1984). Insulin receptor is an insulin-dependent tyrosine protein kinase: copurification of insulin binding activity and protein kinase activity to homogeneity from human placenta. *Proc. Natl Acad. Sci. USA* **81,** 3327–31.

Prigent, S. A., Stanley, K. K., and Siddle, K. (1990). Identification of epitopes on the human insulin receptor reacting with rabbit polyclonal antisera and mouse monoclonal antibodies. *J. Biol. Chem.* **265,** 9970–7.

Pyne, N. J., Cushley, W., Nimmo, H. G., and Houslay, M. D. (1989*a*). Insulin stimulates the tyrosyl phosphorylation and activation of the 52 kDa peripheral plasma-membrane cyclic AMP phosphodiesterase in intact hepatocytes. *Biochem. J.* **261,** 897–904.

Pyne, N. J., Heyworth, C. M., Balfour, N., and Houslay, M. D. (1989*b*). Insulin affects the ability of Gi to be ADP-ribosylated but does not elicit its phosphorylation in intact hepatocytes. *Biochem. Biophys. Res. Comm.* **165,** 251–6.

Rafaeloff, R., Patel, P., Yip, C., Goldfine, I. D., and Hawley, D. M. (1989). Mutation of the high cysteine region of the human insulin receptor α-subunit increases insulin receptor binding affinity and transmembrane signalling. *J. Biol. Chem.* **264,** 15900–4.

Ray, L. B. and Sturgill, T. W. (1988). Insulin-stimulated microtubule-associated protein kinase is phosphorylated on tyrosine and threonine *in vivo*. *Proc. Natl Acad. Sci. USA* **85,** 3753–7.

Reaven, G. M. (1988). Role of insulin resistance in human disease. *Diabetes* **37,** 1595–607.

Reddy, S. S. K. and Kahn, C. R. (1988). Insulin resistance: a look at the role of insulin receptor kinase. *Diabetic Med.* **5,** 621–9.

Rochet, N., Sadoul, J. L., Ferrua, B., Kubar, J., Tanti, J. F., Bougneres, P., Vialettes, B., Van Obberghen, E., Le Marchand-Brustel, Y., and Freychet, P. (1990). Autoantibodies to the insulin receptor are infrequent findings in Type I (insulin-dependent) diabetes mellitus of recent onset. *Diabetologia* **33,** 411–16.

Roome, J., O'Hare, T., Pilch, P. F., and Brautigan, D. L. (1988). Protein phosphotyrosine phosphatase purified from the particulate fraction of human placenta dephosphorylates insulin and growth factor receptors. *Biochem. J.* **256,** 493–500.

Rosen, O. M. (1987). After insulin binds. *Science* **237,** 1452–8.

Rossiter, B. J. F. and Caskey, C. T. (1990). Molecular scanning methods of mutation detection. *J. Biol. Chem.* **265,** 12753–6.

Rossomando, A. J., Payne, D. M., Weber, M. J., and Sturgill, T. W. (1989). Evidence that pp42, a major tyrosine kinase target protein, is a mitogen-activated serine/threonine protein kinase. *Proc. Natl Acad. Sci. USA* **86,** 6940–3.

Roth, J. and Taylor, S. I. (1982). Receptors for peptide hormones: alterations in diseases of humans. *Ann. Rev. Physiol.* **44,** 639–51.

Rothenberg, P. L. and Kahn, C. R. (1988). Insulin inhibits pertussis toxin-catalyzed ADP-ribosylation of G-proteins: evidence for a novel interaction between insulin receptors and G-proteins. *J. Biol. Chem.* **263,** 15546–52.

Rothenberg, P., Kahn, C. R., Backer, J. M., Araki, E., Wilden, P. A., Cahill, D. A., Goldstein, B. J., and White, M. F. (1991). Structure of the insulin receptor substrate IRS-1 defines a unique signal transduction protein. *Nature* **352,** 73–7.

Ruderman, N. B., Kapeller, R., White, M. F., and Cantley, L. C. (1990). Activation of phosphatidylinositol 3-kinase by insulin. *Proc. Natl Acad. Sci. USA* **87,** 1411–15.

Russell, D. S., Gherzi, R., Johnson, E. L., Chou, C. K., and Rosen, O. M. (1987). The protein-tyrosine kinase activity of the insulin receptor is necessary for insulin-mediated receptor down-regulation. *J. Biol. Chem.* **262,** 11833–40.

Saltiel, A. R. (1990). Second messengers of insulin action. *Diabetes Care* **13,** 244–56.

Schaefer, E. M., Siddle, K., and Ellis, L. (1990). Deletion analysis of the human insulin receptor ectodomain reveals independently folded soluble subdomains and insulin binding by a monomeric α-subunit. *J. Biol. Chem.* **265,** 13248–53.

Schechter, Y., Elias, D., Maron, R., and Cohen, I. R. (1984). Mouse antibodies to the insulin receptor developing spontaneously as anti-idiotypes. I. Characterization of the antibodies. *J. Biol. Chem.* **259,** 6411–15.

Schechter, Y., Yaish, P., Chorev, M., Gilon, C., Braun, S., and Levitzki, A. (1989). Inhibition of insulin-dependent lipogenesis and anti-lipolysis by protein tyrosine kinase inhibitors. *EMBO J.* **8,** 1671–6.

Schenker, E. and Kohanski, R. A. (1988). Conformational states of the insulin receptor. *Biochem. Biophys. Res. Comm.* **157,** 140–5.

Seino, S. and Bell, G. I. (1989). Alternative splicing of human insulin receptor messenger RNA. *Biochem. Biophys. Res. Comm.* **159,** 312–16.

Seino, S., Seino, M., Nishi, S., and Bell, G. I. (1989). Structure of the human insulin receptor gene and characterization of its promoter. *Proc. Natl Acad. Sci. USA* **86,** 114–18.

Seino, S., Seino, M., and Bell, G. I. (1990). Human insulin-receptor gene. *Diabetes* **39,** 129–33.

Shia, M. A. and Pilch, P. F. (1983). The β subunit of the insulin receptor is an insulin-activated protein kinase. *Biochemistry* **22,** 717–21.

Shier, P. and Watt, V. M. (1989). Primary structure of a putative receptor for a ligand of the insulin family. *J. Biol. Chem.* **264,** 14605–8.

Shimada, F., Taira, M., Suzuki, Y., Hashimoto, N., Nozaki, O., Taira, M., Tatibana, M., Ebina, Y., Tawata, M., Onaya, T., Makino, H., and Yoshida, S. (1990). Insulin-resistant diabetes associated with partial deletion of insulin-receptor gene. *Lancet* **335,** 1179–81.

Shimoyama, R., Fujita-Yamaguchi, Y., and Boden, G. (1989). Anti-insulin receptor antibodies in human diabetes. *Diabetes Res. Clin. Pract.* **7,** S59–S66.

Shoelson, S. E., White, M. F., and Kahn, C. R. (1988). Tryptic activation of the insulin receptor: proteolytic truncation of the α-subunit releases the β-subunit from inhibitory control. *J. Biol. Chem.* **263,** 4852–60.

Sibley, E., Kastelic, T., Kelly, T. J., and Lane, D. M. (1989). Characterization of the mouse insulin receptor gene promoter. *Proc. Natl Acad. Sci. USA* **86,** 9732–6.

Siddle, K., Soos, M. A., O'Brien, R. M., Ganderton, R. H., and Taylor, R. (1987). Monoclonal antibodies as probes of the structure and function of insulin receptors. *Biochem. Soc. Trans.* **15,** 47–51.

Siddle, K., Soos, M. A., O'Brien, R. M., Ganderton, R. M., and Pillay, T. S. (1988). Monoclonal antibodies to the insulin receptor. In *Clinical applications of monoclonal antibodies* (ed. R. Hubbard and V. Marks), pp. 87–9. Plenum Publishing, New York.

Sinha, M. K., Pories, W. J., Flickinger, E. G., Meelheim, D., and Caro, J. F. (1987). Insulin-receptor kinase activity of adipose tissue from morbidity obese humans with and without NIDDM. *Diabetes* **36,** 620–5.

Sissom, J. and Ellis, L. (1989). Secretion of the extracellular domain of the human insulin-receptor from insect cells by use of a baculovirus vector. *Biochem. J.* **262,** 119–26.

Smith, D. M., King, M. J., and Sale, G. J. (1988). Two systems *in vitro* that show insulin-stimulated serine kinase activity towards the insulin receptor. *Biochem. J.* **250,** 509–19.

Smith, R. M. and Jarett, L. (1990). Partial characterization of mechanism of insulin accumulation in H35 hepatoma cell nuclei. *Diabetes* **39,** 683–9.

Soler, A. P., Thompson, K. A., Smith, R. M., and Jarett, L. (1989). Immunological demonstration of the accumulation of insulin, but not insulin receptors, in nuclei of insulin-treated cells. *Proc. Natl Acad. Sci. USA* **86,** 6640–4.

Sonne, O. (1988). Receptor-mediated endocytosis and degradation of insulin. *Physiol. Rev.* **68,** 1129–96.

Soos, M. A. and Siddle, K. (1989). Immunological relationships between receptors for insulin and insulin-like growth factor-I: evidence for structural heterogeneity of insulin-like growth factor-I receptors involving hybrids with insulin receptors. *Biochem. J.* **263,** 553–63.

Soos, M. A., Whittaker, J., Lammers, R., Ullrich, A., and Siddle, K. (1990). Receptors for insulin and insulin-like growth factor-I can form hybrid dimers: characterization of hybrid receptors in transfected cells. *Biochem. J.* **270,** 383–90.

Stadtmauer, L. and Rosen, O. M. (1986). Increasing the cAMP content of IM-9 cells alters the phosphorylation state and protein kinase activity of the insulin receptor. *J. Biol. Chem.* **261,** 3402–7.

Steele-Perkins, G. and Roth, R. A. (1990). Insulin-mimetic anti-insulin receptor monoclonal antibodies stimulate receptor kinase activity in intact cells. *J. Biol. Chem.* **265,** 9458–63.

Sun, X. J., Rothenberg, P., Kahn, C. R., Backer, J. M., Araki, E., Wilden, P. A., Cahill, D. A., Goldstein, B. J., and White, M. F. (1991). Structure of the insulin receptor substrate IRS-1 defines a unique signal transduction protein. *Nature* **352,** 73–7.

Sung, C. K., Maddux, B. A., Hawley, D. M., and Goldfine, I. D. (1989). Monoclonal antibodies mimic insulin activation of ribosomal protein S6 kinase without activation of insulin receptor tyrosine kinase: studies in cells transfected with normal and mutant insulin receptors. *J. Biol. Chem.* **264,** 18951–9.

Sweet, L. J., Morrison, B. D., and Pessin, J. E. (1987). Isolation of functional $\alpha\beta$ heterodimers from the purified human placental $\alpha_2\beta_2$ heterotetrameric insulin receptor complex: a structural basis for insulin binding heterogeneity. *J. Biol. Chem.* **262,** 6939–42.

Takayama, S., Kahn, C. R., Kubo, K., and Foley, J. E. (1988*a*). Alterations in insulin receptor autophosphorylation in insulin resistance: correlation with altered sensitivity to glucose transport and antilipolysis to insulin. *J. Clin. Endoc. Metab.* **66,** 992–9.

Takayama, S., White, M. F., and Kahn, C. R. (1988*b*). Phorbol ester-induced phosphorylation of the insulin receptor decreases its tyrosine kinase activity. *J. Biol. Chem.* **263,** 3440–7.

Tanti, J. F., Gremeaux, T., Rochet, N., Van Obberghen, E., and Le Marchand-Brustel, Y. (1987). Effect of cyclic AMP-dependent protein kinase on insulin receptor tyrosine kinase activity. *Biochem. J.* **245,** 19–26.

Tavare, J. M. and Dickens, M. (1991). Changes in insulin receptor tyrosine, serine and threonine phosphorylation as a result of substitution of tyrosine-1162 with phenylalanine. *Biochem. J.* **274,** 173–80.

Tavare, J. M., O'Brien, R. M., Siddle, K., and Denton, R. M. (1988). Analysis of insulin-receptor phosphorylation sites in intact cells by two-dimensional phosphopeptide mapping. *Biochem. J.* **253,** 783–8.

Taylor, R. (1989). Aetiology of non-insulin dependent diabetes. *Br. Med. Bull.* **45,** 73–91.

Taylor, S. I., Kadowaki, T., Kadowaki, H., Accili, D., Cama, A., and McKeon, C. (1990). Mutations in insulin-receptor gene in insulin-resistant patients. *Diabetes Care* **13,** 257–79.

Thies, R. S., Ullrich, A., and McClain, D. A. (1989). Augmented mitogenesis and impaired metabolic signaling mediated by a truncated insulin receptor. *J. Biol. Chem.* **264,** 12820–5.

Thies, R. S., Molina, J. M., Ciaraldi, T., Friedenberg, G., and Olefsky, J. M. (1990*a*). Insulin-receptor autophosphorylation and endogenous substrate phosphorylation in human adiopocytes from control, obese, and NIDDM subjects. *Diabetes* **39,** 250–9.

Thies, R. S., Webster, N. J., and McClain, D. A. (1990*b*). A domain of the insulin receptor required for endocytosis in rat fibroblasts. *J. Biol. Chem.* **265,** 10132–7.

Tonks, N. K. and Charbonneau, H. (1989). Protein tyrosine dephosphorylation and signal transduction. *TIBS* **14,** 497–500.

Tonks, N. K., Diltz, C. D., and Fischer, E. H. (1988). Characterization of the major protein-tyrosine-phosphatases of human placenta. *J. Biol. Chem.* **263,** 6731–7.

Tornqvist, H. E. and Avruch, J. (1988). Relationship of site-specific β-subunit tyrosine autophosphorylation to insulin activation of the insulin receptor (tyrosine) protein kinase activity. *J. Biol. Chem.* **263,** 4593–601.

Tornqvist, H. E., Gunsalus, J. R., Nemenoff, R. A., Frackelton, A. R., Pierce, M. W., and Avruch, J. (1988). Identification of the insulin receptor tyrosine residues undergoing insulin-stimulated phosphorylation in intact rat hepatoma cells. *J. Biol. Chem.* **263,** 350–9.

Toyoshige, M., Yanaihara, C., Hoshino, M., Kaneko, T., and Yanaihara, N. (1989). Insulin receptor in human hepatoma PLC/PRF/5 cells and its immunochemical characterization with anti-synthetic peptide antibodies. *Biomed. Res.* **10,** 139–47.

Treadway, J. L., Morrison, B. D., Goldfine, I. D., and Pessin, J. E. (1989). Assembly of insulin/insulin-like growth factor-1 hybrid receptors *in vitro*. *J. Biol. Chem.* **264,** 21450–3.

Treadway, J. L., Morrison, B. D., Wemmie, J. A., Frias, I., O'Hare, T., Pilch, P. F., and Pessin, J. E. (1990). The endogenous functional turkey erythrocyte and rat liver insulin receptor is an $\alpha_2\beta_2$ heterotetrameric complex. *Biochem. J.* **271,** 99–105.

Treadway, J. L., Morrison, B. D., Soos, M. A., Siddle, K., Olefsky, J., Ullrich, A., McClain, D. A., and Pessin, J. E. (1991). Transdominant inhibition of tyrosine kinase activity in mutant insulin/insulin-like growth factor I hybrid receptors. *Proc. Natl Acad. Sci. USA* **88,** 214–18.

Ullrich, A. and Schlessinger, J. (1990). Signal transduction by receptors with tyrosine kinase activity. *Cell* **61,** 203–12.

Ullrich, A., Bell, J. R., Chen, E. Y., Herrera, R., Petruzzelli, L. M., Dull, T. J., Gray, A., Coussens, L., Liao, Y. C., Tsubokawa, M., Mason, A., Seeburg, P. H., Grunfeld, C., Rosen, O. M., and Ramachandran, J. (1985). Human insulin receptor and its relationship to the tyrosine kinase family of oncogenes. *Nature* **313,** 756–61.

Ullrich, A., Gray, A., Tam, A. W., Yang-Feng, T., Tsubokawa, M., Collins, C., Henzel, W., Le Bon, T., Kathuria, S., Chen, E., Jacobs, S., Francke, U., Ramachandran, J., and Fujita-Yamaguchi, Y. (1986). Insulin-like growth factor I receptor primary structure: comparison with insulin receptor suggests structural determinants that define specificity. *EMBO J.* **5,** 2503–12.

Villalba, M., Wente, S., Russell, D. S., Ahn, J., Reichelderfer, C. F., and Rosen, O. M. (1989). Another version of the human insulin receptor kinase domain: expression, purification, and characterization. *Proc. Natl Acad. Sci. USA* **86,** 7848–52.

Waugh, S. M. and Pilch, P. F. (1989). Insulin binding changes the interface region between α subunits of the insulin receptor. *Biochemistry* **28,** 2722–7.

Waugh, S. M., DiBella, E., and Pilch, P. F. (1989). Isolation of a proteolytically-derived domain of the insulin receptor containing the major site for cross-linking/binding. *Biochemistry* **28,** 3448–55.

Wedekind, F., Baer-Pontzen, K., Bala-Mohan, S., Choli, D., Zahn, H., and Brandenburg, D. (1989). Hormone binding site of the insulin receptor: analysis using photoaffinity-mediated avidin complexing. *Biol. Chem. Hoppe-Seyler* **370,** 251–8.

Weiland, M., Brandenburg, C., Brandenburg, D., and Joost, H. G. (1990). Antagonistic effects of a covalently dimerized insulin derivative on insulin receptors in 3T3-L1 adipocytes. *Proc. Natl Acad. Sci. USA* **87,** 1154–8.

White, M. F., Haring, H. U., Kasuga, M., and Kahn, C. R. (1984). Kinetic properties and sites of autophosphorylation of the partially purified insulin receptor from hepatoma cells. *J. Biol. Chem.* **259,** 255–64.

White, M. F., Livingston, J. N., Backer, J. M., Lauris, V., Dull, T. J., Ullrich, A., and Kahn, C. R. (1988*a*). Mutation of the insulin receptor at tyrosine 960 inhibits signal transmission but does not affect its tyrosine kinase activity. *Cell* **54,** 641–9.

White, M. F., Shoelson, S. E., Keutmann, H., and Kahn, C. R. (1988*b*). A cascade of tyrosine autophosphorylation in the β-subunit activates the phosphotransferase of the insulin receptor. *J. Biol. Chem.* **263,** 2969–80.

Whittaker, J. and Okamoto, A. (1988). Secretion of soluble functional insulin receptors by transfected NIH3T3 cells. *J. Biol. Chem.* **263,** 3063–6.

Whittaker, J., Soos, M. A., and Siddle, K. (1990). Hybrid insulin receptors: molecular mechanisms of negative-dominant mutations in receptor-mediated insulin resistance. *Diabetes Care* **13,** 576–81.

Wilden, P. A. and Pessin, J. E. (1987). Differential sensitivity of the insulin receptor kinase to thiol and oxidising agents in the absence and presence of insulin. *Biochem. J.* **245,** 325–31.

Wilden, P. A., Backer, J. M., Kahn, C. R., Cahill, D. A., Schroeder, G., and White, M. F. (1990). The insulin receptor with phenylalanine replacing tyrosine-1146 provides evidence for separate signals regulating cellular metabolism and growth. *Proc. Natl Acad. Sci. USA* **87,** 3358–62.

Williams, J. F., McClain, D. A., Dull, T. J., Ullrich, A., and Olefsky, J. M. (1990). Characterization of an insulin receptor mutant lacking the subunit processing site. *J. Biol. Chem.* **265,** 8463–9.

Xu, Q.-Y., Paxton, R. J., and Fujita-Yamaguchi, Y. (1990). Substructural analysis of the insulin receptor by microsequence analyses of limited tryptic fragments isolated by sodium dodecyl sulfate-polyacrylamide gel electrophoresis in the absence or presence of dithiolthreitol. *J. Biol. Chem.* **265,** 18673–81.

Yamamoto-Honda, R., Koshio, O., Tobe, K., Shibasaki, Y., Momomura, K., Odawara, M., Kadowaki, T., Takaku, F., Akanuma, Y., and Kasuga, M. (1990). Phosphorylation state and biological function of a mutant human insulin receptor Val^{996}. *J. Biol. Chem.* **265,** 14777–83.

Yarden, Y. and Ullrich, A. (1988). Growth factor receptor tyrosine kinases. *Ann. Rev. Biochem.* **57,** 443–78.

Yip, C. C., Hsu, H., Patel, R. G., Hawley, D. M., Maddux, B. A., and Goldfine, I. D. (1988). Localization of the insulin-binding site to the cysteine-rich region of the insulin receptor α-subunit. *Biochem. Biophys. Res. Comm.* **157,** 321–9.

Yonezawa, K. and Roth, R. A. (1990). Various proteins modulate the kinase activity of the insulin receptor. *FASEB J.* **4,** 194–200.

Zhang, B., Tavare, J. M., Ellis, L., and Roth, R. A. (1991). The regulatory role of known tyrosine autophosphorylation sites of the insulin receptor kinase domain: an assessment by replacement with neutral and negatively charged amino acids. *J. Biol. Chem.* **266,** 990–6.

Zick, Y. (1989). The insulin receptor: structure and function. *Crit. Rev. Biochem. Mol. Biol.* **24,** 217–69.

7 | Mechanisms whereby insulin may regulate intracellular events

RICHARD M. DENTON and JEREMY M. TAVARÉ

As made clear in the preceding chapter, it is now beyond reasonable doubt that the initial event in the action of insulin is the binding of the hormone to a single class of receptors on the surface of the cells of target tissues, the most important of which are liver, fat, and muscle. This binding activates the intrinsic tyrosine kinase activity of the intracellular domain of the receptor, resulting in the rapid phosphorylation of tyrosines within first the receptor and then other intracellular proteins (Chapter 6). This chapter will be concerned with what happens beyond this first essential step in the insulin signalling systems which brings about the all-important effects of the hormone on carbohydrate, lipid, and protein metabolism described in Chapter 5. It needs to be emphasized that, at the time of writing, there remain fundamental gaps in our understanding of these systems despite much research effort.

The effects of insulin range over a wide time-scale: from those that occur within minutes (for example, the regulation of many enzymes in carbohydrate and fat metabolism by reversible phosphorylation) to those taking several hours or longer. Any plausible explanation as to the nature of the intracellular signalling systems involved in insulin action must, therefore, take into account not only the wide array of responses but also their differing time courses.

Cells of target tissues typically contain about 10^5 cell-surface insulin receptors but only about 10^3 need to be occupied to elicit a full insulin response. The systems involved in insulin signalling must, therefore, involve a considerable degree of amplification as the end result of insulin action is the manipulation of the activity of many millions of target proteins within the cells; such target proteins include glucose transporters, glycogen synthase, acetyl-CoA carboxylase, and many others.

In general terms, a good deal is known about the actual steps in the pathways of carbohydrate and lipid metabolism which are effected by insulin (Table 1). As explained in Chapter 5, their combined effect is to increase the net uptake of glucose from the blood and increase its conversion to glycogen and triglyceride. At the same time, insulin inhibits the breakdown of triglyceride and glycogen and, under appropriate conditions, the rates of fatty acid oxidation, ketone body formation, and gluconeogenesis in the liver are also restricted. Insulin also has important effects on DNA synthesis and transcription and on protein synthesis, but these are rather less well understood than those of the hormone on carbohydrate and lipid metabolism. In this chapter, we will cover each of these aspects in turn and then consider some of the mechanisms that may link the stimulation of the intrinsic tyrosine kinase activity of the insulin receptor to the alterations in activity of key steps in those metabolic pathways and other intracellular processes. Many aspects of insulin action have been considered in greater detail in a number of recent books (Czech 1985; Belfrage *et al.* 1986; Espinall 1990) and lengthy review articles (Denton 1986; Rosen 1987; Houslay and Siddle 1989; Zick 1989).

7.1 Effects of insulin on carbohydrate and lipid metabolism

Table 1 summarizes the principal effects of insulin on carbohydrate and lipid metabolism which occur within minutes after the exposure of liver, muscle, or adipose tissue to insulin. There are, in addition, important longer-term effects, involving changes in gene expression and maybe translation rates, which result in alterations in the concentration of specific enzymes in these tissues. Particularly striking examples of such effects include the induction of the NADP-linked dehydrogenases of the pentose cycle, acetyl-CoA carboxylase, fatty acid synthase, and liver L-type pyruvate kinase and glucokinase, together with the repression of liver phosphoenolpyruvate carboxykinase (see later).

Table 1 Principal acute effects of insulin on carbohydrate and lipid metabolism

Process	Effect	Tissue*	Mechanisms involved
Glucose transport	Increase	F, M	Translocation of glucose transporters
Glycogen synthesis	Increase	L, F, M	Dephosphorylation of glycogen synthase
Glycogen breakdown	Decrease†	L, F, M	Dephosphorylation of phosphorylase kinase and hence phosphorylase
Glycolysis	Increase†	L	Dephosphorylation of pyruvate kinase and fructose 2,6-bisphosphate kinase (both processes)
Gluconeogenesis	Decrease†	L	
Pyruvate→acetyl-CoA	Increase	(L), F	Dephosphorylation of pyruvate dehydrogenase
Fatty acid synthesis	Increase	F, L, M	Activation and increased phosphorylation of acetyl-CoA carboxylase
Fatty acid oxidation	Decrease	L, (F)	Inhibition of carnitine acyltransferase?
Triacylglycerol synthesis	Increase	F, (L, M)	Stimulation of esterification of glycerol phosphate?
Triacylglycerol breakdown†	Decrease	F	Dephosphorylation of triacylglycerol lipase

The changes listed under 'mechanisms involved' are not necessarily complete.
† These effects are not usually apparent unless tissue cAMP levels are increased by the presence of another hormone.
* Abbreviations: F, white and brown adipose tissue; M, muscle; L, liver.
For further details, including specific references, see Denton (1986).

In this section we will summarize first the recent progress in our understanding of the action of insulin on glucose transport and, in particular, the extent to which it may be explained by the translocation of glucose transporters to the plasma membrane from an intracellular location.

We will then consider the means whereby insulin may alter the activity of intracellular enzymes, with particular emphasis on glycogen synthase, pyruvate dehydrogenase, triglyceride lipase, and acetyl-CoA carboxylase. The action of insulin on the first two enzymes is of interest because their dephosphorylation and activation is most easily observed under basal conditions (i.e. in the absence of other hormones) when there are typically no detectable changes in cAMP levels and dephosphorylation occurs at sites which are not phosphorylated by cAMP-dependent protein kinase. By contrast, the dephosphorylation of other proteins can be explained, at least in part, by a decrease in the activity of cAMP-dependent protein kinase secondary to a lowering of the concentration of cAMP. Such effects of insulin are usually only evident when cell cAMP concentrations are first elevated by other hormones, such as glucagon or β-adrenergic agonists. Examples in carbohydrate metabolism include phosphorylase kinase (and hence phosphorylase) and liver pyruvate kinase; triacylglycerol lipase is the most important example of an enzyme in this group in lipid metabolism. Acetyl-CoA carboxylase is of interest because it is probably the best example of an enzyme whose activity is altered by insulin through increases in its phosphorylation. As will become evident, one of the paradoxes about the effects of insulin is that the hormone brings about both the dephosphorylation of a number of key intracellular proteins while increasing the phosphorylation of others.

7.1.1 Glucose transport

The entry of glucose into most mammalian cells is mediated by an integral membrane transport protein through a facilitated diffusion mechanism. At least five different isoforms (GLUT 1–5) have been identified with distinct tissue distributions (Bell *et al.* 1990; Gould and Bell 1990). All the isoforms appear to have similar structures, in which 12 putative transmembrane domains probably form a hydrophobic channel through which glucose can be transported. In two major mammalian tissues, muscle and adipose tissue, this process is acutely regulated by insulin. The major isoform present in these tissues is GLUT 4 together with a smaller amount of GLUT 1. Most studies on the action of insulin on glucose transport have been carried out using isolated fat cells prepared by collagenase digestion of adipose tissue, as these cells have proved most amenable to the detailed study of the mechanisms involved.

The important separate studies of Kono and Cushman and their respective co-workers, revealed that the binding of insulin to its receptor results in the rapid translocation of glucose transporter proteins from an apparently intracellular membrane location to the plasma membrane (Cushman and Wardzala 1980; Suski and Kono 1980). Early studies involved the use of either the binding of cytochalasin B (a specific inhibitor of glucose transport) to glucose transporters in subcellular membrane fractions, or the reconstitution of glucose transport activity derived from similar subcellular membrane fractions into artificial liposomes. This has been substantiated more recently by the use of photochemical cross-linking of glucose transporters with [^{3}H]cytochalasin B or glucose analogues and the use of specific antisera for Western blotting and immunocytochemistry (Oka and Czech 1984; Blak *et al.* 1988; Clark and Holman 1990; Holman *et al.* 1990).

In unstimulated fat cells, it appears that only a small percentage of the total cellular complement of glucose transporters reside in the plasma membrane. Addition of insulin to cells markedly increases this number, by up to tenfold, depending on how the measurement is made, and concomitantly decreases the number of glucose transporters associated with the intracellular membrane compartment. This translocation event is energy-dependent, is independent of protein synthesis, and occurs with a half-time of 2–3 min. It explains earlier observations that under most conditions insulin increased the maximal rate of glucose transport into fat cells without having a major effect on the K_m. Evidence for translocation has also been obtained in rat heart and diaphragm muscle. Recent studies have indicated that insulin is capable of initiating the translocation of both GLUT 1 and GLUT 4 (Calderhead *et al.* 1990). However, translocation may not fully account for the overall increase in glucose transport in fat cells, since the increase in plasma membrane-associated transporters has consistently been found to be considerably less than the increase in glucose transport (Simpson and Cushman 1986; Calderhead and Lienhard 1988).

Phorbol esters mimic the effects of insulin on translocation of glucose transporters to the plasma membrane of both rat fat cells and Swiss 3T3-L1 cells (e.g. Muhlbacher *et al.* 1988). However, although phorbol esters cause a translocation event of equivalent magnitude to that seen with insulin, they have a considerably smaller overall effect on glucose transport compared to insulin. This strongly suggests that insulin (but not phorbol esters) may be capable of inducing an increase in the inherent activity of glucose transporters once they are inserted into the plasma membrane.

Such an increase in the activity of glucose transporters in the plasma membrane could be brought about through their covalent modification by phosphorylation. However, so far no direct evidence has been found for any changes in phosphorylation of glucose transporters following exposure of cells to insulin (Lawrence *et al.* 1990*a*,*b*). Nevertheless, okadaic acid, which is a specific and potent inhibitor of protein phosphatases 1 and 2A, stimulates glucose transport in both fat and muscle cells (Haystead *et al.* 1989; Tanti *et al.* 1991) which suggests some component of the system is regulated by reversible phosphorylation.

Insulin also promotes the translocation of IGF-II receptors and transferrin receptors to the plasma membrane, from an intracellular membrane compartment that has similar characteristics to that which contains the insulin-sensitive pool of glucose transporters. The time courses and extents of the translocations of glucose transporters, IGF-II receptors, and transferrin receptors are also broadly similar

(Oka *et al.* 1984; Davis *et al.* 1986). By contrast, as discussed in Chapter 6, insulin promotes the translocation of occupied insulin receptors from the plasma membrane to an 'endocytic' subcellular membrane compartment. The half-maximal response of this internalization closely parallels that of the binding of insulin to its receptor. However, the response to insulin of the translocation of glucose transporters and IGF-II receptors is shifted an order of magnitude to the left in comparison with insulin binding, suggesting that the mechanisms involved differ. There are also large differences in the numbers involved. The binding of insulin to about 1000 receptors per fat cell is sufficient to cause the translocation of more than one million glucose transporters.

Overall, it is becoming increasingly evident that the movement of proteins within cells is an important feature of some of the effects of insulin. The interrelationships between these translocation events in both spatial and mechanistic terms will no doubt be revealed in the next few years, but at present our understanding, as with so many other aspects of insulin action, is very rudimentary.

7.1.2 Glycogen synthase

Insulin activates rabbit muscle glycogen synthase through stimulating its dephosphorylation on three specific serine residues close to the carboxyl-terminus (C30, C34, C38) (Cohen 1987) and it is in this tissue that the enzyme has been most extensively studied. Insulin has also been shown to activate glycogen synthase through apparently similar dephosphorylation mechanisms in both rat adipocytes (Lawrence and James 1984) and rat diaphragm (Lawrence *et al.* 1983).

Glycogen synthase can be phosphorylated on nine or more serines by at least seven distinct protein kinases. In the resting state, the enzyme contains about 3 moles of phosphate per mole of enzyme shared between the various phosphorylation sites. Only by the thorough analysis of the extent of phosphorylation of each individual site has it become clear that the major effect of insulin is to promote the dephosphorylation of serines C30, C34, and C38 (Cohen 1987; Poulter *et al.* 1988). On the other hand, agents that promote increases in cAMP concentrations, such as adrenalin and glucagon, inhibit glycogen synthase activity through increasing the phosphorylation state of serines C30, C34, and C38, together with two further sites (Cohen 1987).

Serines C30, C34, and C38 are phosphorylated by a cAMP- and Ca^{2+}-independent protein kinase known as GSK-3 and dephosphorylated by both protein phosphatase 1 and 2A. It can be argued, therefore, that insulin could act through inhibition of GSK-3, activation of either protein phosphatase 1 or 2A, or a combination of these events (Cohen *et al.* 1985; Cohen 1987).

No evidence has been found for any decreases in the activity of GSK-3 which survive cell breakage and conventional assay technique; indeed, evidence has been reported that insulin may increase the activity of this enzyme (Yang *et al.* 1988). Considerable attention has hence been focused on the possibility that insulin may act through increasing phosphatase activity. Phosphatase 1 activity is specifically blocked by a small protein known as inhibitor 1, but only after this inhibitor protein has been phosphorylated by cAMP-dependent protein kinase. The possibility thus arises that insulin may act through the dephosphorylation of inhibitor 1, perhaps via the inhibition of this protein kinase. Some direct evidence in favour of this mechanism has been obtained (Foulkes *et al.* 1982) but only under conditions of raised β-catecholamines (and hence raised cAMP levels). Such a mechanism does not appear to offer an explanation for the dephosphorylation of serines C30, C34, and C38 under basal conditions, since under these conditions the levels of cAMP and the extent of inhibitor-1 phosphorylation are already very low (Kahtra *et al.* 1980; Cohen *et al.* 1985).

Larner and colleagues have argued that insulin acts through low molecular weight mediators released at the plasma membrane, which not only stimulate phosphatase activity but may also inhibit cAMP-dependent protein kinase activity (Larner *et al.* 1978). Similar substances may also be involved in the regulation of pyruvate dehydrogenase, cAMP phosphodiesterase, and other intracellular effects of insulin, and are discussed further in following sections. Spermine (a highly basic polyamine) has been shown to influence the specificity of protein phosphatase 2A so that its activity against serines C30, C34, and C38 is enhanced to a markedly greater extent than its activity against the other phosphorylation sites on the enzyme (Tung *et al.* 1985). Thus an additional possibility is that insulin may act through increasing the concentration of spermine or a related compound, and this is also discussed further in the following section.

Perhaps the most exciting possibility is that the dephosphorylation of glycogen synthase is brought about through the activation of a form of protein phosphatase 1 which is complexed with a 160 kDa subunit known as the 'G-subunit' (Dent *et al.* 1990). This subunit appears to be involved in directing the phosphatase to the glycogen particle (to which glycogen synthase also binds) as well as regulating the activity of the phosphatase. Phosphorylation of the 'G-subunit' on one serine by cyclic AMP-dependent protein kinase appears to cause dissociation of the phosphatase and diminished activity, while phosphorylation on another serine results in an increase in its activity against glycogen synthase. The injection of rabbits with insulin has been shown to increase the phosphorylation of this site, as well as increasing the activity of the protein kinase apparently responsible for its phosphorylation (Dent *et al.* 1990). The protein kinase involved appears to be very similar, if not identical, to ribosomal protein S6 kinase II, to which we will be returning later in this chapter. It should, however, be noted that these observations suggest that increases in protein kinase activity may underlie decreases in the phosphorylation of specific proteins through alterations in protein phosphatase activity.

7.1.3 Pyruvate dehydrogenase

Once pyruvate has been converted to acetyl-CoA by pyruvate dehydrogenase, an exclusively intramitochondrial enzyme complex, there is no means in mammals whereby the acetyl-CoA can be used for the resynthesis of glucose. The enzyme must, therefore, be under particularly tight control and, furthermore, during starvation the enzyme must be greatly inhibited to conserve the restricted reserves of carbohydrate. Regulation of the pyruvate dehydrogenase is achieved in part by end-product inhibition by acetyl-CoA and NADH, and in part by reversible phosphorylation. The phosphorylated form of the enzyme is essentially inactive (Denton and Halestrap 1979; Wieland 1983).

Exposure to insulin of tissues, such as adipose tissue and mammary gland, in which the enzyme plays a predominantly biosynthetic role (since much of the acetyl-CoA formed is used for the synthesis of fatty acids) results in the rapid activation of pyruvate dehydrogenase activity. Such short-term effects are not found in muscle, where pyruvate dehydrogenase has essentially a catabolic role since the acetyl-CoA formed is mainly oxidized via the citrate cycle to CO_2. Most studies have been concerned with the means whereby insulin causes a two- to threefold increase in activity in rat epididymal fat cells (Denton *et al.* 1975, 1989). The increase in activity persists during the preparation and subsequent incubation of intact mitochondria (Denton *et al.* 1984, 1989) and this property has greatly facilitated investigations into the mechanism involved. In particular, it has allowed good evidence to be obtained that the effect of insulin is brought about by the activation of pyruvate dehydrogenase phosphatase rather than inhibition of the kinase. However, no changes in phosphatase activity are detectable in extracts of mitochondria from insulin-treated tissue. This suggests that insulin may cause a change in the concentration of some effector of the phosphatase within mitochondria that then dissociates from the phosphatase during the preparation of extracts.

The activity of pyruvate dehydrogenase phosphatase is known to be regulated by Ca^{2+}, Mg^{2+}, and possibly changes in the $NADH/NAD^+$ ratio, but there is now considerable evidence that insulin does not alter the activity of the phosphatase through changes in these particular regulators (Denton *et al.* 1984, 1989). For example, the effect of insulin still persists within mitochondria made permeable to Mg^{2+} or Ca^{2+} by incubation with the ionophere A23187, or to all substances up to a molecular weight of 1000 to 2000 by treatment with toluene. Further studies of the persistent effect of insulin in these permeabilized mitochondria have shown that insulin causes an increase in the sensitivity of pyruvate dehydrogenase phosphatase to Mg^{2+} with little effect apparent on the maximum velocity.

Spermine has a similar effect on purified preparations of pyruvate dehydrogenase phosphatase. Thus spermine can be linked with the actions of insulin on the dephosphorylation not only of glycogen synthase but also on that of pyruvate dehydrogenase. However, it is unlikely that insulin acts by increasing the concentration of spermine in the appropriate cell compartments since the cell content of spermine would appear to be too high to have a conventional second-messenger role, and insulin appears to have little or no short-term effect on the amount of spermine in cells.

One explanation of these observations on pyruvate dehydrogenase would appear to be that insulin causes a change in the concentration of a spermine-like compound in mitochondria. However, the persistence of the activation in toluene-permeabilized mitochondria suggests that the compound is probably not small. An attractive explanation would be that insulin may promote some change in the interactions between the pyruvate dehydrogenase system and the inner mitochondrial membrane (Denton *et al.* 1989).

However, such an explanation is not easily reconciled with the proposal of Jarett (Jarett and Seals 1979; Gottschalk *et al.* 1986) that the activation of pyruvate dehydrogenase by insulin involves a low molecular weight mediator, apparently similar to that proposed by Larner to be involved in the activation of glycogen synthase. This putative mediator was first proposed following the demonstration that addition of insulin to fat cell plasma membranes resulted in formation of a low molecular weight factor which was capable of increasing pyruvate dehydrogenase activity when added to fat cell mitochondria. The mediator appears to act by stimulating the phosphatase, but characterization has been difficult. One reason for this may be the assay systems used, which would appear to be far from optimal (see Denton *et al.* 1989). More recently it has been suggested by Saltiel and colleagues that the mediator may be a phosphoinositol glycan (Saltiel and Cuatrecasas 1988). We will return to this topic in the final section of this chapter.

So far we have been concerned with the mechanism whereby insulin activates pyruvate dehydrogenase activity within a few minutes. As noted earlier, this effect of insulin appears to be restricted to tissues that carry out fatty acid synthesis, and is exerted through a stimulation of the phosphatase. Insulin also has long-term effects on the pyruvate dehydrogenase system and these have been extensively studied by Randle, Kerbey, and colleagues (Randle 1986; Marchington *et al.* 1987, 1990). These studies have shown that decreases in pyruvate dehydrogenase activity in heart muscle, liver, and other tissues in long-term insulin deficient states, such as starvation and alloxan diabetes, are caused by an increase in kinase activity, perhaps via a kinase activator protein. These effects are reversed by insulin over a period of a few hours.

7.1.4 Acetyl-CoA carboxylase

In mammals, the principal sites of fatty acid synthesis are white and brown adipose tissue, liver, and lactating mammary gland. Insulin stimulates fatty acid synthesis in all four tissues within a few minutes but the largest effects are observed in the adipose tissue of rats and mice fed a high carbohydrate diet. In contrast, hormones that increase cellular cAMP levels, such as β-adrenergic agonists and glucagon, cause a marked decrease in fatty acid synthesis, particularly in adipose tissue and liver (Brownsey and Denton 1987; Hardie 1989). All these changes in fatty acid synthesis appear to be brought about largely through changes in the activity of acetyl-CoA carboxylase, although alterations in rates of glucose transport and pyruvate dehydrogenase are also important as these determine the rate of supply of acetyl-CoA.

It is well established that exposure of rat epididymal fat cells to insulin leads to a two- to threefold increase in acetyl-CoA carboxylase activity, and that this increase is associated both with a greater proportion of the enzyme occurring in its more active polymerized form and with increased phosphorylation of the enzyme (Hardie *et al.* 1984; Borthwick *et al.* 1987, 1990; Brownsey and Denton 1987). This phosphorylation probably occurs on at least two different sites on the enzyme. One site (present within a peptide known as the I-peptide) was first identified by Brownsey and Denton (1982) and exhibits a three- to fivefold increase in phosphorylation with insulin. It has been demonstrated that after purification from extracts of insulin-treated tissue a greater proportion of the enzyme was present in an active polymerized form which exhibited a high degree of phosphorylation within the I-peptide (Borthwick *et al.* 1987). The other site, which exhibits a smaller increase in phosphorylation, has been sequenced and contains a serine residue which is phosphorylated by purified casein kinase II *in vitro* (Haystead *et al.* 1988). These sites of phosphorylation are quite distinct from the sites phosphorylated by cAMP-dependent protein kinase (and other kinases), which result in inhibition of enzyme activity (Haystead *et al.* 1988; Hardie 1989). What has been a matter of some debate is whether the undoubted increases in phosphorylation of acetyl-CoA carboxylase promoted by insulin lead directly to its increased activity. The activity of casein kinase II has been shown to be increased by insulin in rat epididymal fat cells (Diggle *et al.* 1991) but phosphorylation of acetyl-CoA carboxylase by casein kinase II does not appear to alter the activity of the enzyme (Tipper *et al.* 1983; Haystead *et al.* 1988). On the other hand, another insulin-activated acetyl-CoA carboxylase kinase has recently been separated from rat epididymal adipose tissue which appears to phosphorylate the enzyme within the I-peptide with a concomitant increase in activity (Borthwick *et al.* 1990). Thus it should be noted that the 'G-subunit' of phosphatase 1 shares with acetyl-CoA carboxylase the characteristic that phosphorylation on different serines can give rise to opposing changes in activity.

7.1.5 ATP-citrate lyase

ATP-citrate lyase, which catalyses the step immediately preceding that catalysed by acetyl-CoA carboxylase in fatty acid synthesis, also exhibits marked increases in phosphorylation following exposure of fat and other cells to insulin. The role of this phosphorylation is an enigma. Hormones that increase the cell content of cyclic AMP, such as glucagon and β-adrenergic agonists, cause comparable increases in phosphorylation of the enzyme on precisely the same serine while having the opposite effect to insulin on fatty acid synthesis (Pierce *et al.* 1982). No regulatory role for this phosphorylation has been found beyond a small change in ATP affinity (Houston and Nimmo 1985) and a possible subcellular redistribution (Stralfors 1987).

7.1.6 Triacylglycerol lipase

The rate-limiting step in lipolysis is the first reaction, catalysed by triacylglycerol lipase, and regulation of this enzyme is the major means whereby lipolysis is controlled by insulin and other hormones. The release of fatty acids from fat cells appear to involve a specific plasma membrane carrier system, which may also be a site of potential hormonal control.

Insulin has little effect on basal rates of lipolysis from fat cells, but reverses the stimulation caused by counter-regulatory hormones, such as adrenalin, ACTH, and glucagon, which increase the cellular content of cAMP. The counter-regulatory hormones lead to the activation of this enzyme through its phosphorylation (probably on a single serine) by cAMP-dependent protein kinase (Stralfors *et al.* 1985). In general, increases in phosphorylation within intact cells correlate well with increases in lipolysis. The effect of insulin is to reverse this increased phosphorylation of the enzyme and hence to diminish the rate of lipolysis. Under many conditions, the insulin-stimulated decreases in lipolysis correlate well with decreases in activity of cAMP-dependent protein kinase and hence are presumably brought about by decreases in cAMP concentrations (see later discussion). However, there are conditions where the inhibition of lipolysis by insulin cannot be fully explained in terms of the effects of the hormone on cAMP concentrations. For example, in an extensive and careful study Londos *et al.* (1985) have shown that, in more highly stimulated cells, the effect of insulin on lipolysis was substantially greater than that appropriate for the inhibition of cAMP-dependent protein kinase activity. One explanation of these observations is that insulin may also be increasing protein phosphatase activity and so enhancing the dephosphorylation of triacylglycerol lipase. There is at present no direct evidence for the stimulation of the phosphatase involved (probably phosphatase 2A) but such a stimulation could also underlie the dephosphorylation of other key enzymes, such as liver pyruvate kinase.

7.2 Effects of insulin on transcription and translation

Insulin increases the overall rate of protein synthesis in many tissues. This effect appears to be mainly exerted through increases in the initiation phase of translation of mRNA by ribosomes. However, as well as this general effect, insulin also regulates the synthesis of many individual proteins. We have already given a number of examples of such enzymes in carbohydrate and lipid metabolism. These longer-term effects of insulin probably arise mainly from alterations in the level of the mRNA itself, due to changes in the rate of transcription of the corresponding gene.

7.2.1 Transcription

The advances in molecular cloning techniques made during the late 1970s and early 1980s have led to the isolation of an ever increasing number of genes. This, in turn, has provided the necessary methodology to examine the molecular mechanisms underlying transcriptional control.

Table 2 Examples of insulin-induced changes in specific mRNA levels

Change	Tissue	Specific mRNAs
Increase	Liver	Glucokinase, glucose 6-phosphate dehydrogenase, pyruvate kinase, fatty acid synthase, malic enzyme, tyrosine aminotransferase, c-*fos*, c-*myc*, acetyl-CoA carboxylase
	Fat	Glyceraldehyde 3-phosphate dehydrogenase, glycerol 3-phosphate dehydrogenase, pyruvate carboxylase, c-*fos*, adipsin
Decrease	Liver	Phosphoenolpyruvate carboxykinase, carbamoyl phosphate synthetase, fructose 1,6-bisphosphatase

Insulin can both increase and decrease the levels of specific mRNAs for a considerable number of enzymes (Table 2). For example, in liver insulin increases the levels of message for the glycolytic enzymes glucokinase, 6-phosphofructo-1-kinase, pyruvate kinase, 6-phosphofructo-2-kinase, and fructose 2,6-bisphosphatase, in addition to decreasing those for the gluconeogenic enzymes phosphoenolpyruvate carboxykinase (PEPCK) and fructose 1,6-bisphosphatase (Granner and Pilkis 1990). The net effect being an increased flux through the glycolytic pathway.

Insulin often has direct effects on mRNA levels; however, in some cases insulin has only been shown to reverse the effects of experimental diabetes, or act in the presence of glucose or a permissive factor, such as dexamethasone. The mechanism by which insulin promotes these changes is thus likely to be complex, particularly as they often occur over a longer time-scale than the more acute metabolic effects described previously in this chapter. It is clear, however, that the effects of insulin in many of these systems occur directly through the insulin receptor rather than the insulin-like growth factor-I receptor.

With respect to the possible post-receptor mechanisms responsible for insulin's effects on transcription, phorbol esters have been shown to have insulin-like effects on c-*fos* (Stumpo and Blackshear 1986) and PEPCK (Granner *et al.* 1986) mRNA levels. However, at least in the case of the former, insulin appears to act through a mechanism independent of protein kinase C (see further discussion later in this chapter).

The changes in mRNA levels for phosphoenolpyruvate carboxykinase (Sasaki *et al.* 1984), tyrosine aminotransferase (Crettaz *et al.* 1988), c-*fos* (Stumpo *et al.* 1988), and glucokinase (Iynedjian *et al.* 1988) appear to be brought about through changes in the rate of gene transcription. This has been shown through the use of the transcript elongation assay, in which nuclei isolated from control or insulin-treated tissues are incubated with radiolabelled nucleotides and the level of incorporation into specific mRNAs is assessed by hybridization with a complementary DNA probe. However, the possibility that insulin may also act post-transcriptionally through regulating mRNA stability or mRNA efflux from the nucleus cannot be ruled out.

Clearly, an important goal is the identification of the factors that are involved in binding to the regulatory elements that are often located upstream of the initiation site of gene transcription. O'Brien *et al.* (1991) have described a 15 bp insulin-responsive sequence in the PEPCK promotor, and reported a nuclear protein that binds to this region, which may be involved in mediating the insulin effect on transcription. Similarly, Stumpo *et al.* (1988) found an insulin-responsive sequence approximately 300–320 bp upstream of the c-*fos* transcription initiation site. A similar region (20–480 bp) appears to be important in insulin-stimulated glyceraldehyde 3-phosphate dehydrogenase gene expression (Alexander *et al.* 1988) and epidermal growth factor-stimulated c-*fos* gene expression (Fisch *et al.* 1987). The role of c-*fos* is not well understood at present, although intriguingly it is a DNA-binding protein which may be capable of regulating the transcription of other genes (Distel *et al.* 1987).

7.2.2 Translation

Increases in protein synthesis in tissues or cells exposed to insulin are associated with rapid increases in the proportion of ribosomes bound to mRNA to form active polysomes. This indicates that insulin stimulates peptide chain initiation, i.e. that phase of the ribosome cycle in which ribosomes bind to the mRNA and to the initiator tRNA (met-$tRNA_i$) (Kimball and Jefferson 1988). A number of components of the translational machinery involved in initiation undergo reversible phosphorylation both in whole cells and *in vitro* and are, therefore, likely targets for regulation by insulin. Most attention has been focused on initiation factor 2 (eIF2) and ribosomal protein S6.

Phosphorylation of the α-subunit of eIF2 leads to inhibition of the recycling of eIF2 which is essential for its continuing participation in peptide-chain initiation. This mechanism is known to be important in the control of chain initiation by haem in reticulocytes. Thus, the possibility arises that insulin may act through bringing about the dephosphorylation of eIF2, and some evidence has been presented (Towle *et al.* 1984) for such an effect in cultured chondrocytes. However, in more recent studies, the level of phosphorylation of this factor in skeletal muscle of rats in which the level of circulating insulin was altered (for example, by fasting and refeeding) was always found to be low and unchanged by the manipulations, although these influence rates of protein synthesis (Cox *et al.* 1988).

The S6 protein is a component of 40S subunits of eukaryotic ribosomes, and may be involved in the binding of mRNA. Exposure of many different cell types to insulin results in a marked increase in the phosphorylation of this protein on at least five different serines and threonines. However, the relationship between this increase in phosphorylation and rates of protein synthesis remains to be clearly established. The phosphorylation is often transient, whereas the increase in protein synthesis persists, and *in vitro* it has proved difficult to demonstrate any increase in activity of cell-free systems associated with increases in S6 phosphorylation (Tabarini *et al.* 1984; Kimball and Jefferson 1988). However, it should be emphasized that protein synthesis is extremely complex and thus there must always be considerable uncertainty about the extent to which *in vitro* reconstituted systems for protein synthesis reflect events occurring in the cell.

Nevertheless, considerable progress has been made with identifying the protein kinases involved in the increased phosphorylation of S6 in cells exposed to insulin and other hormones that promote protein synthesis (Kimball and Jefferson 1988; Erikson 1991). Substantial increases in S6 kinase activity are apparent in high-speed supernatant fractions prepared from insulin-stimulated cells. However, these increases are only apparent if the initial extraction is made with medium containing protein phosphatase inhibitors. These and other observations suggest that the persistent increase in S6 kinase activities in extracts of insulin-stimulated cells is the result of their increased phosphorylation. We will return to this important topic in the next section.

7.3 Possible mechanisms involved in insulin action

A full understanding of the molecular mechanisms by which insulin brings about the wide diversity of metabolic effects summarized in Fig. 1 is one of the most important challenges in endocrinology. Many of the effects of insulin are accompanied by changes in the phosphorylation state of serines and threonines on specific proteins. Examples of both increases and decreases in protein phosphorylation are listed in Table 3.

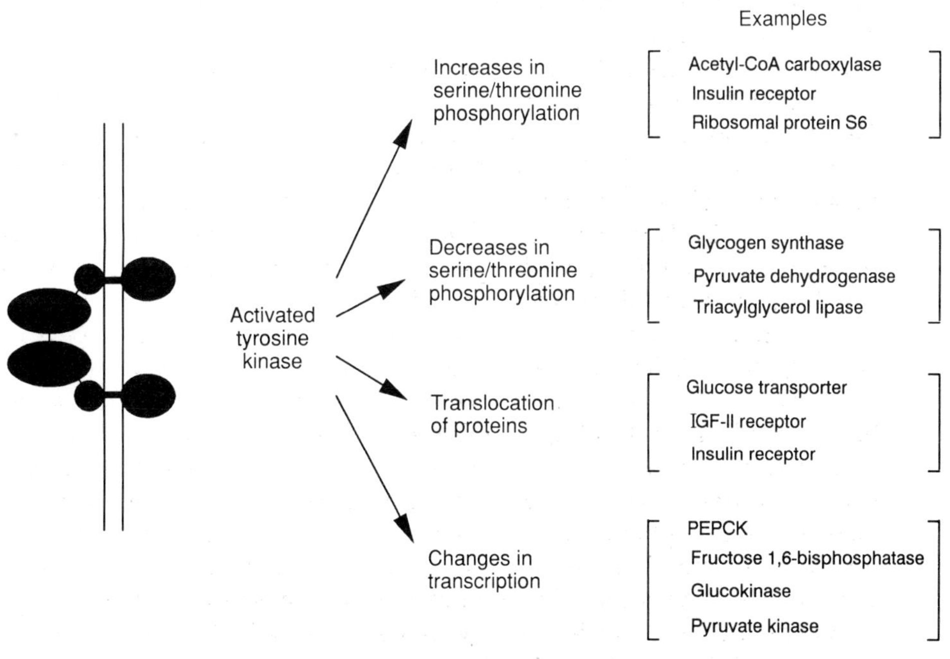

Fig. 1 A summary of some of the established intracellular changes that occur following the binding of insulin to its receptor on the cell surface

Table 3 Examples of proteins exhibiting changes in serine/threonine phosphorylation in intact cells exposed to insulin

Change in phosphorylation	Protein	Cell compartment	Tissue type
Decreases	Glycogen synthase	Glycogen particles	F, L, M
	Phosphorylase*†	Glycogen particles	F, L, M
	Phosphorylase kinase*†	Glycogen particles	F, L, M
	G-subunit of protein phosphatase 1	Glycogen particles	M
	Pyruvate kinase†	Cytoplasm	L
	HMG-CoA reductase*	Microsomes	L
	Phenylalanine hydroxylase†	Cytoplasm	L
	Triglyceride lipase†	Triacylglycerol complex	F
	Pyruvate dehydrogenase	Mitochondria	F, L
	Inhibitor 1†	Cytoplasm	F, L
Increases	Acetyl-CoA carboxylase	Cytoplasm	F, L
	ATP-citrate lyase	Cytoplasm	F, L
	Insulin receptor	Plasma membrane	F, L
	M_r 63 000 protein	Cytoplasm	F
	M_r 61 000 protein	Plasma membrane	F
	M_r 46 000 protein	Cytoplasm	L
	S6-ribosomal protein	Microsomes	F, L
	M_r 22 000 protein	Cytoplasm	F, L
	Cyclic AMP phosphodiesterase	Microsomes	F, L

Abbreviations: F, fat; L, liver; M, muscle.
* Inferred from changes in activity.
† Indicates that the insulin effect is only apparent if cellular cAMP is previously elevated by another hormone.

As already discussed in this chapter, insulin can stimulate the dephosphorylation of several specific proteins, including glycogen synthase and pyruvate dehydrogenase, in the absence of changes in cAMP, probably through increasing protein phosphatase activity. In some cases, however, insulin can only promote dephosphorylation once cAMP levels (and hence increased cAMP-dependent protein kinase activity) have been previously elevated by other hormones, such as glucagon or β-adrenergic agonists.

Insulin-stimulated protein phosphorylation appears to occur largely through increases in the activity of a number of cyclic-nucleotide-independent protein kinases. In addition to stimulating the phosphorylation of a number of well-characterized proteins, such as acetyl-CoA carboxylase and the ribosomal protein S6, insulin also promotes the serine and threonine phosphorylation of several relatively abundant cellular proteins of unknown function. These include two proteins of M_r 22 000 and 46 000 in rat epididymal adipose tissue and liver, respectively, which exhibit particularly large increases in phosphorylation (Denton 1986).

Insulin also promotes the translocation of several proteins between cellular compartments, and has more long-term effects on DNA synthesis and transcription. These processes may also be regulated by changes in protein phosphorylation. We end this chapter by discussing some of the more promising hypotheses for insulin action, which together may explain the complexity behind insulin's effects.

7.3.1 Role of the insulin receptor tyrosine kinase activity

The earliest known event after the binding of insulin to the α-subunit of its receptor is the activation of a protein kinase activity intrinsic to the receptor β-subunit which specifically phosphorylates tyrosine residues on both the β-subunit itself and exogenous substrates. This aspect of insulin action is dealt with in detail in Chapter 6, but the following important aspects are especially relevant to this chapter:

1. The integrity of the insulin receptor protein tyrosine kinase activity appears to be essential for all of insulin's effects.

2. The activity of the tyrosine kinase appears to be decreased in some states of insulin resistance and experimental diabetes.

3. The precise role of the kinase activity is not known at present. For example, it is not established whether the numerous exogenous substrates reported in the literature play a direct role in insulin action or whether they simply reflect the increased activity of the receptor tyrosine kinase. The most important physiological substrate for the tyrosine kinase may be the β-subunit itself. In intact cells, insulin stimulates β-subunit phosphorylation on up to five tyrosine residues. Three of these reside in the tyrosine kinase domain of the β-subunit and are phosphorylated very rapidly both in intact cells and purified receptor preparations. The phosphorylation of these three tyrosines appears to result in the stimulation of the tyrosine kinase activity towards exogenous substrates. This is followed in some, but not all, cells by phosphorylation of two tyrosines in the carboxyl-terminal tail. This sequence of phosphorylation events may initiate a series of conformational changes within the β-subunit, thereby allowing the intracellular domain to interact with other proteins which may be involved in the signalling mechanism, such as substrates for the tyrosine kinase, protein serine kinases, and GTP-binding proteins (see below). There is mounting evidence from the study of mutant insulin receptors that there may be a divergence of signalling pathways at the level of the receptor.

4. Insulin, phorbol esters, and analogues of cAMP also stimulate serine and threonine phosphorylation of the β-subunit in intact cells. Whether these phosphorylation events have an important regulatory role remains to be established, but they clearly indicate that the insulin receptor must interact closely with at least one serine/threonine-specific protein kinase.

7.3.2 Role of cyclic nucleotides and G-proteins in insulin action

As already emphasized, a number of the actions of insulin are undoubtedly brought about via decreases in cellular cAMP levels—at least in liver and adipose tissue previously exposed to a hormone which increases cAMP levels. The effect of insulin probably involves a combination of the activation of phosphodiesterase activity and inhibition of adenyl cyclase. The former may be quantitatively more important, since decreases are rather less evident when cAMP phosphodiesterase activity is blocked by methylxanthines (Houslay 1985, 1986).

Direct inhibitory effects of insulin on adenyl cyclase activity in plasma membranes can be observed, but rather precise conditions are necessary and the effects are quite small. Heyworth and Houslay (1983) found inhibition of adenyl cyclase activity in rat liver plasma membranes only in the presence of glucagon, Mg^{2+}, and an adequate concentration of GTP.

Activation by insulin of particulate phosphodiesterase activity with a low K_m for cAMP has been observed by many groups in both fat and liver cells. What is rather mystifying is that conditions which result in high levels of cAMP in liver and fat cells also lead to persistent activation of cAMP phosphodiesterase activity (Makins and Kono 1980; Houslay 1985, 1986). This is apparently the result of phosphorylation of the phosphodiesterase by cAMP-dependent protein kinase (Degerman *et al.* 1990). Thus under the very conditions in which insulin causes the greatest diminution in cAMP concentrations there may be little or no persistent effect of insulin on particulate cAMP phosphodiesterase activity.

In liver, two distinct membrane-bound cAMP phosphodiesterases have been found to be activated by insulin. One is associated with the plasma membrane, the other with intracellular 'dense vesicles'. However, neither appears to play an essential role in lowering cAMP as decreases can be observed on addition of insulin to liver cells exposed to glucagon under conditions where no persistent changes in these enzymes occur (Houslay 1985, 1986). It is possible that a further distinct phosphodiesterase present in the cytoplasmic fraction of cells is also activated by insulin.

Nevertheless, the liver plasma membrane enzyme shows a number of fascinating properties. Activation can be demonstrated on addition of insulin to liver plasma membrane preparations and this appears to involve increased phosphorylation of the enzyme, perhaps on tyrosine residues (Houslay 1986; Pyne *et al.* 1989). Houslay (1985, 1986) has concluded that a guanine nucleotide-binding protein (G-protein) may be involved in the action of insulin on adenyl cyclase and phosphodiesterase, and that this protein may be distinct from the well-characterized G_s and G_i which are involved in the activation and inhibition, respectively, of adenyl cyclase by other hormones. Direct evidence for such a nucleotide-binding protein is lacking, but insulin has been reported to inhibit the cholera-toxin-induced ADP-ribosylation of a protein of M_r 25 000 in liver plasma membranes. This protein could represent the α-subunit of a G-protein closely associated with the insulin receptor (Heyworth *et al.* 1985).

Further evidence that there may be interactions between the insulin receptor and the G-protein regulatory system has included demonstrations that:

(1) a number of G-proteins undergo tyrosine phosphorylation by purified insulin receptors *in vitro* (Zick *et al.* 1986);

(2) that addition of insulin to liver and fat cell membranes alters the conformation of G_i so it is no longer a substrate for ADP-ribosylation by pertussis toxin (Rothenberg and Kahn 1988); and

(3) that insulin stimulates GTP-ase activity in human platelet membranes (Gauber and Houslay 1987).

In summary, it seems very likely that insulin, like other hormones, regulates adenyl cyclase through the G-protein system, but the G-protein or proteins involved remains to be characterized. The extent to which G-proteins are involved in other actions of insulin remains an important unanswered question. Baldini *et al.* (1991) have recently shown that non-hydrolysable GTP analogues are capable of inducing the translocation of glucose transporters when applied to permeabilized rat fat cells, which represents a strong hint that G-proteins may play a key role in mediating at least some of the other actions of insulin.

Under appropriate conditions, insulin not only causes a decrease in cAMP but also a transient increase in cGMP concentrations (Fain 1980). This increase is only observed in fat and liver cells and as yet no satisfactory role has been found. Other agents, such as carbachol and noradrenalin, can bring about similar changes without having the characteristic effects of insulin. In contrast, the increase in cGMP concentrations observed on incubating fat cells with insulin is greatly diminished if the cells are incubated in Ca^{2+}-free medium, but the metabolic effects of insulin remain essentially unaltered (Pouter *et al.* 1976; Fain 1980).

7.3.3 Role of other small molecular weight mediators, including phosphoinositol glycans

As mentioned in previous sections, the group of Larner obtained evidence in the 1970s that the activation of glycogen synthase may involve the formation of a small molecular weight mediator substance. Initially, evidence for such putative mediators came from studies using crude acid extracts of muscle from insulin-treated animals, which suggested that the mediators may be responsible for the dephosphorylation of glycogen synthase by both inhibiting cAMP-dependent protein kinase activity and stimulating phosphatase activity (Larner *et al.* 1978). In parallel experiments, Jarett and colleagues observed that addition of insulin to fat and other cell plasma membranes resulted in the generation of one or more apparently similar small molecular weight substances which stimulated pyruvate dehydrogenase activity when added to fat cell mitochondria preparations (Gottschalk *et al.* 1986). Subsequently, similar preparations were found to have insulin-mimicking effects on phosphodiesterase and phospholipid methyltransferase activity. The characterization of the active substances in these preparations has proved difficult, but the early studies indicated that the substances were apparently heat- and acid-stable and carried a net negative charge at pH 7 (Larner *et al.* 1978; Gottschalk *et al.* 1986). It was also concluded that the substances were probably peptides or peptide conjugates and that there was probably a family of related but different mediator substances with differing specificities (Larner 1988).

More recently, Saltiel, Cuatrecasas, and colleagues have proposed that these putative mediator substances may, in fact, be phosphoinositol glycans released from plasma membrane-associated glycolipids by the action of a specific insulin-activated phospholipase C (Saltiel *et al.* 1986; Saltiel and Cuatrecasas 1988). Initially, using rat liver membranes or cultured BC3H-1 muscle cells, they found that insulin caused the rapid hydrolysis of one or more novel glycolipids within the cell membrane, resulting in the formation of diacylglycerol and two apparently closely related water-soluble components. These latter components were capable of stimulating microsomal cAMP phosphodiesterase activity and appeared to contain inositol, phosphate, and glucosamine. Subsequent studies by this group, and also by the laboratories of Jarett and Mato, indicated that partially purified preparations of these components were also capable of influencing the activity of pyruvate dehydrogenase, adenyl cyclase, and cAMP-dependent protein kinase in various subcellular assay systems (Villalba *et al.* 1987; Gottschalk and Jarett 1988; Saltiel and Cuatrecasas 1988). However, it should be noted that considerably more studies are required to ensure that the effects observed are a true reproduction of those brought about by the action of insulin on intact cells. Partially purified preparations of the phosphoinositol glycan head groups have also been reported to be capable of eliciting a number of insulin-like actions, including the stimulation of fat synthesis and inhibition of lipolysis when added to intact cells, but glucose transport is not stimulated (Kelly *et al.* 1987; Gottschalk and Jarett 1988; Saltiel and Cuatrecasas 1988).

At the time of writing, there is considerable confusion over the structure of the active glycolipid head groups considerd to be released by insulin. It seems likely that they consist of phosphoinositol linked to glucosamine which in turn is linked through further glycosidic bonds to about four monosaccharide residues (probably galactose) with one or two additional phosphate groups (Merida *et al.* 1988; Saltiel *et al.* 1989).

A closely related structure is also found in the anchor by which certain proteins are attached to the surface cells. Over 30 different proteins are now known to be anchored in this way, including alkaline phosphatase and lipoprotein lipase, and such proteins can be released by the treatment of cells with appropriate inositol-specific phospholipase C preparations. Thus it is of considerable interest that exposure of cells to insulin may also cause the release of alkaline phosphatase and lipoprotein lipase, apparently through the activation of an endogenous phospholipase C acting on the outside of cells (Saltiel *et al.* 1989). In fact, there is mounting evidence that the glycolipid precursor is orientated in the same way as in the protein anchors (Alvarez *et al.* 1988), in which case the glycan head groups would be released outside cells. To act as mediators of insulin action on intracellular processes the head groups would then have to enter the cells across the plasma membrane—and in the case of pyruvate dehydrogenase also penetrate the inner mitochondrial membrane. Evidence for the transfer of the head groups across membranes, presumably by specific transport systems, is presently lacking.

7.3.4 Insulin action may involve cascades of protein kinases

We have already stressed that one of the central themes of insulin action is the increased phosphorylation of a number of specific proteins (Table 3) and that it is becoming increasingly evident that an important early action of insulin is to activate arrays (or cascades) of protein serine and threonine kinases which are responsible for these increases in phosphorylation. Table 4 lists some of the protein serine and threonine kinases that have been reported to be activated by insulin, and emphasizes their distinct substrate specificities. The possible importance of the activation of serine/threonine protein kinases in insulin action was first raised more than a decade ago (see Denton *et al.* 1981) but there has been an explosion in attention given to this possibility in the last couple of years (Czech *et al.* 1988; Weiel *et al.* 1990).

In most cases, the effect of insulin on serine/threonine protein kinase activity has been detected because the activation is stable through cell extraction and partial purification. Such persistence may be a reflection of insulin causing the translocation of the kinase between compartments of the original cells (Denton *et al.* 1981; Denton 1986) or of insulin initiating some form of covalent modification, such as reversible phosphorylation. This latter possibility seems the more likely in the case of the kinases which phosphorylate ribosomal protein S6 (S6 kinases), MAP-2 kinase, myelin basic protein kinase, and Raf-1 kinase, as the insulin stimulation is lost or diminished if the kinases are extracted in the absence of protein phosphatase inhibitors or treated with appropriate protein phosphatases, and the addition of okadaic acid can mimic the effect of insulin.

Two distinct S6 kinases have been described, with apparent molecular weights of 90 kDa and 70 kDa on SDS-polyacrylamide gel electrophoresis. Both have been

Table 4 Examples of insulin-induced increases in serine- and threonine-specific protein kinase activities within intact cells

Source of kinase	**Identity of kinase**	**Substrate for kinase***
Fat cells	Acetyl-CoA carboxylase kinase	Acetyl-CoA carboxylase
Many cell types	S6 kinase	Ribosomal protein S6
Many cell types	MAP-2 kinase	(Microtubule associated protein-2) S6 kinase, Raf-1
Rabbit reticulocytes	Protease activated kinase II	Ribosomal protein S6
3T3-L1 and rat adipocytes BALB/c/3T3 cells	Casein kinase II	Acetyl-CoA carboxylase, 22 kDa protein
Placenta, liver, IM9 cells	Insulin-receptor associated	Insulin receptor
Liver, fat cells	Kemptide kinase	ATP-citrate lyase?
Many cultured cells	Raf-1 kinase	NI

* Although some of the kinases were identified using artificial substrates, only their probable physiological substrates have been included.
NI, not identified.
For further details see Denton (1986) and Weiel *et al.* (1990)

cloned and it is now evident that their accurate molecular weights are in fact close to 83 kDa and 59 kDa respectively (Jones *et al.* 1988; Banerjee *et al.* 1990; Kozma *et al.* 1990).

The former kinase, which is also referred to as S6 kinase II, was first described in amphibian oocytes (Erikson and Maller 1986) but it is now evident that similar enzymes are present in mammalian tissues (Weiel *et al.* 1990; Erikson 1991) and may be involved in the activation of the G-subunit of phosphatase I in muscle (Dent *et al.* 1990). Activation of this kinase is brought about by its phosphorylation on serine and threonine residues by another insulin-activated kinase, known as MAP-2 kinase because of its ability to phosphorylate microtubule-associated protein 2 *in vitro* (Sturgill *et al.* 1988; Boulton *et al.* 1990). It seems unlikely that MAP-2 is a physiological substrate; rather it seems that the physiological substrates include S6 kinase II and possibly also Raf-1.

The smaller S6 kinase ('70K-kinase') may be the more active of the two S6 kinases in mammalian cells (Ballou *et al.* 1991). Again, the kinase is activated by phosphorylation on serine and threonine residues but the insulin-activated kinase involved has yet to be characterized.

It should be noted that the two cascades involving the two S6 kinases (Fig. 2) are essentially incomplete, as some form of 'switch' is needed to link the activation of the sequences of serine/threonine to the activation of the insulin receptor tyrosine kinase. MAP-2 kinase and Raf-1 kinase have been shown to exhibit increased tyrosine as well as serine/threonine phosphorylation in insulin-treated cells. However, neither of these kinases appear to be direct substrates of the insulin receptor (Anderson *et al.* 1990; Blackshear *et al.* 1990). It is also not clear how the other insulin-activated serine/threonine kinases such as casein kinase II, kemptide kinase (Klarlund *et al.* 1990) and acetyl-CoA carboxylase kinase may fit into these putative kinase cascades. Under certain conditions, the addition of insulin to a Triton extract of human placental membranes results not only in the increased

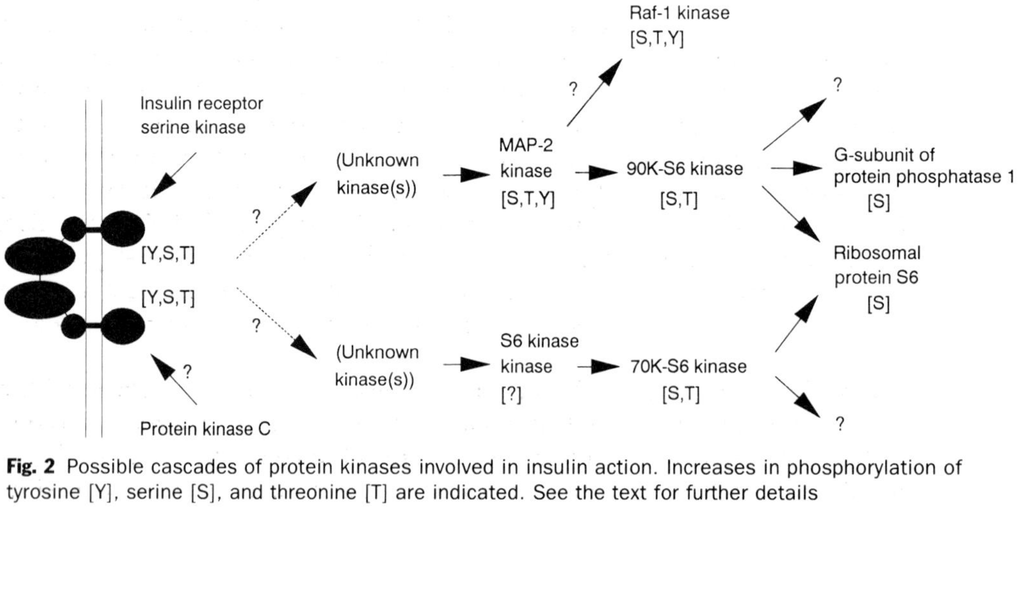

Fig. 2 Possible cascades of protein kinases involved in insulin action. Increases in phosphorylation of tyrosine [Y], serine [S], and threonine [T] are indicated. See the text for further details

tyrosine phosphorylation of the insulin receptor but also the increased phosphorylation of added acetyl-CoA carboxylase, apparently within the same tryptic peptide that shows increased phosphorylation in fat cells exposed to insulin (Denton *et al.* 1986). This observation suggests that it should be possible eventually to reconstruct the insulin signalling system involved in a cell-free system from defined components.

Finally, a brief comment should be made about the possible role of protein kinase C in insulin action. This is a very controversial area (Blackshear 1986; Farese 1990). A number of groups have reported that insulin activates protein kinase C, apparently by increasing the concentration of diacylglycerol. However, others have not been able to find any stimulation. In any case, it is not at all clear what role activation of protein kinase C might play in insulin action. Possibilities include stimulation of glucose transport (but see Shibata *et al.* 1991) and increased threonine phosphorylation of the insulin receptor. However, many of the effects of insulin are still apparent when cells are depleted of the kinase by chronic prior treatment with phorbol esters.

7.4 Final comments

It must be emphasized that the possible mechanisms described above should not be considered to be exclusive of each other—rather that some combination may be involved. It has become clear that the critical question is the identification of the proteins interacting with the insulin receptor which are involved in signal transfer. Such proteins may or may not be substrates for the tyrosine kinase activity of the insulin receptor. Candidates that we have already discussed include G-proteins, serine/threonine protein kinases, and enzymes that may be involved in the formation of low molecular weight mediators such as inositol-specific phospholipase C. Studies of the epidermal growth factor and platelet-derived growth factor receptors have indicated that these other tyrosine kinase receptors are associated with a number of proteins, including phospholipase C, phosphatidylinositol 3-kinase and ras GAP (Anderson *et al.* 1990; Ullrich and Schlessinger 1990) Cantley *et al.* 1991). Moreover, these are all phosphorylated on tyrosines by the two receptors and, in the case of the first two proteins, this may lead to an increase in enzymic activity. However, similar observations have only been reported with the insulin receptor in the case of phosphatidylinositol 3-kinase (Ruderman *et al.* 1990; and the physiological role of this enzyme is a mystery (Cantley *et al.* 1991). As discussed in Chapter 6, it is possible that pp185, which is one of the most prominent intracellular substrates for the insulin receptor tyrosine kinase activity, may play a central linking role between the insulin receptor and other components of the signalling system.

Once insulin is bound, the insulin receptor is rapidly internalized (Table 2). This raises the possibility that at least some of the actions of insulin may be a consequence of the uptake of the occupied receptor into the cell. The polarity of the receptor within the endosome will be with the tyrosine kinase domain, and associated proteins such as serine/threonine kinases, on the cytoplasmic face of the endosome. It is in this way that the phosphorylation of key intracellular proteins may be brought about.

It can be asserted with confidence that insights into the molecular basis of insulin action have an impact far beyond the understanding and possible treatment of diabetes, important though that is. In particular, they will shed light on the mode of action of oncogenes and receptors for a number of important growth factors which, like insulin, have intrinsic protein–tyrosine kinase activities.

References

Alexander, M. C., Lomanoto, M., Nasria, N., and Ranaika, C. (1988). Insulin stimulates glyceraldehyde-3-phosphate dehydrogenase gene expression through *cis*-acting DNA sequences. *Proc. Natl Acad. Sci. USA* **85,** 5092–6.

Alvarez, J. F., Varela, I., Ruiz-Albbusac, J. M. and Mato, J. M. (1988). Localisation of the insulin-sensitive phosphatidylinositol glycan at the outer surface of the cell membrane. *Biochem. Biophys. Res. Comm.* **152,** 1455–62.

Anderson, N. G., Maller, J. L., Tonks, N. K., and Sturgill, T. W. (1990). Requirement for integration of signals from two distinct phosphorylation pathways for activation of MAP kinase. *Nature* **343,** 651–3.

Baldini, G., Hohman, R., Charron, M. J., and Lodish, H. F. (1991). Insulin and non-hydrolyzable GTP analogs induce translocation of GLUT-4 to the plasma membrane in alpha-toxin-permeabilized rat adipose cells. *J. Biol. Chem.* **266,** 4037–40.

Ballou, L. M., Luther, H., and Thomas, G. (1991). MAP2 Kinase and 70K-S6 kinase lie on distinct signalling pathways. *Nature* **349,** 348–50.

Banerjee, P., Ahmad, M. F., Grove, J. R., Kozlosky, C., Price, D. J., and Avruch, J. (1990). Molecular structure of a major insulin mitogen-activated 70-kDa S6 protein kinase. *Proc. Natl Acad. Sci. USA* **87,** 8550–4.

Belfrage, P., Donner, J., and Stralford, P. (1986). *Mechanisms of insulin action*. Elsevier, Amsterdam.

Bell, G. I., Kayano, T., Buse, J. B., Burant, C. F., Takeda, J., Lin, D., Fukumoto, H., and Seino, S. (1990). Molecular biology of mammalian glucose transporters. *Diabetes Care* **13,** 198–208.

Blackshear, P. J. (1986). Protein phosphorylation in cultured cells; insulin, growth factors and phorbol esters. In *Mechanisms of insulin action* (ed. P. Belfrage, J. Donner, and P. Stralfors), pp. 211–28. Elsevier, Amsterdam.

Blackshear, P. J., Haupt, D. M., App, H., and Rapp, U. R. (1990). Insulin activates the Raf-1 protein kinase. *J. Biol. Chem.* **265,** 12131–4.

Blak, J., Gibbs, E. M., Lienhard, G. E., Slot, J. W., and Geuve, H. J. (1988). Insulin-induced translocation of glucose transporters from post-golgi compartments to the plasma membrane of 3T3-L1 adipocytes. *J. Cell. Biol.* **106,** 69–76.

Borthwick, A. C., Edgell, N. J., and Denton, R. M. (1987). Use of rapid gel-permeation chromatography to explore the inter-relationships between polymerisation and activity of acetyl-CoA carboxylase. Effects of insulin and phosphorylation by cyclic-AMP dependent protein kinase. *Biochem. J.* **241,** 773–82.

Borthwick, A. C., Edgell, N. J., and Denton, R. M. (1990). Protein-serine kinase from rat epididymal adipose tissue which phosphorylates and activates acetyl-CoA carboxylase—possible role in insulin action. *Biochem. J.* **270,** 795–801.

Boulton, T. G., Yancopoulos, C. D., Gregory, J. S., Slaughter, C., Moomaw, C., Hsu, J., and Cobb, M. H. (1990). An insulin-stimulated protein kinase similar to yeast kinases involved in cell cycle control. *Science* **249,** 64–7.

Brownsey, R. W. and Denton, R. M. (1982). Evidence that insulin activates acetyl-CoA carboxylase by increased phosphorylation of a specific site. *Biochem. J.* **202,** 77–86.

Brownsey, R. W. and Denton, R. M. (1987). Acetyl-CoA carboxylase. In *The enzymes,* Vol. 18 (ed. E. G. Krebs and P. D. Boyer), pp. 123–46. Academic Press, New York.

Calderhead, D. M. and Lienhard, G. E. (1988). Labelling of glucose transport at the cell surface in 3T3-L1 adipocytes. *J. Biol. Chem.* **263,** 12171–4.

Calderhead, D. M., Kitagawa, K., Tanner, L. I., Holman, G. D., and Lienhard, G. E. (1990). Insulin regulation of the two glucose transporters in 3T3-L1 adipocytes. *J. Biol. Chem.* **265,** 13800–8.

Cantley, L. C., *et al.* (1991). Oncogenes and signal transduction. *Cell* **64,** 281–302.

Clark, A. E. and Holman, G. D. (1990). Exofacial photolabelling of the human erythrocyte glucose transporter with an azitrifluoroethylbenzoyl-substituted bismannose. *Biochem. J.* **269,** 615–22.

Cohen, P. (1987). Role of multisite phosphorylation in the hormonal control of glycogen synthase from mammalian muscle. In *The enzymes,* Vol. 18 (ed. E. G. Krebs and P. D. Boyer), pp. 462–99. Academic Press, Orlando.

Cohen, P., Parker, P. J., and Woodgett, J. R. (1985). The molecular mechanism by which insulin activates glycogen synthase in mammalian skeletal muscle. In *Molecular basis of insulin action* (ed. M. P. Czech), pp. 213–33. Plenum Press, New York.

Cox, S., Redpath, N. T., and Proud, C. G. (1988). Regulation of polypeptide-chain initiation in rat skeletal muscle. Starvation does not alter the activity or phosphorylation state of initiation factor eIF-2. *FEBS Lett.* **239,** 333–8.

Crettaz, M., Muller-Wieland, D., and Khan, C. R. (1988). Transcriptional and post-transcriptional regulation of tyrosine aminotransferase by insulin in rat hepatoma cells. *Biochemistry* **27,** 495–500.

Cushman, S. W. and Wardzala, L. J. (1980). Potential mechanism of insulin action on glucose transport in the isolated rat adipose cell: apparent translocation of intracellular transport systems to the plasma membrane. *J. Biol. Chem.* **255,** 4758–62.

Czech, M. P. (1985). *Molecular basis of insulin action.* Plenum Press, New York.

Czech, M. P., Karlund, J. K., Yagaloff, K. A., Bradford, A. P., and Lewis, R. E. (1988). Insulin receptor signaling. Activation of multiple serine kinases. *J. Biol. Chem.* **263,** 11017–20.

Davis, R. J., Corvera, S., and Czech, M. P. (1986). Insulin stimulates cellular iron uptake and causes the redistribution of intracellular transferrin receptors to the plasma membrane. *J. Biol. Chem.* **261,** 8708–11.

Degerman, E., Smith, C. J., Tornqvist, H., Vasta, V., Belfrage, P., and Manganiello, V. C. (1990). Evidence that insulin and isoprenaline activate the cGMP-inhibited low-*Km* cAMP phosphodiesterase in rat fat cells by phosphorylation. *Proc. Natl Acad. Sci. USA* **87,** 533–7.

Dent, P., Lavoinne, A., Nakielny, S., Caudwell, F. B., Watt, P., and Cohen, P. (1990). The molecular mechanism by which insulin stimulates glycogen synthesis in mammalian skeletal muscle. *Nature* **348,** 302–8.

Denton, R. M. (1986). Early events in insulin actions. *Adv. Cyclic Nucl. Prot. Phos. Res.* **20,** 289–336.

Denton, R. M. and Halestrap, A. P. (1979). Regulation of pyruvate metabolism in mammalian tissues. *Essays in Biochem.* **15,** 37–77.

Denton, R. M., *et al.* (1975). Regulation of mammalian pyruvate dehydrogenase. *Mol. Cell. Biochem.* **9,** 27–53.

Denton, R. M., Brownsey, R. W. and Belsham, G. J. (1981). A partial view of the mechanism of insulin action. *Diabetologia* **21,** 347–62.

Denton, R. M., McCormack, J. G., and Marshall, S. E. (1984). Persistence of the effect of insulin on pyruvate dehydrogenase activity in rat white and brown adipose tissue during the preparation and subsequent incubation of mitochondria. *Biochem. J.* **217,** 441–52.

Denton, R. M., Thomas, A. P., Tavaré, J. M., Borthwick, A. C., Brownsey, R. W., Hopkirk, T. J., and McCormack, J. G. (1986). Mechanisms involved in the stimulation of fatty acid synthesis by insulin. In *Mechanisms of insulin action,* Vol. 7 (eds P. Belfrage, J. Donner, and P. Stralfors), pp. 283–304. Elsevier Biomedical Press, Amsterdam.

Denton, R. M., Midgley, P. J. W., Rutter, G. A., Thomas, A. P., and McCormack, J. G. (1989). Studies into the mechanism whereby insulin activates pyruvate dehydrogenase in adipose tissue. *Ann. NY Acad. Sci.* **573,** 285–6.

Diggle, T. A., Schmitz-Peiffer, C., Borthwick, A. C., Welsh, G. I., and Denton, R. M. (1991). Evidence that insulin activates casein kinase 2 in rat epididymal fat cells and that this may result in the increased phosphorylation of an acid-soluble 22KDa protein. *Biochem. J.* **279,** 545–51.

Distel, R. J., Ro, H., Rosen, B. S., Croves, D. L., and Spiegelman, B. M. (1987). Nucleoprotein complexes that regulate gene expression in adipocyte differentiation: direct participation of *c-fos*. *Cell* **49,** 835–44.

Erikson, E. and Maller, J. L. (1989). *In vivo* phosphorylation and activation of ribosomal protein S6 kinases during *Xenopus* oocyte maturation. *J. Biol. Chem.* **264,** 13711–17.

Erikson, R. L. (1991). Structure, expression and regulation of protein kinases involved in the phosphorylation of ribosomal protein S6. *J. Biol. Chem.* **266,** 6007–10.

Espinall, J. (1990). *Understanding insulin action: principles and molecular mechanisms.* Ellis Horwood, Chichester.

Fain, J. N. (1990). Hormonal regulation of lipid mobilization from adipose tissue. *Biochem. Action. Horm.* **7,** 119–204.

Farese, R. V. (1990). Lipid-derived mediators in insulin action. *Proc. Natl Acad. Sci. USA* **195,** 312–24.

Fisch, T. M., Pryures, R. and Roeder, R. G. (1987). *c-fos* sequences necessary for basal expression and induction by epidermal growth factor, 2-0-tetradecanoyl phorbol-13-acetate and the calcium ionophore. *Mol. Cell. Biol.* **7,** 3490–502.

Foulkes, J. G., Cohen, P., Strada, S. J., Everson, M. W., and Jefferson, L. S. (1982). Antagonistic effects of insulin and β-adrenergic agonists on the activity of protein phosphatase inhibitor-1 in skeletal muscle of the perfused rat hemicorpus. *J. Biol. Chem.* **257,** 12493–6.

Gauber, D. and Houslay, M. D. (1987). Insulin stimulates a novel GTPase activity in human platelets. *FEBS Lett.* **216,** 914–18.

Gould, G. W. and Bell, G. I. (1990). Facilitative glucose transporters: an expanding family. *TIBS* **15,** 18–23.

Gottschalk, W. K. and Jarett, L. (1988). The insulinometric effects of the polar head group of an insulin-sensitive glycophospholipid on pyruvate dehydrogenase in both subcellular and whole cell assays. *Arch. Biochem. Biophys.* **261,** 175–85.

Gottschalk, W. K., Macaulay, S. L., Macaulay, J. O., Kelly, K., Smith, J. A., and Jarett, L. (1986). Characterization of mediators of insulin action. *Ann. NY Acad. Sci.* **488,** 385–405.

Granner, D. and Pilkis, S. (1990). The genes of hepatic glucose metabolism. *J. Biol. Chem.* **265,** 10173–6.

Granner, D. K., Sasaki, K., Chu, S., Koch, T., Cripe, T., and Peterson, D. (1986). Mutihormone control of phosphoenol pyruvate carboxykinase gene transcription; the dominant role of insulin. In *Mechanisms of insulin action* (ed. P. Belfrage, J. Donnier, and P. Stralfors), pp. 365–82. Elsevier, Amsterdam.

Hardie, D. G. (1989). Regulation of fatty acid synthesis via phosphorylation of acetyl-CoA carboxylase. *Prog. Lipid Res.* **28,** 117–46.

Haystead, T. A. J., Campbell, D. G., and Hardie, D. G. (1988). Analysis of sites phosphorylated on acetyl-CoA carboxylase in response to insulin in isolated adipocytes. Comparison with sites phosphorylated by casein kinase-2 and the calmodulin-dependent multiprotein kinase. *Eur. J. Biochem.* **175,** 347–54.

Haystead, T. A. J., *et al.* (1989). Effects of the tumour promoter okadaic acid on intracellular protein phosphorylation and metabolism. *Nature* **337,** 78–81.

Heyworth, C. M. and Houslay, M. D. (1983). Insulin exerts actions through a distinct species of guanine nucleotide regulatory protein: inhibition of adenyl-cyclase. *Biochem. J.* **214,** 547–52.

Heyworth, C. M., Whetton, A. D., Wong, S., Martin, B. R., and Houslay, M. D. (1985). Insulin inhibits the cholera toxin catalysed ribosylation of a Mr-25 000 protein in rat liver plasma membranes. *Biochem. J.* **228,** 593–603.

Holman, G. D., Kozka, I. J., Clark, A. E., Flower, C. J., Saltis, J., Habberfield, A. D., Simpson, I. A., and Cushman, S. W. (1990). Cell surface labeling of glucose transporter isoform Glut4 by bis-mannose photolabel correlation with stimulation of glucose transport in rat adipose cells by insulin and phorbol ester. *J. Biol. Chem.* **265,** 18172–9.

Houslay, M. D. (1985). The insulin receptor and signal generation at the plasma membrane. In *Molecular mechanisms of transmembrane signalling* (ed. P. Cohen and M. D. Houslay), pp. 279–333. Elsevier Biomedical Press, Amsterdam.

Houslay, M. D. (1986). Insulin, glucagon and the receptor-mediated control of cyclicAMP concentrations in liver. *Biochem. Soc. Trans.* **14,** 183–93.

Houslay, M. D. and Siddle, K. (1989). Molecular basis of insulin receptor function. *Br. Med. Bull.* **45,** 264–84.

Houston, B. and Nimmo, H. G. (1985). Effects of phosphorylation on the kinetic properties of rat liver ATP-citrate lyase. *Biochim. Biophys. Acta* **844,** 233–9.

Iynedjian, P. B., Gjinovci, A., and Renold, A. E. (1988). Stimulation by insulin of glucokinase gene transcription in liver of diabetic rats. *J. Biol. Chem.* **263,** 740–4.

Jarett, L. and Seals, J. R. (1979). Pyruvate dehydrogenase activation in adipocyte mitochondria by an insulin generated mediator from muscle. *Science* **206,** 1407–8.

Jones, S. W., Erikson, E., Blenis, J., Maller, J. L., and Erikson, R. L. (1988). A *Xenopus* ribosomal protein S6 kinase has two apparent kinase domains that are each similar to distinct protein kinases. *Proc. Natl Acad. Sci. USA* **85,** 3377–81.

Kahn, C. R. (1985). The molecular mechanism of insulin action. *Ann. Rev. Med.* **36,** 429–51.

Kahtra, B. S., Chiasson, J. L., Shikama, H., Exton, J. H., and Soderling, T. R. (1980). Effects of epinephrine and insulin on the phosphorylation of phosphorylase phosphatase inhibitor 1 in perfused rat skeletal muscle. *FEBS Lett.* **114,** 253–6.

Kelly, K. L., Mato, J. M., Merida, I., and Jarett, L. (1987). Glucose transport and antilipolysis are differentially regulated by the polar head group of an insulin-sensitive glycophospholipid. *Proc. Natl Acad. Sci. USA* **84,** 6404–7.

Kimball, S. R. and Jefferson, L. W. (1988). Cellular mechanisms involved in the action of insulin on protein synthesis. *Diabetes/Metabolism Rev.* **4,** 773–88.

Klarlund, J. K., Bradford, A. P., Milla, M. G., and Czech, M. P. (1990). Purification of a novel insulin-stimulated protein kinase from rat liver. *J. Biol. Chem.* **265,** 227–34.

Kozma, S. C., Ferrari, S., Bassand, P., Siegmann, M., Totty, N., and Thomas, G. (1990). Cloning of the mitogen-activated S6-kinase from rat liver reveals an enzyme of the 2nd messenger subfamily. *Proc. Natl Acad. Sci. USA* **87,** 7365–9.

Larner, J. (1988). Insulin-signalling mechanisms. *Diabetes* **37,** 262–75.

Larner, J., Lawrence, J. C., Walkenbach, R. J., Roach, P. J., Hazen, R. J., and Huang, L. C. (1978). Insulin control of glycogen synthesis. *Adv. Cyclic Nucl. Prot. Phos. Res.* **9,** 425–39.

Lawrence, J. C. and James. C. (1984). Activation of glycogen synthase by insulin in rat adipocytes. Evidence of hormonal stimulation of multisite dephosphorylation by glucose transport-dependent and -independent pathways. *J. Biol. Chem.* **259,** 7975–82.

Lawrence, J. C., Jr, Hiken, J. F., Depaoli-Roach, A. A., and Roach, P. J. (1983). Hormonal control of glycogen synthase in rat hemidiaphragms. *J. Biol. Chem.* **258,** 10710–19.

Lawrence, J. C., Hiken, J. F., and James, D. E. (1990*a*). Phosphorylation of the glucose transporter in rat adipocytes—identification of the intracellular domain at the carboxyl terminus as a target for phosphorylation in intact cells *in vitro*. *J. Biol. Chem.* **265,** 2324–32.

Lawrence, J. C., Hiken, J. F., and James, D. E. (1990*b*). Stimulation of glucose transport and glucose transporter phosphorylation by okadaic acid in rat adipocytes. *J. Biol. Chem.* **265,** 19768–76.

Lodish, H. F. (1988). Anion exchange and glucose transport proteins: structure, function and distribution. *The Harvey Lectures* **82,** 19–46.

Londos, C., Honnor, R. C., and Dhillon, G. S. (1985). CyclicAMP-dependent protein kinase and lipolysis in rat adipocytes. III. Multiple modes of insulin regulation of lipolysis and regulation of insulin response by adenylate cyclase regulators. *J. Biol. Chem.* **260,** 15139–45.

Makins, H. and Kono, T. (1980). Characterization of insulin-sensitive phospho-diesterase in fat cells II. Comparison of enzyme activities stimulated by insulin and isoproterenol. *J. Biol. Chem.* **255,** 7850–4.

Marchington, D. R., Kerbey, A. L., Jones, A. E., and Randle, P. J. (1987). Insulin reverses effects of starvation on the activity of pyruvate dehydrogenase kinase in cultured hepatocytes. *Biochem. J.* **246,** 233–6.

Marchington, D. R., Kerbey, A. L., and Randle, P. J. (1990). Longer-term regulation of pyruvate dehydrogenase kinase in cultured rat cardiac myocytes. *Biochem. J.* **267,** 245–7.

Merida, I., Corrales, F. J., Clements, R., Ruiz-Albusac, J. M., Villalba, M., and Mato, J. M. (1988). Different phosphorylated forms of an insulin sensitive glycosyl-phosphotidylinositol from rat hepatocytes. *FEBS Lett.* **236,** 251–5.

Muhlbacher, C., Karnieli, E., Schaff, P., Obermaier, B., Mushack, J., Rattenhuber, E., and Häring, H. V. (1988). Phorbol esters imitate in rat fat cells the full effect of insulin on glucose-carrier translocation, but not on 3-*O*-methylglucose transport activity. *Biochem. J.* **249,** 865–70.

O'Brien, R. M., Lucas, P. C., Forest, C. D., Magnuson, M. A., and Granner, D. K. (1990). Identification of a sequence in the PEPCK gene that mediates a negative effect of insulin on transcription. *Science* **249,** 533–7.

Oka, Y. and Czech, M. P. (1984). Photoaffinity labelling of insulin-sensitive hexose transporters in intact rat adipocytes. Direct evidence that latent transporters become exposed to the extracellular space in response to insulin. *J. Biol. Chem.* **259,** 8125–33.

Pierce, M. W., Palmer, J. L., Keutmann, H. T., Hall, A., and Avruch, J. (1982). The insulin-directed phosphorylation site on ATP-citrate lyase. *J. Biol. Chem.* **257,** 10681–6.

Poulter, L., Ang, S. G., Gibson, B. W., and Cohen, P. (1988). Analysis of the *in vivo* phosphorylation sites of rabbit skeletal muscle glycogen synthase by fast atom bombardment mass spectrometry. *Eur. J. Biochem.* **175,** 497–510.

Pouter, R. H., Butcher, F. R., and Fain, J. N. (1976). Studies on the rôle of cyclic guanosine 3′,5′-monophosphate and extracellular Ca^{2+} in the regulation of glycogenolysis in rat liver cells. *J. Biol. Chem.* **251,** 2987–92.

Pyne, N. J., Cushley, W., Nimmo, H. G., and Houslay, M. D. (1989). Insulin stimulates the tyrosyl phosphorylation and activation of the 52-kDa peripheral plasma-membrane cyclic AMP phosphodiesterase in intact hepatocytes. *Biochem. J.* **261,** 897–904.

Randle, P. J. (1986). Fuel selection in animals. *Biochem. Soc. Trans.* **14,** 799–806.

Rosen, O. M. (1987). After insulin binds. *Science* **237,** 1452–8.

Rothenberg, P. L. and Kahn, C. R. (1988). Insulin inhibits Pertussis toxin-catalysed ADP-ribosylation of a G-protein. *J. Biol. Chem.* **263,** 15546–52.

Ruderman, N. B., Kapeller, R., White, M. F., and Cantley, L. C. (1990). Activation of phosphatidylinositol 3-kinase by insulin. *Proc. Natl Acad. Sci. USA* **87,** 1411–15.

Saltiel, A. R. and Cuatrecasas, P. (1988). In search of a second messenger for insulin. *Am. J. Physiol.* **255,** C1–C11.

Saltiel, A. R., Fox, J. A., Sherline, P., and Cuatrecasas, P. (1986). Insulin stimulated hydrolysis of a novel glycolipid generates modulators of cAMP phospho-diesterase. *Science* **223,** 967–72.

Saltiel, A. R., Osterman, D. G., Darnell, J. C., Chan, B. L., and Sorbaracazan, L. R. (1989). The role of glycosylphosphoinositides in signal transduction. *Rec. Progr. Hormone Res.* **45,** 353–82.

Sasaki, K., *et al.* (1984). Multihormonal regulation of phosphoenol pyruvate carboxylase gene transcription. *J. Biol. Chem.* **259,** 15242–51.

Shibata, H., Robinson, F. W., Benzing, C. F., and Kono, T. (1991). Evidence that protein kinase-C may not be involved in the insulin action on cAMP phosphodiesterase—studies with electroporated rat adipocytes that were highly responsive to insulin. *Arch. Biochem. Biophys.* **285,** 97–104.

Simpson, I. A. and Cushman, S. W. (1986). Hormonal regulation of mammalian glucose transport. *Ann. Rev. Biochem.* **55,** 1059–89.

Stralfors, P. (1987). Isoproterenol and insulin control the cellular localization of ATP-citrate lyase through its phosphorylation in adipocytes. *J. Biol. Chem.* **262,** 11486–9.

Stralfors, P., Olsson, H., and Belfrage, P. (1985). Hormone sensitive lipase. In *The enzymes*, Vol. XVIIIB (ed. P. D. Boyer and E. G. Krebs), pp. 147–79. Academic Press, New York.

Stumpo, D. J. and Blackshear, P. J. (1986). Insulin and growth factor effects on c-fos expression in normal and protein kinase C-deficient 3T3-L1 fibroblasts and adipocytes. *Proc. Natl Acad. Sci. USA* **83,** 9453–7.

Stumpo, D. J., Stewart, T. N., Gilman, M. Z., and Blackshear, P. J. (1988). Identification of c-fos sequences involved in induction by insulin and phorbol esters. *J. Biol. Chem.* **263,** 1611–14.

Sturgill, T. W., Ray, L. B., Erikson, E., and Maller, J. L. (1988). Insulin-stimulated MAP-2 kinase phosphorylates and activates ribosomal protein S6 kinase II. *Nature* **334,** 715–18.

Suski, K. and Kono, T. (1980). Evidence that insulin causes translocation of glucose transport activity to the plasma membrane from an intracellular storage site. *Proc. Natl Acad. Sci. USA* **77,** 2542–5.

Tabarini, D., Heinrich, J., and Rosen, O. M. (1984). Activation of S6 kinase activity in 3T3-L1 cells by insulin and phorbol ester. *Proc. Natl Acad. Sci. USA* **81,** 7797–801.

Tanti, J. F., Gremeaux, T., Van Obberghen, E., and Le Marchand Brustel, Y. (1991). Effects of okadaic acid, an inhibitor of protein phosphatases-1 and phosphatases-2A, on glucose transport and metabolism in skeletal muscle. *J. Biol. Chem.* **266,** 2099–103.

Tipper, J. P., Bacon, G. W., and Witters, L. A. (1983). Phosphorylation of acetyl-coenzyme A carboxylase by casein kinase I and casein kinase II. *Arch. Biochem. Biophys.* **227,** 386–96.

Towle, C. A., Markin, H. J., Avruch, J., and Treadwell, B. V. (1984). Insulin-promoted decrease in the phosphorylation of protein synthesis initiation factor eIF-2. *Biochem. Biophys. Res. Comm.* **121,** 134–40.

Tung, H. Y. L., Pelech, S., Fisher, M. J., Pogson, C. I., and Cohen, P. (1985). The protein phosphatases involved in cellular regulation. Influence of polyamines on the activities of protein phosphatase-1 and protein phosphatase-2A. *Eur. J. Biochem.* **149,** 305–13.

Ullrich, A. and Schlessinger, J. (1990). Signal transduction by receptors with tyrosine kinase activity. *Cell* **61,** 203–12.

Villalba, M., Kelly, K. L., and Mato, J. M. (1987). Inhibition of cyclic AMP-dependent protein kinase by the polar head group of an insulin-sensitive glycophospholipid. *Biochim. Biophys. Acta* **968,** 69–76.

Weiel, J. E., Ahn, N. G., Seger, R., and Krebs E. G. (1990). Communication between protein tyrosine and protein serine threonine phosphorylation. *Biology and Medicine of Signal Transduction* **24,** 182–95.

Wieland, O. H. (1983). The mammalian pyruvate dehydrogenase complex: structure and regulation. *Rev. Physiol. Biochem. Pharmacol.* **96,** 123–70.

Yang, S. D., Ho, L. T., and Fung, T. J. (1988). Insulin induces activation and translocation of protein kinase FA (a multifunctional protein phosphatase activator) in human platelet. *Biochem. Biophys. Res. Comm.* **151,** 61–9.

Zick, Y. (1989). The insulin receptor—structure and function. *Crit. Rev. Biochem. Mol. Biol.* **24,** 217–69.

Zick, Y., Sagit-Eisenberg, R., Pines, M., Gierschik, P., and Spiegel, A. (1986). Multisite phosphorylation of the α-subunit of transducin by the insulin receptor kinase and protein kinase C. *Proc. Natl Acad. Sci. USA* **83,** 9294–7.

PART V PATHOLOGY OF INSULIN DEFICIENCY

Editors' introduction

Diabetes mellitus is a clinical syndrome characterized by elevation of the fasting blood glucose concentration. It is common in humans but occurs also in other species. In Western countries, up to 5 per cent of the population have some form of the disease, which therefore constitutes a major public health problem. It has become clear that this condition includes two major and quite distinct diseases, type I (insulin-dependent or juvenile-onset) and type II (non-insulin-dependent or maturity onset) diabetes. These disorders differ in their genetics, incidence, and pathology. Both, however, involve a decrease in insulin secretion. In type I diabetes this arises from a loss of β-cells, whereas in type II diabetes the ability of the β-cells to respond to glucose with insulin secretion is defective. In addition, in type II diabetes the efficacy of insulin is also usually reduced (insulin resistance). Both types of diabetes are associated with long-term secondary complications, which include micro- and macrovascular disease, neuropathy, nephropathy, and retinopathy. These complications considerably reduce both the life expectancy and quality of life of diabetics, and provide one of the main spurs for the search for a cure for both diseases.

As described in detail in Chapters 8–10, type I and type II diabetes have a genetic component, although despite intensive research the gene(s) involved have not been identified. It is likely that one or more genes may confer the major susceptibility, while other genes may contribute to the genetic background of the disease. Environmental factors, such as stress, infection and life-style, are also of causative importance.

Type I diabetes

Type I diabetes is an auto-immune disease and β-cell destruction occurs as a consequence of an inappropriate auto-immune response to the β-cell. The evidence consists of:

(1) the presence of excess lymphocytes in and around the pancreatic islets of type I diabetics;

(2) the occurrence in the blood of diabetics of circulating antibodies to islet cells;

(3) an increased risk of the disease in individuals who have particular variants of the genes which encode membrane proteins involved in immune responses.

A group of genes on human chromosome 6, referred to as the HLA gene cluster, constitutes the major histocompatibility complex (MHC). The proteins encoded in this region participate in immunological recognition, both between different types of lymphocytes and between lymphocytes and cells bearing antigens. The MHC encodes three major classes of molecules, class I, class II, and class III antigens. It is the class I and II regions which encode molecules involved in immunological recognition. Class I molecules comprise a single polypeptide that is non-covalently associated with a polypeptide, called β2-microglobulin, which is coded for outside the MHC. The class II antigens are dimers consisting of one α- and one β-chain. The class I MHC molecules include the HLA-A, HLA-B, and HLA-C antigens which are expressed in all nucleated cells in varying amounts. The class II genes are arranged into three main subregions, DP, DQ, and DR, each containing genes for at least one α- and one β-chain. The antigens coded for by the D region are only expressed in certain cells of the immune system. The large number of MHC genes, and the high degree of polymorphism they show, means that within the population there is a very large number of possible combinations and thus great variability between individuals. Certain combinations, however, of MHC antigens occur more frequently than others (linkage disequilibrium).

A given individual's MHC antigens can be determined by tissue-typing, using either the serological or the mixed lymphocyte reaction methods. The former uses an antibody known to react specifically with a given HLA antigen: in the presence of complement, a test cell (usually a lymphocyte) bearing the HLA antigen recognized by the antibody will be killed. The latter method uses B-lymphocytes homozygous for a particular HLA-D antigen. Test T-lymphocytes bearing a different antigen are stimulated to proliferate by these B-lymphocytes whereas test T-cells bearing the same antigen do not react.

The main function of class I MHC molecules is to present foreign antigens to cytotoxic T-cells so that they may be destroyed. Class II molecules present foreign antigens to helper T-cells, which co-operate with B-lymphocytes to induce antibody production. Activated helper T-cells also release lymphokines, which aid macrophages in killing micro-organisms. In auto-immune diseases, the complex processes involved in differentiating between self and foreign antigens are defective, and an individual's proteins are treated as foreign by the immune system.

Almost all auto-immune diseases are associated with particular HLA types. In type I diabetes, both the DR3 and DR4 variants of the DR gene appear to confer an increased risk of the disease. This does not necessarily imply that the DR gene is the diabetes gene, simply that it is associated with the disease in some way. Indeed, individuals heterozygous for DR3 and DR4 have an even greater risk, which implies the involvement of other genetic factors. There is also evidence that the DQ region may be more closely linked to type I diabetes than DR: a common feature of haplotypes associated with the disease is a single amino-acid substitution at position 57 of the DQ β-chain encoded by the DQ gene.

Type II diabetes

There is overwhelming evidence that type II diabetes is a genetic disease, but the gene or genes involved are unknown. In particular, there is no association of type II diabetes with specific HLA types. In some populations with an unusually high prevalence of diabetes, or in the special form of the disease known as maturity-onset diabetes of the young (MODY), the disease is inherited as an autosomal dominant trait. However, for the great majority of type II diabetics, evidence for a dominant mode of inheritance is lacking.

Since defective insulin secretion appears to be the prime defect in type II diabetes, candidate genes presumably include those involved in the uptake and metabolism of glucose by the β-cell. As yet, of these genes that have been investigated none have been directly implicated in disease susceptibility. Secretion of a defective insulin has been described in some diabetics, but this form of the disease is very rare.

Our growing knowledge of the molecular basis of diabetes raises exciting possibilities for future therapy.

8 Introduction to diabetes

ROBERT TURNER and ANDREW NEIL

8.1 The syndrome of diabetes

Diabetes is the name given to the clinical description of patients with a number of symptoms arising from raised glucose levels. The syndrome can occur from many specific 'secondary causes', including pancreatectomy, iron overload of the β-cells resulting from haemochromatosis, excess cortisol production in Cushing's syndrome, and excess growth hormone secretion in acromegaly and insulin-resistant syndromes, including insulin-receptor defects. The majority of patients do not, however, have any of these identified defects and are said to have 'idiopathic' diabetes. The metabolism of glucose involves many organs, in each of which several important metabolic steps occur. Key regulatory points include hepatic regulation of glucose uptake and release, muscle utilization of fuels, control by pancreatic production of insulin and glucagon, and neurogenic controls. Diabetes may arise from abnormalities at one or several sites in the complex feedback loops in this system. Two main types of diabetes can be distinguished. Different nomenclatures for these types have arisen because overlap between the types makes a strict, simple classification covering all patients impossible.

8.1.1 Juvenile-onset and maturity-onset diabetes

One of the main features of juvenile-onset diabetes was the sudden appearance in non-obese children or young adults of a severe disease which only responded satisfactorily to insulin therapy. On the other hand, maturity-onset patients tended to present at an older age, were often obese and then responded to diet, without need for insulin therapy (Table 1). This suggested that the maturity-onset disease was not simply due to insulin deficiency. This concept received support when the initial insulin bioassays in the 1950s showed that these patients often had normal or even high insulin levels, whereas the younger insulin-dependent diabetic patients had low insulin levels. Further confirmation was obtained when the insulin immunoassay was invented by Yalow and Berson (1960).

This nomenclature based on the age of onset of diabetes has fallen out of fashion with the realization that auto-immunity to the islets is the characteristic pathology of type I diabetes, and that this is not confined to juvenile-onset but can occur in maturity-onset diabetes and may present at any age (Spencer *et al*. 1983). However, the distinction based on age is still a useful shorthand description of presentation.

Table 1 Main clinical characteristics of the two types of diabetes

Age of onset	**Juvenile onset** age, 0–30 years non-obese short history (e.g. 2–4 weeks thirst) high glucose levels often ketosis	**Maturity onset** usually 35 years or more usually obese often symptoms for years moderately raised glucose levels rarely ketosis unless major infection
Therapeutic requirement	**Insulin-dependent diabetes** requires insulin therapy insulin sensitive (small doses of insulin needed)	**Non-insulin-dependent diabetes** can be treated with diet or tablets insulin resistant (large doses of insulin needed)
Pathology	**Type I diabetes** islet cell antibodies present β-cells destroyed by immune inflammatory cells	**Type II diabetes** islet cell antibodies absent often amyloid in islets

8.1.2 Non-insulin-dependent diabetes/insulin-dependent diabetes, NIDDM/IDDM

The advent of insulin therapy led to the clinical classification of two types of diabetes, 'insulin-dependent diabetes' and 'non-insulin-dependent diabetes', which relate to the empirical requirement for insulin therapy (National Diabetes Data Group 1979; WHO Study Group 1985). This was introduced because some maturity-onset diabetic patients are of normal weight, have severe disease which requires insulin and resemble juvenile-onset diabetic patients. Although IDDM is defined by the patient's dependence on insulin for survival, in practice this definition is often extended to include patients who require insulin therapy to prevent symptoms. Similarly, the term NIDDM is often restricted to patients who can be maintained symptom-free either by diet or by tablet therapy. This usage is strictly incorrect, since patients who present with NIDDM may later develop more severe diabetes that requires insulin therapy to prevent symptoms. Although withholding insulin would lead to substantial glycosuria and weight loss before a steady state of suboptimal health is reached, insulin is not required to keep these patients alive. In practice, a patient initially treated by diet or tablets, but later transferred to insulin, is often termed an insulin-treated NIDDM patient. Conversely, if any patient requires insulin therapy to prevent symptoms soon after presentation he is often termed IDDM.

This empirical classification is not satisfactory. Dependence on insulin therapy is not easily defined and the classification of middle-aged non-obese patients receiving insulin as 'insulin-dependent' or 'insulin-requiring' may or may not be relevant to the nature of the basic disease processes.

8.1.3 Type I/type II diabetes

This classification relates to the aetiology of the disease. A distinction between two types of diabetes was apparent when it was found that juvenile-onset, insulin-dependent diabetic patients often had islet cell antibodies, which rarely occurred in non-insulin-dependent diabetic patients. Subsequently, they were shown to have specific HLA types (Gorsuch 1987), and the concept that insulin-dependent diabetes was due to an 'auto-immune' disease process became accepted. On the other hand, non-insulin-dependent diabetes had neither of these attributes and it was realized that this was a separate disease with a distinct aetiology.

In practice, physicians often refer to IDDM and type I diabetes as being synonymous, and similarly use interchangeably NIDDM and type II diabetes. Indeed, some journals, e.g. *Diabetologia*, requests that the two terms are always used together. It would be more sensible if type I diabetes were only used in patients in whom there was good evidence for auto-immune disease. However, in practice, the islet cell antibodies detected at diagnosis of juvenile-onset patients are not found in 15 per cent of otherwise typical patients, and in the other 85 per cent the antibodies disappear over the next year. Thus, the key diagnostic feature of type I diabetes is not reliable. Equally, although classical islet cell antibody-positive patients often have HLA-DR3 or 4, or 3/4, the association is not tight and the disease can occur with any DR type. Thus, if islet cell antibodies are not found, it is difficult to make a definitive diagnosis of type I diabetes. Conversely, some classical NIDDM patients have anti-islet cell antibodies, and it is not known whether they have occult type I disease.

'Type Ib' is sometimes used to distinguish a subgroup who present in middle age with NIDDM with islet cell antibodies in conjunction with many other organ-specific antibodies, e.g. against gastric cells, thyroid, and other auto-immune endocrine diseases. It is thought that they may represent a separate, auto-immune polyendocrinopathy syndrome.

8.1.4 WHO classification of diabetes

Most general populations have a continuum of glucose tolerance from normal to abnormal. A defined cut-off level of glycaemia that determines when a patient has diabetes has been accepted by the National Diabetes Data Group (1979) and the WHO Expert Committee on Diabetes in 1980. The criteria chosen relate to two observations:

1. Bimodality of glucose tolerance in some populations. This was first found among Pima Indians (Bennett *et al.* 1971) and subsequently in Micronesian (Zimmet *et al.* 1982) and Mexican American populations (Gardner *et al.* 1984). The diabetic patients were found to have considerably higher blood glucose concentrations than the normal population, and the anti-mode between the two populations was taken to represent the characteristic of diabetes.

2. Increased risk of development of specific diabetic complications above defined glycaemic levels. The development of retinopathy and proteinuria in Pima Indians (Pettit *et al.* 1980) has been shown to be restricted to individuals with a blood glucose level greater than 11.1 mmol/l two hours after a glucose load. In Caucasian

Table 2 Epidemiological association of degree of hyperglycaemia with complications

	OGTT	Blood or plasma	OGTT 2 h glucose (mM)	Fasting plasma glucose (mM)
Retinopathy				
Jarrett and Keen (1976)	50 g	blood	>11.1	
Pettitt *et al.* (1980)	75 g	plasma	>11.1	>7.8
Cardiovascular mortality				
Fuller *et al.* (1980)	50 g	blood	>5.3	
Eschwege *et al.* (1985)	75 g	plasma	>7.0	

The association of marked hyperglycaemia with retinopathy (and with proteinuria in the Pima Indian study) probably reflects a direct relationship. The upper 5 per cent of the distribution of glucose tolerance in the normal population is associated with increased cardiovascular risk, but it is uncertain whether this relates directly to the degree of glycaemia or to an associated factor such as insulin resistance or dyslipidaemia

civil servants retinopathy was confined to individuals with levels above a similar value (Jarrett and Keen 1972) (see Table 2).

Table 3 gives the specific glucose criteria that have been accepted for diabetes and the less severe abnormality, impaired glucose tolerance (IGT). IGT is defined as a glycaemic response to a standard glucose challenge intermediate between normal and diabetic. The criteria are based on the oral glucose tolerance test, which is simple to do but has a considerable day-to-day variation in its result. The clinical significance of IGT is much less clear than for diabetes, since patients with IGT are not predisposed to microvascular disease, but between 1 and 5 per cent of patients per year with IGT have been shown to develop diabetes. Patients diagnosed as having IGT may be in transit to developing diabetes, although the syndrome may arise from different disease processes in which long-term impaired glucose tolerance persists (Yudkin *et al.* 1990). The risk of coronary heart disease may be increased (Al Sayegh and Jarrett 1979).

The WHO definition has been helpful in allowing physicians and epidemiologists in different countries to compare diabetes using the same criteria (Yudkin *et al.* 1990) and allows the identification of individuals at risk of developing microvascular complications. However, with a progressive disease such as diabetes there are theoretical and practical objections to taking a single fixed glucose concentration as the criterion for diabetes, since the disease is present before the blood glucose reaches that level. Furthermore, following dietary control, the blood glucose concentration may subsequently fall below the critical level.

The fasting plasma glucose concentration is stable from day to day in both normal and non-insulin-dependent diabetic patients with a given state of nutrition and otherwise in good health (Holman and Turner 1979). The fasting plasma glucose concentration can be a useful index of the severity of diabetes, and is probably superior to the above definitions of IGT and NIDDM.

Table 3 WHO criteria for diabetes and impaired glucose tolerance: diagnostic values for a 75 g oral glucose tolerance test

	Glucose concentration (mmol/litre)			
	Whole blood		Plasma	
	Venous	Capillary	Venous	Capillary
Diabetes mellitus				
Fasting value	≥6.7	≥6.7	≥7.8	≥7.8
2 h after glucose load	≥10.0	≥11.1	≥11.1	≥12.2
Impaired glucose tolerance				
Fasting value	<6.7	<6.7	<7.8	<7.8
2 h after glucose load	6.7–10.0	7.8–11.1	7.8–11.1	8.9–12.2

8.2 Epidemiology of insulin-dependent diabetes (IDDM)

In all populations, IDDM is much less common than NIDDM, since the risk of developing NIDDM is at least 30 times greater (Melton *et al.* 1983). There is considerable variation in IDDM between geographical and ethnic groups. The disease is less common in Black populations than White, and uncommon in Polynesian and Japanese populations (Diabetes Epidemiology Research International Group 1988). Northern Europe, particularly Finland, Sweden, Scotland, and Norway, has a higher incidence than southern Europe (Rewers *et al.* 1988). There is an inverse correlation with the mean temperature, but the difference in disease incidence may also be due to other factors such as the socio-economic status of the different countries or to the population density. Living indoors in cold countries may increase the exposure to specific viruses potentially involved in the aetiology of the disease. There is evidence that during the past 10 years there has been a doubling of the incidence of the disease in northern Europe (Rewers *et al.* 1988; Metcalf and Baum 1991).

8.2.1 Genetic factors

Twin studies have shown concordance of IDDM in 30–50 per cent of identical twin pairs (Barnett *et al.* 1981). Because a concordance rate of 100 per cent would be expected if the disease were entirely due to genetic factors, environmental factors must also be important.

In Caucasians, the risk of IDDM is greater in those individuals who are HLA-A8 and HLA-B15 (see Chapter 10 for a detailed discussion of immunological aspects of IDDM). Stronger associations have been found with the HLA-DR locus, specifically an increased prevalence in individuals with HLA-DR4 and HLA-DR3, and a reduced prevalence in those with HLA-DR2. Individuals who are compound heterozygotes, HLA-DR3/4, have a greatly increased risk, twenty- to forty-fold higher than the general population. Although only 25 per cent of the general population share the same HLA type, 60 per cent of the diabetic siblings of a patient are HLA identical (Gorsuch 1987). However, although the HLA type is important, it is likely that other genetic factors increase the predisposition to the disease.

Fifteen per cent of patients with IDDM come from families in which there is an affected first-degree relative. Nevertheless, the other 85 per cent of patients may have the same HLA type as those who have a diabetic first-degree relative. Together with the monozygotic twin data, this suggests that there are a large number of genetically predisposed individuals who do not become diabetic. In the Barts–Windsor prospective study of non-affected siblings of 100 patients, by 25 years only 16 per cent of HLA-identical siblings developed diabetes (Tarn *et al.* 1988). Some first-degree relatives of patients develop islet cell antibodies which then disappear, and it is possible that a *forme fruste* of the disease can occur.

8.2.2 Age of onset

IDDM can present from the first few months of life through to the 90s. There is a peak at about 11–14 years of age, with a decline during the teenage years (Spencer and Cudworth 1983). The incidence in adult years is unclear, but is probably about half that in teenage years. Patients presenting in later years who are non-obese and have marked glycaemia are often referred to as IDDM. Some, but not all, have evidence of type I disease (see below).

8.2.3 Speed of onset

Although, clinically, IDDM often presents with a short, 2–3 week history, there is considerable evidence that the disease has a protracted prodromal period (Eisenbarth 1988). The clinical presentation of the disease is most common during winter months, and it is possible that this is induced by secondary infection from viruses which exacerbate the hyperglycaemia.

8.2.4 Pathology

Islets of Langerhans of IDDM patients, examined post-mortem, show a characteristic lymphocytic infiltration, which is part of the auto-immune process. The β-cells are reduced in number and eventually disappear completely; the islets then consist predominantly of α-cells. It is therefore not surprising that patients are insulin-dependent.

8.2.5 Environmental factors

Few specific viral infections have been implicated. Patients with congenital rubella have a markedly increased incidence of the disease, but there is no evidence that rubella is a frequent cause of IDDM. Coxsackie B4 virus can cause diabetes in mice and has been implicated in man in some case reports, but Coxsackie infection is not apparent in most patients (Gamble *et al.* 1985). It is possible that several viruses that affect the pancreas may stimulate an auto-immune process.

8.3 Epidemiology of non-insulin-dependent diabetes (NIDDM)

NIDDM is the most frequent form of diabetes in all populations (Zimmet *et al.* 1982; Melton *et al.* 1983), with 50 per cent of elderly adults affected in some populations. As found for IDDM, some races have a higher incidence of the disease; for example, the Pima Indians of Arizona have a prevalence 10 times that of the general population of the USA (Knowler *et al.* 1978). Diabetes is also prevalent in Mexican Americans (Gardner *et al.* 1984). These two groups may share similar susceptibility genes, acquired from Indian ancestry (Sievers and Fisher 1984). NIDDM is also common in American Blacks, Melanesians, Micronesians, Australian Aborigines, Asian Indians and Arabs (Zimmet 1982). In Mauritius, Chinese Indians and Creoles all have the same high prevalence of the disease. The high incidence of diabetes in Malta may relate to genes from Arabic ancestors. Although it is generally stated that these ethnic groups have a particularly high prevalence, the data probably also reflect an unusually low susceptibility to the disease.

8.3.1 The importance of obesity and environmental factors

Obesity is the major factor that makes diabetes become clinically apparent (West 1978). Since, in every population, the more obese subjects have a greater incidence of diabetes, those countries with the greatest prevalence of obesity have the highest incidence of the disease. During the Second World War, a period of enforced undernutrition of the population, there was a dramatic decrease in the incidence of type II diabetes in Norway, but little change in juvenile-onset disease presenting before 15 years (Westlund and Nicolaysen 1972). The incidence of diabetes is also related to the duration of obesity (West 1978; Modan *et al.* 1986). In addition, those individuals who have truncal obesity are more likely to develop diabetes than those who have peripheral obesity (Ohlson *et al.* 1985).

Obesity is closely associated with insulin resistance, and the increase in insulin resistance is likely to be the major determinant of diabetes occurring in obese subjects. Nevertheless, the majority of obese subjects do not develop diabetes and another factor is also required, which may be a β-cell susceptibility gene.

Diabetes is more common in women who have had several pregnancies. As pregnancy is characterized by increased insulin resistance from placental lactogen secretion, it is possible that pregnancy exacerbates the effect of insulin resistance, so inducing the disease (West 1978). A sedentary life-style is also frequent among affected subjects. A high alcohol intake predisposes to diabetes, but it is not known whether this is due to obesity, to increased cortisol production or to alcohol-induced hepatic or β-cell damage.

8.3.2 Genetic factors

NIDDM in identical twins has been reported to show 56–100 per cent concordance (Barnett *et al.* 1981; Newman *et al.* 1987). The higher figure probably includes selective recruitment, whereas the lower figure is from a cohort studied prospectively, in whom some unaffected twins may yet develop diabetes. It is apparent that the disease has a stronger inherited component than type I diabetes, but the twin studies cannot distinguish between monogenic and polygenic inheritance.

The bimodality of distribution of glucose tolerance in the general population has been taken to indicate that a single major gene is involved. However, it is also possible that bimodality could arise from a polygenic disease, if once the blood glucose level reaches a certain threshold, 'feed-forward' secondary influences become operative (such as hyperglycaemia inducing secondary β-cell dysfunction or insulin resistance).

A single gene disorder provides the simplest model for NIDDM, but it is apparent that expression of the disease usually requires environmental influences, in particular obesity. The WHO study group (1985) have suggested that an autosomal dominant gene underlies NIDDM, although this would require low penetrance in the majority of affected subjects. A recessive gene with a high population frequency cannot be excluded, but is unlikely in view of the lack of a reported high prevalence of diabetes in the offspring of conjugal diabetic parents. Even if a common major gene is present, many secondary modifying genes and environmental influences are likely to be operative. If the major gene affects β-cell function, the occurrence of diabetes may require the simultaneous presence of other factors, such as an insulin-receptor mutation giving 70 per cent efficiency of insulin action; obesity-related insulin resistance; a sedentary life-style; or some combination of these. Non-obese subjects have a higher prevalence of a family history of diabetes than obese subjects (Kobberling *et al.* 1985). This suggests that they have a higher genetic load: it is uncertain whether this is due to homozygosity of a single gene; to compound heterozygosity at a single gene, such as different mutations in the insulin receptor; or to polygenic inheritance.

8.3.3 Age of onset

The incidence of NIDDM increases between 30 and 60 years, and then tends to level off (Palumbo *et al.* 1976). At all ages, the majority of patients are obese but some are normal weight. Those who present in early adult life often have both parents affected with impaired glucose tolerance or diabetes, and may have inherited genes from both parents (O'Rahilly *et al.* 1988).

8.3.4 Speed of onset

The onset of NIDDM is normally insidious and is often exacerbated by an infection. It occurs most frequently in the winter months, possibly due to higher levels of infection and inactivity (UK Prospective Diabetes Study Group 1988).

8.3.5 Pathology

The islets can appear normal in NIDDM, although morphometry has shown that the area of β-cells is decreased to 80 per cent of normal (Clark *et al.* 1988). This on its own would probably not be sufficient to induce diabetes, and a functional defect probably also occurs. Extracellular islet amyloid, caused by deposition of islet amyloid polypeptide (IAPP) between the β-cells and the capillaries, is present in 85 per cent of patients (Clark *et al.* 1987). In both man and monkeys, there is a close association between amyloid deposition and clinical diabetes, and it is possible that extracellular or even intracellular deposition of amyloid may be one factor interfering with insulin production (Clark *et al.* 1990).

IAPP is a normal product of the β-cell and is co-secreted with insulin at about one-tenth the molar concentration (Sanke *et al.* 1991). The reason for amyloid deposition is uncertain. However, the association of diabetes with obesity may be related to increased β-cell stimulation (due to increased food intake and to obesity-associated insulin resistance). Therefore, in obesity, increased release of insulin may be accompanied by increased secretion of IAPP, leading to increased amyloid deposition in susceptible individuals. It is therefore tempting to implicate a major genetic abnormality of IAPP metabolism in NIDDM; however, the IAPP gene itself has a normal sequence (Nishi *et al.* 1990) and the gene is not linked to inheritance of diabetes in families (Cook *et al.* 1991).

8.4 Prognosis of diabetes

When insulin was discovered, it was thought that the problems of diabetes would become purely of historical interest. However, although it is possible to render most patients symptom-free, diabetic patients of all kinds are susceptible to long-term complications. These can be devastating and, at all ages of presentation, diabetic patients have an expected life-span of about 75 per cent of that of the normal age-matched population (Panzram and Zabel-Langhennig 1981). The reduction in life expectancy declines continuously with increasing age at diagnosis (Panzram 1987). Two main types of long-term problems arise, as described in the following sections.

8.4.1 Specific diabetes-related microvascular complications

These relate predominantly to small vessel disease, characterized by progressive obliteration of capillaries. This occurs throughout the body, but is mainly apparent clinically in the kidney and in the eye. The capillaries in the renal glomerulus and in the retina are 'end capillaries' with little communication between each other, so that when one capillary closes the tissue perfused by that capillary becomes ischaemic. The cause of capillary closure is uncertain, but it seems possible that the pericyte cells, which help to maintain the integrity of the endothelial cells lining the capillaries, may be particularly susceptible to hyperglycaemia. One hypothesis is that within these cells enhanced sugar metabolism via aldose reductase leads to sorbitol accumulation and consequent metabolic disturbances. Increased platelet stickiness may also be involved, and increased glycosylation of several different proteins may be a factor. The basement membrane of the capillaries becomes thickened, and this is seen particularly in the mesangium, the central part of the renal glomerulus.

Not all diabetic patients are at risk from microvascular disease. In juvenile-onset diabetic patients, renal failure has normally appeared by age 55 years, and the risk then decreases. Similarly, about 15 per cent of diabetic patients never develop retinopathy. Although the degree and duration of hyperglycaemia are likely to be factors (Dornan *et al.* 1982), unknown susceptibility factors must be operating.

The long-term consequences of the microvascular disease include renal failure and blindness. Diabetes is the most common cause of blindness in middle age; in old age, senile macular degenerations are more common although diabetes is still a contributing factor. Hypertension can also ensue, secondary to the renal dysfunction. In the kidneys, a nephrotic syndrome can occur when the damaged basement membrane of the glomerulus allows excess filtration of albumin. Proteinuria, up to 20 g/day can lead to low plasma albumin levels and peripheral oedema. Nephrotic syndrome and renal failure are predominantly seen in the fifth decade in type I diabetic patients, whereas blindness from microvascular disease can occur in any type of diabetes from 10 years after diagnosis. A major factor is the development of new, fragile vessels in the eye in response to ischaemia of the retina. These can bleed to form an intra-ocular, vitreous haemorrhage which causes sudden loss of vision. In addition, ischaemia of the retina can lead directly to blindness when it affects the macula, the most sensitive central part of the retina, and gives a maculopathy.

Diabetic patients can also develop a severe polyneuropathy (Thomas *et al.* 1982), characterized by the demyelination of segments of many nerves. The accumulative effect is most apparent in the long nerves, giving peripheral loss of sensation and motor weakness. The inability to sense trauma, e.g. a nail in the foot, and abnormal pressure on the foot due to muscle weakness, can lead to foot ulcers and infection. The autonomic nervous system can be involved, with impotence being the main symptom. The neuropathy is in part secondary to ischaemia of the nerves, resulting from microvascular disease, small vessel thromboses and macrovascular disease. In addition, metabolic derangements occur, possibly secondary to increased glucose through the sorbitol pathway. Both accumulation of sorbitol and a secondary deficiency of myoinositol may be involved.

A major factor for the development of all types of complications may be increased glycosylation of proteins. This can lead to advanced glycosylation end-products (AGE) which can induce cross-linking of proteins or macrophage attack with local release of lymphokines, which may cause cellular damage.

8.4.2 Macrovascular disease secondary to accelerated development of atheroma

The major cause of death in both type I and type II diabetes is myocardial infarction from coronary artery atheroma (Pyorola 1982). Patients also have an increased risk of developing peripheral vascular disease, which can lead to amputations and to strokes. Diabetic patients in their 40s or 50s have an increased death rate compared with the non-diabetic population, with about a twofold and 3.5-fold increased risk of cardiovascular death in men and women respectively (Panzram 1987). Diabetes presenting at an earlier age has an even greater effect in females, for it completely nullifies the protective effects which give pre-menopausal women a low cardiovascular risk (Barrett-Connor *et al.* 1991); diabetic pre-menopausal women have a sixteenfold increased risk of heart disease (Dorman *et al.* 1984). The overall mortality in these earlier-onset patients is increased fivefold in men and elevenfold in women, compared with the general population (Dorman *et al.* 1984). It is likely that factors additional to hyperglycaemia are involved, since in some populations, such as the Japanese, macrovascular disease is rare although microvascular diseases still occur. The low saturated fat and cholesterol intake of Japanese populations may be important: other factors that induce atheroma include smoking, hypertension, and abnormal plasma lipids.

8.4.3 Short-term problems/complications

Acute and potentially life-threatening metabolic derangements can occur in diabetes, the most important of which are hypoglycaemia and diabetic ketoacidosis (DKA). DKA is due either to an absolute insulin deficiency, which may occur in new acute-onset type I diabetes when the diagnosis is delayed, or to relative insulin deficiency caused by an infection which induces secretion of counter-regulatory hormones, such as cortisol or glucagon. The annual incidence of DKA ranges from 3 to 8 episodes per 1000 patients (Fishbein 1985) and may account for about 16 per cent of deaths in patients aged under 50 (Tunbridge 1981). Case fatality rates increase with age, from 3 per cent or less in the under-50s to nearly 50 per cent at older ages (Gale *et al.* 1981).

Hypoglycaemia may be induced by insulin or any of the sulphonylurea group of oral hypoglycaemic agents. The annual incidence of hypoglycaemia requiring hospital treatment in insulin-treated patients is about 100 per 1000 patients. The rate for sulphonylurea-induced hypoglycaemia is lower, and a Swedish study reported 4 per 1000. Fatality rates from hypoglycaemia amongst insulin-treated patients under 50 is about 0.2 per 1000. The estimated annual mortality rate is lower for sulphonylurea-induced than for insulin-induced hypoglycaemia (Ferner and Neil 1988). The mortality for glibenclamide-associated hypoglycaemia calculated from reported adverse reactions in Sweden was 0.033 per 1000 per year (Campbell 1984).

8.5 Diabetes therapy

Diabetes therapy has two major aims: normal symptom-free existence and the prevention of long-term complications.

8.5.1 Normal symptom-free existence

Treatment aims primarily to prevent hyperglycaemia-induced thirst, loss of weight, and tiredness. Overtherapy that induces brief episodes of hypoglycaemia, with loss in concentration or even coma, needs to be avoided. Symptoms from hyperglycaemia usually only occur when the blood glucose is over the renal threshold, i.e. above about 10–14 mmol/l: below this range, patients remain symptom-free, even though their blood glucose levels are considerably higher than the normal fasting (4.0–5.5 mmol/l) and postprandial (6–7 mmol/l) levels. This continued hyperglycaemia is likely to be a major factor in the development of complications.

8.5.2 Prevention of long-term complications

Animal studies have shown that if one maintains normoglycaemia by assiduous treatment of diabetes, microvascular complications (nephropathy and retinopathy) can be prevented (Engerman *et al.* 1977). So far, there is little direct evidence that this is applicable to man (Pirart 1978), but this may be because the appropriate long-term studies have not been done. Diabetic complications take years to develop, and their prevention can only be shown by long-term studies. Currently two studies are in progress and are expected to report in approximately 1994:

1. The Diabetes Control and Complication Trial in the USA studies 1400 type I diabetic patients. Half are randomized to a strict insulin regimen, aiming for near-normal glycaemia, whereas the other half have the higher plasma glucose concentrations that are usually obtained with routine insulin therapy. The major end-point is the prevention of progress of retinal eye disease, assessed by repeat photographs of the retina.

2. The UK Prospective Diabetes Study, which includes 5000 maturity-onset diabetic patients presenting between the ages of 25 and 65, compares 'active therapy' to maintain near-normal glucose levels using either sulphonylurea or insulin, with 'diet therapy' alone, in which higher glucose concentrations usually persist. The major end-point is the prevention of macrovascular complications.

8.5.3 Principles of therapy

In all patients, the primary aim is to try to maintain near-normal body weight. This is in order to prevent the insulin resistance that occurs with overnutrition and obesity. Patients frequently remain overweight, however, in spite of dietary advice. In view of the high incidence of heart disease, smoking is discouraged and a low fat, high carbohydrate diet is advised.

8.5.4 Treatment of type I diabetes

Most patients require insulin therapy, although those with modest β-cell deficiency can be treated, like type II diabetic patients, with tablets. Since insulin given by mouth is digested as a dietary protein, it has to be administered by injection.

Normal people produce half their insulin as a low, basal rate and half in response to meals. The insulin response to meals occurs within 5 minutes and lasts for 2–3 hours following each meal. The basal insulin supply can be given to diabetics fairly easily using a long-acting, crystalline insulin which is slowly absorbed. The major difficulty comes in coping with meals, since soluble insulin takes 30 minutes to be absorbed from a subcutaneous injection (i.e. it really needs to be given 30 minutes before a meal) and lasts for 4–6 hours. This long time-course is a major nuisance and a snack has to be taken 2–3 hours after an injection to cover the prolonged insulin absorption. In addition, absorption of insulin varies from injection to injection. The insulin requirements of a patient are also less after exercise and greater when stressed or ill. Therefore, it is not surprising that most patients continue to have high glucose levels, and that there is a risk of hypoglycaemia if one aims for normoglycaemia. If normoglycaemia is to be achieved, patients need to be attentive to their life-style and assess the response to their insulin therapy by measuring their blood glucose. This is done by pricking a finger and placing the blood onto a strip containing the enzyme glucose oxidase: the glucose concentration is determined either by an electronic sensor or by a colour change. Many patients do this regularly four times per day, before meals and before bed, in order to assess the appropriate insulin doses, although others find this unacceptable.

8.5.5 Treatment of type II diabetes

Diet therapy, inducing weight reduction, may be sufficient to reduce the blood glucose to below the renal threshold and to make patients symptom-free, although it is rarely sufficient to induce normal fasting glucose levels. If symptoms persist despite dietary therapies, then most physicians treat with tablets containing sulphonylureas to stimulate insulin secretion. This approximately doubles the β-cell efficiency, but, nevertheless, continued symptom-free hyperglycaemia with a fasting glucose level of 8–10 mmol/l is common. 'Second generation' drugs such as glibenclamide or glipizide are no more effective than the 'first generation' drugs tolbutamide or chlorpropamide. Biguanide therapy with metformin, to improve glucose uptake, is an alternative, but like sulphonylurea it only induces a modest decrease of blood glucose. If symptoms recur on diet and tablet therapy, patients are transferred to insulin therapy. It is possible that therapy should be more aggressive, and tablets or insulin to induce normal glucose levels should be given to all patients. The UK Prospective Diabetes Study is examining whether it would be beneficial to aim for a normal fasting glucose, with either a long-acting insulin or with tablets. The only previous long-term study of therapy, the University Group Diabetes Program (UGDP), raised the possibility that sulphonylurea therapy may induce fatal heart attacks more often than expected; the results were inconclusive, however, and have not affected prescribing habits.

8.5.6 The crystal ball

It is likely that maintenance of a normal blood glucose concentration will eventually be shown to prevent diabetic complications and maintain health. However, to achieve this in most patients, new technology will need to be developed. Possibilities include:

(1) a glucose-sensing needle inserted into the skin, linked to a portable computer which appropriately adjusts the delivery of insulin; and

(2) islet transplantation.

Better means of delivering insulin are required and a shorter-acting insulin, possibly administered by a nasal spray before each meal, could provide insulin with a similar time course to that produced by normal subjects. As diabetes is a common disease, the cost of treating all patients assiduously will be large, but is likely to be considerably less than the large sums spent at present on care of the complications of the disease.

References

Al Sayegh, H. and Jarrett, R. J. (1979). Oral glucose tolerance tests and diagnosis of diabetes: results of a prospective study based on the Whitehall Survey. *Lancet* **ii,** 431–3.

Barnett, A. H., Spiliopoulos, A. J., Pyke, D. A., Stubbs, W. A., Burrin, J., and Alberti, K. G. (1981). Metabolic studies in unaffected co-twins of non-insulin dependent diabetes. *Br. Med. J.* **281,** 1656–8.

Barrett-Connor, E. L., Cohn, B. A., Wingar, D. L., and Edelstein, S. L. (1991). Why is diabetes mellitus a stronger risk factor for fatal ischaemic heart disease in women than in men? *JAMA* **265,** 627–31.

Bennett, P. H., Burch, T. A., and Miller, M. (1971). Diabetes in American (Pima) Indians. *Lancet* **ii,** 125–8.

Bliss, M. (1982). *The discovery of insulin.* The University of Chicago Press, Chicago.

Campbell, I. W. (1984). Metformin and glibenclamide: comparative risks. *Br. Med. J.* **289,** 289.

Clark, A., Cooper, G. J. S., Lewis, C. E., Morris, J. F., Willis, A. C., Reid, K. B. M., and Turner, R. C. (1987*a*). Islet amyloid formed from diabetes associated peptide may be pathogenic in type 2 diabetes. *Lancet* **ii,** 231–4.

Clark, A., Matthews, D. R., Naylor, B. A., Wells, C. A., Hosker, J. P., and Turner, R. C. (1987*b*). Pancreatic islet amyloid and elevated proinsulin secretion in familial maturity-onset diabetes. *Diabetes Research* **4,** 51–5.

Clark, A., Wells, C. A., Burley, I. D., Cruickshank, J. K., Vanhegan, R. I., Matthews, D. R., Cooper, G. J. S., Holman, R. R., and Turner, R. C. (1988). Islet amyloid. Increased A-cells, reduced β-cells and exocrine fibrosis: quantitative changes in the pancreas in Type II diabetes. *Diabetes Research* **9,** 151–9.

Clark, A., Saad, M. F., Nezzer, T., Uren, C., Knowler, W. C., Bennett, P. H., and Turner, R. C. (1990). Islet amyloid polypeptide in diabetic and non-diabetic Pima Indians. *Diabetologia* **33,** 285–9.

Cook, J. T. E., Patel, P. P., Clark, A., Hoppener, J. W. M., Lips, C. J. M., Mosselman, S., and Turner, R. C. (1991). Non-linkage of the islet amyloid polypeptide gene with Type 2 (non-insulin-dependent) diabetes mellitus. *Diabetologia* **34,** 103–8.

Diabetes Epidemiology Research International Group (1988). Geographic patterns of childhood insulin-dependent diabetes mellitus. *Diabetes* **37,** 1133–119.

Dorman, J. S., LaPorte, R. E., Kuller, L. H., Cruickshanks, K. J., Orchard, T. J., Wagener, D. K., Becker, D. J., Cavender, D. E., and Drash, A. L. (1984). The Pittsburg insulin-dependent diabetes mellitus (IDDM) morbidity and mortality study. *Diabetes* **33,** 271–6.

Dornan, T. L., Mann, J. I., and Turner, R. C. (1982). Factors protective against retinopathy in insulin-dependent diabetics free of retinopathy for 30 years. *Br. Med. J.* **285,** 1073–7.

Eisenbarth, G. S. (1986). Type I diabetes mellitus. A chronic autoimmune disease. *New Engl. J. Med.* **314,** 1360–8.

Engerman, R., Bloodworth, J. M. B., and Nelson, S. (1977). Relationship of microvascular disease in diabetes to metabolic control. *Diabetes* **26,** 760–9.

Eschwege, E., Richard, J. L., Thibult, N., Ducimetiere, P., Wernek, J. M. and Rosselin, G. E. (1985). Coronary heart disease mortality in relation with diabetes, blood glucose and plasma insulin levels. The Paris prospective study 10 years later. *Horm. Metab. Res.* (Suppl. 15), 41–6.

Ferner, F. E. and Neil, H. A. W. (1988). Sulphonylurea and hypoglycaemia. *Br. Med. J.* **296,** 249–50.

Fishbein, M. A. (1985). Diabetic ketoacidosis, hyperosmolar nonketotic coma, lactic acidosis and hypoglycaemia. In *Diabetes in America*, Ch. XII, NIH Publication no. 85–1468. US Dept of Health and Human Sciences.

Fuller, J. H., Shipley, A. J., Rose, G., Jarrett, R. J., and Keen, H. (1980). Coronary heart disease risk and impaired glucose tolerance. The Whitehall study. *Lancet* **1,** 1373–6.

Gale, E. A. M., Dornan, T. L., and Tattersall, R. B. (1981). Severely uncontrolled diabetes in the over-fifties. *Diabetologia* **21,** 25–8.

Gamble, D. R., Cumming, H., Odugbesan, O., and Barnett, A. H. (1985). Coxsackie B Virus and juvenile-onset-insulin-dependent diabetes. *Lancet* **2,** 455–6.

Gardner, L. I., Stern, M. P., Haffner, S. M., Gaskill, S. P., Mazuda, H. P., Relethford, J. H., and Eifler, C. W. (1984). Prevalence of diabetes in Mexican-Americans: relationship to percent of gene pool derived from native American sources. *Diabetes* **33,** 86–92.

Gorsuch, A. N. (1987). The immunogenetics of diabetes. *Diabetic Med.* **4,** 510–16.

Holman, R. R. and Turner, R. C. (1979). Maintenance of basal plasma glucose and insulin concentration in maturity-onset diabetes. *Diabetes* **28,** 227–30.

Jarrett, R. J. and Keen, H. (1976). Hyperglycaemia and diabetes mellitus. *Lancet* **ii,** 1009–12.

King, H. and Zimmet, P. (1988). Trends in the prevalence and incidence of diabetes. Non-insulin-dependent diabetes mellitus. *World Health Statistics Quarterly* **41,** 190–6.

Knowler, W. C., Bennett, P. H., Hamman, R. F. (1978). Diabetes incidence and prevalence in Pima Indians: a 19-fold greater incidence than in Rochester, Minnesota. *Am. J. Epidemiol.* **108,** 497–505.

Kobberling, J., Tillil, H., and Lorenz, H.-J. (1985). Genetics of type 2A- and type 2B diabetes mellitus. *Diabetes Res. Clin. Pract.* **1**(1) s311.

Melton, L. J., Palumbro, P. J., and Chu, C.-P. (1983). Incidence of diabetes mellitus by clinical type. *Diabetes Care* **6,** 75–86.

Menser, M. A., Forrest, J. M., and Bransby, R. D. (1978). Rubella infection and diabetes mellitus. *Lancet* **1,** 57–60.

Metcalfe, M. A. and Baum, J. D. (1991). Incidence of insulin-dependent diabetes in children under 15 years in the British Isles during 1988. *Br. Med. J.* **302,** 443–7.

Modan, M., Karasik, A., Halkin, H., Fuchs, Z., Lusky, A., Shitrit, A., and Modan, B. (1986). Effect of past and concurrent body mass index on prevalence of glucose intolerance and type 2 (non-insulin-dependent) diabetes and on insulin response. The Israel study of glucose intolerance, obesity and hypertension. *Diabetologia* **29,** 82–9.

National Diabetes Data Group (1979). Classification and diagnosis of diabetes mellitus and other categories of glucose intolerance. *Diabetes* **28,** 1039–57.

Newman, B., Selby, J. V., King, M. C., Slemenda, C., Fabsitz, R., and Friedman, G. D. (1987). Concordance for type 2 (non-insulin-dependent) diabetes mellitus in male twins. *Diabetologia* **30,** 763–8.

Nishi, M., Bell, G. I., and Steiner, D. F. (1990). Islet amyloid polypeptide (amylin): no evidence of an abnormal precursor sequence in 25 Type 2 (non-insulin-dependent) diabetic patients. *Diabetologia* **33,** 628–30.

Ohlson, L. O., Larsson, B., Svardsudd, K., Welin, L., Eriksson, H., Wilhelmsen, L., Bjorntorp, P., and Tibblin, G. (1985). The influence of body fat distribution on the incidence of diabetes mellitus. 13.5 years of follow-up of the participants in the study of men born in 1913. *Diabetes* **34,** 1055–8.

O'Rahilly, S., Wainscoat, J. S., and Turner, R. C. (1988). Type 2 (non-insulin-dependent) diabetes mellitus. New genetics for old nightmares. *Diabetologia* **31,** 407–14.

Palumbo, P. J., Elveback, L. R., Ch, U. C.-P., Connolly, D. C., and Kurland, L. T. (1976). Diabetes mellitus: Incidence, prevalence, survivorship and causes of death in Rochester, Minnesota, 1945–1970. *Diabetes* **25,** 566–73.

Panzram, G. (1987). Mortality and survival in Type 2 (non-insulin-dependent) diabetes mellitus. *Diabetologia* **30,** 123–31.

Panzram, G. and Zabel-Langhennig (1981). Prognosis of diabetes in a geographically defined population. *Diabetologia* **20,** 587–91.

Pettitt, D. J., Knowler, W. C., Lisse, J. R., and Bennett, P. H. (1980). Development of retinopathy and proteinuria in relation to plasma glucose concentrations in Pima Indians. *Lancet* **ii,** 1050–2.

Pirart, J. (1978). Diabetes and its degenerative complications: A prospective study of 4,400 patients observed between 1947 and 1973. *Diabetes Care* **1,** 168–88 and 252–63.

Pyorala, K. and Laakso, M. (1983). In *Macrovascular disease in diabetes* (ed. J. I. Mann, K. Pyorala, and A. Teuscher), pp. 183–247. Churchill Livingstone, Edinburgh.

Rewers, M., LaPorte, R. E., King, H., and Tuomilehto, J. (1988). Trends in the prevalence and incidence of diabetes: Insulin-dependent diabetes mellitus in childhood. *World Health Statistics Quarterly* **41,** 179–89.

Sanke, T., Hanabusa, T., Nakano, Y., Oki, C., Okai, K., Nishimura, S., Kondo, M., and Nanjo, K. (1991). Plasma islet amyloid polypeptide (Amylin) levels and their responses to oral glucose in Type 2 (non-insulin-dependent) diabetic patients. *Diabetologia* **34,** 129–32.

Sievers, M. L. and Fisher, J. R. (1985). In *Diabetes in America: diabetes data compiled 1984,* Publication No. (NIH) 85–1468: X-1-20. US Department of Health and Human Services.

Spencer, K. M. and Cudworth, A. G. (1983). In *Diabetes in epidemiological perspective* (ed. J. I. Mann, K. Pyorala, and A. Teuscher), pp. 99–111. Churchill Livingstone, Edinburgh.

Tarn, A. C., Thomas, J. M., Dean, B. M., Ingram, D., Schwarz, G., Bottazzo, G. F., and Gale, E. A. (1988). Predicting insulin-dependent diabetes. *Lancet* **16,** 845–50.

Thomas, P. K., Ward, J. D., and Watkins, P. J. (1982). Diabetic neuropathy. In *Complications of diabetes* (ed. K. Keen and J. Jarrett), (2nd edn). Edward Arnold, London.

Tunbridge, W. M. G. (1981). Factors contributing to deaths of diabetics under fifty years of age. *Lancet* **ii,** 569–72.

Turner, R. C., Holman, R. R., Matthews, D. R., O'Rahilly, S. P., Rudenski, A. S., and Braund, W. J. (1986). Diabetes nomenclature: classification or grading of severity? *Diabetic Med.* **3,** 216–20.

UK Prospective Diabetes Study Group IV (1988). Characteristics of newly-presenting Type 2 diabetic patients: male preponderance and obesity at different ages. Multicentre Study. *Diabetic Med.* **5**(2) 154–9.

West, K. M. (1978). *Epidemiology of diabetes and its vascular lesions,* pp. 231–48. Elsevier, Amsterdam.

Westlund, K. and Nicolaysen, R. (1972). Ten year mortality and morbidity related to serum cholesterol: A follow-up of 3,751 men aged 40–49. *Scand. J. Clin. Lab. Invest.* **30** (Suppl. 127), 3–24.

World Health Organisation Study Group (1985). *Diabetes Mellitus, Report of a WHO Study Group,* Technical Report Series 727, pp. 1–113. World Health Organisation, Geneva.

Yalow, R. S. and Berson, S. A. (1960). Immunoassay of endogenous plasma insulin in man. *J. Clin. Invest.* **39,** 1157–75.

Yudkin, J. S., Alberti, K. G. M. M., McLarty, D. G., and Swai, A. B. M. (1990). Impaired glucose tolerance. Is it a risk factor for diabetes or a diagnostic rat-bag. *Br. Med. J.* **301,** 397–401.

Zimmet, P. (1982). Type 2 (non-insulin-dependent) diabetes—an epidemiological overview. *Diabetologia* **22,** 399–411.

Zimmet, P., Taft, P., Guinea, A., Guthrie, W., and Tchoma, L. (1977). The high prevalence of diabetes mellitus on a central pacific island. *Diabetologia* **13,** 111–15.

9 Aetiology of type I diabetes: genetic aspects

RALF WASSMUTH, INGRID KOCKUM, ALLAN KARLSEN, WILLIAM HAGOPIAN, HEIKE BÄRMEIER, SYAMALIMA DUBE and ÅKE LERNMARK

9.1 Background

Ancient literature has indicated that two forms of diabetes could be discerned. One was associated with emaciation, dehydration, polyuria, and lassitude, and the other with stout build, gluttony, obesity, and sleepiness (Bliss 1982). At the beginning of the twentieth century, however, and even after the discovery of insulin in 1921, diabetes was more often than not regarded as a single syndrome. A distinction was later made between juvenile and maturity onset diabetes mellitus but it was not until 1940 that these two forms of diabetes were suggested to be genetically distinct (Rotter *et al.* 1990). It also became clear that diabetes mellitus is not only insulin dependent or non-insulin dependent. A large number of disorders or conditions are associated with glucose intolerance; their causes range from monogenic disorders to diabetes secondary to a pancreatic affliction or to a peripheral insulin resistance induced by drugs or tumours. A recent survey lists more than 60 distinct genetic disorders associated with glucose intolerance and, sometimes, clinical diabetes (Rotter *et al.* 1990). Glucose intolerance and diabetes are therefore symptoms rather than defining diagnostic criteria. In order to properly diagnose insulin-dependent diabetes a set of diagnostic criteria are necessary. Such criteria, e.g. those developed by WHO (1985), are of particular importance in studies aimed at a better understanding of the genetics, aetiology, and pathogenesis of insulin-dependent diabetes.

Investigations during the 1960s and the 1970s provided further evidence that insulin-dependent diabetes showed features distinct from non-insulin-dependent diabetes. One of these was the ability of the β-cells to release insulin. The advent of the radioimmunoassay for insulin made it possible to measure plasma and pancreatic insulin accurately. Patients with insulin-dependent diabetes were clearly insulinopenic while patients with non-insulin-dependent diabetes were able to secrete insulin. In non-insulin-dependent diabetes there seemed to be a decrease in β-cell responsiveness to an increase in plasma glucose (Pfeifer *et al.* 1981). The discovery of proinsulin (Steiner *et al.* 1989) made it possible to study the ability of the endocrine pancreas to release insulin by the radioimmunoassay of circulating

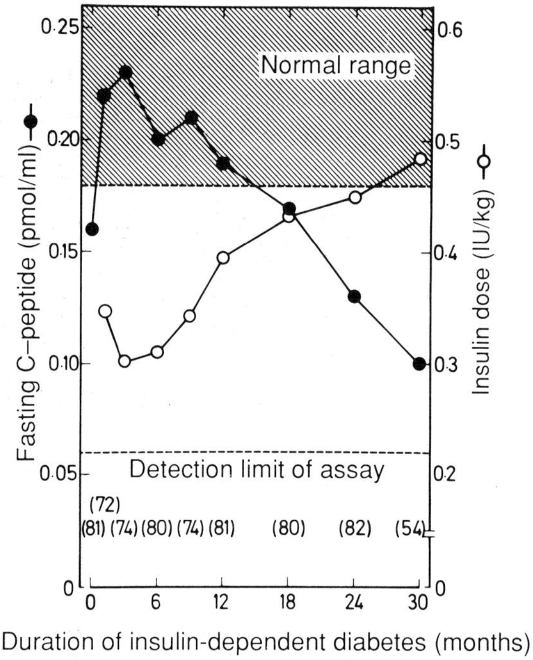

Fig. 1 Effects of the duration of insulin-dependent diabetes on fasting plasma C-peptide and insulin dosage (reproduced from Marner *et al.* 1985)

C-peptide (Horwitz *et al.* 1975). Numerous studies of circulating C-peptide have indicated that the insulin-dependent diabetic patient has a markedly decreased ability to secrete insulin, and that this ability decreases with increasing duration of the disease (Fig. 1).

Another feature that distinguishes insulin-dependent from non-insulin-dependent diabetes is the histopathology of the pancreas. Degeneration and inflammatory reactions in and around the islets of Langerhans had been reported years before the discovery of insulin (Opie 1901). These early reports were 're-discovered' and extended by Gepts (1965) and Gepts and LaCompte (1981). In a series of insulin-dependent diabetic patients with a short history of disease, Gepts demonstrated, first, that the β-cells (but not the other endocrine cells) were reduced or lacking in the pancreatic islets, and secondly, that in many, but not all, of the new patients this loss of β-cells was associated with insulitis (Von Meyenburg, 1940), i.e. an infiltration of mononuclear cells in and around the pancreatic islets (Fig. 2). The presence of insulitis was taken as evidence that insulin-dependent diabetes was an auto-immune disorder. It was hypothesized that the destruction of β-cells was immunologically mediated. This hypothesis was supported by later demonstrations that insulin-dependent diabetes often occurs in patients with other organ-specific auto-immune disorders (reviewed by MacCuish and Irvine 1975). In addition, newly diagnosed insulin-dependent diabetic patients show anti-pancreatic cellular hypersensitivity (Nerup *et al.* 1971; MacCuish *et al.* 1974) associated with certain HLA types often found in diseases of auto-immune character (Singal and Blajchman

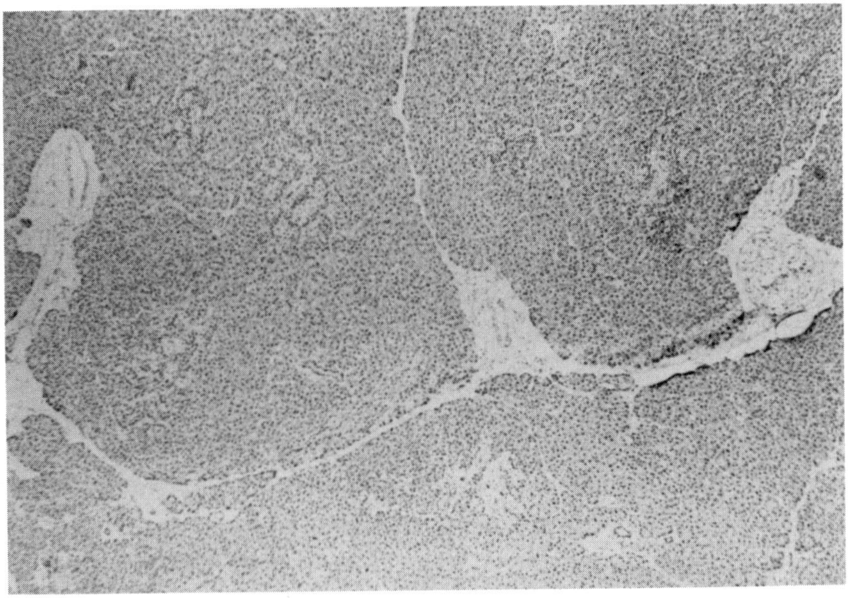

Fig. 2 Insulitis in the pancreas of a 13-month-old patient with newly diagnosed insulin-dependent diabetes mellitus. Note that all islets seen in this overview are affected. Haematoxylin and eosin staining.

1973; Nerup *et al.* 1974) as well as the presence of islet cell antibodies (Bottazzo *et al.* 1974; MacCuish *et al.* 1974; Lendrum *et al.* 1975). Taken together these data strongly suggested an auto-immune aetiopathogenesis of insulin-dependent diabetes. These features were not found among patients with non-insulin-dependent diabetes.

In 1979, clinical criteria for distinguishing between insulin-dependent and non-insulin-dependent diabetes were proposed, to allow comparisons between studies (WHO Report 1976; NDD Group 1979; WHO Study Group 1985). Since neither the genetics, aetiology, nor pathogenesis of insulin-dependent and non-insulin-dependent diabetes are known, the proposed diagnostic criteria should allow a better comparison between different investigations of these diseases throughout the world.

Recent investigations seem to favour the use of the WHO criteria (WHO Study Group 1985). A unified set of diagnostic criteria is of particular importance when studying the inheritance of diabetes mellitus. It should be kept in mind, however, that the criteria are advisory, rather than fixed, in order to promote a better understanding of diabetes mellitus. The 1985 WHO report also recommends that the terms type I and type II should be regarded as completely synonymous with insulin-dependent (IDDM) and non-insulin dependent (NIDDM) diabetes mellitus, respectively (WHO Study Group 1985).

The use of defined criteria to classify different types of diabetes is thus a relatively recent event. This should be kept in mind when reviewing earlier literature on the genetic aspects of insulin-dependent diabetes. The inheritance of this disease is particularly complicated to understand. First, only 10–13 per cent of new patients have a first degree relative with the disease (Mason *et al.* 1987; Tillil and Köbberling 1987; Dahlquist *et al.* 1989). Secondly, the disease may occur at any age (Andres 1971; Wilson *et al.* 1985; Tattersall 1986). Depending on the age at diagnosis, the clinical onset may be mild and the disease may subsequently progress to insulin-dependent diabetes, especially in patients with islet cell antibodies (Di Mario *et al.* 1983; Landin-Olsson *et al.* 1989). Similarly, secondary failure of oral hypoglycaemic agents is often found to be associated with islet cell antibodies or HLA types associated with insulin-dependent diabetes (Irvine *et al.* 1980; Gleichmann *et al.* 1984; Groop *et al.* 1986). These phenomena and recent findings of transient immune markers such as islet cell antibodies (Gorsuch *et al.* 1981; Spencer *et al.* 1984; Landin-Olsson *et al.* 1989; Karjalainen 1990) and of subclinical β-cell dysfunction with or without persistent immune markers (Bärmeier *et al.* 1990; McCulloch *et al.* 1990), underline the difficulty in obtaining reliable information on the phenotypic expression of the genotype(s) of insulin-dependent diabetes. In addition, indirect evidence suggests that the phenotypic expression may be influenced by both environmental factors and gender. The understanding of the genetic aspects of insulin-dependent diabetes has also improved with the introduction of molecular biological methods.

9.2 Studies in twins

The genetic inheritance of a disease may often be best understood from studies of monozygotic twins. Our understanding of insulin-dependent diabetes in twins derives largely from studies of twins in the UK (Tattersall and Pyke 1972; Pyke 1979). These investigators first compared twins with insulin-dependent diabetes—developing their disease before the age of 35—with non-insulin-dependent diabetic twins who developed their disease when older than 45 years (Table 1). This analysis showed that while 89 per cent of the non-insulin-dependent patients were concordant for the disease, this was so only for 55 per cent of twins with insulin-dependent diabetes. Additional references to twins studies can be found in Rotter *et al.* (1990).

The question traditionally asked in twin studies is 'What is the concordance rate?'. This is because the higher the concordance rate the more likely it is that the disease

Table 1 Diabetes mellitus in identical twins (data from Pyke 1979)

	Number of pairs		
Disease	**Concordant**	**Discordant**	**Total**
IDDM	73	59	132
NIDDM	47	6	53

is due to genetic mechanisms, whereas the lower the concordance rate the more likely it is that environmental factors are important. Recent studies on UK twins indicate that the concordance rate for insulin-dependent diabetes may actually be below 50 per cent, more precisely around 30 per cent (Olmos *et al.* 1988; Rotter *et al.* 1990). The reason for this low concordance rate is not clear. One possible explanation is that both twins are born with a propensity to develop insulin-dependent diabetes. The disease does not develop however, unless there is an aetiological agent in the environment which triggers pathogenetic events which result in β-cell destruction and consequent loss of insulin production.

Monozygotic twins show a heterogeneous pattern with respect to the clinical onset of insulin-dependent diabetes. Twins who are both HLA-DR3 and HLA-DR4 positive have a higher concordance rate than twins with just one of these HLA types (Barnett *et al.* 1981). Older twins have less chance of becoming concordant, and the longer the time that has elapsed since the diagnosis in the first twin, the less is the possibility of the pair becoming concordant. Impaired β-cell function (Heaton *et al.* 1987) and a number of other functional and immune abnormalities have been reported in the non-diabetic twin, but none of them seem to predict insulin-dependent diabetes (Millward *et al.* 1986; Heaton *et al.* 1988; Johnston *et al.* 1989; Beer *et al.* 1990). It is therefore possible that the diabetogenic process in these individuals has remitted.

The data on monozygotic twins are often criticized because there may be a bias in the selection of twins; thus concordant pairs are more often reported than discordant pairs, and the twins may not have been followed for long enough. An important aspect of recent investigations is that immune abnormalities have been detected but shown to be transient. Islet cell antibodies, for example, were detected but found to subsequently disappear (Millward *et al.* 1986). It is possible, therefore, that both twins may have been born with the disease but the pathogenesis of the disorder progressed differently, allowing factors in the environment to protect one twin from developing the disease (rather than inducing the disease in the other twin).

The concept that insulin-dependent diabetes may be entirely genetic and that the environment promotes protection rather than acts to trigger the disease is consistent with recent observations in the spontaneously diabetic BB rat and in the NOD mouse. These observations will be discussed below.

9.3 Studies in families

Studies in families are important in revealing patterns of inheritance. Numerous studies on the mode of inheritance of insulin-dependent diabetes have been published and reviewed (Köbberling and Tattersall 1982; Rich 1990; Rott *et al.* 1990; Wassmuth and Lernmark 1990). It has not, however, been possible to establish clearly a mode of inheritance of IDDM. Almost every possible model has been proposed and insulin-dependent diabetes has been referred to as the geneticist's nightmare (Köbberling and Tattersall 1982). A large number of genetic markers and traits have been tested, including markers for blood group types and HLA transplantation antigens. None of these markers has shown linkage to insulin-dependent diabetes, in family studies.

The difficulties in studying IDDM in families are well illustrated in a number of investigations (Table 2). First, when the age at onset is below 20 years only 10–13 per cent of new patients have a family history of diabetes (Mason *et al.* 1987; Tillil and Köbberling 1987; Dahlquist *et al.* 1989). This finding has been confirmed in studies comparing the frequency of diabetes among similar relatives to controls (Rotter *et al.* 1990). In a recent case-control study, the 13 per cent frequency of IDDM among parents and siblings was confirmed but NIDDM and organ-specific auto-immunity were also identified as familial risk factors (Dahlquist *et al.* 1989).

Table 2 Family studies of insulin-dependent diabetes mellitus (data from Tillil and Köbberling 1987; Thomson *et al.* 1988; Rotter *et al.* 1990)

	Lifetime recurrence risk of IDDM if age at onset of proband is <25 years (%)
Identical twins	25–50
First-degree relatives	19–13
parents	2–3
children	5–7
Siblings	6–8
HLA-identical	16
HLA-haplo-identical	5
HLA-non-identical	1

Families affected by insulin-dependent diabetes have been subjected to linkage analysis with a variety of gene markers (Table 3). Linkage analysis is the most powerful technique for determining the position of a disease gene on the chromosomal map. If the disease gene and the gene marker used are close to each other on the chromosome, they tend to be inherited together. The closer the marker the less likely it is that crossing-over will occur to separate the marker from the gene. Using molecular genetics, several gene mutations causing diseases such as cystic fibrosis and neurofibromatosis have been identified, cloned, and sequenced.

Several genes at different loci on the human genome have been studied to determine their linkage or association to IDDM. In association studies the frequency of a marker among patients is compared to its frequency in the control population. In this approach, it is asked whether a gene is distributed with the disease throughout the population. The association with the HLA complex on chromosome 6 was first reported in 1973 for HLA-B15 (Singal and Blajchman 1973)

Table 3 Linkage and association between insulin-dependent diabetes and gene markers

Chromosome	Marker	Association	Linkage	Reference
1	NT			
2	NT			
3	NT			
4	NT			
5	NT			
6	HLA-B,-DR,-DQ	+	+	Wassmuth (1990)
7	T-cell receptor β-chain constant region	+	−	Hoover (1986) Millard (1987)
8	NT			
9	NT			
10	NT			
11	Insulin gene 5′ flanking region	+	−	Bell (1984)
12	NT			
13	NT			
14	IgG heavy chain allotype	+	−	Field (1984)
15	NT			
16	NT			
17	NT			
18	Kidd blood group (JK^b)	− (HLA-DR)	−	Barbosa (1982)
19	Lewis blood group (Lea-b-)	+	−	Vague *et al.* (1978)
20	NT			
21	NT			
22	NT			
X	No evidence for gender transmission			
Y	No evidence for gender transmission			

NT, not tested.

and in 1974 for B8 (Nerup *et al.* 1974). These initial observations have since been confirmed and extended in numerous reports.

The approach used to detect genes involved in the development of insulin-dependent diabetes usually involves two steps. First, a polymorphic gene marker is tested for association with IDDM. If a significant association is found, data on genetic linkage are sought in family studies, using families with multiple affected members (Fig. 3). DNA probes for more than 4000 loci or genes have been developed and are available for testing by methods such as restriction fragment length polymorphism (RFLP) or polymerase chain reaction analysis. This 'candidate gene' approach however is likely to be unproductive since we cannot predict which molecules or genes may control insulin-dependent diabetes.

A novel means of obtaining linkage information and chromosomal location is by the use of genes which are highly polymorphic and appear in the genome in a way that is unique to each individual. These DNA sequences represent regions of variable number of tandem repeats (VNTR) which are composed of nucleotide repeats of varying lengths. Probes specific for VNTR sequences generate DNA fingerprints which may be used to study both association and linkage. Such studies are feasible in man (Hyer *et al.* 1991). The complicated mode of inheritance in IDDM would,

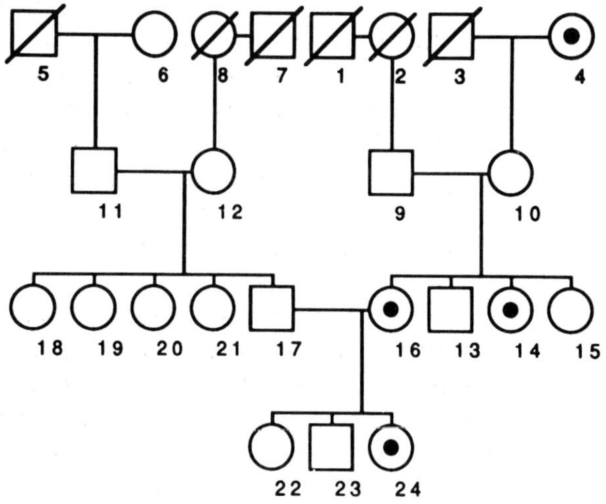

Fig. 3 Pedigree of a family with multiple affected members in different generations. Family members are numbered. Crossed-out members are dead, while members with IDDM are indicated by a dot. Analysis of several generations may disclose chromosomal markers which are linked to the disease.

however, require a large number of families with multiple affected members in order to link polymorphic markers to the disease. The outcome of such an investigation may be to locate the gene(s) necessary for diabetes on one or several of the human chromosomes. One such marker, HLA, has already been localized to chromosome 6. Current investigations of the HLA genes are focused on the identification of those sequences that confer the highest risk for the development of insulin-dependent diabetes. As will be discussed below certain HLA genes are necessary but not sufficient for the development of IDDM.

9.4 Studies in the population

Studies of genetic markers for disease susceptibility may also be carried out using an epidemiological approach. Patients are compared with controls for the presence or absence of a specific marker. The results are expressed as a risk, and the strength of the association expressed as an odds ratio (OR); formulae and a discussion of their use are found elsewhere (Green 1982; Wassmuth *et al.* 1990). In IDDM such studies are necessary since non-familial cases comprise nearly 90 per cent of new patients. The studies are, however, complicated by the fact that the incidence rate of the disease varies with age and is also influenced by gender and geographical location. Indeed, a control in one study may be a patient in a subsequent investigation. In most studies to date the patients and the controls have been ascertained at the same time. Ideally, the patients and controls should be followed prospectively to obtain an estimate of relative risk: this is, however, seldom achieved. Complicating factors include the overall low incidence rate, seasonal variation, variable age at onset, and the low recurrence rate of disease in families with members already affected. In populations with known incidence rates, it is possible to determine the absolute risk (AR) as a percentage of individuals being factor-positive or negative.

The markers most commonly used when studying insulin-dependent diabetes are those in the HLA region on the short arm of chromosome 6 (Fig. 4). The genetics of the HLA region is complicated by the fact that these alleles show a reduced rate of crossing-over and tend to be inherited together, a phenomenon termed linkage disequilibrium. Common sets of alleles or extended haplotypes (supratypes) are shown in Table 4. Any polymorphic marker from the HLA region will therefore reveal a positive or negative association with insulin-dependent diabetes. The HLA supratypes, however, show differences between ethnic groups (Dawkins *et al.* 1988), which may influence the rates obtained in various studies. Since nearly all genes in the HLA region code for proteins that are involved in the immune response, it has been hypothesized that the combination of these genes confers susceptibility to IDDM. For example, low levels of complement associated with the C4AQ0 or C4BQ0 alleles may influence the immune response. Similarly, tumour necrosis factor (TNF) α and β gene polymorphism may be important since this cytokine potentiates β-cell cytotoxicity of interleukin-1β (Mandrup-Poulsen *et al.* 1987). The recent discovery of the ABC (ATP-binding cassette) gene centromeric to HLA-DQ is of particular interest because of its location and possible functional importance in the process of antigen presentation (Spies *et al.* 1990; Trowsdale *et al.* 1990). In conclusion, although there is an increased risk of first-degree relatives acquiring insulin-dependent diabetes, nearly 90 per cent of new patients have unaffected family members. Current genetic analysis of insulin-dependent diabetes suggests that HLA-region encoded genes are required but not sufficient for the disease. Additional genes (how many is still unclear) are thought to be needed. A simple explanation would be the presence of a single recessive gene located on a chromosome other than chromosome 6. Homozygosity for this gene would be required along with the presence of the HLA susceptibility gene(s).

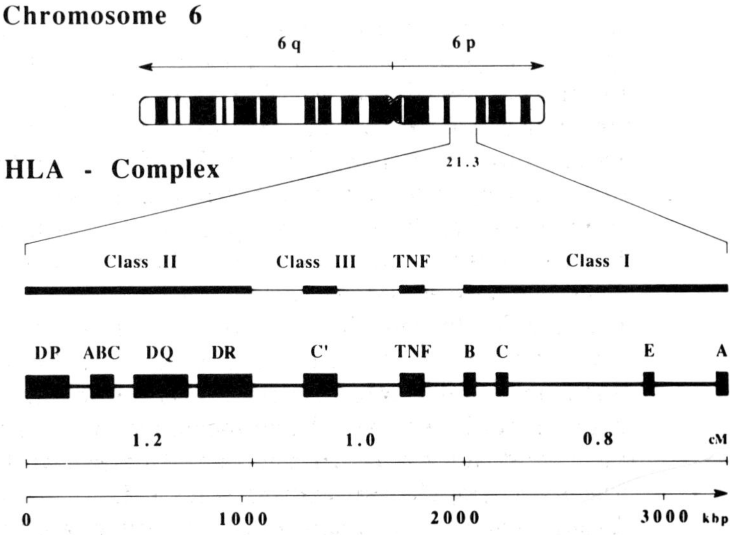

Fig. 4 The HLA-region on the short arm of chromosome 6.

Table 4 Extended HLA-region haplotypes which are in linkage disequilibrium

Centromeric end		Telomeric end	Association to IDDM
	DQw2-DR3-C4B1-C4-AQ0-BfS-B8-A1		positive
	DQw2-DR3-C4BQ0-C4A3-BfF1-B18-A1		positive
	DQw8-DR4-C4B1-C4A3-BfS-B15-?		positive
	DQw8-DR4-C4B3-C4A4-BfS-B15-?		positive
	DQw7-DR4-C4B3-C4A4-BfS-B15-?		neutral
	DQw1-DR2-C4B1-C4A3-BfS-B7-A3		negative

In the absence of environmental influences (e.g. freedom from specific pathogens) these double-marker-positive individuals would be bound to develop insulin-dependent diabetes. Environmental factors may reduce rather than enhance the degree of penetrance. This model is based on recent investigations in the BB rat and NOD mouse and serves as an hypothesis which can also be tested in man.

9.5 Association studies

The identification of HLA as a marker for IDDM has resulted in numerous investigations. Initially, HLA typing was carried out by serological techniques. These techniques have largely been replaced by genotyping using cloned probes or oligonucleotides against defined genes. Typing sera and gene probes are standardized and an international classification and nomenclature is established in histocompatibility testing workshops (Dupont 1988; Bodmer *et al.* 1990).

As indicated above, the first association between HLA and IDDM was with HLA-B15 and B8. The strongest association with the typing sera for the HLA-B class I molecules was with B8/15 heterozygosity (Nerup *et al.* 1974). The association between HLA-B and IDDM was shortly thereafter shown to be secondary to a strong association with HLA-DW3 and DW4 (Thomsen *et al.* 1975). In these studies, a cellular detection system was used to determine the HLA-D specificities. Studies with typing sera for HLA-DR (R for related) specificities, DR3 and DR4, supported the conclusion that DR was more closely associated with risk of diabetes than HLA-B (Table 5). Platz *et al.* (1981) demonstrated that HLA-B8-positive patients with insulin-dependent diabetes were more often positive for HLA-DR3 than were healthy B8-positive controls. Subsequently, numerous investigations have confirmed the strong association between HLA-DR3 and/or DR4 and insulin-dependent diabetes (Wassmuth *et al.* 1990).

About 90–95 per cent of patients with insulin-dependent diabetes are either DR3 and/or DR4 positive, compared to 50–60 per cent of controls. Again, the highest known risk seems to be conferred by heterozygosity of HLA-DR3/4 which is

Table 5 HLA specificities associated with insulin-dependent diabetes

HLA specificity	IDDM		Controls		OR
	N	%	N	%	
DR3	67	50.8	13	15.3	5.7
DR4	115	87.1	38	44.7	8.4
DR3/4	55	41.7	6	7.1	9.4
DQw2	74	56.1	20	23.5	4.1
DQw8	110	83.3	32	37.6	8.3
DQw2/w8	60	45.5	4	4.7	16.9
4 kb	73	55.3	20	23.5	4
12 kb	116	87.9	44	51.8	6.8
4/12 kb	62	47	7	8.2	9.9

present, dependent on the ethnic background, in 30–40 per cent of patients compared to 3–8 per cent of controls. Similarly, the highest risk for Chinese patients with onset of insulin-dependent diabetes below the age of 11 was in HLA-DR3/9 individuals (Hawkins *et al.* 1987) and for Japanese patients in DR4/9 (Kobayashi *et al.* 1986). As will be shown below, recent data suggest that heterozygosity both of HLA-DQw2/8 (Wassmuth *et al.* 1990) and an HLA-DQ β-chain polymorphism (Michelsen *et al.* 1990; Reijonen *et al.* 1990) confers higher risks than HLA-DR3/4.

The discovery of additional genes for HLA class II molecules in the HLA-D region resulted in the demonstration, first by RFLP analysis (Owerbach *et al.* 1983) and later by direct sequence analysis (Todd *et al.* 1987, 1988), that the DQ region may be closer to IDDM than DR. Using a similar approach to that of Platz *et al.* (1981), it was demonstrated that DR4-positive patients are more often positive for DQw8 than healthy DR4-positive controls (Wassmuth *et al.* 1990). Diabetes risk would therefore be closer to DQ than DR. This constitutes another example of linkage disequilibrium and a lesson that care has to be taken not to speculate about a possible functional role of HLA-DQ class II molecules in IDDM aetiology and pathogenesis until the 'diabetes locus' has been properly identified. Further association studies to map the 'diabetes susceptibility locus' accurately will therefore require a large number of population-based patients and controls. Alternatively (and assuming that this hypothetical locus is the same in ethnic groups other than Caucasians, e.g. Mexican Americans or US Blacks), transracial gene mapping may also be an effective approach (Jenkins *et al.* 1990; Todd 1990).

From sequence comparisons of DQ β-chain genes, Todd *et al.* (1988) concluded that haplotypes associated with insulin-dependent diabetes (DR4-DQw8, DR3-DQw2, DR1-DQw5, DR2-DQw1.AZH) differed at amino-acid position 57 from uniform haplotypes which are negatively associated (DR4-DQw7, DR2-DQw1.2, DR2-DQw1.12). In a hypothetical model of the class II DQ heterodimer, this amino acid is located at the end of the α-helix. Although the crystallographic structure of a class II molecule is yet to be determined, it is inferred from class I molecules that the position 57 amino acids would be accessible for both peptide binding and T-cell receptor recognition (Todd *et al.* 1988). Positively associated haplotypes have alanine, serine, or valine in this position, while aspartic acid (Asp) is present in negatively associated haplotypes. Numerous studies have been conducted to test this hypothesis. On the grounds that as many as 2 per cent of Norwegian, 9 per cent of Polish, 22 per cent of Chinese and 56 per cent of Japanese IDDM patients develop IDDM despite being Asp/Asp homozygous (see (Wassmuth and Lernmark 1990)) the hypothesis may be refuted. Other observations which also do not lend support to the position 57 idea are:

(1) HLA-DR7/DQw2 (non-Asp in position 57) is negatively associated with diabetes, and

(2) HLA-DQw7 (Asp in position 57) is positively associated with IDDM in DQw8-positive individuals.

It is therefore concluded that genotyping solely for aspartic acid at position 57 is insufficient to assess the disease risk for IDDM.

DR and DQ genotyping in Swedish (Wassmuth *et al.* 1990) and US (Baisch *et al.* 1990) IDDM patients and controls has allowed analyses of complete DR-DQ haplotypes. These analyses reveal a pattern of haplotypes (such as DR3-DQw2 in DQw8-positive individuals or DR1-DQw7 in DQw8-positive individuals) which are positively associated with IDDM. Some haplotypes may confer risk despite the fact that they contain a protective DQ type. Susceptibility to IDDM may therefore require either alleles from both the DR and the DQ locus or there may be a dominant susceptibility provided by a gene which is close to DQw8. Negative association with IDDM also shows a dominant pattern of inheritance. Examples are DQw1.2 (Baisch *et al.* 1990) and DQw1.18 (Wassmuth *et al.* 1990) which are both negatively associated with IDDM even when they occur together with HLA-DQw8. It will therefore be necessary to subject the samples to complete DR-DQ genotyping in order to assess susceptibility to IDDM. It may also prove useful to further explore gene polymorphism associated with the HLA-DQ region (Michelson and Lernmark 1987; Reijonen *et al.* 1990) and newly discovered genes such as the ATP-binding cassette (ABC) genes (Spies *et al.* 1990; Trowsdale *et al.* 1990) which have been mapped to the HLA-D region of chromosome 6.

9.6 Linkage studies and sib pair analysis

The pattern of inheritance of IDDM is complex and, as pointed out above, does not fit available models of genetic inheritance. Linkage analysis of a non-Mendelian disease such as IDDM has resulted in conflicting data. In a joint study of Caucasians with IDDM (Thomson *et al.* 1988), it was concluded that DR3 predisposes in a recessive-like and DR4 in a dominant and intermediate-like fashion. After removal of DR3 and DR4, predisposing effects of DR1 and DRw8 were observed — probably explained by the DQw-types discussed above. The risk estimates for siblings, based on an overall sibling risk of 6 per cent, showed a 13 per cent, 5 per cent, or 2 per cent risk for siblings sharing two, one, or no haplotypes, respectively, with the proband. Again, the highest risk, 19 per cent, was conferred by HLA-DR3/4 heterozygous pairs. This analysis, which confirms and extends several previous publications (reviewed and discussed by Risch 1987, 1989) is taken to indicate

(1) that the mode of inheritance of IDDM remains largely unexplained;

(2) a single locus model is rejected; and

(3) additional non-HLA linked determinants are more important than HLA since the lifetime risk for a sibling is five times higher than that predicted by HLA. Using a maximum-likelihood affected pair method, Risch (1989) obtained strong evidence for linkage with HLA.

Family studies are also for analysing possible distorted patterns of inheritance. It is conceivable that homozygosity at two different loci for diabetes would be lethal. Effects on the fetus during pregnancy would add to factors that could affect the pathogenesis after birth. Exposure to antigenic stimulus during pregnancy may also affect the ability of the immune system to develop tolerance to self-antigens. Several authors have reported that the risk for IDDM was significantly higher in sibships with an IDDM father compared to those with an affected mother (Warram *et al.* 1984; MacDonald *et al.* 1986; Vadheim *et al.* 1986; Risch 1989). In one study it was observed that HLA-DR4-positive fathers were more likely to transmit this allele to their children than were DR4-positive mothers (Vadheim *et al.* 1986). Similarly, among DR3/4-positive probands and affected siblings significantly more affected children had received DR4 from their father (p) than from their mother (m); DR3(m)/DR4(p) was more common than DR3(p)/DR4(m) (Deschamps *et al.* 1990). In these studies the normal ratio is expected to be 50 per cent; healthy DR3- and DR4-positive parents and their children in non-diabetic families need to be evaluated before these data are fully conclusive. These and similar analyses may provide important information about possible effects of HLA on the pathogenesis of IDDM, for example during pregnancy.

The increased risk to children of fathers with IDDM, confirmed in a recent population-based case-control study (Dahlquist *et al.* 1989) also suggests that diabetic mothers somehow protect their children from diabetes. It may be hypothesized that IDDM is established already *in utero*. The propensity to develop the disease would thus be due to environmental factors *preventing* rather than environmental insults inducing the disease. There are several observations that support this hypothesis. First-degree relatives of IDDM patients more often have organ-specific auto-antibodies without clinical disease. Subclinical auto-antibodies are also frequent. The prevalence of islet cell antibodies (4 per cent) exceeds that of the disease (about 0.2 per cent) by a factor of 20. This means that roughly 1 out of 20 islet cell antibody-positive individuals will develop IDDM. Auto-antibodies may be present early in life (Ivarsson *et al.* 1988; Landin-Olsson *et al.* 1989). This is in contrast to the general population where auto-antibodies develop with increasing age. Their increased frequency may be due to a loss of tolerance rather than to the induction of auto-immunity by environmental factors. It is therefore possible that auto-antibodies may appear early in life if the fetal environment is not supportive of tolerance development.

In analysing families with and without IDDM, it was observed that the presence of auto-antibodies was associated with HLA-DR3, but only in the families with IDDM (Hägglöf *et al.* 1986). In the control families the frequency of HLA-DR3 was the same but there was no association between the presence of auto-antibodies and HLA-DR3. These data suggest that auto-antibodies may be determined by a gene lying outside the HLA locus. It may be speculated that auto-immunity is inherited as a separate entity. This could, for example, be involved in the generation of tolerance. Tolerance is the failure of the immune system to recognize self-antigen. It is thought that tolerance is induced *in utero* and during the first few months of postnatal or neonatal life. Tolerance is likely to be due to an active process or immunization which induces suppressor cells or causes suppressor mechanisms which are dependent on active removal of self-reactive T-helper cells. A genetic defect in IDDM might involve the mechanism by which such self-reactive T-helper cells are removed in the thymus. It is postulated that this 'auto-immunity' gene would be inherited as a recessive gene. HLA is the 'secondary' gene since it is permissive, i.e. the HLA type is necessary but not sufficient for diabetes to develop.

An explanation of why diabetic mothers frequently have offspring who do not develop IDDM is suggested by recent observations in transplantation biology (Claas *et al.* 1988). In investigations of kidney transplant donors and recipients it is found that the graft is accepted better if the maternal HLA haplotype of the patient is also present in the graft HLA haplotype. In other words, an improved graft acceptance is observed if the donor is matched with the HLA type of the mother of the recipient rather than the father. Similar observations have also been reported for bone marrow transplantation.

A possible explanation of these observations is that cells expressing the mother's HLA types are entering the bood circulation of the fetus. These HLA-bearing cells are antigenic to the fetus and induce tolerance. Tolerance develops not only against the HLA-type inherited from one of the mother's chromosomes but also towards the non-inherited HLA-type. In transplantation situations the immune system of the recipient will fail to react with the graft since tolerance has developed against the maternal HLA types not inherited by the fetus. Therefore, when auto-immunity is first initiated and a β-cell auto-antigen is presented by the maternal haplotypes, non-reactivity would persist. In contrast, a paternal HLA-haplotype would provide an initiating signal to the immune system since tolerance has not been induced. The increasingly detailed analyses of HLA-DQ specificities and genomic sequences which are rapidly becoming available and the identification of the major β-cell auto-antigen as glutamic acid decarboxylase (Baekkeskov *et al.* 1990) may help to resolve this issue.

9.7 A second gene for diabetes

Two animals, the BB rat and the NOD mouse develop IDDM in a manner comparable to the human disease (Mordes *et al*. 1987). In contrast to man, both BB rats and NOD mice are available as inbred lines. During the past few years, it has been observed that the incidence rate of IDDM in these rodent models has increased. Although this may be the result of inbreeding, recent observations suggest that an additional phenomenon may have been equally important (Leiter 1989). In both BB rat and NOD mouse colonies initially kept under standard animal breeding conditions, the incidence rate of IDDM increased when the animals were bred under specific pathogen-free conditions. These data suggest that the 'cleaner' the animal the higher the frequency of IDDM. Auto-immunity would therefore appear to be associated with the absence of appropriate stimuli for induction of tolerance. This hypothesis is testable since animals reared in a pathogen-free environment can be subjected to treatment with external antigens. Two such experiments have already been performed. First, BB rats infected with T-lymphocytotrophic virus showed a marked decrease in the frequency of IDDM (Dryberg *et al*. 1988). Secondly, both BB rats (Rabinovitch *et al*. 1984) and NOD mice (Sadelain *et al*. 1990) were found to be fully protected against developing IDDM when injected with complete Freund's adjuvant before weaning. Although the mechanism of protection in these experiments remains to be clarified, it is suggested that early immunostimulation in genetically susceptible animals is sufficient to remove the propensity to develop IDDM. Similar experiments in man will be difficult to perform. More effective HLA typing methods are rapidly becoming available, however, and response patterns to common viruses and vaccines will eventually be available. It is conceivable that protection from IDDM may be induced by vaccination; the problem to date is that we cannot distinguish protective from inducing environmental factors.

9.8 Perspectives

The clinical onset of IDDM is associated with a number of immune abnormalities. Using islet cell antibodies as a marker for IDDM, it has been established that diabetes exists as a subclinical disease. Subclinical diabetes may last for years before prediabetes (significant β-cell destruction) and, finally, clinical IDDM develop. A wealth of data collected on twins, families, and in the population has shown that the disease is familial. The subclinical disease may often remit. Although IDDM is genetically linked to certain HLA-DR/DQ class II molecules, it needs to be clarified whether these molecules determine the propensity to develop an immune response to certain antigens, failure to maintain tolerance, or the ability to produce disease-associated auto-antibodies. The observation that HLA-DQ confers a higher risk for IDDM than HLA-DR is an important advance. Continued molecular analysis of the HLA-DR/DQ region and extensive population-based association studies (best carried out as large multicentre studies) and transracial gene mapping should allow the HLA-D locus 'diabetes' sequences to be finally identified. Studies in both the BB rat (Markholst *et al.* 1990) and the NOD mouse (Leiter 1989) suggest the presence of another diabetes gene outside the MHC. The availability of a human linkage map and a growing number of polymorphic DNA probes should make it possible to eventually identify non-MHC diabetes genes. In one such study in man, it was possible to exclude a candidate locus on chromosome 11 (Hyer *et al.* 1991). The growing number of DNA probes allows a near saturation of the human chromosome map. Probes that segregate with the disease can therefore be used first to localize the disease to a certain chromosome. A knowledge of the chromosome will allow further detailed mapping of the locus using technologies that have resulted in the recent spectacular cloning and sequencing of genes such as that for cystic fibrosis (Rommens *et al.* 1989). It would seem that progress in this field is dependent more on the availability of suitable families than on techniques in molecular genetics.

A major challenge to understanding IDDM is the effect of the environment. The HLA class II molecules interact with the environment by virtue of their ability to present antigen to the immune system. We need to understand in what way these molecules bind and present β-cell-specific peptides. Perhaps there is a diabetogenic sequence to be found in glutamic acid decarboxylase (64K antigen), the major autoantigen detected by IDDM sera? It is possible that individuals prone to develop IDDM are poor presenters of this self-antigen during the induction of tolerance. Destruction, due to virus (rubella, mumps, Coxsackie, rotavirus, etc.) of β-cells late in life may therefore initiate a disease process which is perpetuated by a genetic propensity to develop auto-immunity. Recent data in spontaneously diabetic BB rats and NOD mice, however, suggest that environmental factors may be protective. For example, the penetrance of diabetes in inbred animals often reaches 100 per cent in specific pathogen-free environments. Similar phenomena may also occur in man, and research needs to be focused on the possible protective interaction of external factors, both before and after birth. Further studies on the molecular mechanisms of the human immune response, both in terms of tolerance and the response to external antigens, should help to reveal the complex pattern of genetic inheritance of IDDM.

References

Andres, R. (1971). Aging and diabetes. *Med. Clin. North Am.* **55,** 835–46.

Baekkeskov, S., Aanstoot, H. J., Christgau, S. (1990). Identification of the 64K autoantigen in insulin-dependent diabetes as the GABA-synthesizing enzyme glutamic acid decarboxylase. *Nature* **347,** 151–6.

Baisch, J. M., Weeks, T., Giles, R., Hoover, M., Stastny, P., and Capra, J. D. (1990). Analysis of HLA-DQ genotypes and susceptibility in insulin-dependent diabetes mellitus. *New Engl. J. Med.* **322,** 1836–82.

Barbosa, J., Rich, S., Dunsworth, T., and Swanson, J. (1982). Linkage disequilibrium between insulin-dependent diabetes and the Kidd blood group JK-b allele. *J. Clin. Endocrinol.* **55,** 193–5.

Bell, G. I., Aorita, S., and Koran, J. H. (1984). A polymorphic locus near the human insulin gene is associated with insulin-dependent diabetes mellitus. *Diabetes* **33,** 176–83.

Bliss, M. (1982). *The discovery of insulin.* McClelland and Stewart, Toronto.

Bodmer, S. G., March, E., Parham, P., *et al.* (1990). Nomenclature for factors of the HLA system, 1989. *Immunol. Today* **11,** 3–10.

Bottazzo, G. F., Florin-Christensen, A., and Doniach, D. (1974). Islet cell antibodies in diabetes mellitus with autoimmune polyendocrine deficiencies. *Lancet* **ii,** 1279–83.

Claas, F. H. J., Gijbels, Y., Velden-de Munck, J. V. D., and Rood, J. J. V. (1988). Induction of B cell unresponsiveness to noninherited maternal HLA antigens during fetal life. *Science* **241,** 1815–17.

Dahlquist, G., Blom, L., Tuvemo, T., Nyström, L., Sandström, A., and Wall, S. (1989). The Swedish childhood diabetes study – Results from a nine year case register and one year case-referent study indicating that Type 1 (insulin-dependent) diabetes mellitus is associated with both Type 2 (non-insulin-dependent) diabetes mellitus and autoimmune disorders. *Diabetologia* **32,** 2–6.

Dawkins, R. L., Uko, G., Pestell, R., and McCann, V. J. (1988). Association of MHC supratypes and complotypes with endocrine disorders. In *Immunogenetics of endocrine disorders* (ed. N. R. Fand), pp. 433–48. Alan R. Liss, New York.

Deschamps, I., Hors, J., Clerget-Darpour, F., *et al.* (1990). Excess of maternal HLA-DR3 antigens in HLA-DR3,4 positive type 1 (insulin-dependent) diabetic patients. *Diabetologia* **33,** 425–30.

Di Mario, U., Irvine, W. J., Borsey, D. Q., Kyner, J. L., Weston, J., and Galfo, C. (1983). Immune abnormalities in diabetic patients not requiring insulin at diagnosis. *Diabetologia* **25,** 392–5.

Dupont, B. (1988). *Histocompatibility Testing 1987.* Springer-Verlag, New York.

Dyrberg, T., Schwimmbeck, P. L., and Oldstone, M. B. A. (1988). Inhibition of diabetes in BB rats by virus infection. *J. Clin. Invest.* **8,** 928–31.

Field, L. L., Anderson, C. E., Neiswanger, K., Hodge, S. E., Spence, M. A., and Rotter, J. I. (1984). Interaction of HLA and immunoglobulin antigens in Type 1 (insulin-dependent) diabetes. *Diabetes* **27,** 504–8.

Gepts, W. (1965). Pathologic anatomy of the pancreas in junvenile diabetes mellitus. *Diabetes* **14,** 619–33.

Gepts, W. and LaCompte, P. (1981). The pancreatic islets in diabetes. *Am. J. Med.* **70,** 105–15.

Gleichmann, H., Zörcher, B., Greulich, B., *et al.* (1984). Correlation of islet cell antibodies and HLR-DR phenotypes with diabetes mellitus in adults. *Diabetologia* **27,** 90–2.

Gorsuch, A. N., Spencer, K. M., Lister, J., *et al.* (1981). Evidence for a long prediabetic period in Type 1 (insulin-dependent) diabetes mellitus. *Lancet* **ii,** 1363–5.

Green, A. (1982). The epidemiologic approach to studies of association between HLA and disease. I. The basic measures, concepts and estimation procedures. *Tissue Antigens* **19,** 245–58.

Groop, L. C., Pelkonen, R., Koskimies, S., Bottazzo, G. F., and Doniach, D. (1986). Secondary failure to treatment with oral antidiabetic agents in non-insulin-dependent diabetes. *Diabetes Care* **9,** 129–33.

Hägglöf, B., Rabinovitch, A., Mackay, P., *et al.* (1986). Islet cell and other organ-specific autoantibodies in healthy first degree relatives to insulin-dependent diabetic patients. *Acta Paediatr. Scand.* **75,** 611–18.

Hawkins, B. R., Lam, K. S. L., Ma, J. T. C., *et al.* (1987). Strong association of HLA-DR3/DRw9 heterozygosity with early onset insulin-dependent diabetes mellitus in Chinese. *Diabetes* **36,** 1297–300.

Heaton, D. A., Millward, B. A., Gray, P., *et al.* (1987). Evidence of β cell dysfunction which does not lead on to diabetes: a study of identical twins of insulin dependent diabetics. *Br. Med. J.* **294,** 145–6.

Heaton, D. A., Millward, B. A., Gray, I. P., *et al.* (1988). Increased proinsulin levels as an early indicator of β-cell dysfunction in non-diabetic twins of Type 1 (insulin-dependent) diabetic patients. *Diabetologia* **31,** 182–4.

Horwitz, D. L., Starr, J. J., Mako, M. E., Blackard, W., and Rubenstein, A. H. (1975). Proinsulin, insulin and C-peptide concentrations in human portal and peripheral blood. *J. Clin. Invest.* **55,** 1278–83.

Hoover, M. L., Angelini, G., Ball, E., *et al.* (1986). HLA-DQ and T-cell receptor genes in insulin-dependent diabetes mellitus. *Cold Spring Harbor Symp. Quant. Biol.* **L1,** 803–9.

Hyer, R. N., Julier, C., Buckley, J. D., *et al.* (1991). High-resolution linkage mapping for susceptibility genes in human polygenic disease: Insulin-dependent diabetes mellitus and chromosome II. *Am. Soc. Hum. Gen.* **48,** 243–57.

Irvine, W. J., Gray, R. S., and Steel, J. M. (1980). Islet cell antibody as a marker for early stage type I diabetes mellitus. In *Immunology of diabetes*, pp. 117–54. Teviot Scientific Publications, Edinburgh.

Ivarsson, S. A., Marner, B., Lernmark, Å., and Nilsson, K. O. (1988). Nonislet pancreatic autoantibodies in sibship with permanent neonatal insulin-dependent diabetes mellitus. *Diabetes* **37,** 347–50.

Jenkins, D., Mijovic, C., Fletcher, J., Jacobs, K. H., Bradwell, A. R., and Barnett, A. H. (1990). Identification of susceptibility loci for type 1 (insulin-dependent) diabetes by trans-racial gene mapping. *Diabetologia* **33,** 387–95.

Johnston, C., Millward, B. A., Hoskins, P., Leslie, R. D. G., Bottazzo, G. F., and Pyke, D. A. (1989). Islet-cell antibodies as predictors of the later development of type 1 (insulin-dependent) diabetes. *Diabetologia* **32,** 382–6.

Karjalainen, J. (1990). Islet cell antibodies as predictive markers for IDDM in children with high background incidence of disease. *Diabetes* **39,** 1144–50.

Kobayashi, T., Sugimoto, T., Itoh, T., *et al.* (1986). The prevalence of islet cell antibodies in Japanese insulin-dependent and non-insulin-dependent diabetic patients studied by indirect immunofluorescence and by a new method. *Diabetologia* **35,** 335–40.

Köbberling, J. and Tattersall, B. (1982). *The genetics of diabetes mellitus*. Academic Press, London.

Landin-Olsson, M., Karlsson, A., Dahlquist, G., Blom, L., Lernmark, Å., and Sundkvist, G. (1989). Islet cell and other organ-specific autoantibodies in all children developing Type 1 (insulin-independent) diabetes mellitus in Sweden during one year and in matched control. *Diabetologia* **32,** 387–95.

Leiter, E. (1989). The genetics of diabetes susceptibility in mice. *FASEB J.* **3,** 2231–41.

Lendrum, R., Walker, G., and Gamble, D. R. (1975). Islet-cell antibodies in juvenile diabetes mellitus of recent onset. *Lancet* **ii,** 880–3.

MacCuish, A. C. and Irvine, W. J. (1975). Autoimmunological apects of diabetes mellitus. *Clin. Endocr. Metab.* **4,** 435–71.

MacCuish, A. C., Barnes, E. W., Irvine, W. J., and Duncan, L. J. P. (1974*a*). Antibodies to pancreatic islet-cells in insulin-dependent diabetics with coexistent autoimmune disease. *Lancet* **ii,** 1529–31.

MacCuish, A. C., Jordan, J., Campbell, C. J., Duncan, L. J. P., and Irvine, W. J. (1974*b*). Cell-mediated immunity to human pancreas in diabetes mellitus. *Diabetes* **23**, 693–7.

McCulloch, D. K., Klaff, L. J., Khan, S. E., *et al*. (1990). Nonprogression of subclinical β-cell dysfunction among first degree relatives of IDDM patients: 5-yr follow-up of the Seattle family study. *Diabetes* **39**, 549–56.

MacDonald, M. J., Gottschall, J., Hunter, J. B., and Winter, K. L. (1986). HLA-DR4 in insulin-dependent diabetic parents and their diabetic offspring: a clue to dominant inheritance. *Proc. Natl Acad. Sci. USA* **83**, 7049–53.

Mandrup-Poulsen, T., Bendtzen, K., Dinarello, C. A., and Nerup, J. (1987). Human tumor necrosis factor potentiates human interleukin 1 mediated rat pancreatic β cell cytotoxicity. *J. Immunol.* **139**, 4077–82.

Markholst, H., Andreasen, B., Eastman, S., and Lernmark, Å. (1990). Diabetes segregates as a single locus in crosses between inbred BB rats prone or resistant to diabetes. *J. Exp. Med.*, **174**, 297–88.

Marner, B., Agner, T., Binder, C., Lermark, Å., Nerup, J., Mandrup-Poulsen, T., and Walldorf, S. (1985). Increased reduction in fasting C-peptide is associated with islet cell antibodies in Type 1 (insulin-dependent) diabetic patients. *Diabetologia* **28**, 875–80.

Mason, D. R., Scott, R. S., and Darlow, B. A. (1987). Epidemiology of insulin-dependent diabetes mellitus in Canterbury, New Zealand. *Diabetes Res. Clin. Pract.* **3**, 21–9.

Michelsen, B. and Lernmark, Å. (1987). Molecular cloning of a polymorphic DNA endonuclease fragment associates insulin-dependent diabetes mellitus with HLA-DQ. *J. Clin. Invest.* **79**, 1144–52.

Michelsen, B., Wassmuth, R., Ludvigsson, J., Lernmark, Å, Nepom, G. T., and Fisher, L. (1990). HLA heterozygosity in insulin-dependent diabetes is most frequent at the DQ locus. *Scand. J. Immunol.* **31**, 405–13.

Millward, B. A., Alviggi, L., Hoskins, P. J., *et al*. (1986). Immune changes associated with insulin-dependent diabetes may remit without causing the disease, a study of identical twins. *Br. Med. J.* **292**, 793–6.

Millward, B. A., Welsh, K. I., Leslie, R. D. G., Pyke, D. A., and Demaine, A. G. (1987). T cell receptor beta chain polymorphism are associated with insulin-dependent diabetes. *Clin. Exp. Immunol.* **70**, 152–7.

Mordes, J. P., Desemone, J., and Rossini, A. A. (1987). The BB rat. *Diabetes/Metabolism Rev.* **3**, 725–50.

NDD Group (1979). Classification and diagnosis of diabetes mellitus and other categories of glucose tolerance. *Diabetes* **28**, 1039–57.

Nerup, J., Andersen, O. O., Bendixen, G., Egeberg, J., and Poulsen, J. E. (1971). Anti-pancreatic cellular hypersensitivity in diabetes mellitus. *Diabetes* **20**, 424–7.

Nerup, J., Platz, P., Anderssen, O. O. (1974). HL-A antigens and diabetes mellitus. *Lancet* **ii**, 864–6.

Olmos, P., Aherne, R., Heaton, D. A., Millward, B. A., Risley, D., and Pyke, D. A. (1988). Significance of concordance rates in identical twins of insulin dependent diabetics. *Diabetologia* **31**, 747–50.

Opie, E. (1901). On the relation of chronic interstitial pancreatitis to the island of Langerhans and to diabetes mellitus. *J. Exp. Med.* **5**, 393–9.

Owerbach, D., Lernmark, Å., Platz, P., *et al*. (1983). HLA-D region β-chain DNA endonuclease fragments differ between HLA-DR identical healthy and insulin-dependent diabetic individuals. *Nature* **303**, 815–17.

Pfeifer, M. A., Halter, J. B., and Porte, D. J. (1981). Insulin secretion in diabetes mellitus. *Am. J. Med.* **70,** 559–88.

Platz, P., Jakobsen, B. K., Morling, M., *et al.* (1981). HLA-D and DR-antigens in genetic analysis of insulin-dependent diabetes mellitus. *Diabetologia* **21,** 108–15.

Pyke, D. A. (1979). Diabetes: The genetic connections. *Diabetologia* **17,** 333–43.

Rabinovitch, A., MacKay, P., Ludvigsson, J., and Lernmark, Å. (1984). A prospective analysis of islet cell cytotoxic antibodies in insulin-dependent diabetic children: transient effects of plasmapheresis. *Diabetes* **33,** 224–8.

Reijonen, H., Ilonen, J., Knip, M., Michelsen, B., Åkerblom, H. K. (1990). HLA-DQ beta-chain restriction fragment length polymorphism as a risk marker in type 1 (insulin-dependent) diabetes mellitus: a Finnish family study. *Diabetologia* **33,** 357–62.

Rich, S. S. (1990). Mapping genes in diabetes. *Diabetes* **39,** 1315–19.

Risch, N. (1987). Assessing the role of HLA-linked and unlinked determinants of disease. *Am. J. Hum. Genet.* **40,** 1–14.

Risch, N. (1989). Genetics of IDDM: Evidence for complex inheritance with HLA. *Genetic Epidemiol.* **6,** 143–8.

Rommens, J. M., Iannuzzi, M. C., Kerem, B.-S., *et al.* (1989). Identification of the cystic fibrosis gene: chromosome walking and jumping. *Science* **245,** 1059–65.

Rotter, J. I., Vadheim, C. M., and Rimoin, D. L. (1990). Genetics of diabetes mellitus. In *Ellenberg and Rifkin's Diabetes Mellitus* (ed. H. Rifkin and D. Porte), pp. 378–413. Elsevier, New York.

Sadelain, M. W. J., Qin, H. Y., Lauzon, J., and Singh, B. (1990). Prevention of type 1 diabetes in NOD mice by adjuvant immunotherapy. *Diabetes* **39,** 583–9.

Sadelain, M. W. J., Qin, H. Y., Sumoski, W., Parfrey, N., Singh, B., and Rabinovitch, A. (1990). Prevention of diabetes in the BB rat by early immunotherapy using Freund's adjuvant. *J. Autoimmunity* **3,** 671–80.

Singal, D. P. and Blajchman, M. A. (1973). Histocompatibility (HL-A) antigens, lymphocytotoxic antibodies and tissue antibodies in patients with diabetes mellitus. *Diabetes* **22,** 429–32.

Spencer, K. M., Tarn, A., Dean, B. M., Lister, J., and Bottazzo, G. F. (1984). Fluctuating islet-cell autoimmunity in unaffected relatives of patients with insulin-dependent diabetes. *Lancet* **i,** 764–6.

Spies, T., Bresnahan, M., Bahram, S., *et al.* (1990). A gene in the human major histocompatibility complex class II region controlling the class I antigen presentation pathway. *Nature* **348,** 744–7.

Steiner, D. F., Bell, G. I., and Tager, H. S. (1989). Chemistry and biosynthesis of pancreatic protein hormones. In *Endocrinology,* 2nd edn, Vol. 2 (ed. L. J. DeGroot), pp. 1263–89. W.B. Saunders Company, Philadelphia.

Tattersall, R. B. (1986). Diabetes in the elderly—a neglected area? *Diabetologia* **27,** 167–73.

Tattersall, R. B. and Pyke, D. A. (1972). Diabetes in identical twins. *Lancet* **ii,** 1120–5.

Thomsen, M., Platz, P., Andersen, O. O., *et al.* (1975). MLC typing in juvenile diabetes and idiopathic Addison's disease. *Transplant Proc.* **22,** 125–47.

Thomson, G., Robinson, W. P., Kuhner, M. K., *et al.* (1988). Genetic heterogeneity, modes of inheritance, and risk estimates for a joint study of Caucasians with insulin-dependent diabetes mellitus. *Am. J. Hum. Genet.* **43,** 799–816.

Tillil, H. and Köbberling, J. (1987). Age-corrected empirical genetic risk estimates for first-degree relatives of IDDM patients. *Diabetes* **36,** 93–9.

Todd, J. A. (1990). Genetic control of autoimmunity in type 1 diabetes. *Immunol. Today* **11,** 122–9.

Todd, J. A., Bell, J. I., and McDevitt, H. O. (1987). HLA DQ β gene contributes to susceptibility and resistance to insulin-dependent diabetes melitus. *Nature* **329,** 599–604.

Todd, J. A., Acha-Orbea, H., Bell, J. L., *et al.* (1988). A molecular basis for MHC class II-associated autoimmunity. *Science* **240,** 1003–9.

Trowsdale, J., Hanson, I., Mockridge, I., Beck, S., Townsend, A., and Kelly, A. (1990). Sequences encoded in the class II region of the MHC related to 'ABC' superfamily of transporters. *Nature* **348,** 741–4.

Vadheim, C. M., Rotter, J. I., Maclaren, N. K., Riley, W. J., and Andersen, C. E. (1986). Preferential transmission of diabetic alleles within the HLA complex. *New Engl. J. Med.* **315,** 1314–18.

Vague, P., Melis, C., Mercier, P., Vialettes, B., and Lassmann, V. (1978). The increased frequency of the Lewis negative blood group in a diabetic population. *Diabetologia* **15,** 33–6.

von Meyenburg, H. (1940). Über 'Insulitis' bei Diabetes. *Schweiz. Med. Wochenschr.* **21,** 554–61.

Warram, J. H., Krolewski, A. S., Gottlieb, M. S., and Kahn, C. R. (1984). Differences in risk of insulin-dependent diabetes in offspring of diabetic mothers and diabetic fathers. *New Engl. J. Med.* **311,** 149–52.

Wassmuth, R. and Lernmark, Å. (1990). The genetics of susceptibility to diabetes. *Clin. Immunol. Immunopathol.* **53,** 358–99.

Wassmuth, R., Kockum, I., Nepom, G. T., Holmberg, E., Ludvigsson, J., and Lernmark, Å. (1990*a*). Complete HLA DR and DQ genotyping and insulin dependent diabetes mellitus: towards a new concept of disease susceptibility and resistance. Submitted.

Wassmuth, R., Sasaka, T., Ikeda, N. Y., and Lernmark, Å. (1990*b*). An HLA-DQ-related RFLP provides the missing link between Caucasian and Japanese IDDM patients. *Diabetes* **39** (Suppl. 1), 219A.

WHO Report (1976). Announcements: Review of the notation for allotypic and related markers of human immunoglobulins. *J. Immunol.* **117,** 1056–8.

WHO Study Group (1985). *Diabetes mellitus,* WHO Technical Report Series 727, 7–113. WHO, Geneva.

Wilson, R. M., Van der Minne, P., Deverill, I., *et al.* (1985). Insulin dependence: Problems with the classification of 100 consecutive patients. *Diab. Med.* **2,** 167–72.

10 | Aetiology of type I diabetes: immunological aspects

MICHAEL R. CHRISTIE

Section contributed by JOHN W. SEMPLE, MICHA J. RAPOPORT, and TERRY L. DELOVITCH

10.1 Type I diabetes as an auto-immune disease

Insulin-dependent (type I) diabetes is the result of the specific loss of insulin-secreting pancreatic β-cells. It is now generally accepted that the disease is auto-immune in nature, the destruction of the β-cells occurring as the direct result of an aberrant auto-immune reaction against one or more components of these cells. The main pieces of evidence that point to such a mechanism are as follows:

1. The disease clearly has a genetic component (reviewed in Chapter 9) and at least one of the susceptibility genes maps to the major histocompatibility complex (MHC) and specifically to genes that encode molecules involved in the regulation of immune responses. Immune responsiveness is thus likely to be important in disease pathogenesis.
2. Histological studies of pancreata from diabetic patients have shown that lymphocytic infiltration of islets is a common feature in recent onset diabetes (Gepts and De Mey 1978). Although it can be argued that inflammation is a natural biological response to tissue damage, perhaps induced by exogenous cytotoxic factors, the lymphocytes invading the islet could equally be directly responsible for β-cell destruction. Indeed, lymphocytes from diabetic patients have been shown to be reactive to islet antigen and to show cytotoxicity to islets *in vitro* (Nerup *et al.* 1974; Charles *et al.* 1983).
3. Antibodies to islet cell components are present at high frequency in newly diagnosed diabetic patients, and circulating antibodies can be demonstrated to be cytotoxic to islet cells *in vitro* (Baekkeskov *et al.* 1986).
4. Type I diabetes has, for some time, been known to be associated with other auto-immune disorders, particularly those that are endocrine in nature (Nerup and Lernmark 1981). Thus, type I diabetes is found in approximately 15 per cent

of patients with Addison's disease and in around 10 per cent of patients with thyroid disorders. Antibodies that react with other tissues, such as thyroid, adrenal medulla, or gastric parietal cells, or organ-specific molecules such as thyroglobulin or thyroid microsomal antigen, are also found at high frequency in type I diabetic patients. These observations suggest that one component of disease pathogenesis is likely to be disturbed immune regulation, and that this defect may be common to a number of auto-immune disorders.

5. Suppression of immune responses by the administration of general immunosuppressive agents, such as cyclosporin A, causes a partial or complete remission in recent onset diabetic patients, as judged by a lower requirement for exogenous insulin therapy to maintain normoglycaemia (Feutren *et al.* 1986).
6. Animal models of type I diabetes with features that are common to the human disease clearly have an auto-immune component to β-cell destruction. Thus, disease can be prevented by a number of procedures that block immune function, and diabetes can be transferred to non-diabetic recipients by lymphocytes from diabetic donors.

Recent molecular and functional characterization of a number of components of the immune response, including the T-cell receptor and MHC class I and class II antigens, has substantially advanced our understanding of immunoregulation. This has led to considerable interest in determining the molecular mechanisms of immunological self-tolerance and how self-tolerance fails in auto-immune disease. In this chapter we review our current understanding of the immune system and its abnormalities in type I diabetes, including lessons learnt from animal models of the disease. We then discuss how this knowledge might be applied to develop better therapies for the disease.

10.2 Mechanisms of immunological self-tolerance

The role of the immune system is to distinguish foreign from self; foreign antigens must be eliminated while maintaining the integrity of self-structures. Failure of the immune system to maintain tolerance to self can result in an auto-aggressive reaction against the body's own tissues that can lead to auto-immune disease. A knowledge of mechanisms employed by the immune system to avoid reactivity to self-antigens is necessary if we are to understand how self-tolerance fails in auto-immune disease.

Recognition of foreign antigen is achieved by two clonally expressed receptor systems; the immunoglobulins on B-lymphocytes and receptors on T-lymphocytes (T-cells). Surface immunoglobulin on B-lymphocytes is capable of binding intact antigen whereas processing of antigen and presentation of antigen fragments by MHC class I or class II molecules is required for recognition by the antigen receptor of T-lymphocytes. Binding of antigen by a specific lymphocyte clone is frequently not sufficient for activation and proliferation of lymphocytes. For example, activation and proliferation of B-lymphocytes, and secretion of antibody, requires interaction with and 'help' (for example, signalling molecules such as secreted lymphokines) from CD4-positive, antigen-specific, T-helper cells (T_h cells). Helper factors are also required for the activation and proliferation of cytotoxic T-cells which recognize foreign antigen, such as viral peptides, bound to MHC class I molecules on the surface of infected cells. This requirement for collaboration between components of the immune system itself provides a limitation on the proliferation of lymphocyte clones. Thus, B-cell activation and antibody secretion requires interaction between antigen-specific T-helper cells and B-lymphocytes, both of which are present at very low frequency in the total lymphocyte population. Furthermore, most peripheral tissues do not express MHC class II molecules, so do not have the capacity to present their own self-antigen to T-helper cells, providing a further limitation on self-reactivity. It has been suggested that aberrant expression of class II antigens, for example by thyroid or pancreatic β-cells, may contribute to the development of auto-immunity to these tissues in thyroid disease and diabetes, respectively (Bottazzo *et al.* 1983). Other factors are clearly important to ensure non-reactivity to self-antigens. The major mechanisms that appear to be operating to maintain self-tolerance are:

(1) deletion of self-reactive lymphocyte clones during ontogeny (clonal deletion);
(2) inactivation of self-reactive lymphocyte clones (clonal anergy); and
(3) suppression of self-reactive lymphocytes by T-suppressor cells.

10.2.1 Clonal deletion

Burnett (1959) originally suggested that clones of self-reactive lymphocytes are deleted as a mechanism for tolerance induction. It is unclear to what extent clonal deletion acts as a mechanism for self-tolerance in B-lymphocytes; self-reactive B-lymphocyte clones are easily isolated from normal, non-autoimmune individuals. However, clonal deletion is now established as a major mechanism for self-tolerance in T-lymphocytes.

During T-cell development, precursors of T-lymphocytes migrate from the bone marrow to the thymus, where they mature into distinct T-cell subsets. Only a small proportion (approximately 2 per cent) of incoming prothymocytes will exit the thymus as mature T-cells; the majority die in the thymus. It is proposed that T-cells undergo two selective processes during ontogeny (Boehmer *et al.* 1989). First, thymocytes are 'educated' to recognize antigen in association with self-MHC molecules (positive selection). Thymocytes appear to be programmed to die unless they receive a signal, probably an increase in cytosolic calcium, after interaction with self-MHC molecules expressed on thymic epithelium. The mature T-cell repertoire is thus skewed to recognize foreign antigen in association with self-MHC molecules. Secondly, T-lymphocytes bearing T-cell receptors with high affinity for self-antigens in association with MHC molecules expressed on dendritic cells in the thymus are deleted (negative selection).

Evidence for clonal deletion as a mechanism for maintaining tolerance to self comes from experiments analysing T-cell receptor usage with specific monoclonal antibodies, in mice expressing different self-molecules (Marrack and Kappler 1987). A number of mouse strains fail to express the MHC class II molecule IE due to mutations in the genes encoding this protein. Analysis of T-cells in the periphery of mice that either express or lack IE molecules have shown that IE-positive mice have very low numbers of T-cells expressing the T-cell receptor V_β 17a, despite the presence of the gene for the receptor in the genome. In contrast, approximately 10 per cent of peripheral T-cells express this receptor in IE-negative mice. Analysis of T-cell responses to MHC molecules *in vitro* has demonstrated that a large proportion of T-cells expressing V_β 17a T-cell receptors show reactivity to IE. It was concluded that T-cells bearing V_β 17a are deleted during development in mice that express IE.

These observations have been extended to another antigen that displays differential expression in mouse strains; mls (minor lymphocyte stimulating). Generally, MHC-identical T-cells will not cross-react *in vitro*. T-cells from MHC-identical mice strains will react, however, with lymphocytes bearing different mls antigens. The responses to mls are MHC-restricted. Kappler *et al.* (1988) and MacDonald *et al.* (1988) found that mice expressing the a haplotype of mls (mls^a) have very few T-cells bearing V_β 8.1 and V_β 6, which are precisely the receptors borne on T-cells reactive to mls^a/MHC in *in vitro* studies. The inference from such experiments is that high-affinity interaction between the immature T-lymphocyte and MHC/self-antigen complexes in the thymus generates a signal that causes the deletion of those self-reactive T-lymphocytes.

10.2.2 Tolerance to transgene products expressed specifically on pancreatic β-cells

Evidence for a role for clonal deletion in maintaining tolerance to self-antigens has been confined to antigens that are expressed within the thymus. It is not clear to what extent clonal deletion acts as a mechanism for antigens expressed specifically in non-thymic tissues. The recent advances in genetic manipulation that allow one to direct the expression of exogenous proteins to specific tissues by transgenic mouse technology have led to experiments to define the mechanisms that maintain tolerance to peripheral self-antigens.

Hanahan and co-workers have analysed immune responses to a viral oncoprotein, large T antigen, in transgenic mice expressing this protein under the control of the insulin promoter/enhancer region (Hanahan *et al.* 1989). These mice express large T antigen specifically in the pancreatic β-cells, develop β-cell tumours, and eventually die of hypoglycaemia as a result of excessive insulin production. Two distinct types of T antigen expression were observed in the transgenic mice. In one class of mouse, expression of the protein occurred early in development, at embryonic day 10, and continued throughout adult life. In the second class of mouse, there was a delay in the expression of the transgene until adulthood when, initially, scattered β-cells were found to express the protein, with a subsequent increase in the number of T antigen-positive islets with age. The specific pattern of large T antigen expression is thought to depend on the site of incorporation of the transgene into the genome.

The time of onset of antigen expression was found to influence immune responses to the large T antigen. Transgenic mice expressing the protein early during development did not develop an antibody response on immunization with purified T antigen; i.e. they were immunologically tolerant. In contrast, those mice showing delayed antigen expression were found to be non-tolerant of antigen. Furthermore, a proportion of these mice spontaneously developed antibodies to the protein without immunization. Immunohistochemical analysis of pancreata from non-tolerant mice revealed lymphocytic infiltration into pancreatic islets, composed predominantly of B-lymphocytes, CD4-positive and CD8-positive T-lymphocytes, and macrophages. Thus, tolerance to an antigen expressed specifically by a relatively low number of cells at a site outside of the thymus can be achieved, provided that the antigen is expressed early in development. Failure to express the protein until adulthood leads to non-tolerance and auto-immunity. Thus, delayed antigen expression may represent one potential mechanism whereby tolerance to self is avoided.

A number of groups have studied the transgenic expression of MHC class I or class II molecules specifically by the pancreatic β-cells (Miller *et al.* 1989). Transgenic mice were created with class I or class II MHC genes linked to the insulin promotor. These mice were found to express strongly the transgene products specifically in the pancreatic β-cells, although in some mice expression in the renal tubules was observed. Thymic MHC expression was not detected. Diabetes occurred in all mice at an early age. The onset of diabetes, however, was not accompanied by lymphocytic infiltration into the pancreatic islet, suggesting that β-cell destruction was not immune-mediated. In addition, neonatal thymectomy, a procedure that interferes with maturation of T-cells and thus blocks T-cell-dependent immune responses, did not prevent diabetes. Since cultured pancreatic islets from transgenic mice showed aberrant insulin secretory responses, it appears that transgenic hyperexpression of MHC molecules by β-cells can result in functional impairment, leading to β-cell damage and diabetes *in vivo*. Other transgene products expressed under the control of similar promotors do not result in diabetes (Harrison *et al.* 1989). Furthermore, transgenic mice expressing class II molecules under the control of a different insulin promotor construct, where the level of class II expression was low and comparable to that found on splenic B-lymphocytes, did not develop diabetes (Bohme *et al.* 1989). Thus, it appears that β-cell impairment is a relatively specific property of an inappropriately high level of expression of MHC molecules. Since pancreatic β-cells are found to hyperexpress MHC class I molecules in the diabetic pancreas, this finding may have relevance to β-cell damage in diabetes.

Transgenic mice with hyperexpression of MHC molecules on pancreatic β-cells were found to be immunologically tolerant of the transgene product, even though the MHC molecules were not expressed in the thymus. Therefore, immunological tolerance to peripheral self-antigens may not necessarily involve clonal deletion in the thymus. In support of this idea, mice with pancreatic expression of IE molecules were found to have a high proportion of V_{β} 17a- and V_{β} 5-positive T-cells which are normally deleted in IE-expressing mice (Burkly *et al.* 1990). However, T-cells from those mice that express V_{β} 5 T-cell receptors showed poor stimulation in response to antibodies to these T-cell receptors, suggesting that some IE-reactive T-cells, although present, are functionally inactivated as a result of pancreatic IE expression.

In contrast to transgenic mouse models with hyperexpression of MHC molecules, lymphocytes from mice with pancreatic IA expression close to levels seen in normal class II-expressing cells were found to be immunologically intolerant to the MHC antigen when tested in *in vitro* assays (Bohme *et al.* 1989). Nevertheless, the IA-expressing pancreatic β-cells were not destroyed, despite the presence of IA-reactive T-cells. There are a number of possible explanations for the survival of the β-cells in these mice. For example, IA-reactive T-cells may not normally have access to islet antigen, or *in vivo* the activation of IA-reactive T-cells may be inhibited by T-suppressor cells. Alternatively, pancreatic β-cells may be unable to provide co-stimulatory signals required for activation of IA-reactive lymphocytes. In support of the latter hypothesis, the transgenic expression of the cytokine, interferon-γ in pancreatic β-cells has been shown to cause the loss of tolerance to β-cell antigens, as assessed by an *in vitro* assay, and the transgenic mice were shown to develop diabetes by an immune-mediated mechanism (Sarvetnick *et al.* 1990). It is known that interferon-γ has co-stimulatory activity *in vitro* allowing T-cell activation. The production of interferon-γ within the pancreatic islet may result in the activation of T-cells reactive to islet antigen, leading to auto-immunity and diabetes. In IA-transgenic mice, the failure of β-cells to secrete co-stimulatory signals that are necessary for T-cell activation, including interferon-γ, may allow the survival of β-cells, despite the presence of T-cells that are reactive to the transgenic antigen. From these observations one can envisage that a physiological event that leads to a high local cytokine release, such as an inflammatory reaction to an infection, might also, under certain circumstances, lead to an auto-immune response to self-antigen.

10.2.3 Idiotypic regulation

The relative abundance of each T- or B-cell clone is very low, probably too low to induce immunological tolerance to the respective antigen receptors. The polymorphic regions on the antigen receptors of T- and B-lymphocytes can be considered to be antigenic and the immune system may become responsive to the variable domains on the receptors following expansion of specific clones during an immune response. Antigenic determinants on the antigen-binding site of B- or T-cell receptors are termed 'idiotypes', and T- or B-cells bearing receptors specific for the idiotype are called anti-idiotypic. Jerne (1974) has suggested that idiotype–anti-idiotype interactions between antigen receptors may form a network of connections that is central to the regulation of immune responses (Fig. 1). Thus, proliferation

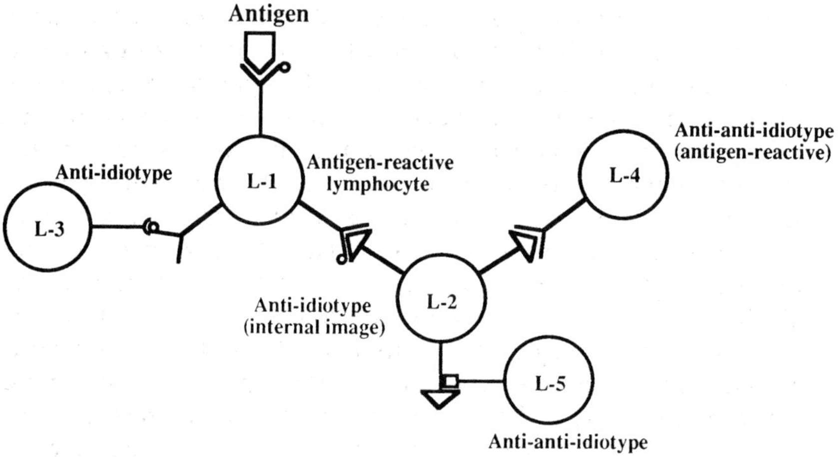

Fig. 1 Idiotypic networks and immune regulation. Antigen-reactive lymphocytes (L-1) bearing receptors for antigen may be functionally connected to other lymphocytes through receptor–receptor interactions. Anti-idiotypic lymphocytes (L-2, L-3) recognize distinct determinants on the antigen receptor; L-2 has an idiotype with an internal image of antigen and binds to the region on L-1 involved in antigen binding. Anti-idiotypic lymphocytes themselves interact with other sets of lymphocytes (L-4, L-5) to extend the network. Idiotype–anti-idiotype interactions are thought to be involved in immunoregulation, including the prevention of responses to self-antigens (see text)

of lymphocytes responsive to antigen may be accompanied by the stimulation of anti-idiotypic lymphocytes which may act to regulate the immune response to antigen itself. Anti-idiotypic lymphocytes might consist of cells with receptors that bear an 'internal image' of the antigen (L-2 in Fig. 1) which could potentially stimulate a response. Additional sets of anti-idiotypic lymphocytes recognizing determinants distinct from those binding the epitope on the stimulating antigen (L-3 in Fig. 1) might inhibit the response. The idiotype–anti-idiotype network is thought to pre-exist the presence of antigen. Addition of antigen is envisaged as causing a perturbation of the network, overcoming the suppressive effects of anti-idiotypes, resulting in proliferation of antigen-reactive lymphocytes.

Elements of immune networks have been shown to exist for a number of antigens. For example, immunization of mice with insulin was shown to induce at least two antibody specificities: antibodies to insulin itself and antibodies that have the capability of binding to the insulin receptor and mimicking the biological effects of the insulin molecule (Schechter *et al.* 1984). The latter antibodies were shown to be anti-idiotypic to insulin antibodies and probably possess an 'internal image' of insulin that structurally resembles the insulin receptor binding site on the insulin molecule. When the experiments were performed with an insulin molecule that had been modified at residues that interact with the insulin receptor, then immunization no longer induced anti-idiotypic antibodies that had insulin receptor binding activity (Schechter *et al.* 1984).

Evidence that idiotypic networks can have a regulatory function comes from observations that exogenous administration of an anti-idiotype *in vivo* will either stimulate or suppress lymphocytes that share the relevant idiotypes. However, most observations have been restricted to relatively oligoclonal responses. It is unclear to what extent idiotypic networks operate to regulate the polyclonal responses to foreign antigen. However, idiotypic networks may play a role in controlling immune responses to self-antigens. For example, induction of an anti-idiotypic response to autoreactive T-lymphocytes has been shown to be very effective at blocking auto-immune disease in a number of animal models (Cohen 1989). Furthermore, these networks may also regulate the activity of a distinct subset of B-lymphocytes that may play a role in preventing immune responses to self. B-lymphocytes secreting antibodies to self (natural antibodies) can readily be demonstrated in normal healthy individuals but these are clearly not pathogenic. The autoreactive lymphocytes frequently express the cell-surface marker, CD5, and secrete antibodies that are 'polyreactive', having the capability to bind to many structurally unrelated self-antigens and also to many foreign antigens. These antibodies also display a high degree of idiotypic connectivity, i.e. the antibodies will frequently react with other natural antibodies produced by the same individual. The frequency of B-lymphocytes that secrete natural antibodies is very high in the neonate and these lymphocytes display a high degree of connectivity, implying that idiotypic networks, perhaps stimulated by self-antigen, are important in the development of the immune system (Coutinho *et al.* 1989). These natural antibody-secreting B-lymphocytes persist into adulthood, albeit at a lower frequency, and presumably have some physiological function. Cohen and Cooke (1986) have suggested that the function of these antibodies is to prevent an immune response to epitopes on foreign antigens that cross-react with structures on self-molecules and that could lead to an auto-immune response. The probability that at least one of the many potential epitopes on an invading organism resembles a structure on a self-molecule is likely to be high. The self-perpetuating secretion of non-pathogenic auto-antibodies that bind such self-mimicking structures, and block immune responses to those regions, may provide a means to focus the immune system on those epitopes on foreign antigens that would not result in a cross-reactive, potentially pathogenic, auto-immune response.

10.2.4 Suppression

Non-responsiveness to antigen can also be achieved through the active suppression of immune responsiveness mediated by a subset of T-cells. Immune responses to β-cell-specific antigens may be regulated by suppression. For example, Jensen *et al.* (1984) have shown that some mouse strains that are normally non-responsive to porcine insulin can become responsive upon depletion of irradiation-sensitive, insulin-specific suppressor T-cells.

The precise mechanisms by which suppression operates are poorly understood due to difficulties both in the cloning of suppressor T-cells and in the isolation of soluble factors involved in intracellular communication between cells that mediate immune suppression. Suppression is thought to be initiated by a subset of CD4-

positive T-cells (suppressor-inducers). Effector cells are CD8-positive T-cells that are particularly sensitive to X-ray irradiation or cyclophosphamide treatment. The nature of the target structure for suppressor T-cells is not clear. Some suppressor cells (anti-idiotypic T-cells) may recognize idiotypic determinants on peptide fragments of T-cell receptors expressed in association with MHC molecules by the antigen-specific T-cell targets of suppression. Other suppressor cells may be antigen non-specific and recognize the state of activation of the T-cell (Cohen 1989), perhaps via a cell-surface activation marker.

Two model systems, the autologous mixed lymphocyte reaction (AMLR) in man and the syngeneic mixed lymphocyte reaction (SMLR) in mouse, have provided a means to study immunosuppression *in vitro*. In these reactions, subsets of CD4-positive T-cells respond to self MHC class II molecules expressed by non-T lymphoid cells (Weksler *et al.* 1981). Amongst those cells responding are those that are capable of mediating suppression of both T- and B-cell reactions *in vitro*. These suppressor–inducer cells may be identified as a distinct subset of T-cells with monoclonal antibodies recognizing the surface antigen 2H4.

The AMLR has been shown to be defective in a number of human auto-immune diseases, including systemic lupus erythematosus (SLE) and multiple sclerosis. Likewise, the SMLR is deficient in a number of auto-immune-prone mouse strains including NZB, the F_1 generation NZB × NZW and MRL/lpr (Weksler *et al.* 1981). Although the AMLR and SMLR have to be regarded as *in vitro* phenomena, they may provide indicators of the ability to generate suppressor T-cells. The low responses in a number of auto-immune states may therefore be indicative of suppressor T-cell defects that may well contribute to disease pathogenesis.

10.3 Animal models

10.3.1 The NOD mouse

The non-obese diabetic (NOD) mouse was derived from a cataract-developing substrain (CTS) of outbred ICR/JC1 mice by selective breeding. CTS sublines with differences in fasting blood glucose were established and, from these, the diabetes-susceptible NOD mouse strain and a non-diabetic non-obese normal (NON) strain were developed. Many colonies of NOD mice are now established which differ in diabetes incidence and time of onset. A typical incidence of diabetes in the NOD mouse is 70 per cent for female mice and 15 per cent for males, the disease developing between 80 and 200 days of age for most mice. The sex difference is related to effects of sex hormones; castration increases the incidence of diabetes in males and reduces that in females, whereas testosterones prevents diabetes and oestradiol increases incidence in castrated animals. Environmental factors, including diet, can influence the rate of development of the disease and there are also substrain differences in disease incidence. This is most clearly illustrated by the low incidence of diabetes in the NOD/Wehi colony compared with the NOD/Lt strain housed in the same unit (Baxter *et al.* 1989).

The NOD mouse shows many characteristics that are typical of type I diabetes in humans (Lampeter *et al.* 1989). The disease is polygenic, and at least one of the susceptibility genes maps to the MHC. The NOD mouse expresses a unique IA MHC molecule and no IE. Insulitis is present in both males and females to a similar degree, despite the differences in disease incidence, starting with a peri-insulitis at 6–8 weeks, progressing to infiltration of the islet and β-cell destruction (Fig. 2). The predominant infiltrating cells are monocytes, B-lymphocytes, and CD4-positive T-cells, with lower numbers of CD8-positive T-lymphocytes present. No evidence of immunoglobulin deposition on islets has been reported, although insulin antibodies and antibodies reacting with cell-surface antigens of islets have been detected in the serum of most animals. In addition to insulitis, NOD mice frequently have lymphocytic infiltration in salivary, thyroid, and adrenal glands, indicating that immune dysregulation is not restricted to responses directed at pancreatic β-cell antigens.

The NOD mouse is characterized by a high proportion of T-lymphocytes in lymphoid organs. The importance of T-cells in mediating the disease has been illustrated by the prevention of diabetes by thymectomy, by *in vivo* treatment with anti-T-cell antibodies, or by cyclosporin treatment. Splenocytes or purified spleen T-cells from diabetic donors are able to transfer disease to irradiated recipients. Both CD4- and CD8-positive T-cells are required for transfer of insulitis and diabetes and both T-cell subsets must be from diabetic animals. T-cells from young, non-diabetic donors are ineffective, which suggests that both T-cell subsets need to be primed with islet antigen (Miller *et al.* 1988).

Attempts have been made to characterize T-cells that might be important in disease pathogenesis in this animal model, generating T-cell lines by stimulation with islet antigen. Haskins *et al.* (1989) have produced a panel of CD4-positive T-cell clones that respond specifically to islet antigen in association with MHC molecules of the NOD haplotype. Co-transplantation of T-cell clones and NOD islets into a (CBA × NOD) F_1 recipient caused destruction of the islet graft, whereas pituitary grafts (acting as controls) were not affected when transplanted with the same T-cell clones, providing evidence that these T-cells might be disease related. However, attempts to transfer diabetes by injecting the T-cell clones into (CBA × NOD) F_1 mice were not successful, although histological examination did reveal periductal and perivascular infiltration of lymphocytes in the pancreata of most mice, which might be regarded as the first signs of insulitis.

Reich *et al.* (1989*a*) have produced both CD4- and CD8-positive clonal T-cell lines by stimulating lymphocytes from islets of recently diabetic NOD mice. In disease transfer studies, injection of a mixture of CD4- and CD8-positive clones into irradiated NOD or (NOD × Balb/c) F_1 mice produced intense insulitis, whereas injection of single clones were ineffective. These results provide further evidence that primed, islet-specific T-cells from both CD4 and CD8 subsets are required for disease transfer. Further characterization of such islet-specific T-cell clones, particularly with regard to antigen specificity and T-cell receptor usage, should provide important information on the mechanism of β-cell destruction in the NOD mouse model.

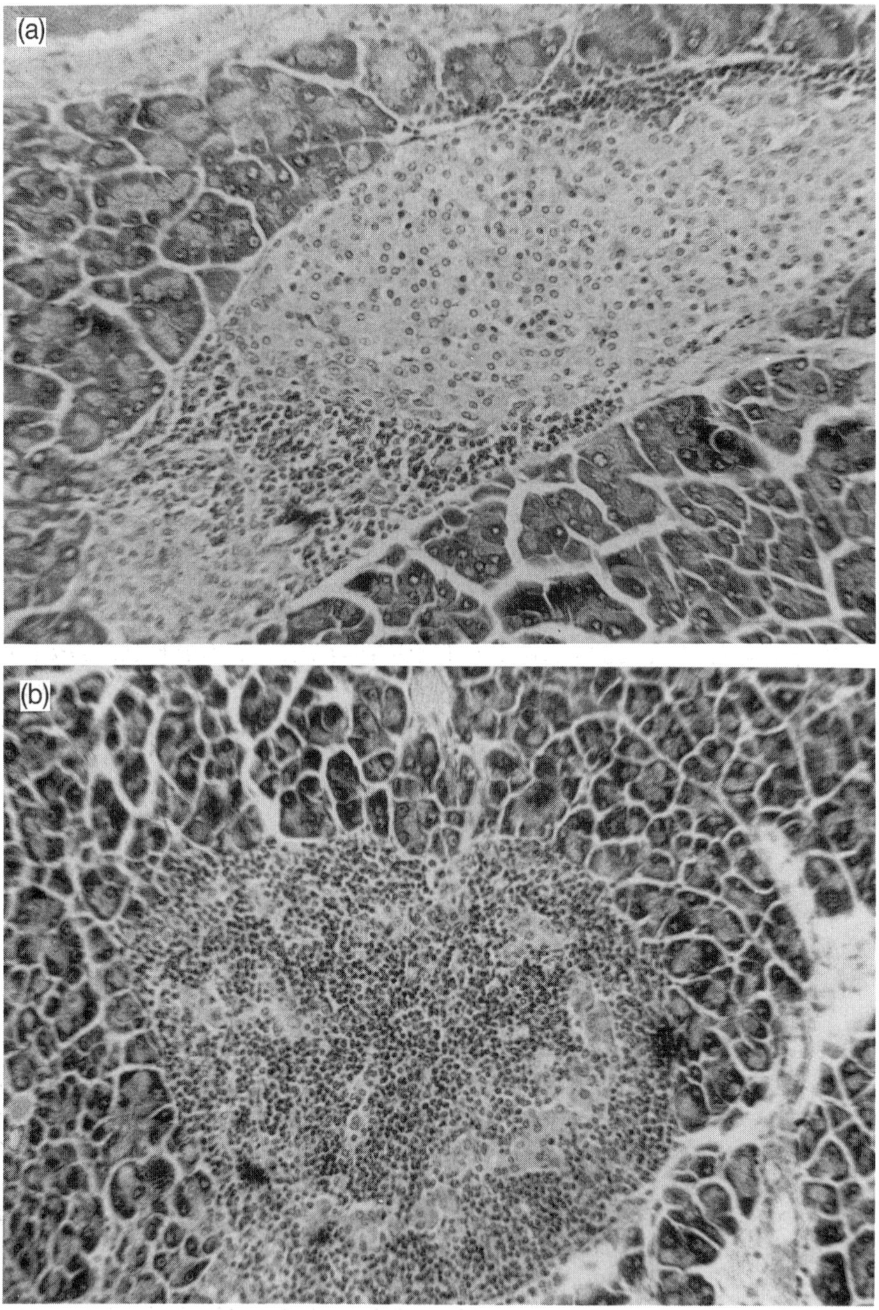

Fig. 2 Progressive insulitis in the NOD mouse. (a) Early lymphocytic infiltration of islets in the NOD mouse is characterized by an accumulation of mononuclear cells around still-intact islets (large islet) followed by infiltration into the islet interior (small islet at lower left of figure). (b) Late stage insulitis with heavy lymphocytic infiltration and extensive β-cell damage. (Photographs kindly provided by Drs Danny Zipris and Terry Delovitch, University of Toronto, Canada)

In spite of sex and inter-colony differences in the incidence of diabetes, most NOD mice, including non-diabetic animals, have evidence of lymphocytic infiltration into islets. Virtually all animals of this strain may have immune abnormalities that cause an auto-immune response against β-cell antigens, but in a proportion of these the auto-aggressive reaction does not achieve sufficient severity to reduce β-cell function to a level where metabolic abnormalities are apparent. Harada and Makino (1984) reported that administration of the immunosuppressive agent, cyclophosphamide, accelerated progression to diabetes in NOD mice but was without effect in normal strains of mice. Cyclophosphamide is also effective at promoting diabetes in NOD colonies with a low disease incidence (Charlton *et al.* 1989). The agent is known to cause a significant depletion of lymphoid cells, and a temporary suppression of immune responses. However, effector cells recover function a week after treatment whereas cells with suppressor function recover more slowly. This means that, depending on the dose and time of administration, cyclophosphamide can enhance immune responses due to reduced suppression. Cyclophosphamide treatment of NOD mice may effectively deplete a suppressor cell activity that may be acting, particularly in colonies with a low incidence of diabetes, to down-regulate auto-aggressive immune responses against the β-cell.

Serreze and Leiter (1988) have shown that SMLR may be abnormal in NOD mice. The SMLR is a CD4-positive T-cell response to self-MHC antigens, and T-cells responding may participate in a number of regulatory functions, including suppression. NOD T-cells generated during an SMLR were found to be defective in inducing suppression of T-cell responses to foreign MHC molecules. The reduced SMLR in NOD mice may point to low suppressor cell activity in this mouse strain, as found for many other auto-immune-prone mice. The low SMLR in NOD mice was accompanied by a very low production of interleukin-2; addition of exogenous interleukin-2 restored the ability of NOD T-cells to respond in an SMLR and to generate suppressor activity. The results suggest that there is a defect in the activation of suppressor T-cells in NOD mice that may be related to deficiencies in cytokine secretion.

In addition to the deficiency in interleukin-2 secretion observed in the SMLR, NOD mice have been shown to be defective in the secretion of other cytokines, including interleukin-1 and tumour necrosis factor. Administration of tumour necrosis factor-α and, in one report, interleukin-1 to NOD mice from 4–8 weeks of age has been shown to reduce the severity of lymphocytic infiltration in the pancreatic islet and to prevent or delay diabetes onset (Satoh *et al.* 1989; Jacob *et al.* 1990). In addition, the administration early in life of agents such as Freund's adjuvant or a streptococcal preparation, OK-432, which stimulate the secretion of a variety of cytokines, has also been shown to prevent the onset of diabetes in the NOD mouse (Sadelain *et al.* 1990; Shintani *et al.* 1990). Preliminary studies suggest that this treatment suppresses the generation of effector cells, perhaps by stimulating a population of T-cells with suppressor activity that are antigen non-specific (Sadelain *et al.* 1990).

Analysis of cloned T-cell lines originating from infiltrated islets of diabetic NOD mice have shown that, in addition to T-cells with specificity for islet antigen, T-cells that are autoreactive to NOD MHC are present in the inflamed islet. Injection of these autoreactive T-cells into NOD mice inhibited insulitis and diabetes development (Reich *et al.* 1989*b*). Thus, cells capable of mediating suppressive effects on the effector cells responsible for β-cell destruction appear to be present in the insulitis of the diabetic NOD mouse. These T-cells appear to be reactive to NOD MHC and may perhaps be equivalent to those that respond in an SMLR. Progression to diabetes in the NOD mouse model may therefore depend on the relative activity of suppressor and effector T-lymphocytes.

10.3.2 The BB rat

The BB rat is another animal strain that develops spontaneous diabetes with features similar to human type I diabetes (Prud'homme *et al.* 1985). Onset of the disease occurs in both male and female animals between 60 and 150 days of age, and the animal is also prone to thyroiditis. As in the NOD mouse, diabetes in the BB rat is associated with lymphocytic infiltration into the pancreatic islet and the presence of antibodies reacting with islet cell components. An important role for T-cells in disease pathogenesis is indicated by the prevention of disease by neonatal thymectomy, by treatment with anti-lymphocyte serum, and by the administration of cyclosporin. The disease can be transferred to non-diabetic animals by splenocytes from diabetic rats which have been activated by concanavalin A.

Studies with monoclonal antibodies to mononuclear cell subsets found within the inflamed islet of the prediabetic BB rat suggested that the earliest cell type infiltrating the islet is the macrophage (Hanenberg *et al.* 1989). Macrophage infiltration is accompanied by hyperexpression of MHC class I molecules on endocrine cells. T-cells and natural killer (NK) cells were observed at a later stage, and subsequently large numbers of B-lymphocytes appeared. Macrophages were the predominant cell type at all stages of infiltration and evidence that this cell type is critical for diabetes onset was provided by the observation that depletion of macrophages by silica administration prevented insulitis (including T- and B-cell infiltration) and diabetes. Depletion of T-cells and NK-cells with specific antibodies also prevents diabetes (Like *et al.* 1986). Thus, β-cell destruction in the BB rat appears to involve a collaboration between macrophages, T-cells, and NK-cells. The final effector cell is not known, although in *in vitro* experiments macrophages, NK-cells, and T-lymphocytes can all induce cytotoxic effects on isolated pancreatic β-cells.

The BB rat has severe abnormalities in immune regulation (Prud'homme *et al.* 1985). *In vitro*, BB rat splenocytes show low proliferative responses to concanavalin A and produce low levels of interleukin-2. T-cells respond poorly in allogeneic and syngeneic mixed lymphocyte reactions. *In vivo*, rejection of skin grafts is poor and the animals are particularly vulnerable to infections. Both CD4- and CD8-positive T-cells are severely depleted in the periphery (lymphopenic). Bone marrow transfer studies and thymus grafting experiments suggest that there is a bone marrow abnormality, rather than a thymic defect, that causes the abnormal T-cell activity

(Francfort *et al.* 1985). A subset of mature lymphocytes that express the T-cell marker, RT6, is absent from the diabetes-prone BB rat. Experiments from rat strains that have a low incidence of diabetes (<1 per cent) have suggested that RT6-expressing cells may be important in the regulation of autoreactive lymphocytes. Diabetes-resistant (DR) BB rat lines were derived from the diabetes-prone BB rat but bred for resistance to the disease. DR BB rats are non-lymphopenic and have normal numbers of RT6-positive T-cells. However, the specific depletion of the RT6-positive subset by administration of monoclonal antibody induces diabetes in more than 50 per cent of rats, suggesting that this subset includes regulatory cells that are acting to suppress effector cells involved in auto-immune β-cell destruction.

10.4 β-Cell destruction in human type I diabetes

10.4.1 Humoral and cellular immune abnormalities

Over the past few years evidence has been accumulating that an aberrant immune response to pancreatic β-cells may be directly responsible for the loss of β-cell function that causes type I diabetes. Cellular hypersensitivity to islet antigens was demonstrated in a lymphocyte migration test (Nerup *et al.* 1974). Lymphocytes from diabetic patients have been shown to be cytotoxic to islets or insulinoma cells *in vitro* (Charles *et al.* 1982) and abnormalities in T-cell subsets have been described (reviewed by Drell and Notkins 1987). In addition to cellular immune abnormalities, antibodies to islet cell components are a common feature of the disease. Bottazzo *et al.* (1974) and MacCuish *et al.* (1974) first described islet cell antibodies (ICA) in sera from patients with insulin-dependent diabetes and polyendocrine disease, by indirect immunofluorescence on frozen sections of human pancreatic tissue. Subsequent studies have demonstrated a high prevalence of ICA in newly diagnosed diabetic patients. Furthermore, ICA have been detected in individuals several years before diabetes development, suggesting that despite the acute onset of clinical symptoms, type I diabetes may be preceded by a prolonged period of auto-immunity to pancreatic β-cells. Further analyses of humoral immunity to islet cell components have shown that other antibody activities, distinct from ICA, are present, which recognize β-cell antigens, such as insulin and an islet cell protein of M_r 64 000, the 64K antigen. These antibody activities also precede diabetes onset by several years (Baekkeskov *et al.* 1987; Castano and Eisenbarth 1990). Thus, the presence of immune abnormalities, particularly humoral abnormalities such as ICA, may be useful in the prediction and diagnosis of the disease. Like the disappearance of lymphocytic infiltration after the onset of diabetes, ICA titres decline during the first 1–3 years after disease onset, concomitant with the loss of β-cell function, although ICA may persist in patients with polyendrocrine disorders.

In addition to antibodies to islet cell antigens, a proportion of diabetic patients possess antibodies to antigens that are not specific to islet cells (Drell and Notkins 1987). These include ubiquitous components such as DNA, RNA, tubulin, and actin, and also antigens that are expressed specifically in non-pancreatic tissues, such as the thyroid, adrenal medulla or gastric parietal cells. Some of these immune abnormalities may be epiphenomena; for example, antibodies might be produced as a response to antigens released from damaged islet cells. Additionally, antibody production might be enhanced by the secretion of helper factors during the primary response to a specific target antigen. However, responses to thyroid, adrenal medulla or gastric parietal cell antigens are unlikely to be generated by such mechanisms. These antigens are associated with other auto-immune diseases, such as auto-immune thyroiditis, Graves disease, and Addison's disease, and patients with type I diabetes have a higher risk of developing these disorders than the general population (Nerup and Lernmark 1981). ICA and insulin antibodies are also sometimes found in patients with other auto-immune disorders. Thus, a distinct set of self-antigens, including insulin, thyroglobulin, and thyroid microsomal antigen, appear to be particularly prone to stimulating an auto-immune response. Dysregulation of responses to one or more of these antigens is a general feature of auto-immune disease.

Histological examination of pancreata from young diabetic patients of recent onset provided early evidence that immunological abnormalities might contribute to the severe deficiency of insulin secretion that causes type I diabetes (Gepts 1965). These studies demonstrated that a high proportion (68 per cent) of acute onset juvenile diabetic patients had inflammatory infiltrates in or around the pancreatic islet. Subsequent analyses of pancreata from patients with a duration of diabetes ranging from a few days to many years (Gepts and De Mey 1978) suggested that lymphocytic infiltration was short lived, being observable in recent onset cases but only rarely in patients having diabetes for more than six months. Immunocytochemical staining with antibodies to pancreatic hormones showed a progressive and specific loss of the pancreatic β-cells in the disease. Islets from patients with long-term diabetes were frequently devoid of β-cells, but cells expressing glucagon, somatostatin, or pancreatic polypeptide survived. The finding that lymphocytic infiltration was only found in pancreata containing β-cells suggested that the lymphocytic infiltration represents an immune reaction directed specifically against the pancreatic β-cells.

Since lymphocytic infiltration of the pancreatic islet in type I diabetes may be directly responsible for the disease, characterization of the lymphocytes invading the islet might provide clues to the mechanisms whereby the β-cells are destroyed. A detailed examination of the pancreatic islet during the destructive process has only been possible in a few rare cases. Sibley *et al.* (1985) analysed pancreatic graft biopsies in four patients with long-term (17–27 years) type I diabetes who had received pancreatic grafts from HLA-identical twins or siblings. All four patients recovered normal glucose metabolism after transplantation, but impaired graft function developed after 6–12 weeks and three of the grafts failed after one year, probably as a result of disease recurrence. Immunohistochemical analysis of graft biopsies demonstrated a predominanace of T-cells expressing CD8 antigen (cytotoxic/suppressor phenotype). Lower numbers of CD4-positive T-cells (helper phenotype) and monocytes were present. B-lymphocytes were not detected in the insulitis, and islets did not react with antibodies to immunoglobulin or complement. In only one case, did islet cell antibodies appear in the serum after transplantation. These results strongly suggest a role for T-cell-mediated β-cell destruction in the recurrence of the disease after pancreas grafting.

Bottazzo *et al.* (1985) performed an immunohistochemical analysis of the pancreas of a 12-year-old girl who died within 24 h of diabetes diagnosis. The majority of cells infiltrating the islet were CD8-positive T-cells but other T-cell subpopulations were also present. Increased expression of MHC class I antigens was observed on islet cells, which may increase their susceptibility to cell-mediated cytotoxicity. Some cells were positive for both insulin and HLA class II antigens, raising the possibility that β-cells may have the potential for presenting antigen to T-helper cells. The capillary epithelium was strongly positive for MHC class II antigens, and these cells may represent a major site of antigen presentation to T-helper cells. A few scattered B-lymphocytes were present and some islet cells reacted with antibodies to immunoglobulin and complement. The patient was positive for ICA. This study thus provides evidence for both antibody- and cell-mediated immunity in β-cell destruction.

Functional characterization of peripheral blood T-lymphocytes from diabetic patients, and comparison with the normal situation, has yielded inconsistent results about the total number of circulating T-cells or the relative numbers of T-cell subsets (Drell and Notkins 1987). However, consistent increases in the proportion of activated T-cells, as indicated by increased expression of HLA-DR antigens or interleukin-2 receptors, have been observed. There is also evidence of decreased suppressor-cell activity that may be analogous to the suppressor-cell defects seen in animal models of diabetes and other human or animal auto-immune states. Thus, T-cells from diabetic patients respond poorly in AMLR and interleukin-2 restores the proliferative response (Chandy *et al.* 1984). Interleukin-2 secretion was shown to be abnormal from T-cells of diabetic patients (Zier *et al.* 1984). Cytokine deficiencies may thus be a common feature in human and animal auto-immune disease, and may be a direct cause of defective immunoregulation responsible for disease.

10.4.2 Action of cytokines on islet cells

As well as being of importance in immunoregulation, cytokines are known to have effects on non-lymphoid cells. Some cytokines have cytotoxic and anti-viral activity. Since secretion of cytokines is high during an active immune response, such as that occurring during inflammation, the infiltration of mononuclear cells into the pancreatic islet in type I diabetes is likely to be accompanied by a high local concentration of these molecules. There has, therefore, been considerable interest in investigating the direct effects of cytokines on β-cells *in vitro*.

In vitro experiments have demonstrated the importance of cytokines in the regulation of the immune response against the pancreatic β-cell. Interferon-γ and tumour necrosis factor-α have been shown to upregulate the expression of MHC class I molecules on β-cells and to induce the expression of class II molecules. The latter are normally not expressed by pancreatic β-cells (Campbell *et al.* 1985; Pujol-Borrell *et al.* 1987) and their presence would facilitate the presentation of β-cell antigen to T-cells. If accompanied by co-stimulatory factors, it could also result in the activation of autoreactive T-cells and auto-immunity. Whether MHC class II molecules are induced on pancreatic β-cells *in vivo* is controversial. Double-staining experiments have demonstrated some cells in the pancreas of type I diabetic patients that coexpress insulin and MHC class II molecules (Bottazzo *et al.* 1985). However, electron microscopic examination of MHC class-II-expressing cells from pancreatic islets have detected phagocytic cells which contain the remnants of insulin secretory granules, presumably due to uptake of damaged β-cells (Pipeleers *et al.* 1987). Coexpression of insulin and MHC class II molecules is thus not sufficient to demonstrate class II expression by the endocrine cells. Several studies have shown an increase in MHC class I molecules on β-cells in the diabetic pancreas, and cytokine regulation of MHC class I antigen expression may be relevant to these findings. Interferon-γ and tumour necrosis factor-α have also been shown to induce the expression of intercellular adhesion molecule-1 (ICAM-1) on pancreatic β-cells (Campbell *et al.* 1989). Adhesion molecules are important in cell–cell interactions in immune responses (Springer 1990); ICAM-1 is an adhesion receptor for lymphocyte function-associated antigen 1 (LFA-1) expressed by T-cells. By inducing the expression of both HLA class I molecules and ICAM-1 by β-cells, cytokines such as interferon-γ and tumour necrosis factor may facilitate the adhesion of antigen-specific T-cells, increasing their susceptibility to immune-mediated destruction.

A number of studies have demonstrated that cytokines can have direct cytotoxic effects on pancreatic β-cells. Mandrup-Poulsen and co-workers have produced evidence for an effect of interleukin-1 (IL-1) on islet cell function. Low concentrations of IL-1 were shown to increase insulin release and insulin synthesis by isolated islets, whereas prolonged exposure to the cytokine inhibits β-cell function, due to its cytotoxic action (Spinas *et al.* 1986). This cytotoxicity is apparently selective for β-cells and is potentiated by tumour necrosis factor (Mandrup-Poulsen *et al.* 1987*a*). Pukel *et al.* (1988) showed a dose-dependent cytotoxic action of interferon-γ on monolayer cultures of rat pancreatic islet cells. Interleukin-1, tumour necrosis factor and lymphotoxin were inactive on their own in this system, but in combination produced synergistic cytotoxic effects.

The mechanism for the cytotoxic action of cytokines is unknown. A number of agents, however, have been shown to inhibit cytokine-induced cytotoxicity. For example, glucose confers protection against interleukin-1-mediated damage, suggesting that the metabolic state of the β-cell might influence its susceptibility to cytokines (Mandrup-Poulsen *et al.* 1987*b*). In addition, the oxygen free-radical scavengers, dimethylthiourea and citiolone, were found to protect against islet cell damage by cytokines (Sumoski *et al.* 1989). Cytokines have been shown to stimulate free-radical production in macrophages, primarily through increased arachidonic acid metabolism (Nathan and Tsunawaki 1986), and similar pathways may operate in β-cells. Pancreatic β-cells appear to be particularly sensitive to oxygen free-radical production, due to a low activity of endogenous free-radical scavengers (Malaisse *et al.* 1981). Alloxan, an agent which generates reactive oxygen radicals, also has a cytotoxic action which is relatively specific to pancreatic β-cells. Like cytokine toxicity, this action is inhibited by glucose and free-radical scavengers. Glucose may confer protection against cytotoxicity by alloxan and IL-1 because its metabolism produces reducing equivalents that might neutralize the oxidizing effect of the free radicals.

These *in vitro* studies suggest an alternative mechanism for β-cell damage to the classical view of cell- and antibody-mediated immune destruction. However, *in vivo*, experimental evidence for such a mechanism is lacking. Sutton *et al.* (1989) have investigated the susceptibility of islets to non-specific cytotoxicity by secreted products of an inflammatory response by transplanting a mixture of syngeneic and allogeneic islets beneath the kidney capsule of normal rats. In this system there was total destruction of allogeneic islets without any evidence of damage to syngeneic islets, in spite of the high local concentrations of cytokines which probably occur during the response to the allograft. Nevertheless, cytokine effects on islets are likely to contribute to islet destruction in type I diabetes either directly, or by the induction of accessory molecules important in the immune response.

10.4.3 Environmental factors

No more than 36 per cent of non-diabetic, identical twins of diabetic patients will subsequently develop diabetes (Olmos *et al.* 1988). Since these individuals presumably share diabetes-susceptibility genes, this finding has been taken as evidence that type I diabetes is not a purely genetic disease, and that environmental factors are also important in disease pathogenesis. Furthermore, in spite of evidence for a long period of auto-immunity to β-cells before disease onset, the clinical symptoms of diabetes appear rapidly and show seasonal variation in appearance. Thus, diabetes onset is most frequent in spring and autumn months and is often preceded by a flu-like illness. Viral infections may therefore precipitate the disease: either by stimulating an already ongoing auto-immune response, due to increased lymphokine secretion during the immune response to viral antigens; or by exerting additional stress on an already severely depleted mass of β-cells by increasing insulin requirements or inhibiting β-cell function. No specific environmental factor has been clearly identified as an agent participating in the auto-immune pathogenesis of diabetes. Nevertheless, the study of human type I diabetes and animal models of the disease has implicated a number of factors that might be important, including viral, bacterial, chemical, and dietary components.

There is considerable evidence from spontaneous animal models of diabetes that environmental factors may have beneficial effects on the diabetes-prone individual. Animals such as the NOD mouse or BB rat may be 'preprogrammed' for auto-immunity due to immunoregulatory defects. If these animals are bred in clean environments diabetes still develops, in some instances at a higher incidence than in the normal animal facility. Exposure to environmental pathogens at a critical stage of development may, to some extent, restore immunoregulation, perhaps by stimulating antigen non-specific suppressor cell activity. Thus, infection with certain viruses or treatment with immune stimulators, such as adjuvants, have clearly been shown to inhibit auto-immune effector cell activity and prevent diabetes in the NOD mouse (Oldstone 1988; Sadelain *et al.* 1990; Shintani *et al.* 1990).

Environmental agents may have a deleterious effect on β-cell function, as a result of direct cytotoxicity to these cells (Yoon *et al.* 1987). β-Cell damage may also result in the release of sequestered antigen from sites not normally exposed to the immune system, resulting in immune stimulation and auto-immunity. Auto-immune responses may be further enhanced during an inflammatory reaction to tissue damage or viral infection by the induced expression of MHC class II molecules on β-cells, allowing increased presentation of self-antigens. A number of chemicals are known to have relatively specific cytotoxic effects on β-cells, including alloxan, streptozotocin and the rodenticide, Vacor. Viruses such as Coxsackie B and encephalomyocarditis virus can infect pancreatic β-cells and induce diabetes in animals (Yoon *et al.* 1987). In humans, pancreatic Coxsackie B4 infection has been associated with diabetes onset and virus isolated from such patients can induce diabetes on inoculation of mice. There is no evidence that Coxsackie infection induces an auto-immune reaction to the β-cell. Rather, the virus may act as a causative or precipitating factor by the direct killing of the β-cells. However, there is evidence that β-cell damage can be associated with auto-immunity since multiple low doses of the β-cell toxin, streptozotocin, can induce an auto-immune form of diabetes in certain strains of mice and rats, and Vacor poisoning in humans has been reported to be accompanied by the presence of auto-antibodies to islet cells (Yoon *et al.* 1987).

Auto-immune responses may also be induced as a result of cross-reactivity of epitopes on foreign antigen with regions on self-molecules. For example, a major antigen associated with an animal model of rheumatoid arthritis, rat adjuvant arthritis, that is induced by injection of killed *Mycobacterium tuberculosis* in adjuvant, has been shown to be the 65 kDa mycobacterial heat-shock protein, hsp 65. T-lymphocytes reactive to hsp 65 from arthritic animals can transfer the disease to normal animals, indicating that the auto-antigen in adjuvant arthritis is a self-molecule, bearing a determinant that is cross-reactive to epitopes on the mycobacterial hsp 65. Heat-shock proteins are major targets of immune responses to mycobacterial infection but are also amongst the most highly conserved proteins in nature (Young and Elliott 1989). There is therefore considerable potential for cross-reactivity between foreign and self heat-shock proteins and the auto-antigen in adjuvant arthritis may well be rat hsp 65. The relevance of the observations to human disease is demonstrated by the finding that T-lymphocytes isolated from synovial fluid of patients with rheumatoid arthritis show reactivity to both mycobacterial and human heat-shock proteins.

Interestingly, there is evidence that hsp 65 may also play a role in diabetes development in the NOD mouse (Elias *et al.* 1990). Both antibodies and T-cells that recognize hsp 65 of *M. tuberculosis* appear before the onset of spontaneous diabetes in the NOD mouse, and T-cells reactive to the heat-shock protein can transfer insulitis and diabetes to young, non-diabetic NOD mice. Furthermore, immunization of NOD mice with hsp 65 in the presence of adjuvant induced early diabetes, whereas administration of the protein in a form that induces tolerance prevented diabetes. The implication from these experiments is that a protein that is cross-reactive with hsp 65 of *M. tuberculosis* might also be an auto-antigen in NOD mice.

Other components of micro-organisms have been shown to be powerful stimulators of T-cells. These proteins, which have been called superantigens, include enterotoxins such as staphylococcal enterotoxin B. Superantigens stimulate a large proportion of the T-cell population, and the T-cells responding often share particular V_β elements of their receptors. These observations suggest that the superantigen may bind directly to the V_β domains and stimulate T-cell responses. It is possible that, with such a large proportion of the T-cell repertoire responding, the activated T-cells include those that are autoreactive (White *et al.* 1989). Bacterial infection and enterotoxin production is indeed associated with certain auto-immune diseases, such as Reiter's syndrome and ankylosing spondylitis.

Diet is an additional factor that might be important in disease pathogenesis. Some dietary components may directly influence β-cell function; for example, nitrosamines may be directly cytotoxic to β-cells. Alternatively, dietary factors may influence immune function. In the BB rat, replacing the normal chow diet with a defined, semi-purified diet prevented diabetes, reduced the incidence of insulitis, increased thymus weight, and increased peripheral lymphocyte numbers (Scott *et al.* 1985). Similarly, a decrease in diabetes incidence was found in NOD mice fed on a defined diet (Coleman *et al.* 1990). Supplementing the defined diet with a natural ingredient mouse diet (OG 96) or brewer's yeast, a component of OG 96, restored the high incidence of diabetes in the colony. The diabetogenic component of OG 96 was found to be soluble in chloroform/methanol, suggesting that this might be a lipid component. Studies of dietary influences in humans are, of course, more difficult to interpret, but diet may well play a role in influencing disease onset in human type I diabetes as well.

10.5 Islet cell antibodies

The majority of patients with type I diabetes possess antibodies that bind specifically to islets on frozen pancreatic tissue. This assay for islet cell antibodies (ICA) is currently the most frequently used procedure for analysing anti-islet auto-immunity in diabetes. With current procedures, approximately 80 per cent of recent onset patients possess ICA, while less than 5 per cent of healthy controls are positive (Baekkeskov *et al.* 1986). The frequency of antibodies in control populations is usually higher than the incidence of disease (0.1–0.5 per cent) so the possession of these antibodies *per se* cannot be taken as completely predictive of disease development. It is possible that ICA-positive individuals who do not develop diabetes may have some degree of β-cell destruction which does not achieve the severity required for the appearance of abnormalities in glucose metabolism. Also, the nature of the ICA test gives the potential for binding to multiple islet cell components, some of which may be irrelevant to disease pathogenesis. Nevertheless, the presence of ICA does provide a very good marker for subsequent disease development in first-degree relatives of diabetic patients, particularly when antibody titre is taken into account (Bonifacio *et al.* 1990).

A number of procedures have been developed for the detection of antibodies to islet cell components (Lernmark 1987); indirect immunofluorescence on sections of frozen human pancreas is the method most frequently used. Rapidly frozen pancreas from blood group O donors are employed. Cut sections are dried on slides and incubated with serum. The sections are rinsed and a fluoresceinated second antibody is used as a detection agent. A fluorescent reaction which is specific to islets indicates the presence of ICA. ICA levels can be quantified by end-point titration.

Establishment of workshops for the inter-laboratory comparison of antibody assays revealed considerable inconsistency between laboratories in assessment of serum ICA levels. This has, to some extent, been resolved by serum exchange and standardization of the assay with control sera (Bonifacio *et al.* 1987). The quality of the pancreas substrate is likely to be an important determinant of sensitivity and reproducibility. Monkey or rat pancreas have been shown to also express islet cell antigens, and may be useful when human tissue is not available. Other variations in the assay technique have been developed. Islet cell antibodies that fix complement are often reported. In this assay, antibodies bound to the pancreas section are incubated in the presence of complement, activating the complement cascade of reactions. The complex is detected by a fluorescent antibody to complement component C3. The complement-fixing ICA assay is less sensitive as it tends to detect only those antibodies that are of high titre as measured by conventional procedures.

ICA display, typically, a cytoplasmic fluorescence on the pancreas section. The antibodies have been shown to react with all cell types within the islet, rather than specifically with β-cells, so it is not clear what role these antibodies have in disease pathogenesis. Antibodies reacting with the surface of isolated pancreatic islet cells (islet cell-surface antibodies, ICSA) are also detected in serum from diabetic patients. These antibodies may be detected by indirect immunofluorescence, radioligand binding (e.g. iodinated protein A), or in a cell ELISA (enzyme-linked immuno-absorption assay). Due to the problem of isolating viable human islet cells, isolated rat or mouse islets (or in some instances insulinoma cells) are generally employed as a source of tissue. It is then necessary to preabsorb the sera with rat or mouse tissue to remove any antibodies that react with rodent-specific, rather than islet-specific determinants. An additional cause of artefactual binding may be the frequent presence of circulating antibodies to bovine serum albumin. This protein is often included in islet preparation media and may bind to the surface of islet cells. The frequency of ICSA at onset of diabetes is reported to be 50–70 per cent with normal islet cells and 20–30 per cent with insulinoma tissue (Lernmark 1987). Approximately 2 per cent of control individuals are positive.

Analysis of the cell specificity of ICSA by cell sorting, electron microscopy, and immunocytochemistry has indicated that most antibodies in young, recent onset type I diabetic patients react specifically with β-cells (Winkel *et al.* 1982). In a few cases, antibodies were also present that reacted with purified A or PP cells. Older type I diabetic patients, non-insulin-dependent diabetics, and one ICSA-positive normal healthy control individual possessed antibodies that reacted exclusively to A or PP cells. Thus, there can be heterogeneity in the cell specificity (and therefore antigen specificity) of ICSA, although most young type I diabetic patients possess antibodies that are β-cell specific.

Antibodies from diabetic patients have been shown to have effects on pancreatic β-cell function *in vivo*. Incubation of islet cells with immunoglobulin fractions from diabetic patients have been shown to inhibit insulin secretion (Kanatsuna *et al.* 1983), although stimulatory effects have also been observed (Wilkinson *et al.* 1988*a*). Johnson *et al.* (1990) suggested that the inhibitory effect may be due to inhibition of a glucose transporter. Immunoglobulin fractions from diabetic patients were found to inhibit 3-*O*-methyl glucose uptake into islets, with high K_m transport being most affected. The inhibitory activity was removed by preincubation of sera with liver or islets but not with red blood cells. Since liver and islets both possess similar glucose transporters, this protein, or a protein associated with it, may represent one target for ICSA.

10.5.1 Gangliosides as targets of ICA

Identification of the antigens detected by islet cell antibodies could lead to a more specific and reproducible assay that is simpler to perform and quantify. Although the target antigens for ICA have yet to be identified, recent studies suggest that at least part of the reactivity may be directed against ganglioside antigens. Nayak *et al.* (1985) performed a number of chemical treatments of human pancreatic tissue to determine their effect on subsequent reactivity with antibodies in patients' sera. Treatment of the pancreas with the protease, pronase, did not affect its reactivity with sera from three diabetic patients, suggesting that the antigen recognized was not a protein. However, periodate treatment abolished reactivity, and antibody binding could be restored by borohydride reduction. Neuraminidase treatment or extraction of the pancreas with chloroform/methanol also prevented binding. Antigens on the section of pancreas recognized by a monoclonal antibody to gangliosides, showed a similar sensitivity to these treatments. Since gangliosides are known to be sensitive to periodate oxidation and neuraminidase, and may be solubilized by chloroform/methanol, the authors hypothesized that the antigen recognized by ICA may be a ganglioside.

To further test this possibility, ganglioside expression was analysed in pancreatic islet cells. Gangliosides were partially purified by chloroform/methanol extraction,

C18 cartridge chromatography and thin layer chromatography (Colman *et al.* 1988). The predominant gangliosides in human pancreatic islets migrated on TLC with a GM3 standard and as two bands with similar mobility to a GM2 standard. Total human pancreas expressed predominantly GM3 and GD3 gangliosides. Preincubation of diabetic patients' sera with a glycolipid fraction of human pancreas blocked the ability to bind to pancreas sections in an ICA assay. Further chracterization by TLC or HPLC indicated that the blocking activity was localized to a monosialoganglioside fraction with a mobility intermediate between GM2 and GM1. This blocking activity was not recovered from human brain or liver, providing indirect evidence that the antigen recognized by ICA is a monosialoganglioside which might show restricted tissue expression.

In another study, Gillard *et al.* (1989) analysed ganglioside expression in a pancreatic β-cell line, RIN m5F. This cell line had previously been shown to bind antibodics from diabetic patients in a cell ELISA. Subclones of the cell line which displayed strong antibody binding in this assay contained predominantly GM3, GD3, and GT3 gangliosides, whereas these gangliosides were reduced in subclones with low binding activity. Reactivity of purified gangliosides GM3, GD3, and GT3 to serum antibodies from diabetic patients or normal individuals was analysed by ELISA. Increased binding by diabetes-associated antibodies was only observed with GT3 (isolated from salmon eyes); high binding to brain GM2 was seen in both controls and diabetic patients, whereas binding to brain GM3 was low in the two groups.

This study shows that direct binding of antibodies from a proportion of diabetic patients to gangliosides can be demonstrated, although the ganglioside recognized is different from that suggested by the indirect approach of blocking ICA activity, described above. Type I diabetic patients may well have antibodies to different gangliosides, some of which may be islet cell or β-cell specific. Although the fluorescence seen in the ICA assay is typically cytoplasmic, gangliosides are known to be expressed on the surface of cells, and RIN cells employed in the study of Gillard *et al.* express antigens recognized by ICSA. Thus, gangliosides clearly have the potential to be targets of ICSA. Further characterization of glycolipid antigens, particularly in terms of their tissue-specific expression and subcellular localization is obviously important to determine their potential role in the pathogenesis of type I diabetes.

10.6 Immunity to insulin

J. W. SEMPLE, M. J. RAPOPORT, and T. L. DELOVITCH

Many diabetic patients develop circulating insulin auto-antibodies (IAAs) either at or before the clinical onset of diabetes, and the majority of patients develop insulin antibodies (IAs) upon initiation of insulin treatment. While it is not clear whether insulin is the primary auto-antigen that elicits auto-immune diabetes, insulin auto-antibodies have been proposed to be particularly useful in predicting the onset of disease (Castano and Eisenbarth 1990). Furthermore, the availability of insulin in a highly purified form has made the protein useful as a model antigen for studying immune responses to β-cell proteins.

10.6.1 Humoral immunity to insulin in animals

Animal models have yielded considerable information with respect to the immunogenicity of insulin. Insulin is a small protein of irregular tertiary structure having two interchain disulphide bonds (A7–B7 and A20–B19) and one intrachain disulphide bond (A6–A11) (Blundell *et al.* 1972). Amino-acid differences between species occur predominantly within the A-chain loop that is comprised of residues A6 to A11. Bovine (BI), ovine (SI), equine (EI), and porcine (PI) insulins differ in sequence only in their A-loop region, whereas human insulin (HI) differs not only in the A-loop but also at residue B30. These amino-acid polymorphisms elicit insulin antibodies with different specificities in various animal species.

Early studies on the immune response to insulin were performed in rabbits. Both highly purified (monocomponent) PI and HI exhibited very low antibody formation, and their ability to stimulate antibody production in rabbits was indistinguishable (Schernthaner *et al.* 1983). BI, on the other hand, induced a higher rabbit antibody response due to the larger number of amino-acid differences between rabbit and beef insulin.

Studies of immune responsiveness to insulin in different inbred guinea-pig strains demonstrated that the genetic control of IA formation was MHC-linked (Barcinski and Rosenthal 1977; Thomas *et al.* 1981). Strain 2 guinea-pigs respond to conformational determinants in the A-loop region (A8–A10) whereas strain 13 guinea-pigs respond to isolated B-chains, particularly in the B5–B19 region. These observations are consistent with the determinant selection hypothesis of immune responsiveness, i.e. the antigen-presenting cell selects the determinants to be presented to the T-cell in association with class II MHC molecules of a given haplotype.

Most of our knowledge of the humoral immunity to insulin has been obtained in mice. Keck (1975) originally demonstrated that inbred strains of mice could be classified as either genetic responders or non-responders to insulin. Mice of the H-2^k haplotype yielded no detectable antibody response to BI, whereas H-2^d mice produced a strong antibody response. Studies on the reactivity of a panel of 18 mouse monoclonal antibodies against BI and HI, implicated the presence of residues A4, the A-chain loop, B3, and B28 to B30 as immunodominant epitopes recognized by these antibodies (Schroer *et al.* 1983). These epitopes are in close proximity in the three-dimensional structure of insulin (Tainer *et al.* 1985).

10.6.2 Cellular immunity to insulin in animals

T-cell responsiveness to insulin in mice has been investigated by several laboratories. Jensen *et al.* (1984) demonstrated that T-cell reactivity to insulin was MHC-associated. T-helper (T_h) cells and B-cells from H-2^d, but not H-2^k mice, clearly respond to BI, PI, and HI. H-2^b mice are genetic low responders to PI, but PI-specific T_h-cells can be detected if their spleen cells are depleted of CD8-positive T-suppressor (T_s) cells. It was proposed that T_s-cells recognize A-chain-loop-associated epitopes of PI, and their dominant effect results in the low responder phenotype of these mice (Jensen *et al.* 1984). Whiteley *et al.* (1988) have recently demonstrated that HI transgenic mice which have circulating physiological concentrations of HI are tolerant to HI, whereas H-2 identical normal mice are not. This tolerance was shown to be due not to clonal deletion of T_h-cells but to a functional inhibition of T_h-cells. This supports the concept that clonal anergy to insulin can be induced by peripheral antigen-specific mechanisms.

10.6.3 Insulin processing and presentation by antigen-presenting cells

Studies on insulin processing and presentation of antigen fragments to helper T-cells by antigen-presenting-cells have shown that PI-specific murine T-cell clones recognize an antigen fragment consisting of the intact A-chain-loop disulphide linked to a portion of the B-chain (Naquet *et al.* 1987). Isolated A-chains or B-chains of insulin do not stimulate a T- or B-cell response *in vitro* or *in vivo*. These results indicate that the insulin fragment recognized by T-cells is a heterodimeric disulphide-linked peptide. This type of structure has not been found in T-cell epitopes of other soluble protein antigens. Competitive inhibition of binding, and presentation experiments using insulin and isolated A- and B-chains, have revealed that a heterodimeric insulin peptide bound to MHC class II is presented by an antigen-presenting cell to a T_h-cell via interaction of the B-chain residues with the MHC class II molecules and of the A-chain residues with the T-cell receptor (Delovitch *et al.* 1989).

The nature of the minimal insulin peptide(s) that binds to the groove of an MHC class II molecule and stimulates T_h-cells has not been identified (Delovitch *et al.* 1989). Insulin may have specific processing requirements for presentation. For example, the activity of insulin degrading enzyme (IDE), a neutral metalloendoproteinase, may be essential for the processing of insulin into immunogenic peptides by an antigen-presenting cell (Delovitch *et al.* 1989). However, since IDE-generated peptides cannot stimulate T_h-cells when presented on glutaraldehyde-fixed antigen-presenting cells, additional factor(s) appear necessary to further process insulin into a minimal peptide. Additional processing enzymes may be required to generate fragments that can be presented to helper T-cells. Alternatively, MHC class II molecules may influence the processing of intermediate insulin peptides bound to the antigen-binding site on the molecule. MHC-directed processing of insulin may explain, in part, the genetic control of immune responsiveness to insulins and may have a direct bearing on the HLA-linked control of IAA production in humans.

10.6.4 Humoral immunity to insulin in humans

Humoral immunity to insulin may occur both as a response to self-insulin, detectable before initiation of insulin therapy, or as a response to the therapeutic administration of exogenous insulin. These responses elicit insulin auto-antibodies and insulin antibodies, respectively. These two types of antibodies differ significantly in their immunopathogenesis and in the clinical problems that arise from their presence in diabetic and/or prediabetic patients.

Insulin auto-antibodies (IAAs) are present in about 40 per cent of newly diagnosed type I diabetic patients, up to 26 per cent in selected high-risk prediabetic populations, and have been detected in between 0.35 per cent and 8 per cent of the normal population (Wilkin 1990). Titres of IAAs are inversely correlated with age, and higher IAA titres in childhood diabetes correlate directly with the length of the subclinical phase that precedes disease onset (Karjalainen *et al.* 1989; Castano and Eisenbarth 1990). IAAs have also been found in several non-diabetic states, including insulin auto-immune syndrome, auto-immune thyroiditis, connective tissue diseases, drug-induced IAAs, mumps, and chicken pox (Wilkin 1990).

Analysis of the HLA haplotype association of IAAs indicated a positive association of IAA production and HLA-DR3 and/or DR4 (Atkinson *et al.* 1986), the same haplotypes that confer an increased risk for type I diabetes. This association was observed in first-degree relatives of diabetic patients and in IAA-positive normal controls, suggesting that IAAs are restricted to persons genetically susceptible to diabetes. In selected high-risk groups for diabetes, IAAs are found together with other diabetes-related auto-antibodies, including those against proinsulin, glucagon, and C-peptide, but mainly with islet cell antibodies (ICAs). Although the data regarding the latter association are highly variable, the majority of ICA-positive high-risk individuals are not IAA-positive.

The relatively low frequency of IAAs in diabetic populations limits the usefulness of IAA alone as a predictive marker for type I diabetes. In addition, the presence and titre of IAAs are not suitable indicators for the management of diabetes (Karjalainen *et al.* 1988; Sochett and Daneman 1989). IAA-positive patients neither differ clinically in metabolic control from IAA-negative patients, nor can IAAs predict the outcome of immune suppression therapy with cyclosporin (Mandrup-Poulsen *et al.* 1990). Despite the uncertainty concerning the value of IAA as a reliable serological marker for IDDM, the combined presence of both IAAs and ICAs in a IDDM high-risk individual enhances the predictive value of ICA alone (Wilkin 1990).

Recently, a different approach was reported that may increase the value of IAAs as a serological marker for IDDM. By using epitope-restricted IAAs instead of 'whole' IAA determinations, Wilkin *et al.* (1988*b*) demonstrated that they could improve the association between IAAs and diabetes. Therefore, increasing the specificity without reducing the sensitivity of detection of IAAs improved their diagnostic value. Nell *et al.* (1989) demonstrated that the IAA response in their diabetic patients was oligoclonal and restricted to two different insulin epitopes.

One epitope is human insulin B-chain specific, and the other is a cross-reactive epitope found on human and animal insulins that requires the presence of both A- and B-chain residues. It is not yet clear which of the different epitope-restricted IAAs is the most discriminating. In addition, there appears to be a linear relationship between first-phase insulin secretion, the level of IAA, and the duration to disease onset in ICA-positive first-degree relatives (Castano and Eisenbarth 1990). This finding shifts the focus from the role of IAA as a predictor of type I diabetes to that of a barometer of the subclinical process preceding the disease.

The administration of heterologous insulin triggers an insulin antibody response in the majority of type I and type II diabetic patients. Multiple factors determine the immunogenicity of an insulin preparation. These include the species of origin, purity, mode of administration, contaminants, physiochemical properties, the aggregatability of the insulin molecules, and the different vehicles used to modify the activity profiles of the various preparations (Haeften 1989). During the past decade, introduction of monocomponent, highly purified insulins and semisynthetic or recombinant human insulins has decreased significantly the intensity of the insulin antibody response in insulin-treated patients. Despite this decrease, no obvious improvement in the management of diabetes or in the reduction of long-term chronic diabetic complications was apparent (Haeften 1989). The insulin antibody response is polyclonal, as compared to the oligoclonal IAA response, the latter being restricted to a small number of insulin epitopes (Wilkin 1990). The lack of a general consensus regarding the HLA haplotype association of an insulin antibody response may be due to ethnic differences. However, it seems that high levels of insulin antibodies are associated with the expression of HLA-DR4 and/or DR7, and low levels with HLA-DR3 and/or DR5 (Reeves *et al.* 1984).

The presence of insulin antibodies in a diabetic patient may disturb glucose homeostasis in two opposing ways. First, insulin antibodies may mediate hypoglycaemia by binding insulin secreted during a meal and releasing it postprandially. When the bound insulin dissociates from the insulin antibody, it inappropriately raises the plasma free insulin, resulting in postprandial hypoglycaemia (Taylor *et al.* 1989). Binding of insulin to insulin antibodies can also result in its decreased clearance from the blood and so predispose to hypoglycaemia. Secondly, low levels of insulin antibodies have also been shown to potentiate insulin action *in vitro* by cross-linking insulin receptor complexes. Whether such a mechanism operates *in vivo* is not yet known. Anti-idiotypic antibodies to insulin antibodies, with insulin-like activity, can also cause hypoglycaemia (Taylor *et al.* 1989). It is possible, as the evidence from animal studies suggests, that the occurrence of such insulinomimetic antibodies is a part of the idiotypic–anti-idiotypic insulin antibody network triggered by the presence of insulin antibodies (Cohen *et al.* 1984).

Anti-idiotypic antibodies to insulin antibodies may appear after immunization with insulin in mice (Schechter *et al.* 1984) or spontaneously in prediabetic patients (Maron *et al.* 1983), and may even precede the appearance of IAAs. Such anti-idiotypic antibodies represent an internal image of insulin and can mimic its biological activities (Cohen *et al.* 1984; Taylor *et al.* 1989). While the role and prevalence of these antibodies in IDDM patients are unclear, they have been implicated as an additional serological marker for incipient diabetes (Maron *et al.* 1983). Anti-insulin receptor antibodies may also contribute to the development of IDDM by causing peripheral insulin resistance (Cohen *et al.* 1984).

A certain degree of insulin resistance is a common feature of almost all diabetic syndromes (Kahn 1986). Immunological insulin resistance resulting from a high titre of insulin antibodies, particularly in patients treated with beef insulin, has been reported to occur in less than 0.1 per cent of insulin-treated patients. Altered pharmacokinetics and bioavailability of insulin due to IA–insulin complexes, as well as increased sequestration of these complexes in the reticuloendothelial system, are probably the mechanisms mediating immunological insulin resistance. However, multiple mechanisms may underlie insulin resistance in diabetic patients, and more than one could be responsible for insulin resistance in the same individual (Kahn 1986; Kofler *et al.* 1989). Therefore, evaluation of the role of immunological insulin resistance in an insulin-antibody-positive diabetic patient could prove difficult. Fortunately, due to the introduction of highly purified insulin and recombinant human insulin, the occurrence of insulin antibodies in diabetics has decreased significantly. It is expected, therefore, that immunological insulin resistance, already uncommon in the past, will become exceedingly rare.

10.6.5 Cellular immunity to insulin in humans

Whether or not the presence of T-cells reactive to self-insulin can be considered as a marker of type I diabetes is still an open question, since no large-scale prospective studies have addressed this question. Naquet *et al.* (1988) examined T-cell autoreactivity to human insulin (HI) in insulin-treated diabetics and in their first-degree relatives, some of whom were HLA-identical siblings. CD4-positive T-cells were found to proliferate in response to HI in both groups. Although the highest T-cell response to HI was found in the diabetic group, the T-cell repertoire for HI was similar in both groups. Both groups recognized two conformational insulin epitopes formed by the interaction between A- and B-chain residues. Interestingly, the T-cell responses observed in HLA-DR4-positive individuals were restricted by the HLA-DQ3.2 allele, the same that confers susceptibility to diabetes. Based on this data, albeit limited to a small number of subjects, it was suggested that autoreactive T-cells to HI may be present in diabetic as well as non-diabetic subjects, and do not necessarily signify an auto-immune reaction directed towards the β-cells.

The regulation of the helper T-cell response to HI by CD8-positive T-cells was examined in an IDDM patient with immunological insulin resistance (Naquet *et al.*1989). Administration of sulphated beef insulin to the patient resulted in the emergence of CD8-positive insulin-specific T-cells, which suppressed the CD4-positive proliferative response to HI and decreased the titre of insulin antibodies in the patient, thus abrogating the insulin-resistant state. It is possible, therefore, that CD8-positive cells induce peripheral tolerance to insulin and that the net balance of activities of CD4-positive and CD8-positive T-cells regulates the level of immune responsiveness to insulin in diabetic patients. In addition, Epstein–Barr virus-transformed B-lymphocytes derived from the immunological insulin-resistant patient process insulin differently than B-cells from non-insulin-resistant patients and normal individuals (Semple and Delovitch 1991). Therefore, altered antigen processing may modify the intensity of the insulin antibody response in insulin-treated diabetic patients. Other workers showed that various parameters of the cellular immune response, such as the mitogen-induced proliferation of mononuclear cells and the percentage of NK-cells, proved superior to the insulin antibody response in reflecting alterations of immune responsiveness to different insulin preparations (Eicher *et al.* 1988). Thus, T-cell autoreactivity to insulin may reflect a normal physiological state, and also have a significant role in diabetic pathological states such as immunological insulin resistance. Further studies should elucidate the clinical significance of human T-cell autoreactivity to insulin.

10.7 Auto-immunity to an islet 64K antigen

Attempts have been made to identify other proteins expressed by the pancreatic islets that might play a role in the specific destruction of islet cells in insulin-dependent diabetes. One successful approach has been to label endogenous proteins with [^{35}S]methionine, solubilize islet proteins with non-ionic detergent, and incubate the extract with sera from newly diagnosed insulin-dependent diabetic patients or from normal healthy individuals. Antigens specifically recognized by antibodies in sera are detected by SDS-polyacrylamide gel electrophoresis and autoradiography. Using these techniques, a protein of M_r 64 000 (the 64K antigen) was identified that specifically bound to antibodies in sera from insulin-dependent diabetic patients (Baekkeskov *et al.* 1988). Expression of this protein was found to be tissue restricted; a strong expression of the protein was observed in pancreatic islets but the protein was not detected in liver, kidney, thyroid, adrenals, pituitary, thymus, or spleen. Furthermore, preparation by cell sorting of subpopulations of islet cells enriched in β-cells and non-β-cells indicated that the protein was only found in the β-cells (Christie *et al.*1990*a*). Further studies have shown that antibodies to the detergent-solubilized molecule are present in approximately 80 per cent of new onset diabetic patients, a frequency similar to that of ICA (Christie *et al.* 1988). Antibodies could be detected several years before disease onset, in some cases preceding the appearance of ICA (Baekkeskov *et al.* 1987; Atkinson *et al.* 1990) and, in contrast to ICA, may persist for as long as β-cell function is detectable (Christie *et al.* 1990*b*). Disease specificity of the antibody response was suggested by the finding that antibodies could not be detected in patients with other autoimmune disorders, such as Hashimoto's thyroiditis, Graves disease, or systemic lupus erythematosus (Baekkeskov *et al.* 1987). Recent findings have demonstrated that type I diabetic patients may possess at least two distinct antibody activities that bind epitopes on different proteolytic fragments of the antigen (Christie *et al.*

1990*c*). One of the antibody activities appears to bind to an epitope that is hidden on the detergent-solubilized molecule, but exposed on proteolytic cleavage. The significance of this finding is unclear, but epitopes that are not accessible to the immune system are likely to escape tolerance induction. Inappropriate degradation, processing, and presentation of the antigen may subsequently result in auto-immunity. Analysis of the frequency of the distinct antibody markers indicated that each marker was found in approximately 80 per cent of newly diagnosed diabetic patients; since the activities were distinct over 90 per cent of recent onset diabetic patients possessed at least one antibody activity. Thus, analysis of distinct antibody activities directed to the 64K antigen provides a highly sensitive and specific diagnostic marker for insulin-dependent diabetes.

Antibodies that immunoprecipitate a 64K islet antigen have also been detected in the NOD mouse and the BB rat animal models of type I diabetes (Atkinson and Maclaren 1988; Baekkeskov *et al.* 1988). These antibodies appear at a very early age in both animals and in some animals may disappear after the onset of disease. The antibodies were detected only very rarely in control rat or mouse strains. The NOD mouse and BB rat may thus prove to be very important in the characterization of auto-immune responses to the 64K antigen and elucidation of its role in disease pathogenesis. However, despite the similarity in molecular weight, it remains to be established that the antigens detected by antibodies in the animal models are identical to that associated with human type I diabetes.

A recent study (Baekkeskov *et al.* 1990) has identified the human diabetes-associated 64K antigen as glutamic acid decarboxylase (GAD). This enzyme is involved in the synthesis of the neurotransmitter γ-aminobutyric acid (GABA). The enzyme is expressed in GABA-ergic neurons of the central nervous system and in the β-cells of the pancreatic islet (but not in other islet cells). It has been proposed that GABA secretion by β-cells may play a role in regulating hormone secretion from pancreatic α- and δ-cells, which express receptors for the neurotransmitter (Rorsman *et al.* 1989).

The key to the identification of the antigen came from observations on a rare disorder of the central nervous system, stiff-man syndrome. The disease is characterized by progressive muscle rigidity and spasms. Symptoms are alleviated by drugs that enhance GABA-ergic neurotransmission, suggesting that GABA-ergic pathways are impaired. In a study of 33 patients with stiff-man syndrome, Solimena *et al.* (1990) identified a subpopulation of patients with antibodies to GABA-ergic neurons; immunoprecipitation and Western blotting analysis suggested that GAD was the predominant antigen. All patients who were positive for antibodies to GABA-ergic neurons also had antibodies reacting with pancreatic islets, and a proportion of these (6/20) had insulin-dependent diabetes. Other organ-specific antibodies were also present in most patients, including antibodies to gastric parietal cells, thyroid microsomal antigen, and thyroglobulin. Stiff-man syndrome was also associated with other auto-immune diseases; three of the 20 patients had Graves disease, two had hypothyroidism, two pernicious anaemia. Thus, stiff-man syndrome might itself be the result of auto-immunity against GABA-ergic neurons, and against GAD in particular.

The association of stiff-man syndrome with insulin-dependent diabetes and the similarities in properties of the antigens associated with the two diseases, in particular the tissue-restricted expression of GAD, prompted studies of whether the diabetes-associated antigen was in fact GAD. Baekkeskov *et al.* (1990) were able to demonstrate that antibodies to purified brain GAD, antibodies in sera from patients with stiff-man syndrome, and antibodies in sera from insulin-dependent diabetic patients all cross-reacted with islet or brain 64K antigen. Furthermore, the 64K islet antigen showed a similar pattern on two-dimensional gel electrophoresis as brain GAD, and GAD activity could be measured in immunoprecipitates of brain or islet extracts with sera from diabetic, but not control, individuals. These studies thus provided strong evidence that the diabetes-associated 64K antigen was islet GAD.

The finding that GAD, an enzyme expressed by GABA-ergic neurons and in pancreatic β-cells, is an auto-antigen in two pathological states associated with auto-immunity to these tissues suggests that auto-immunity to GAD may play a role in the pathogenesis of both diseases. However, it is clear that auto-immunity to the protein itself is not sufficient for disease development. Thus, although most patients with stiff-man syndrome have high circulating antibody levels to glutamate decarboxylase, only a proportion will develop diabetes. Conversely, stiff-man syndrome is extremely rare in diabetic patients. There are likely to be differences in the immune responses, perhaps in the site of the response or the epitope specificity that determines disease specificity; the failure of diabetes-associated anti-GAD antibodies to bind GAD on sections of brain or on Western blots, in contrast to antibodies in stiff-man syndrome, does suggest that there are differences in antibody specificity. There may also be specific isoforms of the enzyme that are expressed only by the β-cells. Further characterization of the β-cell 64K antigen, and of T- and B-cell responses to it, are clearly important to elucidate the role of GAD immunity in the pathogenesis of insulin-dependent diabetes.

10.8 Prospects for therapy

Since the discovery of insulin in 1921, there has been an effective treatment for type I diabetes that has allowed the patient to maintain normoglycaemia and lead a relatively normal life. However, even with good metabolic control and considerable improvements in the quality and delivery of insulin, diabetes is associated with secondary complications that affect, primarily, the kidney and circulatory system and result in a lower life expectancy. Thus, there is a continuing search for improved treatments for the diabetic patient. With the now very strong evidence that type I diabetes is an auto-immune disease, interest has turned to the employment of immunosuppressive agents as a potential treatment. For ethical reasons, immunosuppressive therapy has only been employed after onset of overt diabetes, when the residual β-cell mass is likely to be low. A number of agents have been tried, including azathioprine, prednisone, anti-lymphocyte serum, and cyclosporin;

results with cyclosporin have been most encouraging. In two multicentre, double-blind trials (Feutren *et al.* 1986; Canadian–European Randomized Control Trial Group, 1988), 23–24 per cent of patients receiving cyclosporin still did not require insulin after 9–12 months of treatment, compared with 6–10 per cent of the group receiving placebo. Cyclosporin was most effective in patients with a relatively short duration of diabetes, who probably have a greater β-cell reserve. However, even at the relatively low doses employed (7.5–10 mg/kg), the treatment was associated with moderate toxic effects, as indicated by decreased creatinine clearance and structural damage seen in kidney biopsies after 1 year of treatment. Continual administration of cyclosporin is likely to be required to maintain remission from insulin, and the long-term toxic effects of the agent in a diabetic population have yet to be reported. Since diabetic patients are already at risk for kidney disease as a secondary complication of diabetes, even slight renal toxicity in these patients could be very serious. Nevertheless, the results of the cyclosporin trials are sufficiently encouraging to prompt the search for safer, more specific therapies.

It has been proposed that problems associated with general immunosuppressive therapy might be overcome by directing immune intervention specifically at autoreactive T-cells responsible for β-cell destruction. Our current understanding of the molecular requirements for initiating and sustaining an immune response to an antigen suggest a number of potential sites for immune intervention by blocking the presentation of auto-antigen to autoreactive T-cells (Fig. 3).

A number of cell types have been shown to act as antigen-presenting cells, including macrophages, dendritic cells, and B-lymphocytes. The role each kind of antigen-presenting cell plays during an immune response is not clear. However, antigen-specific B-lymphocytes probably represent the most efficient cells presenting antigen, due to the presence of high-affinity surface receptors that can efficiently bind and internalize the antigen. Therefore, elimination of antigen-specific B-lymphocytes, for example by administering antigen coupled to a toxin, will have

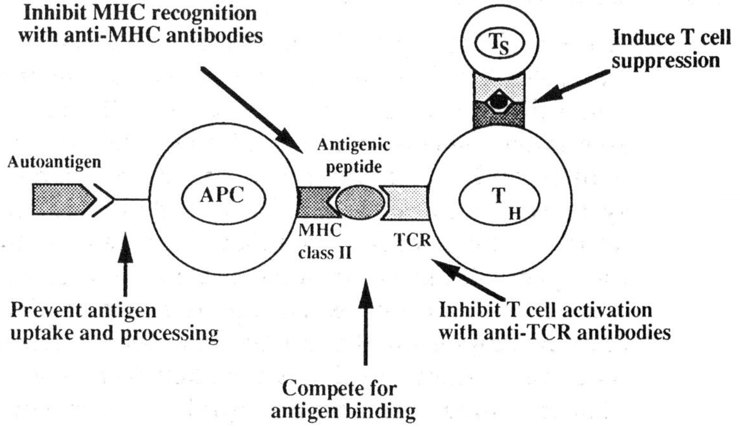

Fig. 3 Antigen-specific immunotherapy of type I diabetes. Novel therapies for type I diabetes may include procedures designed to specifically block activation of autoantigen-specific T-helper cells (T_H) by antigen presented in association with MHC class II molecules on an antigen-presenting cell (APC). Immune intervention may involve inhibiting the processing of antigen and presentation of antigen fragments to specific T-cell receptors (TCR) or inducing suppression of responses via T-suppressor cells (T_s)

inhibitory effects on the immune response, both by preventing efficient uptake and processing of auto-antigen and by preventing specific antibody production.

Inhibition of antigen recognition by antigen-specific T-lymphocytes has been achieved in a number of ways, all of which have been shown to be effective *in vivo* at preventing auto-immune disease in animal models. If the specific MHC class II molecules that present auto-antigen are known, then T-cell recognition might be prevented by administering antibodies to the specific class II molecule. Blocking T-cell-receptor binding to MHC with antibodies to MHC class II molecules has prevented auto-immune disease in a number of animal models, including NOD diabetes. Alternatively, peptides with high affinity for the appropriate MHC class II molecule may also block the immune response. This approach has proved effective in experimental auto-immune encephalomyelitis, an animal model of multiple sclerosis (EAE; Hood *et al.* 1989). EAE is induced by injecting rats with myelin basic protein, a component of the myelin sheath of nerve fibres, which results in cellular auto-immunity to the endogenous protein, degradation of the nerve fibres, and paralysis, symptoms similar to those seen in multiple sclerosis. Peptide fragments of the protein are also effective. EAE is T-cell dependent and MHC restricted; only animals with particular MHC haplotypes develop the disease, and high-affinity antibodies to MHC molecules block the disease. Regions of the molecule recognized by pathogenic T-cells have been identified and these differ according to the MHC haplotype of the animal. For example, a peptide comprising amino acids 1–9 is the disease-causing determinant of myelin basic protein in IA^u rats, whereas amino acids 89–101 are important in IA^s rats. Since residues on the peptides that interact with MHC are likely to be different from those interacting with the T-cell receptor, it was argued that, by modifying specific residues, it should be possible to design peptides that still show strong binding to MHC but that no longer activate T-cells. Such peptides might compete with encephalitogenic peptides for MHC binding. This was found to be true in EAE; some amino-acid substitutions on the N-terminal peptide of myelin basic protein effectively inhibited T-cell responses to the protein *in vitro* and prevented EAE *in vivo*.

A major problem with therapies directed against MHC molecules is that such therapies are still relatively non-specific. Since individuals express relatively few MHC molecules, blocking the T-cell receptor interaction at the level of MHC, either with antibodies to MHC or with competing peptides, removes a significant proportion of the antigen-presenting capacity of the individual, which may leave them, to some extent, immunodeficient. Strategies at preventing T-cell activation at the level of the antigen receptor would be preferable. For EAE, T-cells reacting with encephalitogenic peptides have been characterized and these show a relatively restricted T-cell receptor usage; the specific receptors that are used depend on the MHC haplotype of the animal (Hood *et al.* 1989). Treatment of animals with antibodies specific for the Vβ region of the specific T-cell receptors employed did indeed reduce the incidence of the disease; complete abrogation may not have been achieved due to the usage of T-cell receptors to which antibodies were not available. Recovery from ongoing disease could also be achieved by a similar therapy. Thus, antibodies to T-cell receptors may be useful for reversing auto-immune disease as well as its prevention.

All the above approaches probably require continuing treatment in order to be effective. Depletion of T-cell subpopulations in mice treated with antibodies to the T-cell receptor lasted several weeks; competing peptide therapy or anti-MHC antibody treatment would probably be effective for a much shorter period. The restoration of immunoregulatory function, which appears to be defective in type I diabetes, would be the most effective way of preventing diabetes in susceptible individuals. Animal models have provided evidence that the single administration of an immunostimulatory agent early in life may restore antigen non-specific suppressor cell function and prevent subsequent development of diabetes (Sadelain *et al.* 1990). Alternatively, Cohen and co-workers have developed a novel procedure, termed T-cell vaccination, whereby antigen-specific T-cell suppression can be induced (Cohen 1989). In this procedure, the single injection of as few as 100 autoreactive T-cells was found to induce T-cell suppression and prevent auto-immune disease in EAE. For effective vaccination, it was necessary to activate the T-cells *in vitro*, either with antigen or with a T-cell mitrogen such as concanavalin A, and to treat the activated T-cells with a chemical cross-linker, such as glutaraldehyde or formaldehyde. The latter treatment increases the effectiveness of the vaccine and prevents proliferation of the autoreactive T-cells *in vivo*. Identification and cloning of the auto-immune T-cells was not essential. Thus, administration of unselected, concanavalin A-activated and cross-linked splenic T-cells from NOD mice into prediabetic NOD mice reduced the incidence of diabetes in these mice (Cohen 1989). Characterization of the regulatory T-cells induced by T-cell vaccination suggests that anti-idiotypic T-cells and antigen non-specific regulatory T-cells that recognize the state of activation of the autoreactive T-cell are important in the down-regulation of the auto-immune response.

A better understanding of the role of T-cell immunity to β-cell antigens is required before similar procedures can be employed as a preventative therapy for type I diabetes. In particular, auto-antigens associated with diabetes in man and in animal models need to be identified and purified. The availability of suitable quantities of antigen will allow the generation of T-cell lines specific for antigen. Analysis of the T-cell-receptor genes expressed by antigen-specific T-cell clones will be important, since a restricted T-cell repertoire for autoreactive T-cells may be essential for successful employment of immune intervention at the level of the T-cell receptor. Knowledge of the T-cell subsets invading the pancreas in human type I diabetes is clearly lacking. Finally, specific, high-capacity screening assays for anti-β-cell auto-immunity need to be developed to identify individuals at risk of developing diabetes and to allow treatment to be initiated early, before the majority of the β-cells are destroyed.

References

Atkinson, M. A. and MacClaren, N. K. (1988). Autoantibodies in nonobese diabetic mice immunoprecipitate 64000-Mr islet antigen. *Diabetes* **37**, 1587–90.

Atkinson, M. A., MacClaren, N. K., Riley, W. G., Winter, W. E., Fish, D. D. and Spillar, R. P. (1986). Are insulin autoantibodies markers for insulin-dependent diabetes mellitis? *Diabetes* **35**, 894–8.

Atkinson, M. A., MacClaren, N. K., Sharp, D. W., Lacy, P. E. and Riley, W. J. (1990). 64000 Mr antibodies as predictors of insulin-dependent diabetes. *Lancet* **335**, 1357–60.

Baekkeskov, S., Christie, M. R., and Lernmark, A. (1986). Islet cell antibodies. In *Immunology of endocrine diseases* (ed. A. M. McGregor), pp. 73–88. MTP Press, Lancaster.

Baekkeskov, S., Landin, M., Kristensen, J. K., Srikanta, S., Bruining, G. J., Mandrup-Poulsen, T., de Beaufort, C., Soeldner, J. S., Eisenbarth, G., Lindgren, F., Sundquist, G., and Lernmark, A. (1987). Antibodies to a 64000-Mr human islet cell protein precede the clinical onset of insulin-dependent diabetes. *J. Clin. Invest.* **79**, 926–34.

Baekkeskov, S., Markholst, H., and Christie, M. (1988). The 64-kD islet cell target antigen of humoral immunity associated with insulin-dependent diabetes in man and the BB-rat. In *Frontiers in diabetes research. Lessons from animal diabetes II* (ed. E. Shafrir and A. E. Renold), pp. 40–5. John Libbey, London.

Baekkeskov, S., Aanstoot, H. K., Christgau, S., Reetz, A., Solimena, M., Cascalho, M., Folli, F., Richter-Olesen, H., and De Camilli, P. (1990). Identification of the 64k autoantigen in insulin-dependent diabetes as the GABA-synthesizing enzyme glutamic acid decarboxylase. *Nature* **347**, 151–6.

Barcinski, M. A. and Rosenthal, A. S. (1977). Immune response gene control of determinant selection. I. Intramolecular mapping of the immunogenic sites on insulin recognized by guinea pig T and B cells. *J. Exp. Med.* **145**, 726–39.

Baxter, A. G., Adams, M. A., and Mandel, T. E. (1989). Comparison of high- and low-diabetes-incidence NOD mouse strains. *Diabetes* **38**, 1296–300.

Blundell, T., Dodson, G., Hodgkin, D., and Mercola, D. (1972). Insulin: The structure of the crystal and its reflection in chemistry and biology. *Adv. Prot. Chem.* **26**, 279–402.

Boehmer, H. von, Teh, H. S., and Kieselow, P. (1989). The thymus selects the useful, neglects the useless and destroys the harmful. *Immunology Today* **10**, 57–61.

Bohme, J., Haskins, K., Stecha, O., van Ewijk, W., LeMeur, M., Gerlinger, P., Benoist, C., and Mathis, D. (1989). Transgenic mice with I-A on islet cells are normoglycemic but immunologically intolerant. *Science* **244**, 1179–83.

Bonifacio, E., Lernmark, A., Dawkins, R. L., and coworkers (1987). Serum exchange and the use of dilutions have improved precision of measurement of islet cell antibodies. *J. Immunol. Methods* **106**, 83–8.

Bonifacio, E., Bingley, P. J., Shattock, M., Dean, B. M., Dunger, D., Gale, E. A. M., and Bottazzo, G. F. (1990). Quantification of islet cell antibodies and prediction of insulin-dependent diabetes. *Lancet* **335**, 147–9.

Bottazzo, G. F., Florin-Christensen, A., and Doniach, D. (1974). Islet cell antibodies in diabetes mellitus with autoimmune polyendrocrine deficiencies. *Lancet* **ii**, 1279–83.

Bottazzo, G. F., Pujol-Borrel, R., Hanafusa, T. and Feldmann, M. (1983). Hypothesis: role of aberrant HLA-DR expression and antigen presentation in the induction of endocrine autoimmunity. *Lancet ii*, 1115–19.

Bottazzo, G. F., Dean, B. M., McNally, J. M., MacKay, E. H., Swift, P. G. F., and Gamble,

D. R. (1985). *In situ* characterisation of autoimmune phenomena and expression of HLA molecules in the pancreas in diabetic insulitis. *New Engl. J. Med.* **313,** 353–60.

Burkly, L. C., Lo, D., and Flavell, R. A. (1990). Tolerance in transgenic mice expressing major histocompatibility molecules extrathymically on pancreatic cells. *Science* **248,** 1364–8.

Burnett, F. M. (1959). *The clonal selection theory of immunity.* Cambridge University Press.

Campbell, I. L., Wong, G. H. W., Schrader, J. W., and Harrison, L. C. (1985). Interferon γ enhances the expression of the major histocompatibility class I antigens on mouse pancreatic beta-cells. *Diabetes* **34,** 1205–9.

Campbell, I. L., Cutri, A., Wilkinson, D., Boyd, A. W., and Harrison, L. C. (1989). Intercellular adhesion molecule 1 is induced on isolated endocrine islet cells by cytokines but not by reovirus infection. *Procl. Natl Acad. Sci. USA* **86,** 4282–6.

Canadian–European Randomized Control Trial Group (1988). Cyclosporin-induced remission of IDDM after early intervention: association of 1 y of cyclosporin treatment with enhanced insulin secretion. *Diabetes* **37,** 1574–82.

Castano, L. and Eisenbarth, G. S. (1990). Type 1 diabetes: A chronic autoimmune disease of human mouse and rat. *Ann. Rev. Immunol.* **8,** 647–79.

Chandy, K. G., Charles, A. M., Kershnar, A., Buckingham, B., Waldeck, N., and Gupta, S. (1984). Autologous mixed lymphocyte reaction in man: XV. Cellular and molecular basis of deficient autologous mixed lymphocyte response in insulin-dependent diabetes mellitus. *J. Clin. Immunol.* **4,** 424–8.

Charles, M. A., Suzuki, M., Waldeck, N., Dodson, L. E., Slater, L., Ong, K., Kershnar, A., Buckingham, B., and Golden, M. (1983). Immune islet killing mechanisms associated with insulin-dependent diabetes: *in vitro* expression of cellular and antibody-mediated islet cell cytotoxicity in humans. *J. Immunol.* **130,** 1189–94.

Charlton, B., Bacelj, A., Slattery, R. M., and Mandel, T. E. (1989). Cyclophosphamide-induced diabetes in NOD/WEHI mice: evidence for suppression in spontaneous autoimmune diabetes mellitus. *Diabetes* **38,** 441–7.

Christie, M., Landin Olsson, M., Sundkvist, G., Dahlquist, G., Lernmark, A., and Baekkeskov, S. (1988). Antibodies to a Mr-64000 islet cell protein in Swedish children with newly diagnosed Type 1 (insulin-dependent) diabetes. *Diabetologia* **31,** 597–602.

Christie, M. R., Pipeleers, D. G., Lernmark, A., and Baekkeskov, S. (1990*a*). Cellular and subcellular localization of a Mr-64000 protein autoantigen in insulin-dependent diabetes. *J. Biol. Chem.* **265,** 376–81.

Christie, M. R., Daneman, D., Champagne, P., and Delovitch, T. L. (1990*b*). Persistence of antibodies to 64000-Mr islet cell protein after onset of Type 1 diabetes. *Diabetes* **39,** 653–6.

Christie, M. R., Vohra, G., Champagne, P., Daneman, D., and Delovitch, T. L. (1990*c*). Distinct antibody specificities to a 64-kD islet-cell antigen in Type 1 diabetes as revealed by trypsin treatment. *J. Exp. Med.* **172,** 789–94.

Cohen, I. R. (1989). Physiological basis of T-cell vaccination against autoimmune disease. *Cold Spring Harbor Symp. Quant. Biol.* **54,** 879–84.

Cohen, I. R. and Cooke, A. (1986). Natural antibodies might prevent autoimmune disease. *Immunol. Today* **7,** 363–4.

Cohen, I. R., Elias, D., Maron, R., and Shechter, Y. (1984). Immunization to insulin generates anti-idiotypes that behave as antibodies to the insulin hormone receptor and cause diabetes mellitus. In *Idiotypy in biology and medicine* (ed. H. Kohler, J. Urbain, and P. A. Cazenave), pp. 385–400. Academic Press, Orlando Florida.

Coleman, D. L., Kuzava, J. E., and Leiter, E. H. (1990). Effect of diet on incidence of diabetes in nonobese diabetic mice. *Diabetes* **39,** 432–6.

Colman, P. G., Nayak, R. C., Campbell, I. L., and Eisenbarth, G. S. (1988). Binding of cytoplasmic islet cell antibodies is blocked by human pancreatic glycolipid extract. *Diabetes* **37,** 645–52.

Coutinho, A., Bandeira, A., Pereira, P., Portnoi, D., Holmberg, D., Martinez, C., and Freitas, A. A. (1989). Selection of lymphocyte repertoires: the limits of clonal versus network organization. *Cold Spring Harbor Symp. Quant. Biol.* **54,** 159–70.

Delovitch, T. L., Lazarus, A. H., Phillips, M. L., and Semple, J. W. (1989). Antigen binding and processing by B cell antigen presenting cells: Influence on T- and B-cell activation. *Cold Spring Harbor Symp. Quant. Biol.* **54,** 333–41.

Drell, D. W. and Notkins, A. L. (1987). Multiple immunological abnormalities in patients with Type 1 (insulin-dependent) diabetes mellitis. *Diabetologia* **30,** 132–43.

Eicher, H. L., Lauritano, A. A., Woertz, L. L., Selam, J. L., Gupta, S., and Charles, M. A. (1988). Cellular alterations associated with human insulin therapy. *Diabetes Research* **8,** 111–15.

Elias, D., Markovits, D., Reshev, T., van der Zee, R., and Cohen, I. R. (1990). Induction and therapy of autoimmune diabetes in the non-obese diabetic (NOD/Lt) mouse by a 65-kDa heat shock protein. *Proc. Natl. Acad. Sci. USA* **87,** 1576–80.

Feutren, G., Papoz, L., Assan, R., Vialettes, B., Karsenty, G., Vexiau, P., Du Rostu, H., Rodier, M., Sirmai, J., Lallemand, A., and Bach, J.-F. (1986). Cyclosporin increases the rate of and length of remissions in insulin-dependent diabetes of recent onset. *Lancet* **ii,** 119–23.

Francfort, J. W., Naji, A., Silvers, W. K., and Barker, C. F. (1985). The influence of T-lymphocyte precursor cells and thymus grafts on the cellular immunodeficiencies of the BB rat. *Diabetes* **34,** 1134–8.

Gepts, W. (1965). Pathologic anatomy of the pancreas in juvenile diabetes. *Diabetes* **14,** 619–33.

Gepts, W. and De Mey J. (1978). Islet cell survival determined by morphology: an immunocytochemical study of the islets of Langerhans in juvenile diabetes mellitus. *Diabetes* **27** (Suppl. 1), 251–61.

Gillard, B. K., Thomas, J. W., Nell, L. J., and Marcus, D. M. (1989). Antibodies against ganglioside GT3 in the sera of patients with Type 1 diabetes mellitus. *J. Immunol.* **142,** 3826–32.

Haeften, T. W. van (1989). Clinical significance of insulin antibodies in insulin-treated diabetic patients. *Diabetes Care* **12,** 641–8.

Hanahan, D., Jolicoeur, C., Alpert, S., and Skowronski, J. (1989). Alternative self or nonself recognition of an antigen expressed in a rare cell type in transgenic mice. Implication for self tolerance and autoimmunity. *Cold Spring Harbor Symp. Quant. Biol.* **54,** 821–36.

Hanenberg, H., Kolb-Bachofen, V., Kantwerk-Finke, G., and Kolb, H. (1989). Macrophage infiltration precedes and is a prerequisite for lymphocytic insulitis in pancreatic islets of pre-diabetic BB rats. *Diabetologia* **32,** 126–34.

Harada, M. and Makino, S. (1984). Promotion of spontaneous diabetes in non-obese diabetes-prone mice by cyclophosphamide. *Diabetologia* **27,** 604–6.

Harrison, L. C., Campbell, I. L., Allison, J., and Miller, J. F. A. P. (1989). MHC molecules and beta-cell destruction. Immune and nonimmune mechanisms. *Diabetes* **38,** 815–18.

Haskins, K., Portas, M., Bergman, B., Lafferty, K., and Bradley, B. (1989). Pancreatic T-cell clones from nonobese diabetic mice. *Proc. Natl Acad. Sci. USA* **86,** 8000–4.

Hood, L., Kumar, V., Osman, G., Beall, S. S., Gomez, C., Funkhouser, W., Kono, D. H., Nickerson, D., Zaller, D. M., and Urban, J. L. (1989). Autoimmune disease and T-cell immunologic recognition. *Cold Spring Harbor Symp. Quant. Biol.* **54,** 859–74.

Jacob, C. O., Aiso, S., Michie, S. A., McDevitt, H. O., and Acha-Orbea, H. (1990). Prevention of diabetes in nonobese diabetic mice by tumor necrosis factor (TNF): Similarities between TNF-alpha and interleukin 1. *Proc. Natl Acad. Sci. USA* **87,** 968–72.

Jensen, P. E., Pierce, C. A., and Kapp, J. A. (1984). Regulatory mechanisms in immune responses to heterologous insulins. II. Suppressor T cell activation associated with nonresponsiveness. *J. Exp. Med.* **160,** 1012–26.

Jerne, N. K. (1974). Towards a network theory of the immune system. *Ann. Immunol. (Paris)* **125C,** 373–89.

Johnson, J. H., Crider, B. P., McCorkle, K., Alford, M., and Unger, R. H. (1990). Inhibition of glucose transport into rat islet cells by immunoglobulins from patients with new-onset insulin-dependent diabetes mellitus. *New Engl. J. Med.* **322,** 635–59.

Kahn, C. R. (1986). Insulin resistance: a common feature of diabetes mellitus. *New Engl. J. Med.* **315** (4), 252–3.

Kanatsuna, T., Baekkeskov, S., Lernmark, A., and Ludvigsson, J. (1983). Immunoglobulin from insulin-dependent diabetic children inhibits glucose-induced insulin release. *Diabetes* **32,** 520–4.

Kappler, J. W., Staerz, U., White, J., and Marrack, P. (1988). Self-tolerance eliminates T cells specific for mls-modified products of the major histocompatibility complex. *Nature* **332,** 35–40.

Karjalainen, J., Knip, M., and Akerblum, H. K. (1988). Insulin antibodies at the clinical manifestation of Type 1 (insulin dependent) diabetes-A poor predictor of clinical course and antibody response to exogenous insulin. *Diabetologia* **31,** 129–33.

Karjalainen, J., Salmela, P., Ilonen, J., Surchel, H. M., and Knip, M. (1989). A comparison of childhood and adult Type 1 diabetes mellitus. *New Engl. J. Med.* **320,** 881–6.

Keck, K. (1975). Ir. gene control of insulin and A-chain loop as carrier determinants. *Nature* **254,** 78–9.

Koler, M., Ramirez, L. C., and Raskin, P. (1989). Insulin resistance and diabetes: Mechanism and possible intervention. *Diabetes Res. Clin. Pract.* **7,** 83–98.

Lampeter, E. F., Sigmore, A., Gale, E. A. M., and Pozzilli, P. (1989). Lessons from the NOD mouse for the pathogenesis and immunotherapy of Type 1 (insulin-dependent) diabetes mellitis. *Diabetologia* **32,** 703–8.

Lernmark, A. (1987). Islet cell antibodies. *Diabetic Med.* **4,** 285–92.

Like, A. A., Biron, C. A., Weringer, E. J., Byman, K., Srcynski, E. and Guberski, D. L. (1986). Prevention of diabetes in BioBreeding/Worcester rats with monoclonal antibodies that recognize T lymphocytes or natural killer cells. *J. Exp. Med.* **164,** 1145–59.

MacCuish, A. C., Jordan, J., Campbell, C. J., Duncan, L. J. P., and Irvine, W. J. (1974). Antibodies to islet cells in insulin-dependent diabetes with co-existent auto-immune disease. *Lancet* **ii,** 1529–33.

MacDonald, H. R., *et al.* (1988). T-cell receptor Vβ use predicts reactivity and tolerance to Mls-a encoded antigens. *Nature* **332,** 40–5.

Malaisse, W. J., Malaisse-Lagae, F., Sener, A., and Pipeleers, D. G. (1982). Determinants of the selective cytotoxicity of alloxan to the pancreatic B cell. *Proc. Natl Acad. Sci. USA* **79,** 927–30.

Mandrup-Poulsen, T., Bendtzen, K., Dinarello, C. A., and Nerup, J. (1987*a*). Human tumor necrosis factor potentiates interleukin 1-mediated rat pancreatic β-cell cytotoxicity. *J. Immunol.* **139,** 4077–82.

Mandrup-Poulsen, T., Spinas, G. A., Prowse, S. J., Hansen, B. S., Jorgensen, D. W., Bendtzen, K., Nielsen, J. H., and Nerup, J. (1987*b*). Islet cytotoxicity of interleukin 1. Influence of culture conditions and islet donor characteristics. *Diabetes* **36,** 641–7.

Mandrup-Poulsen, T., Molvig, J., Andersen, H. U., Helquist, S., Spinas, G., Munck, M. and the Canadian–European Randomized Control Trial Group (1990). Lack of predictive value of islet cell antibodies, insulin antibodies and HLA-DR phenotype for remission in cyclosporine-treated IDDM patients. *Diabetes* **39**, 204–10.

Maron, R., Elias, D., Bruining, B. M., Van Hood, G. J., Shechter, Y., and Cohen, I. R. (1983). Autoantibodies to the insulin receptor in juvenile onset diabetes mellitus. *Nature* **303**, 817–18.

Marrack, P. and Kappler, J. (1987). The T cell receptor. *Science* **238**, 1073–9.

Miller, B. J., Appel, M. C., O'Neil, J. J., and Wicker, L. S. (1988). Both the Lyt-2+ and L3T4+ T cell subsets are required for the transfer of diabetes in nonobese diabetic mice. *J. Immunol.* **140**, 52–8.

Miller, J. F. A. P., Morahan, G. A., and Allison, J. (1989). Immunological tolerance: new approaches using transgenic mice. *Immunol. Today* **10**, 53–7.

Naquet, P., Ellis, J., Singh, B., Hodges, R. S., and Delovitch, T. L. (1987). Processing and presentation of insulin. I. Analysis of immunogenic peptides and processing requirements for insulin A-loop-specific T cells. *J. Immunol.* **139**, 3955–63.

Naquet, P., Ellis, J., Tibensky, D., Kenshole, A., Singh, B., Hodges, R., and Delovitch, T. L. (1988). T-cell autoreactivity to insulin in diabetic and related non-diabetic individuals. *J. Immunol.* **140**, 2569–78.

Naquet, P., Ellis, J., Kenshole, A., Semple, J. W., and Delovitch, T. L. (1989). Sulfated beef insulin treatment elicits CD8+ T cells that may abrogate immunologic insulin resistance in Type 1 diabetes. *J. Clin. Invest.* **84**, 1479–87.

Nathan, C. F. and Tsunawaki, S. (1986). Secretion of toxic oxygen products by macrophages: regulatory cytokines and their effects on the oxidase. *Ciba Foundn Symp.* **8**, 211–30.

Nayak, R. C., Omar, M. A. K., Rabizadeh, A., Srikanta, S., and Eisenbarth, G. S. (1985). Cytoplasmic islet cell antibodies. Evidence that the target antigen is a sialoglycoconjugate. *Diabetes* **34**, 617–19.

Nell, J. L., Hulbert, C. and Thomas, W. J. (1989). Human insulin autoantibody fine specificity and H and L chain use. *J. Immunol.* **142**, 3063–9.

Nerup, J. and Lernmark, A. (1981). Autoimmunity in diabetes mellitus. *Am. J. Med.* **70**, 135–41.

Nerup, J., Andersen, O. O., Bendixen, G., Egeberg, J., Gunarsson, R., Kromann, H., and Poulsen, J. E. (1974). Cell mediated immunity in diabetes mellitus. *Proc. R. Soc. Med.* **67**, 506–11.

Oldstone, M. B. A. (1988). Prevention of Type 1 diabetes in nonobese diabetic mice by viral infection. *Science* **239**, 500–2.

Olmos, P., A'Hern, R., Heaton, D. A., Millward, B. A., Risley, D., Pyke, D. A., and Leslie, R. D. G. (1988). The significance of the concordance rate for Type 1 (insulin-dependent) diabetes in identical twins. *Diabetologia* **31**, 747–50.

Pipeleers, D. G., In't Veld P. A., Pipeleers-Marichal, M. A., Gepts, W., and Van de Winkel, M. (1987). Presence of pancreatic hormones in islet cells with MHC-class II antigen expression. *Diabetes* **36**, 872–6.

Prud'homme, G. I., Colle, E., Fuks, A., Goldner-Sauve, A., and Guttmann, R. (1985). Cellular immune abnormalities and autoreactive T lymphocytes in insulin-dependent diabetes in rats. *Immunol. Today* **6**, 160–2.

Pujol-Borrell, R., Todd, I., Doshi, M., Bottazzo, G. F., Sutton, R., Gray, D., Adolf, G. R., and Feldman, M. (1987). HLA class II induction in human islet cells by interferon γ plus tumour necrosis factor or lymphotoxin. *Nature* **326**, 304–6.

Pukel, C., Baquerizo, H., and Rabinovitch, A. (1988). Destruction of rat islet cell monolayers by cytokines. Synergistic interactions of interferon-gamma, tumor necrosis factor, lymphotoxin and interleukin 1. *Diabetes* **37,** 133–6.

Reeves, W. G., Barr, D., Douglas, C. A., Gelsthorpe, K., Hanning, I., Skene, A., Wells, L., Wilson, R. M., and Tattersal, R. B. (1984). Factors governing the human immune response to injected insulin. *Diabetologia* **26,** 266–71.

Reich, E.-P., Sherwin, R. S., Kanagawa, O., and Janeway, C. A. (1989*a*). An explanation for the protective effect of the MHC class II I-E molecule in murine diabetes. *Nature* **341,** 326–8.

Reich, E.-P., Scaringe, D., Yagi, J., Sherwin, R. S., and Janeway, C. A. (1989*b*). Prevention of diabetes in NOD mice by injection of autoreactive T-lymphocytes. *Diabetes* **38,** 1647–51.

Rorsman, P., Berggren, P. O., Bokvist, K., Erikson, H., Mohler, H., Ostenson, C. G., and Smith, P. A. (1989). Glucose-inhibition of glucagon secretion involves activation of GABA-A-receptor chloride channels. *Nature* **341,** 233–6.

Sadelain, M. W. J., Quin, H.-Y., Lauzon, J., and Singh, B. (1990). Prevention of diabetes in NOD mice by adjuvant immunotherapy. *Diabetes* **39,** 583–9.

Sarvetnick, N., Shizuru, J., Liggett, D., Martin, L., McIntyre, B., Gregory, A., Parslow, T., and Stewart, T. (1990). Loss of pancreatic islet tolerance by beta-cell expression of interferon-γ. *Nature* **346,** 844–7.

Satoh, J., Seino, T., Tanaka, S.-I., Shintani, S., Ohta, S., Tamura, K., Sawai, T., Nobunaga, T., Oteki, T., Kumagai, K., and Toyota, T. (1989). Recombinant human tumor necrosis factor α suppresses autoimmune diabetes in nonobese diabetic mice. *J. Clin. Invest.* **84,** 1345–8.

Semple, J. W. and Delovitch, T. L. (1991). Altered processing of human insulin by B-lymphocytes from an immunologically insulin-resistant Type-1 diabetic patient. *J. Autoimmun* **4,** 277–90.

Schechter, Y., Elias, D., Maron, R., and Cohen, I. R. (1984). Mouse antibodies to the insulin receptor developing spontaneously as anti-idiotypes. I. Characterization of the antibodies. *J. Biol. Chem.* **259,** 6411–15.

Schernthaner, G., Borkenstein, M., Fink, M., Mayr, W. R., Menzel, J., and Scober, E. (1983). Immunogenicity of human monocomponent or pork monocomponent insulin in HLA-DR-typed insulin dependent Type 1 diabetic individuals. *Diabetes Care* **6,** 43–8.

Schroer, J. A., Bender, T., Feldmann, R. J., and Kim, K. J. (1983). Mapping epitopes on the insulin molecule using monoclonal antibodies. *Eur. J. Immunol.* **13,** 693–700.

Scott, F. W., Mongeau, R., Kardish, M., Hatina, G., Trick, K. D., and Wojcinski, Z. (1985). Diet can prevent diabetes in the BB rat. *Diabetes* **34,** 1059–62.

Serreze, D. V. and Leiter, E. H. (1988). Defective activation of T suppressor cell function in nonobese diabetic mice. Potential relation to cytokine deficiencies. *J. Immunol.* **140,** 3801–7.

Shintani, S., Satoh, J., Seino, H., Goto, Y., and Toyota, T. (1990). Mechanism of action of a streptococcal preparation (OK-432) in prevention of autoimmune diabetes in NOD mice. Suppression of generation of effector cells for pancreatic B cell destruction. *J. Immunol.* **144,** 136–41.

Sibley, R. K., Sutherland, D. E. R., Goetz, F., and Michael, A. F. (1985). Recurrent diabetes mellitus in the pancreas iso- and allograft. A light and electron microscopic and immunohistochemical analysis of four cases. *Lab. Invest.* **53,** 132–44.

Sochett, E. and Daneman, D. (1989). Relation of insulin autoantibodies to presentation and early course of IDDM in children. *Diabetes Care* **12,** 517–23.

Solimena, M., Folli, F., Aparisi, R., Pozza, G., and De Camilli, P. (1990). Autoantibodies to

GABA-ergic neurones and pancreatic beta-cells in stiff-man syndrome. *New Engl. J. Med.* **322,** 1555–60.

Spinas, G. A., Mandrup-Poulsen, T., Molvig, J., Baek, L., Bendtzen, K., Dinarello, C. A., and Nerup, J. (1986). Low concentrations of interleukin-1 stimulate and high concentrations inhibit insulin release from isolated rat islet of Langerhans. *Acta Endocrin.* **113,** 551–8.

Springer, T. A. (1990). Adhesion receptors of the immune system. *Nature* **346,** 425–34.

Sumoski, W., Baquerizo, H., and Rabinovitch, A. (1989). Oxygen free radical scavengers protect rat islet cells from damage by cytokines. *Diabetologia* **32,** 792–6.

Sutton, R., Gray, D. W. R., McShane, P., Dallmann, M. J., and Morris, P. J. (1989). The specificity of rejection and the absence of susceptibility of pancreatic beta cells to non-specific immune destruction in mixed strain islet grafted beneath the renal capsule in the rat. *J. Exp. Med.* **170,** 751–62.

Tainer, J. A., Getzoff, E. D., Paterson, Y., Olson, A. J., and Lorner, R. A. (1985). The atomic mobility component of protein antigenicity. *Ann. Rev. Immunol.* **3,** 501–35.

Taylor, S. I., Barbetti, F., Accili, D., Roth., J., and Gorden, P. (1989). Syndromes of autoimmunity and hypoglycemia. Autoantibodies directed against insulin and its receptor. *Endocrinol. Metab. Clin. North Am.* **18,** 123–43.

Thomas, D. W., Danho, W., Bullesbach, E., Fohles, J., and Rosenthal, A. S. (1981). Immune response gene control of determinant selection III. Polypeptide fragments of insulin are differentially recognized by T but not by B cells in insulin-immune guinea pigs. *J. Immunol.* **126,** 1095–100.

Weksler, M. E., Moody, C. E., and Kozak, R. W. (1981). The autologous mixed-lymphocyte reaction. *Adv. Immunol.* **31,** 271–312.

White, J., Herman, A., Pullen, A. M., Kubo, R., Kappler, J. W., and Marrack, P. (1989). The Vβ-specific superantigen Staphylococcal Enterotoxin B: stimulation of mature T cells and clonal deletion in neonatal mice. *Cell* **56,** 27–35.

Whiteley, P. J., Jensen, P. E., Pierce, O. W., Abruzzini, A. R., and Kapp, J. A. (1988). Helper T-cell clones that recognize autologous insulin are stimulated in nonresponder mice by pork insulin. *Proc. Natl Acad. Sci. USA* **85,** 2723–7.

Wilkin, T. J. (1990). Insulin autoantibodies as markers for Type 1 diabetes. *Endocrine Rev.* **11,** 92–104.

Wilkin, T. J., Hammonds, P., Mirza, J., Bone, A. J., and Webster, K. (1988*a*). Graves disease of the beta-cell: Glucose dysregulation due to islet cell stimulating antibodies. *Lancet* **ii,** 1151–8.

Wilkin, T., Mirza, I., and Armitage, M. (1988*b*) Insulin autoantibody polymorphism with greater discrimination for diabetes in humans. *Diabetologia* **31,** 670–4.

Winkel, M. van de, Smets, G., Gepts, W., and Pipeleers, D. G. (1982). Islet cell surface antibodies from insulin-dependent diabetics bind specifically to pancreatic beta cells. *J. Clin. Invest.* **70,** 4–9.

Yoon, J. W., Kim, C. J., Pak, C. Y., and McArthur, R. G. (1987). Effects of environmental factors on the development of insulin-dependent diabetes mellitus. *Clin. Invest. Med.* **10,** 457–69.

Young, R. A. and Elliott, T. J. (1989). Stress proteins, infection and immune surveillance. *Cell* **59,** 5–8.

Zier, K. S., Leo, M. M., Spielman, R. S., and Baker, L. (1984). Decreased synthesis of interleukin-2 (IL-2) in insulin-dependent diabetes mellitus. *Diabetes* **33,** 552–5.

11 Aetiology of type II diabetes

EROL CERASI

Type II diabetes is certainly the most common form of the disease, even in Scandinavia where the prevalence of type I diabetes is remarkably high. The past decades have seen an increase in type II diabetes in Western societies, which cannot be accounted for solely by improved diagnostic procedures or increased longevity (Everhart *et al.* 1985). In addition, the traditionally low prevalence of type II diabetes in the Far East and in developing countries is rapidly becoming a feature of the past, due to the ongoing improvement in standards of living. It seems certain, therefore, that humanity will enter the twenty-first century with at least twice as many type II diabetics as today, unless curative measures can be established during this decade.

It is doubtful whether a cure for a disease can be discovered without knowledge of its aetiology, and to investigate the aetiological factors a disease has to be well defined and delineated. There is no shortage of formal definitions, nor of diagnostic criteria, for type II diabetes (see Chapter 8). However, in the 'real world' of clinical medicine, categorizing a diabetic patient as type II, and type II alone, is not always as straightforward. According to the prevailing clinical praxis in any given country, varying proportions of 'type II' diabetics are treated with insulin (Criffiths *et al.* 1986). The investigator cannot always ascertain whether such patients are true secondary failures to oral antidiabetic agents, or type II diabetics given unnecessary insulin treatment due to paramedical factors such as the poor compliance of the patient or an erroneous decision made by an unskilled physician; or whether they belong to the category of so-called slow onset type I diabetics (see Chapters 8 and 10). Even the seemingly obvious recognition of diabetes induced by drugs such as steroids or thiazides may give rise to an aetiological dilemma: perhaps such diabetogenic drugs simply allow the expression of diabetes in predisposed subjects, in whom some, but not all, the aetiological factors were already operative prior to drug administration. These and other confounding factors make the delineation of 'pure' type II diabetics as objects of study for the clarification of the aetiology of the disease quite difficult. It has become common wisdom to state that type II diabetes is a heterogeneous disorder of many aetiologies; it is questionable, however, whether we possess the objective criteria to support or refute this statement. This difficulty in deciding whether type II diabetes is one or many diseases, has major implications for any discussion of the aetiology of type II diabetes.

11.1 Heredity in type II diabetes

The suspicion that diabetes is a genetic disorder stems from the clinical observation of clustering of the disease in families. Although some early studies did suggest that diabetes in the adult may be inherited as an autosomal dominant disease (Grunner 1957), the lack of precision in identifying the patients as type I or type II diabetics, together with other difficulties discussed below, usually led to confusing conclusions; this was elegantly summarized by J. V. Neel (1965) with the now classical characterization of diabetes as 'a geneticist's nightmare'. Two-and-a-half decades later, despite the clear recognition of the various types of diabetes, the clarification of some of the genetic mechanisms of type I diabetes, and the application of techniques of molecular biology to family and population studies, the picture remains sufficiently blurred to justify renewed use of the 'nightmare' image (O'Rahilly *et al.* 1988).

Type II diabetes manifests itself late in life. Furthermore, for each diagnosed diabetic there are almost four subjects with impaired glucose tolerance and one undiagnosed patient (Harris *et al.* 1987; Harris 1989). Therefore, the outcome of population or family studies, by necessity, depends on the criteria and tools used to define diabetes or abnormalities of glucose homeostasis, and on the age of the subjects studied. In the following, the genetics of diabetes will be discussed as a function of the parameters studied, from pure clinical assessments to the use of molecular probes.

11.1.1 Inheritance of clinical type II diabetes

One indication of the importance of genetic factors for the expression of the diabetic phenotype is the vast differences in the prevalence of type II diabetes between distinct ethnic groups, especially when they live in the same geographical area (Ekoe 1986). It has also been shown that in high-prevalence groups (e.g. American Indians) admixture of genes from low-prevalence populations (e.g. Caucasoids) reduces the occurrence of type II diabetes in proportion to the degree of gene mixing (for a full discussion, see Zimmet *et al.* 1989). Obviously, such epidemiological studies provide no information regarding the mode of inheritance of the disease.

In some populations where the prevalence of type II diabetes is extremely high (Pima Indians, Naurus, Mexican Americans), the distribution of fasting or postprandial blood levels shows bimodality, possibly suggestive of a single dominant gene for diabetes (Bennett *et al.* 1976; Zimmet and Whitehouse, 1978; Rosenthal *et al.* 1985); however, in Caucasians the distribution of glucose levels is unimodal (Harris 1989). In several large family studies, either the presence of type II diabetes or glucose intolerance in each generation, or the high frequency of diabetes among parents and siblings, suggested a mode of inheritance most akin to autosomal dominant (Grunnet 1957; Ohlsén *et al.* 1971; Köbberling and Tillil 1982).

Three special cases further illustrate the genetic connection of type II diabetes. The first relates to early onset type II diabetes. O'Rahilly *et al.* (1987) called attention to the fact that patients who develop type II diabetes at a younger age (25–40 years) often have parents who both have diabetes or impaired glucose tolerance. Although these patients had none of the immunological stigmata of type I diabetes, the disease was usually severe and required insulin treatment, which led the authors to suggest that a 'double dose' of diabetogenic genes was inherited from the parents. Indeed, the finding of diabetes or glucose intolerance in 69 per cent of the siblings of patients with early onset type II diabetes by O'Rahilly *et al.* (1987), and in 53 per cent of siblings in another series of patients studied by Köbberling and Tillil (1989) using a different methodology, supports the idea that in this form of the disease diabetogenic genes are acquired from heterozygous parents by Mendelian inheritance. When the selection criteria were reversed, subjects whose parents were both diabetic showed a high incidence of diabetes or impaired glucose tolerance in most but not all studies (Kahn *et al.* 1969; Radder and Terpstra 1975; Tattersall and Fajans, 1975; Ganda *et al.* 1985; Viswanathan *et al.* 1985). It is interesting to note, however, that these diabetic offspring usually failed to present the clinical characteristics of early onset type II diabetes. The reason for this contradiction is not clear.

The second example also concerns young individuals with type II diabetes; however, unlike the early-onset diabetics studied by O'Rahilly *et al.* (O'Rahilly *et al.* 1987; O'Rahilly and Turner 1988) their disease is mild and seldom requires insulin treatment. Called MODY (for maturity onset diabetes of the young), this form of the disease comprises from 5 to 19 per cent of all type II diabetics, depending on the population studied and whether comprehensive screening of family members was performed (Asmal *et al.* 1981; Panzram and Adolph 1983; Mohan *et al.* 1985; Winter *et al.* 1987). Fajans (1990) has recently summarized a 30-year experience of MODY; most pedigrees were large, covered several generations, and presented multiple cases of type II diabetes or glucose intolerance in each generation. Analysis of pedigrees from many countries and different ethnic groups convincingly demonstrates that MODY is inherited as an autosomal dominant trait (reviewed by Fajans 1990).

The third and most compelling proof for the genetic basis for type II diabetes comes from studies on twins. Barnett *et al.* (1981) summarized data on 360 pairs of monozygotic twins where one twin was diabetic, where information on the type of diabetes was available, and where data (clinical or glucose tolerance) on the other twin existed. In the whole group, 60 per cent of the twins were concordant for diabetes, that is, were both diabetic. In the 215 pairs where one twin had type I diabetes (according to clinical criteria), concordance was only 47 per cent, whereas among the 145 pairs where one twin had type II diabetes, concordance was almost twice as high (80 per cent). If restricted to the data of Pyke and collaborators on 200 pairs of British twins (the best controlled twin study), concordance was 54 per cent for type I diabetes, and 91 per cent for type II diabetes (Barnett *et al.* 1981). Thus, even allowing for selection bias and possible misclassification of diabetics, the twin studies undeniably prove that type II diabetes has a genetic cause with almost 100 per cent penetrance at high age. It is important to note that a similar degree of concordance of type II diabetes has recently been reported in another racial group (Committee on Diabetic Twins, Japan Diabetes Society 1988): among 56 pairs of Japanese monozygotic twins, in 46 both twins had type II diabetes (83 per cent concordance).

To summarize, the evidence suggesting that type II diabetes is a genetic disorder is overwhelming, and at least in the case of monozygotic twins, penetrance of the trait at advanced age may reach 100 per cent. In genetically isolated populations with extraordinarily high prevalences of type II diabetes, or in special forms of the disease (e.g. MODY) or in special circumstances (both parents diabetic) inheritance seems to be autosomal dominant. For the vast majority of type II diabetics, however, there is no convincing evidence to suggest dominant inheritance. It is not clear if this is due to basic pathogenic differences (e.g. between MODY and ordinary type II diabetes), or to the fact that diabetogenic genes are enriched in the special forms of diabetes. While the latter may be correct for populations like the Pima Indians, neither the MODY (Fajans 1990) nor the early onset type II diabetes (O'Rahilly and Turner 1988) families seem to represent groups with a high degree of inbreeding.

11.1.2 Inheritance of impaired glucose tolerance (IGT)

A major difficulty in the interpretation of studies dealing with the genetics of type II diabetes originates from the fact that the early stages of the disease are asymptomatic, and unless specific testing, such as glucose tolerance tests, is used, phenotypic expression may be grossly underestimated (Harris *et al.* 1987; Harris 1989). However, there is no consensus that impaired glucose tolerance (IGT) should be regarded as a phenotypic expression of the diabetes genes. Indeed, it is not clear if IGT is an obligatory stage in the development of type II diabetes; certainly not all subjects with IGT develop clinical diabetes within the observation period (Harris 1989). Nevertheless, each year up to 5 per cent of subjects with IGT may develop overt diabetes (Sartor *et al.* 1980; Keen *et al.* 1982; Sicree *et al.* 1987; Ohlson *et al.* 1988). It is therefore important to ascertain whether glucose tolerance *per se* is under genetic control.

Despite the widespread use of glucose tolerance tests as a diagnostic tool, the importance of genetic factors in determining glucose tolerance has not been studied systematically. Familial and ethnic influences on glucose tolerance have been described (King *et al.* 1989; Leonetti *et al.* 1989), but it has not been possible to delineate the mode of inheritance of IGT (Rotter and Rimoin 1981). In 13 monozygotic twins discordant for type II diabetes, borderline or impaired glucose tolerance was found in the majority (Barnett *et al.* 1981; Committee on Diabetic Twins, Japan Diabetes Society 1988). No follow-up data are available on these twins. Hence it is not clear whether IGT was the beginning of type II diabetes, which would emphasize the concordance of diabetes, or alternatively a reflection of the inheritance of a trait akin to, but distinct from, diabetes.

The role of genetic factors in determining various characteristics of the plasma glucose curves during an intravenous glucose tolerance test (IVGTT) or prolonged intravenous glucose infusion has been evaluated in two studies from Stockholm. One comprised 279 non-diabetic subjects belonging to 52 families and included 24 monozygotic and 29 dizygotic twin pairs (Lindsten *et al.* 1976); in the second (Iselius *et al.* 1982), 601 subjects were selected from 96 families where the probands were diabetic, and from 59 non-diabetic families. In both studies, complex mathematical genetic analyses showed that the response of blood glucose to a glucose challenge is significantly influenced by genetic factors. Genetic heritability was always larger than 'cultural heritability', the magnitude of the genetic component (h^2) varying between 0.20 and 0.54 according to the mathematical parameter studied. Interestingly, no difference was found between the children belonging to diabetic and non-diabetic families (Iselius *et al.* 1982); this is in contrast to the finding of 22 per cent IGT in the offspring of one diabetic parent by Leslie *et al.* (1986). It is possible that the selection and size of the populations studied may explain the difference.

Two conclusions seem reasonable. First, in healthy subjects, glucose tolerance is partly controlled by genetic factors. Whether this means that subjects with a tendency to higher glucose levels are at higher risk of developing IGT is not known. Several studies have indicated, however, that in high-risk populations a high 2 h oral glucose tolerance test (OGTT) value is positively correlated with the future development of diabetes. Secondly, IGT occurs at high frequency in first-degree relatives of diabetics. While it is possible that people inherit 'mild' diabetes genes that produce IGT only, it seems more reasonable to assume that the same genetic factors are responsible for IGT and type II diabetes, the former simply being an early stage manifestation of the latter. The finding of IGT in the discordant twin pairs, and the very high degree of concordance for type II diabetes in the same twin material at advanced age, support the latter interpretation.

11.1.3 Inheritance of regulatory functions in glucose homeostasis

Since the phenotypic expression of the diabetic syndrome is variable and occurs late in life when environmental factors play an increasingly complex role, it would be more logical to study the genetic control of some key functions that regulate glucose homeostasis, and which, therefore, may be involved in the pathophysiology of type II diabetes. Two such functions, β-cell secretory capacity and peripheral responsiveness to insulin, are obvious candidates. Unfortunately, however, these functions are difficult to quantitate, and the need to use time- and manpower-consuming procedures for their evaluation has discouraged most researchers. Therefore systematic studies on their genetic aspects are scarce.

Genetic control of insulin release

The qualitative and quantitative characteristics of insulin secretion in response to stimuli are determined by complex mechanisms (see Chapter 4). Although it is natural to expect that genetic control of this response will be as evident as that found for events of similar complexity (e.g. body growth), the fact that β-cell function is measured at a distance and indirectly, via plasma insulin or C-peptide levels, introduces a factor of uncertainty. In addition, many external factors, such as physical fitness (O'Rahilly *et al.* 1988), insulin sensitivity (Hollenbeck and Reaven 1987), or age (Chen *et al.* 1988), have profound effects on β-cell function, which reduces the utility of the *in vivo* insulin response as the phenotypic expression of a given genotype in family or population studies. It is therefore astonishing that, despite these limitations, some studies (Cerasi and Luft 1974; Niskanen *et al.* 1990) have found the insulin response to intravenous or oral glucose to be reproducible over extensive time periods (although others could not verify this; Smith *et al.* 1988).

Initial evidence for genetic regulation of insulin release in man was obtained in the 1960s by the observation that the shape and magnitude of the plasma insulin response to a glucose infusion was strikingly similar in monozygotic twin pairs (Cerasi and Luft 1967). These studies were later repeated on a larger sample of non-diabetic monozygotic and dizygotic twins. The variability of the insulin response within a pair of monozygotic twins was found to be much smaller than the variability between dizygotic twins or non-twin siblings, thus suggesting the existence of a strong genetic influence on this function (Lindsten *et al.* 1976). Finally, in two large family studies, complex segregation analysis of the insulin response to glucose infusion showed that, in quantitative terms, the influence of genetic factors was highly significant (Lindsten *et al.* 1976; Iselius *et al.* 1982). Indeed, not only the total insulin curve but also the kinetics of the response (*i.e.* the magnitudes of the first- and second-phase responses) were found to be under genetic control (Iselius *et al.* 1985). Thus, despite the multitude of factors that can influence insulin secretion, there seems to be a major genetic effect on β-cell function in normal man, which can be evidenced by measuring the plasma hormone levels during standard challenge tests.

More abundant data is available on the insulin response of family members of type II diabetics. Although these studies were not designed to quantitate eventual genetic effects, they certainly indicate that β-cell function is modified in some non-diabetic, glucose-tolerant relatives of diabetic patients. Thus, the insulin response to oral glucose was severely impaired in five monozygotic twins discordant for type II diabetes (Barnett *et al.* 1981). In the offspring of MODY patients (Mohan *et al.* 1986) or first-degree relatives of type II diabetics (Berntorp *et al.* 1986) a high proportion of subjects with normal glucose tolerance had a diminished insulin response to glucose administration. The proportion of decreased insulin responses was high also in the offspring of conjugal diabetic patients (Soeldner *et al.* 1986; Snehalatha *et al.* 1986). An obvious difficulty in the interpretation of such findings is that subjects with decreased insulin secretion but normal glucose tolerance may be in a dynamic state evolving towards glucose intolerance (see Chapter 8); therefore, a low insulin response could be the expression of a diabetogenic process rather than that of genes controlling insulin secretion and predisposing to diabetes.

Genetic control of insulin action

The overall insulin sensitivity of a person is the sum of specific tissue sensitivities, each being determined by a series of complex biological events (see Chapter 5). Notwithstanding this complexity, it is possible to demonstrate that the effect of insulin on glucose disposal in man is indeed influenced by genetic factors. Bogardus *et al.* (1989) performed hyperinsulinaemic–euglycaemic clamps on 245 non-diabetic Pima Indians and analysed the maximal glucose utilization data for population distribution mixtures. Their data were best fitted with the sum of three normal distributions, suggesting to the authors that in Pima Indians insulin action (or insulin resistance) is determined by a single gene with a co-dominant mode of inheritance. Although this hypothesis is attractive, either much larger population studies or segregation analysis of data from numerous pedigrees (see, for example, Iselius *et al.* 1985) would be necessary to verify it. Furthermore, the applicability of these observations, obtained in a highly inbred population, to more heterogeneous populations remains to be assessed.

Iselius *et al.* (1985) analysed the sensitivity to physiologically elevated endogenous insulin concentrations, with the help of mathematical modelling, in data from Swedish families comprising 601 subjects. Using both path and segregation analysis, the authors concluded that sensitivity to endogenous insulin was controlled by multifactorial heritability, the genetic component (h^2) varying between 0.35 and 0.42, without evidence for a major locus. However, the same study demonstrated, as expected, that environmental factors do play a major role in determining insulin sensitivity; thus, cultural heritability by path analysis was larger ($c^2 = 0.56$) than for genetic heritability ($h^2 = 0.35$).

Thus, these two studies clearly suggest that *in vivo* insulin sensitivity, at least to a certain degree, reflects the influence of genetic factors. However, despite the large number of subjects studied, the mode of inheritance of insulin sensitivity could not be defined. This is probably the consequence of the complexity of the phenomenon studied, the limited precision of the *in vivo* methods used, and the existence of important environmental factors that can override the effect of genetic factors on insulin sensitivity.

Two studies indicate that genetic factors play a role in impairing insulin sensitivity in subjects at risk of developing type II diabetes. In 45 Pima Indian sibships (116 non-diabetic subjects) hyperinsulinaemic–normoglycaemic clamp studies demonstrated that maximally stimulated peripheral glucose uptake shows significant familial aggregation (Lilioja *et al.* 1987); after correction for age, sex, and obesity, ~34 per cent of the variability in insulin action could be accounted for by family affiliation. It is interesting to note the similarity between this number and the heritability coefficients (0.35–0.42) found by Iselius *et al.* (1985) in a totally different population, using unrelated measurements and genetic analysis techniques. Eriksson *et al.* (1989) used insulin clamps combined with labelled glucose administration to study glucose metabolism in first-degree relatives of type II diabetics. In 13 of the relatives with normal glucose tolerance, total insulin-stimulated glucose metabolism was reduced to the same extent as in relatives with IGT or with type II diabetes. The major determinant of the reduction was non-oxidative glucose metabolism. Contrasting with these findings, a recent study (Osei 1990) showed augmented basal glucose utilization in first-degree relatives of type II diabetics. The significance of these differences is not immediately apparent.

In summary, there is sufficient evidence to suggest that genetic factors are important for determining the insulin sensitivity of a subject. The genetic link between reduced insulin sensitivity and type II diabetes is less convincing; as is the case for reduced β-cell function, such an impairment could reflect an early metabolic derangement during the development of diabetes, rather than a primary genetic trait.

11.1.4 The search for molecular probes of type II diabetes

In an era when the molecular basis for many hereditary diseases has been elucidated, it is disappointing that the gene(s) for diabetes, or at least genes with linkage to type II diabetes, have remained elusive. Several reports have been published on restriction fragment length polymorphism (RFLP) studies on the genes for insulin, the insulin receptor, and glucose transporters, and various mutations have been described. In the following, a concise review of the main findings will be presented.

The insulin gene

Type II diabetic patients do produce insulin; therefore, gross deletions in the insulin gene would not be expected. However, more discrete mutations in the preproinsulin gene might have several functional consequences, for example:

1. alteration of the efficiency of transcription and/or translation of the hormone, thus limiting the availability of insulin;
2. leading to derangements in the cellular sorting and trafficking of the protein, thus reducing the efficiency of proinsulin cleavage and of regulated insulin secretion; or
3. changing the configuration of the hormone and impairing receptor binding and activation (Steiner *et al.* 1989).

When a point mutation (PheB25 to Leu) in the receptor binding domain of insulin was first discovered to cause diabetes (Tager *et al.* 1979) it was felt that such defects would explain the disease in many type II diabetes patients. However, to-date only six families with similar mutations have been described (PheB25 to Leu, PheB24 to Ser, and ValA3 to Leu; reviewed by Steiner *et al.* 1990). All affected individuals had mild hyperglycaemia (70–340 mg/dl) and marked hyperinsulinaemia (33–440 μU/ml). They were all heterozygous for the insulin gene. The modified insulins bound to insulin receptors with less than 5 per cent efficiency, and had accordingly reduced biological activity and prolonged plasma half-life (Steiner *et al.* 1990).

Neither a markedly augmented plasma half-life of insulin, nor major fasting hyperinsulinaemia are hallmarks of randomly selected type II diabetics, and therefore it is likely that the point mutations described above are uncommon causes of diabetes. This is supported by the studies of Sanz *et al.* (1986), who assessed the incidence of PheB24 or PheB25 mutations in 213 type II diabetics by RFLP analysis using *Mbo*II, which recognizes this site; only one cleavage defect was found, which, furthermore, was not associated with an abnormal insulin.

A second type of point mutation in the proinsulin gene impairs the processing of the prohormone. Four families have been described with minimal changes in glucose homeostasis and with markedly elevated circulating levels of proinsulin-like material (reviewed by Steiner *et al.* 1990). In three instances, the mutated insulin had a substitution for Arg at the Lys–Arg cleavage site of the C-peptide/A-chain junction; more interesting is the family with a HisB10 to Asp mutation (Chan 1987), which is distant from the proinsulin cleavage sites. Further studies have shown that HisB10 is essential for the correct sorting of the nascent proinsulin to the regulated, Golgi-associated pathway, thus permitting normal processing and secretion of the hormone. A mutation at this site directs a proportion of the prohormone to the constitutive pathway, escaping regulation (Carroll *et al.* 1988; Gross *et al.* 1989).

Although type II diabetics show hyperproinsulinaemia as well as fasting hyperinsulinaemia, these are corrected by normalization of the blood glucose levels, and there is no evidence for unregulated release of insulin (see Chapter 8). It is therefore again unlikely that mutations at B10 or at the proinsulin cleavage sites are frequent causes of type II diabetes. However, modifications in the 5′ untranslated region, as well as other discrete mutations in the insulin gene, could modify β-cell function in a subtle manner and eventually lead to diabetes. It was therefore logical to screen diabetic families and populations for such mutations by performing RFLP studies.

Although initial studies seemed to demonstrate an association between 5′-flanking region polymorphism of the insulin gene and type II diabetes (Owerbach and Nerup 1982; Rotwein *et al.* 1983), in later studies comprising large diabetic pedigrees and MODY families, no linkage could be found between the insulin locus and diabetes (Owerbach *et al.* 1983; Hitman *et al.* 1984; Andreone *et al.* 1985; Elbein *et al.* 1988*a*). Similarly, when more subtle effects of insulin gene modifications were investigated, no linkage could be established between 5′-flanking region polymorphism and various parameters of insulin secretion (Permutt *et al.* 1985; Samanta *et al.* 1987; Elbein *et al.* 1988*b*). One exception is the study of Cocozza *et al.* (1988): in 64 healthy subjects, a quantitative correlation was found between the plasma C-peptide response to oral glucose, and the presence of a 1.6 kb insertion at the 5′ end of the insulin gene. Furthermore, in class 3/3 and class 1/1 homozygous subjects defined by RFLP studies, insulin secretion during a hyperglycaemic clamp was found to be threefold higher in class 1/1 individuals (Cocozza *et al.* 1988). If confirmed, these studies may suggest that the 5′-flanking region of the insulin gene could influence insulin secretion in a direct or indirect manner.

The insulin-receptor gene

Severe insulin resistance syndromes, like leprechaunism, are very rare disorders in which the central role played by the insulin receptor has been recognized for many years (see Chapter 6). Recent advances in the molecular biology of the insulin receptor has enabled the detailed description of mutations associated with these syndromes (reviewed by Seino *et al.* 1990*a*). The insulin resistance of type II diabetes, however, is far less pronounced than that of these syndromes; nevertheless, as illustrated by some of the parents of patients with point-mutations in the α-subunit of the receptor (Kadowaki *et al.* 1988; Klinkhamer *et al.* 1989), coexpression of both normal and defective insulin receptors leads to mild insulin resistance, fully compatible with type II diabetes. The human insulin-receptor gene can now be analysed in detail in clinical material (Seino *et al.* 1990*b*). Future studies of mutations, especially in the exons coding for the α-subunit, may reveal whether genetic modifications of the insulin receptor are important for the pathogenesis of common type II diabetes.

RFLP studies of the insulin-receptor gene have so far been disappointing, no clear indication being obtained on the possible link between insulin-receptor alleles and diabetes. On the one hand, a significant association was found between some alleles and type II diabetes in Black Americans (McClain *et al.* 1988), and gestational diabetes in Black and Caucasian women (Ober *et al.* 1989). Moreover, in Chinese Americans (Xiang *et al.* 1989), the occurrence of specific insulin-receptor haplotypes was protective against type II diabetes. In contrast, several other investigators failed to detect correlations between polymorphism of the insulin-receptor gene and type II diabetes (Elbein *et al.* 1986, 1987), gestational diabetes (Elbein *et al.* 1988*b*), or MODY (Elbein *et al.* 1987). Furthermore, the insulin-receptor cDNA sequence in two Pima Indians with type II diabetes and insulin resistance (Moller *et al.* 1989) was found to be normal. It seems, therefore, that if linkage does exist between the insulin-receptor locus and type II diabetes, its demonstration necessitates the study of much larger populations and family materials, and the use of techniques capable of revealing more subtle structural modifications (Seino *et al.* 1990*b*).

The glucose transporters

The insulin-resistance characteristic of type II diabetes is of the so-called post-receptor defect type, which involves the glucose transport system of peripheral tissues. The past few years have witnessed an explosive development in the characterization of the glucose transporter family, and molecular probes have become available for the four main types of transporters (reviewed by Bell *et al.* 1990). However, genetic studies of the family of glucose transporters are still in their infancy.

Restriction fragment length polymorphisms have been identified for the two main types of glucose transporters, GLUT-1 (Shows *et al.* 1987) and GLUT-4 (Bell *et al.* 1989). The ubiquitous glucose transporter GLUT-1, thought to be mainly responsible for basal glucose transport (Bell *et al.* 1990), has been subjected to intensive investigation. Only one study (Li *et al.* 1988) found linkage between genetic variation of GLUT-1 and type II diabetes. In contrast, linkage could be rejected in many studies performed on various ethnic groups with type II diabetes (O'Rahilly *et al.* 1989; Serjeantson *et al.* 1989; Xiang *et al.* 1989; Kaku *et al.* 1990), and in MODY (Vinik *et al.* 1988) and gestational diabetes (Ober *et al.* 1989).

GLUT-4, the adipocyte/muscle insulin-responsive glucose transporter, may be regarded as a more appropriate candidate for linkage studies in insulin-resistant and/or diabetic subjects. However, so far only limited information on GLUT-4 RFLPs exist; the data seem to rule out linkage with type II diabetes (Bell *et al.* 1990).

GLUT-2, the liver/β-cell glucose carrier (Johnson *et al.* 1990*a*,*b*) is a most attractive candidate since it controls glucose output from the liver and glucose entry into the β-cell (and hence insulin release); no polymorphisms of this transporter have been reported yet. Finally, detailed structural studies of the various transporters in type II diabetes, like those described above for mutations in the insulin and insulin-receptor genes, have yet to be performed.

Other candidate genes

In addition to modifications in proinsulin, the insulin receptor, and the glucose transporters, the malfunction of three other proteins could, in theory, impair β-cell function and thus glucose homeostasis. The first, glucokinase, is a low-affinity isoenzyme of the hexokinase family, and is mainly expressed in liver and pancreatic β-cells where it controls the cellular glucose metabolism (Matschinsky 1990). As discussed in Chapter 4, it has been proposed that glucokinase is the glucose sensor of the β-cell and, therefore, the key element in the regulation of insulin release. It has also been demonstrated that the expression of this enzyme is regulated differently in liver and islet cells (Iynedjian *et al.* 1989), insulin being the inducer in the former, and glucose in the latter. Modifications in the function or regulation of glucokinase may therefore impair, selectively or in combination, hepatic glucose metabolism and pancreatic insulin secretion.

Another protein recently implicated both in defective insulin output and the decreased insulin sensitivity of type II diabetes is amylin (Porte and Kahn 1989; Leighton and Cooper, 1990). It is not yet known whether this protein plays a primary or secondary role in diabetes. Nevertheless, it is indeed possible that discrete genetic changes in its structure or regulation ascribe a pathogenic role to this physiological substance.

Finally, a recently described, islet cell-specific protein, REG, seems to be important for β-cell regeneration (Terazono *et al.* 1990; Watanabe *et al.* 1990). Deficient expression of the *reg* gene could therefore diminish the capacity of the β-cells to adapt to increased insulin demand in situations of insulin resistance, and thus promote the development of type II diabetes.

General remarks on molecular probes of type II diabetes

It is apparent from the previous paragraphs that attempts to find a molecular probe for type II diabetes have so far been unsuccessful, despite impressive efforts.

Often, the blame for this lack of success is put on the heterogeneity of the disease, and the fact that diabetes expresses itself very late in life. However, other complex diseases with late phenotypic expression, e.g. multiple endocrine neoplasia type 1, have been amenable to molecular genetic analysis (Byström *et al.* 1990).

Has the choice of the probes used so far been inadequate? The chain insulin → insulin receptor → glucose transporter is certainly involved in the pathophysiology of type II diabetes. The question that can be raised, however, is whether the cDNA probes used adequately define the functions believed to be impaired in diabetes. For instance, no evidence exists to suggest that in the common type II diabetic patient the insulin molecule is defective. Insulin release is deficient; however, the probability that structural changes in the insulin gene would lead to impairment in insulin release is not large. The exact mechanisms that control insulin secretion and that may be defective in the type II diabetic β-cell are not known (see Chapter 4). Probes for some of the important β-cell components (glucokinase, GLUT-2) are available while others are emerging, therefore coming years will certainly witness studies in type II diabetics with such probes. Success will depend on the correctness of the guess of the main pathophysiological mechanism in type II diabetes, and whether such a derangement is a primary (genetic) or an acquired defect. The role of insulin resistance in the pathophysiology of type II diabetes is discussed in detail below; the feeling of this author is that much of the resistance is secondary to the diabetic metabolic state, and therefore hardly expected to be reflected by gross modifications in the structure of insulin receptors or glucose transporters. This obviously does not exclude the possibility that very discrete modifications in the structure of the insulin receptor or glucose transporters may make them sensitive to the metabolic milieu of diabetes and promote insulin resistance.

11.2 Environmental versus genetic factors in the aetiology of type II diabetes

In contrast to the situation in type I diabetes (see Chapter 10), no data exist to suggest that toxic agents or viral infections can induce type II diabetes. However, such a possibility is difficult to exclude *a priori*. Indeed, in rats it is possible to induce a syndrome very similar to type II diabetes by administering streptozotocin in the neonatal period; once adult, these animals present mild hyperglycaemia, reduced insulin output, and some insulin resistance (Blondel *et al.* 1989). Furthermore, the intrauterine metabolic situation may have a long-lasting effect on the offspring. Thus, Petitt *et al.* (1985) demonstrated in a large group of Pima Indians with gestational diabetes that the offspring, studied 5–19 years later, were more obese and had higher blood glucose levels after a glucose challenge than offspring of non-diabetic women. Obviously, it is difficult to dissociate the genetic component from the intrauterine metabolic impact in this example. However, the example still serves as a caveat for the presence of unknown environmental factors acting during the perinatal period which may be responsible for the future development of type II diabetes.

The best-studied environmental factor in the development of type II diabetes is insulin resistance induced by Western life habits, mainly linked to obesity and physical inactivity. Epidemiological examples abound of populations with a low prevalence of type II diabetes in which, on migration and/or change of life-style, the disease frequency sharply increases. Noteworthy are the examples of the Nisei (Japanese–Americans) compared to the parent population in Japan (Leonetti and Fukimoto 1989), and that of Yemenite Jews on arrival and after several years of settlement in Israel (Medalie *et al.* 1978). Obesity, especially abdominally located fat excess, has been shown in many epidemiological and prospective studies to be a significant risk factor for type II diabetes (for example, see Ohlson *et al.* 1988).

Data exist, however, that ascribe a minimal direct role to the environment in the aetiology of type II diabetes. First, genetic control of insulin sensitivity has been suggested from studies in non-diabetic Caucasians (Iselius *et al.* 1985) and Pima Indians (Bogardus *et al.* 1989), and in family members of diabetic patients (Lilioja *et al.* 1987; Eriksson *et al.* 1989). Secondly, obesity itself (and the factors that lead to it) seems to be greatly influenced by genetic factors. Thus, Stunkard *et al.* (1986) demonstrated in 540 Danes adopted as infants, that their adult body weight was highly correlated to that of the biological parents, while no relationship existed with that of the foster parents. These results are compatible with those of an older study (Mayer 1965): 80 per cent of the offspring of two obese parents become obese, against only 14 per cent in the offspring of two lean parents. Furthermore, the rate of energy expenditure, which is an important determinant for the development of obesity and insulin resistance, seems also to be under some genetic control. In an elegant study in American Indians, Ravussin *et al.* (1988) found a significant (albeit weak, $r = 0.15$) negative correlation between 24 h energy expenditure adjusted for body composition, age, and sex, on the one hand, and change in body weight over a 2 year period, on the other hand. In 94 siblings from 36 families, 24 h energy expenditure showed significant familial aggregation, even after correction for the family effect on body size. Therefore, without negating the important impact of life-style, obesity and insulin resistance cannot be taken as pure environmental factors. The quantitative contributions of the genetic versus environmental factors to obesity seem difficult to determine.

Finally, that genetic factors may far outweigh environmental factors in the aetiology of type II diabetes is suggested by concordance studies in twins. In the study by Barnett *et al.* (1981), 39 out of 42 concordant diabetic monozygous twins were living apart at the time type II diabetes was diagnosed. In addition, body weight was different at diagnosis by more than 5 kg in 16 of 21 twins concordant for diabetes; in five pairs, one twin was obese and the other lean when diabetes had developed. Thus, at least in this limited sample of monozygotic twins concordant for diabetes, obesity and environmental factors seem to play a minor role in the development of type II diabetes. To what extent these conclusions can be extrapolated to the general population of type II diabetics is not clear. Nevertheless, much of the data at hand seems to indicate that the most important factor in the aetiology of type II diabetes is the genetic background of the individual.

11.3 Pathophysiology of type II diabetes

It is usually accepted that in type II diabetes both insulin deficiency and insulin resistance prevail. Opinions are divided, however, as to which of these factors is a primary aetiological factor. Analysis is complicated by the fact that hyperglycaemia *per se* impairs both insulin secretion and insulin action. In addition, only few longitudinal, prospective studies exist to permit certitude in assigning a pathogenic role to either changes in insulin release or in insulin sensitivity.

11.3.1 Insulin secretion in type II diabetes

To the uninitiated, perusal of the literature concerning whether type II diabetics are hypo- or hyperinsulinaemic, is a most confusing experience. Particularly among epidemiologists, it is widely held that as glucose intolerance increases, plasma insulin levels become increasingly elevated until major hyperglycaemia occurs, at which time insulin levels fall towards normal, thus producing a bell-shaped curve (Zimmet *et al.* 1989). On the other hand, authors more immediately interested in β-cell physiology usually claim a major impairment of insulin secretion. There are several reasons for these conflicting interpretations. The assessment of insulin secretion in man is difficult. This difficulty stems mainly from two factors. First, insulin release *in vivo* is estimated from peripheral insulin levels, with the assumption that the hepatic extraction and peripheral clearance of insulin are constant; however, it has been shown convincingly that this is not the case (Ferrannini *et al.* 1983). Methods for calculating the insulin secretion rate by C-peptide infusion (Shapiro *et al.* 1988) and by the use of mathematical models (Cerasi and Luft 1967) have been described; however, their use has been exceptional.

A second difficulty relates to the correlation between stimulus and response. The fact that an insulin response cannot be quantified unless it is related to the magnitude of the stimulus applied to the β-cell has been often ignored. Because the sensitivity of the β-cell to intravenous glucose is modest, ignoring the magnitude of the stimulus (within reasonable limits) may not be critical in studies where glucose was injected. However, in most studies glucose was given orally; under these conditions the synergistic interaction between glucose and 'incretins' is very large, and minute differences in blood glucose may have extraordinary effects on the plasma insulin concentration. Thus, in a dose–response study for oral glucose loads it could clearly be shown that at a critical blood glucose range, increasing the level by as little as 1 mmol/l in healthy subjects may be sufficient to increase the insulin response by 50–70 per cent (Cerasi *et al.* 1973). Considered against this background, it becomes apparent that insulin levels in a hyperglycaemic patient, if similar to that of a normoglycaemic subject, in reality indicate severe impairment of insulin secretion.

An additional problem in assessing insulin secretion in diabetics concerns the nature of the product measured by conventional immunoassays. It is known that hyperglycaemic individuals secrete a larger proportion of proinsulin and proinsulin split-products than normal subjects, the proportion varying between 20 and 60 per cent of total insulin-like reactivity (Temple *et al.* 1989). This seems to be due to the combined effects of hyperglycaemia and increased secretory demand, since the proportion of proinsulin increases in normal subjects rendered acutely insulin resistant with dexamethasone (Ward *et al.* 1987) but not in obese persons (Saad *et al.* 1990), and much higher levels are registered in type II diabetics (Ward *et al.* 1987; Saad *et al.* 1990). Thus, true insulin levels in diabetics are probably much lower than accepted. It seems imperative in future epidemiological and clinical studies either to use exclusively anti-insulin antibodies which do not cross-react with proinsulin and its split-products, or to define the degree of cross-reactivity (some antibodies show >60 per cent cross-reactivity).

Bearing in mind the above reservations, some aspects of insulin release in type II diabetics will now be described.

Basal insulin secretion

Insulin secretion is controlled by a variety of factors, including nutrient substrates, hormones, and neurotransmitters, glucose being the dominating regulator. There is no evidence to indicate that basal insulin secretion is regulated differently from stimulated release. Rather, the rate of basal release is simply dependent on the basal glucose concentration.

In type II diabetics basal insulin levels are variable but usually within the range of the control population (Turner and Holman 1976). However, since these patients are hyperglycaemic, their 'basal' state in reality corresponds to a situation of glucose stimulation. This has been shown by inducing normoglycaemia with insulin infusion in type II diabetics; the 'basal' plasma insulin values then diminished markedly (McCarthy *et al.* 1977). Furthermore, from comparison with the steady-state plasma insulin level obtained in control subjects rendered hyperglycaemic by glucose infusion, it can be concluded that the basal insulin secretion in diabetics is diminished to the same extent as the stimulated release. Based on these considerations it has been suggested that the basal insulin level in type II diabetes is the result of β-cell adaptation to hyperglycaemia, the degree of fasting hyperglycaemia in its turn being a function of the adaptive capacity of the β-cell, the feedback regulatory system thus arriving at equilibrium (Matthews *et al.* 1985).

Because of the lower clearance rate of proinsulin, the ratio of proinsulin to insulin is higher at the basal state than during acute stimulations of the β-cell. Therefore, the true basal plasma insulin levels in type II diabetics are proportionately more reduced than apparent from the above studies.

Acute insulin response to glucose

The most pronounced defect of the insulin response to glucose in type II diabetes is seen in the first minutes of stimulation. The insulin secretory pattern (even in

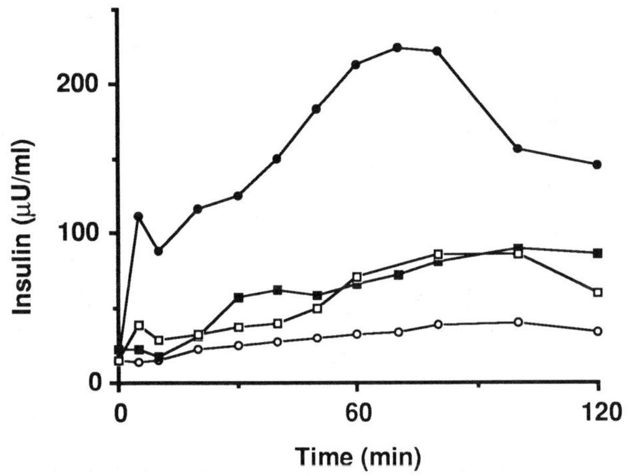

Fig. 1 Patterns of the plasma insulin response to a hyperglycaemic clamp (0–60 min, approximately 18 mM) in normal subjects (●), in subjects with impaired glucose tolerance (□), in mild non-insulin dependent diabetics (■; fasting blood glucose <8 mM), and in more advanced diabetics (○). All were moderately obese (body mass index 27–30). Note that the reduced first-phase insulin response of glucose-intolerant subjects disappears in hyperglycaemic patients and that the second-phase response is further reduced as hyperglycaemia advances

patients with modest hyperglycaemia) is characteristic, the decreased and delayed appearance of the first-phase response being striking (Fig. 1). This pattern is observed both in overtly diabetic subjects and in persons with glucose intolerance only; however, the higher the fasting blood glucose of the patient, the flatter the insulin response to glucose (reviewed by Cerasi 1988). Indeed, in several patients glucose may initially induce a fall in the plasma insulin level.

Reduced acute insulin response to glucose is also observed in a variety of diabetes-like syndromes in animals, such as *Macaca nigra*, the New Zealand white rabbit, Chinese hamsters, C57BL/KsJ-db mice, *Acomys cahirinus*, and non-ketotic BB rats (reviewed by Shafrir 1990). It is of interest that in diabetes induced in rats by experimental manipulations, such as neonatal streptozotocin injection or partial pancreatectomy, the effect of glucose on acute insulin release is also markedly impaired (Shafrir 1990).

Only a few studies exist on the relationship between the time course of glucose-induced insulin release and glucose concentration in diabetes. As shown in Fig. 2, in type II diabetics, the first-phase insulin response to glucose shows decreased maximal capacity (V_{max} reduction) (Cerasi 1988). The second-phase response was found to increase with very high blood glucose levels, the dose–response curve thus showing a shift to the right (increased apparent K_m). A similar K_m change was noted for the insulin response to oral glucose, which corresponds mainly to second-phase release (Fig. 2; Cerasi *et al.* 1973). However, saturation of the insulin response to glucose is obtained only with difficulty in man, and therefore statements regarding the kinetic nature of the secretory defect in diabetes (K_m v. V_{max} changes) should be accepted with caution. Nevertheless, findings very similar to those in type II diabetes were obtained in islets isolated from *Acomys cahirinus*: full glucose dose–response studies demonstrated that, compared to rat islets, the V_{max} of the first-phase response was reduced, while the second-phase response showed an increased K_m (Gutzeit *et al.* 1974).

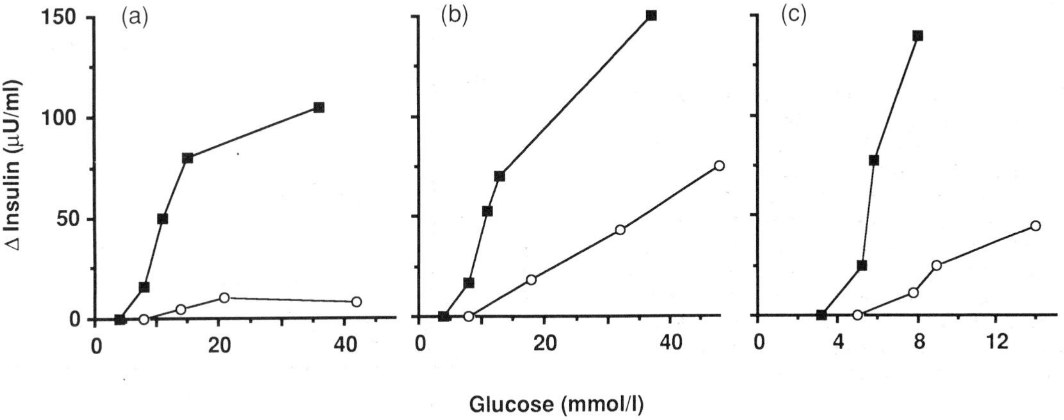

Fig. 2 The *in vivo* glucose–insulin dose-relationship in normal (■) and non-insulin dependent diabetic (○) subjects. Increasing hyperglycaemic stimuli were obtained either by intravenous glucose infusion (a and b) or by oral ingestion (c) of increasing glucose loads. Blood glucose and incremental plasma insulin values at 10 min (a) or 60 min (b) following glucose infusion, or 30 min after oral glucose administration (c) are shown. Responses at (a) reflect mainly first-phase insulin release, while those at (b) and (c) correspond to second-phase secretion (Data adapted from Cerasi *et al.* 1972, 1973)

In addition to these changes in the gross aspects of insulin release, a more subtle malfunction of the β-cell exists in type II diabetes. It has thus been shown that the normal oscillatory nature of insulin secretion is modified in diabetics, the 13–14 min cycle being replaced by irregular bursts of secretion (O'Rahilly *et al.* 1988).

The insulin response to glucose is also diminished in subjects with impaired glucose tolerance (Cerasi 1988). Furthermore, the transition from IGT to manifest type II diabetes seems to require such an impaired insulin response. Kadowaki *et al.* (1984) followed a large group of subjects with glucose intolerance for several years; the majority of those who developed clinical diabetes initially had decreased insulin responses to glucose, the insulin values being a predictive factor for the development of diabetes independently of other risk factors. Subjects who had glucose intolerance and high insulin responses did not progress to diabetes. Similar results were obtained in other studies (reviewed by Cerasi 1988). The question remains, however, whether the transition from glucose intolerance to overt diabetes is caused by progressive reduction of the already low insulin response, or whether a developing insulin resistance tips the balance toward diabetes.

A more controversial question is how glucose intolerance and mild type II diabetes develop. Two main ideas have been proposed:

1. low responsiveness of the β-cell is a primary factor that precedes glucose intolerance;
2. hypersecretion of insulin, induced by insulin resistance or other factors, leads gradually to fatigue of the β-cell, with consequent reduction of the insulin secretion and development of glucose intolerance.

Indirect evidence exists to suggest that glucose intolerance develops in subjects with already diminished insulin secretion. Thus, in identical twins of diabetic patients, the insulin response of the healthy twin was found to be as decreased as in the diabetic twin (Cerasi and Luft 1967; Barnett *et al.* 1981). Similarly, in the offspring of conjugal diabetic parents, where the prevalence of glucose intolerance can be very high, the insulin response was markedly diminished in many subjects (Soeldner *et al.* 1968; Snehalatha *et al.* 1986). Of even greater interest is the demonstration that women with a history of gestational diabetes, but who were normoglycaemic at the time of study, had a reduced insulin response to glucose administration (Ward *et al.* 1985). Since these women were previously diabetic during the stress of pregnancy, and it is known that a high proportion later developed type II diabetes (Harris 1988), the finding of impaired β-cell function at a time of normal glucose tolerance strongly suggests that decreased insulin secretion precedes the appearance of glucose intolerance and hyperglycaemia.

We have described the presence of subjects with a low insulin response to glucose in the normal population also: there was complete overlap between the responses of 90 per cent of patients with mild diabetes and 15–20 per cent of control subjects with normal glucose tolerance (Cerasi and Luft 1967). Intravenous and oral glucose dose–response studies showed that 'low insulin responders' had insulin response curves intermediary between those of the control 'high responders' and the diabetics; indeed, the first-phase response to intravenous glucose in this group showed a right-shift (increased K_m) rather than the decreased V_{max} observed in type II diabetics (Cerasi *et al.* 1972). Low insulin responders were observed at similar frequencies in all age groups, including children (Cerasi and Luft 1970).

The low insulin response in subjects with normal glucose tolerance is not the consequence of β-cell adaptation to increased insulin sensitivity since, measured both directly and indirectly, peripheral sensitivity to insulin was not different from controls, and the mean K_g of the IVGTT was lower in these subjects than in high insulin responders (Cerasi and Luft 1974). In one study (Nordlander *et al.* 1977), the plasma glycerol and non-esterified fatty acid responses to muscular exercise were higher in low insulin responders than in control subjects, suggesting that despite normal glucose tolerance there may be metabolic consequences to the reduction of insulin secretion. However, the most marked consequence of a low insulin response seems to be predisposition to later development of glucose intolerance and diabetes. Two 10-year follow-up studies in Stockholm showed that clinical diabetes occurred only in low insulin responders, and that the frequency of glucose intolerance was fivefold greater in these subjects (Cerasi 1985). Thus, a low insulin response to glucose in a normal subject seems to be a risk factor for the later development of glucose intolerance, as is the low insulin reponse of glucose-intolerant subjects for transition to diabetes.

Most animal models for type II diabetes are hyperinsulinaemic (see below). Two models seem to mimic quite well the 'low insulin syndrome' described above. Our group has studied extensively a semi-desert rodent, the spiny mouse (*Acomys cahirinus*). The majority of these animals show a low insulin response to glucose both *in vivo* and *in vitro*. When obesity (and severe insulin resistance) develops in these animals, the insulin response shows only minimal adaptation, and hyperglycaemia ensues (Gutzeit *et al.* 1979). The *in vitro* insulin release patterns are quite similar in hyperglycaemic and normoglycaemic *Acomys*, and reminiscent of the responses found in diabetic man (lack of first-phase and reduction of second-phase release). A minority of the animals show normal insulin responses (with conserved first-phase release); the frequency of glucose intolerance in these is not known. Another attractive model in this context is the monkey *Macaca nigra*, in which the acute insulin response to glucose shows a gradual decrease during the development of diabetes (Howard 1986). In this model there seems to be a continuous loss of β-cells due to amyloid deposition; it is therefore difficult to evaluate its relevance for the transition from a low insulin response to glucose intolerance in man.

The idea that type II diabetes might develop out of the exhaustion of hyperstimulated β-cells originates from epidemiological studies, where basal or post-glucose plasma insulin levels were found to be elevated in glucose intolerant and mildly diabetic subjects, but diminished (usually back to the normal level) in more advanced diabetics (see former chapters). One important longitudinal study is that of Sicree *et al.* (1987), who investigated a large group of Nauruans over a 6-year period. Sixteen per cent of their subjects with initially normal glucose tolerance progressed to IGT, and a further 6.5 per cent became diabetic. In both groups, at the initial testing (with normal glucose tolerance) the 2-hour plasma insulin values were 25–60 per cent higher than in subjects who remained normal. However, minimal (but statistically not significant) differences did exist also in the 2-hour glucose values, and in the absence of the entire OGTT glucose and insulin curves, it is difficult to accept these findings as evidence for hyperfunction of β-cells prior to the development of diabetes (see discussion above regarding *in vivo* assessment of β-cell function).

Under short-term conditions of induced insulin resistance or hyperglycaemia, normal β-cells seem to be fully capable of adjusting to the increased secretory demands. Thus, healthy subjects given nicotinic acid for 2 weeks developed insulin resistance; however, their insulin response to glucose and to arginine increased by 30–90 per cent, and they remained normoglycaemic (Kahn *et al.* 1989). Chen *et al.* (1989) induced mild hyperglycaemia during 5 days in normal rats by glucose infusion; *in situ* hybridization studies demonstrated that the islet area and number doubled, as did the proinsulin mRNA signal intensity. Obviously such findings do not exclude the possibility that the β-cells may become exhausted if the increased demand is extended over time. The most common situation in which insulin resistance and β-cell hyperfunction occur is obesity; more than 75 per cent of obese subjects remain normoglycaemic and hyperinsulinaemic throughout life. It is, however, not known whether in a subset of the population the β-cells are more vulnerable to deleterious action by chronic stimulation (genetic factors?).

In stark contrast to the situation in man, most animal models of type II diabetes are markedly hyperinsulinaemic, at least until the development of severe hyperglycaemia (reviewed by Shafrir 1990). A most convincing example is that of rhesus monkeys, studied longitudinally in detail by Hansen *et al.* (1988). These animals frequently become very obese, and many progress to IGT and diabetes. While the rate of progression seems to be quite variable, they all proceed through a phase of severe hyperinsulinaemia prior to the impairment of glucose tolerance and appearance of hyperglycaemia, when insulin secretion falls to levels near or above control. This model offers the unique advantage of longitudinal follow-up in individual primates; to what extent it reflects the natural history of type II diabetes in man remains to be established.

Priming effect of glucose

In addition to acutely stimulating insulin release, glucose has a priming effect on the β-cell, which expresses itself as a potentiation of the insulin response to subsequent stimulation (Nesher and Cerasi 1987). This action of glucose can be shown by sequential glucose applications; the insulin response to the second stimulus usually becomes two- to threefold greater. In a series of experiments (Cerasi 1975), in which a 1-hour glucose infusion was repeated following a 50- to 70-minute rest period, it could be demonstrated that the insulin response to the second stimulus was greatly amplified in subjects with type II diabetes or IGT, the sensitivity for the potentiating effect of glucose being equal to or greater than that in control subjects (Fig. 3). A more striking demonstration of the conservation of glucose-induced potentiation in a functionally deficient islet is offered by *Acomys*: repeated stimulation of *Acomys* islets by glucose resulted in the marked amplification of the response; the kinetics of the insulin secretion was also modified, a prominent first-phase response becoming evident (Nesher *et al.* 1985). Thus, while the acute insulin-releasing effect of glucose is greatly reduced in type II diabetes, the ability of the sugar to generate potentiation is retained and may even partly correct the secretory defect.

Is the deficient insulin response *in vivo* specific for glucose?

Most type II diabetics who fail to secrete insulin on glucose administration do respond to non-glucose secretagogues such as arginine, glucagon, isoproterenol, etc. This observation has led to the suggestion that the β-cell defect in type II diabetes is restricted to non-recognition of glucose (Robertson 1989). However, the situation is somewhat more complex and deserves some analysis. Most non-glucose secretagogues show complex synergistic interactions with glucose. These interactions can be described as those that increase the sensitivity of the β-cell to glucose (reduction of the glucose K_m), and those that amplify the action of glucose (increased V_{max}). Sulphonylurea drugs are the best-known representatives of the first group, while arginine represents the latter group.

The plasma insulin response to arginine in type II diabetes is seemingly normal. This is, however, an artefact due to the hyperglycaemia of the patient; indeed, when control subjects were hyperglycaemia-matched with patients by a glucose

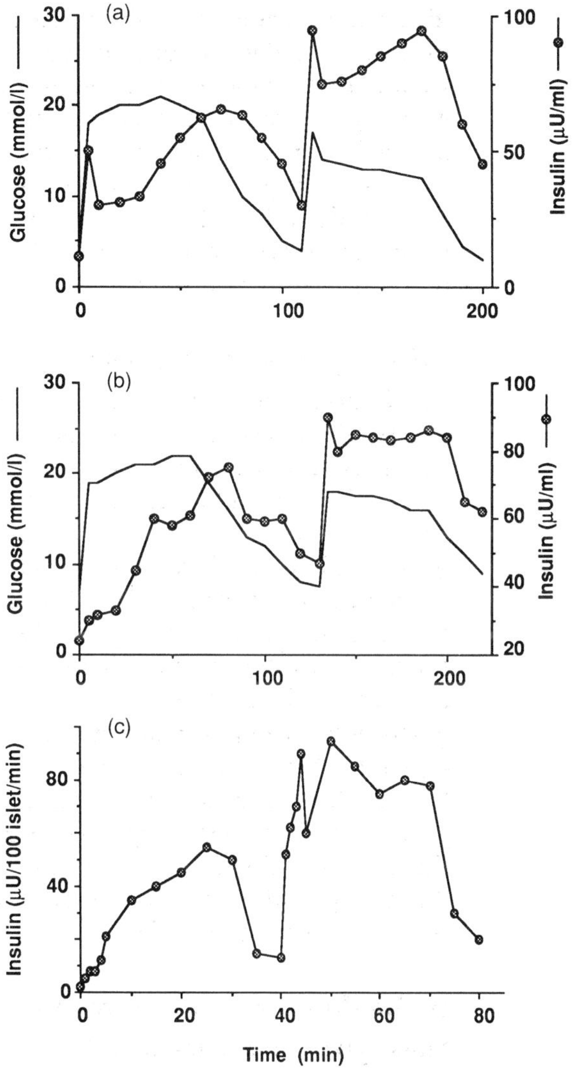

Fig. 3 Amplification of the insulin response by glucose priming *in vivo* (a and b) and *in vitro* (c). (a) A glucose infusion was given at 0–60 min and 110–170 min to control subjects. (b) Glucose infusions were given at 0–60 and 130–190 min to subjects with mild non-insulin-dependent diabetes. Note that the higher insulin values were obtained in the face of lower glycaemias both in control and diabetic subjects. While the kinetics of the insulin response was not modified by glucose priming in normal subjects (a), an almost normal first-phase response appeared in the diabetics (b). (c) Islets isolated from *Acomys cahirinus* were perfused at 0–30 and 40–70 min with 16.7 mM glucose, while the glucose level at 30–40 min and 70–80 min was reduced to 3.3 mM. Note the 'diabetic' pattern of the insulin release during the stimulatory period and the emergence of an almost normal first-phase response following the priming period. Note also the general similarities between the patterns presented in (b) and (c) (Data adapted from Cerasi 1975 and Nesher *et al.* 1985)

infusion, it became evident that the diabetic response was only 10–15 per cent of the matched control one (Nesher *et al.* 1984). Similar results were obtained by performing arginine infusions during normo-, hypo-, and hyperglycaemic clamps, and by superimposing the effect of arginine on a range of glucose loads; the maximal insulin response to combined arginine–glucose stimulation is reduced in type II diabetes, as is the glucose dose–response relationship that describes the effect of arginine at various glucose levels (reviewed by Cerasi 1988).

The above results may be taken to indicate that the diabetic β-cell defect includes the arginine stimulus, and therefore is not specific for glucose. However, arginine is inactive in the absence of glucose, and in kinetic terms its effect can be described as augmentation of the V_{max} of glucose-induced insulin release. Indeed, arginine augmented the insulin or C-peptide level by a factor of 2–4 at all glucose concentrations, equally in controls and patients (Cerasi 1988). Thus, instead of regarding glucose as the potentiator of arginine-induced insulin release, if one takes the physiologically correct approach of arginine as the amplifier of the glucose effect, it appears that arginine exerts a quantitatively and qualitatively normal action in the diabetic β-cell. A normal potentiation by arginine of the diminished glucose-induced insulin release in type II diabetes results in a low insulin response. In low insulin responders with normal glucose tolerance, β-cell sensitivity to arginine was found to be normal (Assan *et al.* 1981). No similar study exists in type II diabetics; therefore, although evidence at hand suggests that the diabetic β-cell responds normally to arginine, discrete sensitivity changes in the recognition of the amino acid cannot be excluded.

In the presence of sulphonylureas, both first- and second-phase glucose-stimulated insulin release is amplified. Most of the effect is observed at medium to high levels of glucose, both *in vitro* and *in vivo* (reviewed by Pfeiffer 1983). In moderately hyperglycaemic type II diabetics, near-normal acute insulin responses to intravenously administered sulphonylurea were observed; however, in combination with a glucose dose–response study, the ability of sulphonylureas to increase the sensitivity of the β-cell for glucose was found to be reduced (Cerasi *et al.* 1979). It is difficult to conclude from this type of acute study whether the diabetic β-cell is intrinsically less responsive to sulphonylureas, or whether the lower secretory response is a reflection of a reduced effect of glucose on insulin release. Under therapeutic conditions, sulphonylureas seem to retain their stimulatory effect over extended periods. Thus, in two studies where patients were treated with sulphonylureas for up to 3 months, insulin secretion was increased by 50–100 per cent (Hosker *et al.* 1985; Gutniak *et al.* 1987). In a recent study, type II diabetics were given gliclazide, and several parameters of β-cell function (response to glucose infusion and to meals, urinary C-peptide secretion, etc.) monitored over 6 months. The insulinogenic effect of glucose was doubled by gliclazide; the effect remained unmodified over the 6-month period (Della Casa *et al.* 1991). Thus, the diabetic β-cell seems capable of responding to sulphonylureas both under acute and chronic conditions.

In animal models of diabetes, the β-cell response to non-glucose stimuli are usually quite complex; therefore a detailed review of the findings is beyond the scope of this chapter (see Shafrir 1990). In the spiny mouse, arginine, tolbutamide theophylline (and iso-butyl methylxanthine) fail to amplify substantially the insulin secretion (Shafrir 1990).

To summarize, much of the data available suggests that the deficient insulin response in type II diabetes is specific for glucose. However, because most other insulin secretagogues depend on glucose in order to stimulate secretion, and the response to glucose itself is defective in diabetes, the insulin response to non-glucose stimuli is also reduced in type II diabetes.

The mechanism of the β-cell defect in type II diabetes

It is quite clear that, in contrast to type I diabetes, and despite the fact that some reduction in β-cell mass as well as other morphological changes may be found, the deficient insulin secretion of type II diabetes cannot be accounted for by gross structural changes in the islet.

Due to the difficulty in obtaining islets from diabetics, much of the information on the nature of the diabetic insulin secretory defect in man is derived from *in vivo* studies. The striking feature of the defect is its restriction to the acute initiatory signal of glucose, while modulatory factors are recognized normally. Among the latter are the ability of glucose to amplify the secretion by the mechanism of time-dependent potentiation and the synergistic effect of arginine on glucose-induced insulin secretion. Furthermore, when the glucose–insulin dose–response relationships in obese and lean type II diabetics with similar degrees of hyperglycaemia were compared, the obese patients showed a threefold increase in the V_{max} of the response. This was identical to the difference in V_{max} between matched obese and lean controls (Nesher *et al.* 1987). Thus, the modulatory effect of obesity on β-cell function also seems to be sensed normally in type II diabetes.

The maintenance of the normal priming effect by glucose in type II diabetes (see above) is quite significant for analysis of the biochemical defect of the β-cell. *In vivo* (Efendic *et al.* 1979) and *in vitro* (Grill *et al.* 1978) studies have shown that glucose induces priming even when its acute insulin-releasing effect is blocked, and demonstrated that the calcium requirements of time-dependent potentiation are different from those of the acute release (Nesher *et al.* 1988). As these observations indicate, glucose utilizes different mechanisms to generate an acute stimulus–secretion coupling signal and a signal that modulates the secretory rates with time. Therefore, the defect in type II diabetes seems to be restricted to the ability of glucose to initiate rapidly insulin release; other functions of glucose (insulin synthesis, slow amplification of release) are conserved, at least in the early stages of the disease.

Although the exact biochemical derangement that leads to glucose non-sensing in the diabetic β-cell is still unknown, several possible mechanisms have been proposed. Islet cyclic AMP, which is augmented by glucose, is regarded as one of the components of the stimulus–secretion coupling system. In islets from low insulin responding *Acomys cahirinus*, the effect of glucose on cyclic AMP levels is markedly reduced, while non-glucose stimulators of adenylate cyclase, such as forskolin, are capable of generating major cyclic AMP responses in these islets (Nesher *et al.* 1989). Thus, in this model of type II diabetes the islet adenylate cyclase does not seem to be defective; rather, the coupling of the glucose stimulus to cyclic AMP generation is deficient. However, the *Acomys* islet defect is not restricted to the generation of cyclic AMP, since phosphodiesterase inhibitors, forskolin, and dibutyryl cyclic AMP all failed to restore the glucose-induced first-phase insulin release, despite adequate elevations in the intracellular cyclic AMP concentration (Nesher *et al.* 1989). From these studies it appears that cyclic AMP defects, although participating in the response, are not the main cause of deficient insulin release.

Glucose itself has recently attracted interest as a possible culprit in the β-cell derangement of type II diabetes. Indeed, several studies have shown that near-normalization of the blood glucose profiles in type II diabetics by various means greatly improves their insulin response to glucose (Kosaka *et al.* 1980; Glaser *et al.* 1988; Gumbiner *et al.* 1990). In a recent study of diabetics with secondary failure to sulphonylurea treatment, we could demonstrate that a 2-week period of near-normoglycaemia induced by continuous subcutaneous insulin infusion led to doubling the insulin and C-peptide responses to meals and intravenous glucagon, and the appearance of a small first-phase insulin response to glucose infusion (Glaser *et al.* 1988). This is shown in Fig. 4. Furthermore, some of the patients could be controlled adequately by oral agents for several months after cessation of the insulin treatment. Similar findings were obtained in various animal models of type II diabetes (reviewed by Shafrir 1990). Thus, chronic hyperglycaemia seems to be deleterious for β-cell function.

Hyperglycaemia *in vivo* is accompanied by several changes in the metabolic and hormonal milieu of the β-cell; it is therefore difficult to ascribe this deleterious effect to glucose *per se*. However, several *in vitro* experiments using isolated islets or the isolated perfused pancreas have confirmed that prolonged exposure to high glucose levels reduces the insulin response (Grodsky 1989; Purrello *et al.* 1990). We have recently used a primary rat islet cell culture, maintained for several weeks on extracellular-matrix-coated dishes, to investigate in some detail the mechanism of this 'toxic' effect of high glucose concentrations (Kaiser *et al.* 1991). When maintained for 1–3 weeks at glucose concentrations above 11 mM, there was a time- and dose-dependent reduction in the acute insulin response to glucose (Fig. 5), which expressed itself as reduction of the V_{max} for glucose. The response to other nutrient secretagogues (glyceraldehyde, α-ketoisocaproic acid) was similarly reduced. By contrast, agents like sulphonylureas, carbamylcholine, or IBMX retained their ability to initiate or modulate insulin release (Kaiser *et al.* 1990). These findings are thus quite similar to those observed *in vivo*, and give credence to the 'glucose toxicity' theory (Rossetti *et al.* 1990). However, similar results were obtained when islet

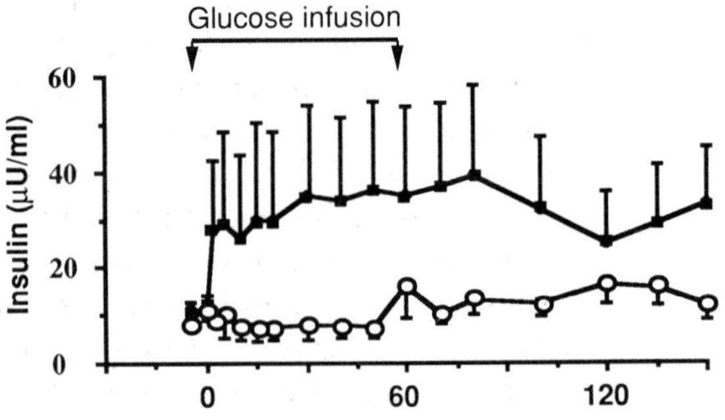

Fig. 4 Correction of the hyperglycaemia improves β-cell function in NIDDM. In a group of patients with secondary failure to sulphonylurea treatment, glucose infusion (0–60 min) failed to generate a significant plasma insulin response (o). Following 2 weeks of intensive insulin treatment and near-normalization of the blood glucose levels, a significant insulin response to glucose infusion (■) appeared. Note that the improved response is still far below normal (compare with Fig. 1) (Adapted from Glaser *et al.* 1988)

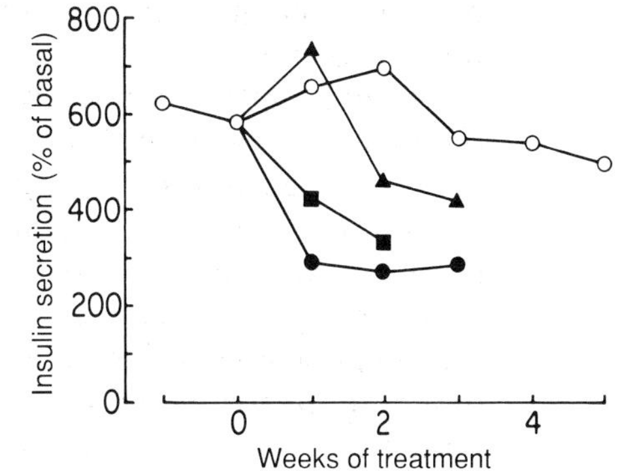

Fig. 5 Induction of 'glucose toxicity' *in vitro*. Rat islets were maintained as monolayer cultures on extracellular-matrix-coated dishes for the times indicated. The control group remained in the normal culture medium (RPMI, 11.1 mM glucose; o) The experimental groups were transferred at zero time to 16.7 (▲), 25 (■), or 33 (●) mM glucose. Osmolarity was corrected with sucrose. At weekly intervals, the cultures were washed, incubated for 2 h in Krebs–Ringer-bicarbonate buffer with 3.3 mM glucose, and then stimulated with 16.7 mM glucose. Results show percental stimulation in relation to the basal (3.3 mM glucose) secretion rate

cultures were maintained at normal glucose levels but in the presence of IBMX, thus inducing chronic hypersecretion of insulin (Kaiser *et al.* 1991). It seems therefore that 'glucose toxicity' is not due to glucose itself but to a metabolic desensitization due to prolonged induction of high secretion rates. Our results also indicate that this desensitization is not due to non-enzymatic glycosylation of β-cell proteins involved in exocytosis. Down-regulation of islet-specific glucose transporters (GLUT-2) can also be excluded since insulin responses to glyceraldehyde and α-ketoisocaproic acid, substrates not transported by the glucose carrier, are also reduced, and a recent report demonstrates that, at least *in vivo*, prolonged hyperglycaemia in rats causes up-regulation of islet GLUT-2 mRNA levels (Chen *et al.* 1990). Finally, the maintenance of a normal response to sulphonylurea, believed to act by closing the ATP-sensitive K^+ channels of β-cells, may indicate that a gross defect in this channel is not involved in hyperglycaemia-induced desensitization of secretion (see Chapter 4). However, the possibility that the link between channel activity and metabolism is disturbed still remains.

Whatever its mechanism, it is unlikely that 'glucose toxicity' is the sole cause of the deficient insulin secretion in type II diabetes. This conclusion is based on the fact that phasic insulin release is markedly reduced in subjects with IGT, in whom minimal hyperglycaemia (at levels insufficient to cause *in vitro* desensitization of the islet) occurs only during short periods following meals, and that even strict normalization of the blood glucose level for several weeks in diabetic patients fails to normalize the insulin secretion (the 80–90 per cent reduction of β-cell function is usually ameliorated to a 60–70 per cent reduction, see preceding chapters). Thus, chronic hyperglycaemia must be seen as an aggravating rather than a causative factor for defective β-cell function in type II diabetes.

It has been known for many years that an amyloid-like material accumulates in the islets of long-standing type II diabetics. The recent identification of amylin in diabetic β-cells and its cloning (Nishi *et al.* 1990) has raised a major interest in this molecule, and intensive research is presently ongoing into its possible role in diabetes. Amylin is co-packed with insulin in β-cell granules, and co-secreted in a regulated manner (Mitsukawa *et al.* 1990). Porte and Kahn (1989) have speculated that amylin might be involved in the processing of proinsulin, and directly or indirectly be causally related to the hyperinsulinaemia and deficient insulin release of type II diabetics. Recent experiments by Nagamatsu *et al.* (1990), however, negate effects of exogenous amylin on islet insulin mRNA levels or pronisulin biosynthesis rates, or on insulin secretion and proinsulin to insulin conversion. It is therefore unclear what role, if any, amylin plays in the aetiology of the diabetic β-cell dysfunction.

Three other possibilities, although speculative, merit mention as possible causes for impaired insulin release in type II diabetes. All relate to early steps in β-cell glucose metabolism; since insulin release is controlled by hexose utilization, they are plausible aetiological candidates (see also Chapter 4). The first relates to the glucose transporter. Johnson *et al.* (1990*b*) have recently shown that new-onset type I diabetics have circulating antibodies against the glucose transporter of islet cells (presumably anti-GLUT-2 antibodies) which block the islet glucose uptake. Combined with the earlier demonstration of an *in vitro* inhibitory effect of type I diabetic sera on insulin release (Kanatsuna *et al.* 1983), these studies definitely suggest that exogenous or intrinsic interference with the function of GLUT-2 may impair insulin secretion. This is emphasized by the preliminary finding that over-expression of glucose transporters in the RIN β-cell line improves its defective insulin response to glucose (Shibasaki *et al.* 1990). It seems, therefore, worthwhile to explore the structure and function of islet glucose transporters in spontaneously diabetic animals.

Once transported into the β-cell, glucose is phosphorylated by glucokinase which, as discussed in Chapter 4, may be the glucose sensor of the islet. It has also been proposed (Matschinsky 1990) that discrete modifications of the function of glucokinase might have drastic effects on β-cell glucose metabolism and lead to the deficient insulin release of type II diabetes. No experimental evidence exists to substantiate this speculation.

Finally, Khan *et al.* (1990*a*, *b*) measured cycling of glucose 6-phosphate back to glucose in islets from control and diabetic animals (induced by neonatal streptozotocin administration to rats, or in hyperglycaemic ob/ob mice). They found a severalfold increase in the rate of glucose cycling in diabetic islets, and proposed that the ensuing reduction in glycolytic flux could explain the reduced insulin response to glucose in these models of diabetes. Whether such a mechanism is common to other models of type II diabetes, not to mention the disease in man, is not known.

11.3.2 Insulin resistance in type II diabetes

The expectation that type II diabetics should demonstrate reduced biological efficiency of insulin stems from the finding that these patients usually have normal to supranormal levels of circulating insulin-like immunoreactivity. This precondition can be challenged on two accounts: the circulating material is largely proinsulin and its split-products, which have markedly reduced insulin-like bioactivity (Temple *et al.* 1989); and the level of insulin (including proinsulin) is grossly inadequate in relation to the patient's hyperglycaemia, which is the dominating β-cell stimulus (as discussed above). Nevertheless, type II diabetics do show reduced responsiveness also to exogenously administered insulin. This has been repeatedly demonstrated by using the technique of normoglycaemic–hyperinsulinaemic clamp (reviewed by DeFronzo 1988). In this technique, blood glucose levels are acutely normalized (e.g. by overnight insulin infusion), and maintained at constant euglycaemia in the presence of various insulin infusion rates by a matched glucose infusion. Total body glucose utilization, calculated from the glucose infusion rate, shows a 30–60 per cent decrease in type II diabetics compared to weight-matched controls at all exogenous insulin levels, indicating that the V_{max} for insulin-stimulated glucose disposal is reduced in diabetics (DeFronzo 1988). This 'post-receptor defect' is found in lean as well as obese diabetics and in subjects with IGT (for further details and the possible mechanisms involved, see Chapters 6 and 7).

The major peripheral site implicated in the insulin resistance of type II diabetes has been the subject of debate. Although most authors favour skeletal muscle as the prime site where both insulin-dependent and insulin-independent glucose uptake occur (e.g. Edelman *et al.* 1990), Gerich *et al.* (1990) arrived at the conclusion that impaired muscle glucose clearance accounts for less than 10 per cent of the reduction in total body glucose consumption. Several factors confound the interpretation of glucose clearance studies, whether performed with or without isotope dilution techniques, which may explain these and other controversies in the literature. One is whether studies are performed at normoglycaemia, or at the patient's habitual hyperglycaemia. Glucose disposal is increased by the mass action effect of hyperglycaemia, and, in type II diabetics total glucose utilization is normal or above normal (Gerich *et al.* 1990), although if calculated per glucose concentration, the clearance values are found to be depressed. In a non-steady-state protocol, where IVGTTs were performed with the addition of increasing doses of insulin in hyperglycaemic patients, we found no difference between lean type II diabetics and lean controls, or between obese type II diabetics and obese non-diabetics, in the dose–response relationship between plasma (exogenous) insulin levels and the K_g of IVGTT, despite the fact that a marked right-shift in the curves was observed in obese subjects (diabetic and non-diabetic alike) (Nesher *et al.* 1987). This is illustrated in Fig. 6. Another confounding factor in clamp studies relates to the prolonged infusion of insulin. It has indeed been shown that, while prolonged continuous hyperinsulinaemia induces insulin resistance, pulsatile administration of the hormone enhances sensitivity to insulin (Ward *et al.* 1990). This may be one explanation for our inability to demonstrate a diabetes-specific insulin insensitivity, since the hormone was administered as intravenous bolus injections in repeated doses (Nesher *et al.* 1987).

Both glycogen synthesis and glucose transport have been implicated in the reduced glucose disposal of type II diabetes. In man (Shulman *et al.* 1990) and rats rendered mildly diabetic by partial pancreatectomy (Rossetti and Giaccari 1990),

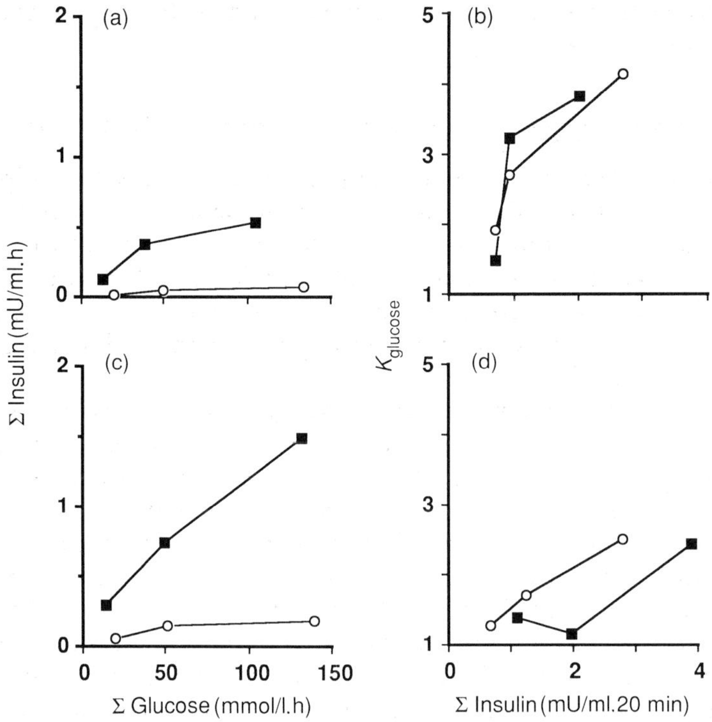

Fig. 6 Pancreatic responsiveness to glucose (a and c) and peripheral sensitivity to insulin (b and d) in normal (■) and NIDDM (○) subjects during non-steady-state conditions. Pancreatic responsiveness was established by giving increasing intravenous loads of glucose and measuring the integrated plasma insulin responses during 60 min (plotted against corresponding integrated blood glucose levels). For peripheral sensitivity to insulin, 0.3 g/kg glucose was given intravenously together with 25, 50, and 100 mU/kg insulin; the plasma insulin levels that resulted were integrated over the initial 20 min and plotted against the disappearance rate of blood glucose ($K_{glucose}$ of IVGTT). Panels (a and b) present results obtained in lean subjects, (c and d) those of obese persons. Note that insulin responses were about threefold higher in obese subjects (normal and diabetic alike) than in their respective lean counterparts, and that the insulin effect on K values was lower and shifted to the right in obese groups. Note also the lack of major differences between diabetics and non-diabetics regarding insulin sensitivity, while pancreatic responsiveness is grossly reduced (Adapted from Nesher *et al.* 1987)

muscle glycogen synthesis was found to be markedly reduced, which accounted for most of the diminished total body glucose disposal at supraphysiological plasma insulin concentrations. At lower insulin levels, glucose transport seemed to be the rate-limiting step which is reduced in type II diabetics (Edelman *et al.* 1990; Rossetti and Giaccari 1990). However, a recent study by Henry *et al.* (1990) seems to negate the determining role of insulin for the intracellular fate of glucose. Indeed, these authors demonstrated convincingly, by matching the rate of peripheral glucose uptake in type II diabetics under conditions of either major hyperglycaemia or major hyperinsulinaemia, that equivalent proportions of glucose are directed to the oxidative and non-oxidative pathways. Thus, the intracellular availability of glucose, rather than the hormonal milieu, determines the further metabolism of the hexose. This indicates that the main problem in type II diabetes may indeed reside at the level of glucose uptake.

11.3.3 Glucose transporters and type II diabetes

From the discussion above it appears that the specific peripheral metabolic dysregulation of type II diabetes is more intimately related to the handling of glucose by muscle and other cells, than to malfunctioning of the insulin signal *per se*. Notwithstanding possible, obesity-unrelated defects in various aspects of the insulin–insulin-receptor interaction (see Chapter 6), the term insulin resistance seems to be a misnomer when applied to the reduction of peripheral glucose utilization in diabetes.

Since many studies (see above) suggest that glucose metabolism in tissues such as muscle is regulated by the rate of uptake of the sugar, several groups have recently investigated the effect of diabetes on the cellular levels of specific glucose transporters. In rats with mild hyperglycaemia induced by streptozotocin, both in adipocyte and skeletal muscle the total, as well as plasma membrane-associated, quantity of glucose transporters, assessed by the cytochalasin B binding assay, was found to be reduced by 30–50 per cent, and was restored by insulin treatment of the rats (Karnieli *et al.* 1987; Barnard *et al.* 1989). Cytochalasin B does not differentiate between the various tissue-specific glucose transporters. GLUT-4, the dominating type of glucose transporter in muscle and adipose tissue, is adversely influenced by diabetes: in streptozotocin-diabetic rats, the steady-state mRNA level and total cellular content of transporter protein was dramatically reduced in adipocytes and, to a lesser extent, in skeletal muscle (Berger *et al.* 1989; Garvey *et al.* 1989; Sivitz *et al.* 1989; Strout *et al.* 1990). In contrast, no change was found in the constitutive GLUT-1 mRNA or protein levels. Restoration of normoglycaemia by insulin treatment (Berger *et al.* 1989; Garvey *et al.* 1989; Sivitz *et al.* 1989) or vanadate administration (Strout *et al.* 1990) normalized the GLUT-4 levels. Thus, in this animal model of diabetes a differential effect on muscle/adipocyte specific GLUT-4 is clearly evident. It is much less clear whether the situation in human type II diabetes is analogous to that of streptozotocin-diabetic rats. In a recent study, Pedersen *et al.* (1990) found no change in the mRNA and protein levels of either GLUT-4 or GLUT-1 in biopsies from skeletal muscle in type II diabetic patients. In contrast, a preliminary report (Elton *et al.* 1990) does claim a decrease in muscle GLUT-4 content in apparently similar groups of type II diabetic patients. Thus, further work will be necessary to establish under which conditions the biosynthesis of cell-specific glucose transporters is disturbed in type II diabetes.

11.3.4 Autoregulation of glucose transport

In hyperglycaemic animals and man, the complexity of the metabolic and hormonal changes that intervene make the interpretation of the data described above difficult. It has been known for some time, however, that glucose deprivation, in the absence of other modifications, may influence cellular glucose uptake (Martineau *et al.* 1972). This direct role of glucose has recently been extensively investigated. Using either *in vitro* incubations of isolated rat soleus muscles or cultures of L8 myocytes and myotubes, it could be shown that the rate of uptake of the glucose analogues 2-deoxyglucose and 3-*O*-methylglucose is down-regulated by the presence of glucose during a preincubation (Sasson and Cerasi 1986; Sasson *et al.* 1987). This down-regulatory effect was time-dependent, being completed within 3–5 hours, and dose-dependent, such that the transport rate was reduced by 2–3 per cent for each mmol/l glucose above 3–4 mmol/l. Furthermore, the down-regulatory effect of glucose on glucose uptake was paralleled, in quantitative terms, by down-regulation of total muscle glucose utilization (Sasson *et al.* 1987). These effects were entirely reversible, muscle adapting its hexose uptake with precision to the level determined by the glucose concentration of the preceding few hours (Sasson and Cerasi 1986; Cerasi *et al.* 1989).

Is this direct autoregulatory effect of glucose mediated by changes in glucose transporter biosynthesis, as described above for diabetic rodents? Glucose does influence the mRNA levels for the transporter specific to myocyctes in culture, GLUT-1 (Wertheimer *et al.* 1989; Walker *et al.* 1990). In our hands, the down-regulation of GLUT-1 mRNA by glucose is very similar in its dose- and time-dependence to the down-regulation of glucose transport. This could suggest that the two events are causally related; however, reversal of down-regulation of transport is complete within 3–4 hours following glucose withdrawal, while GLUT-1 mRNA levels are augmented only after a lag period of 8–12 hours (Sasson and Cerasi 1986; Wertheimer *et al.* 1991). Therefore, while chronic hyperglycaemia in animals and high glucose levels *in vitro* may indeed deplete the muscle cell glucose transporters, the acute regulatory effect of glucose must employ another mechanism. This has been shown recently to be the case for GLUT-1 in myocytes in culture (Greco-Perotto *et al.* 1992; Walker *et al.* 1990). In muscle, as in adipocytes, glucose transporters are located in intracellular, microsomal vesicles, and the plasma membrane, the two compartments being in a dynamic equilibrium. It is accepted that, at least partly, insulin stimulates glucose uptake by translocating glucose transporters from the intracellular pool to the plasma membrane (Hirshman *et al.* 1990). As shown in Fig. 7, we have found that glucose, in the absence of hormones and other substrates, induces internalization of glucose transporters, thus depleting the plasma membrane (Greco-Perotto *et al.* 1992). This translocation seems to be reversible; furthermore, the relative changes in the plasma membrane GLUT-1 were quantitatively similar to the changes in glucose uptake, suggesting that the autoregulatory effect of glucose does not involve modulation of the intrinsic, glucose transporting efficiency of the individual glucose carriers.

It is as yet not entirely clear how the above autoregulatory mechanism operates *in vivo* under conditions where muscle cells are confronted with varying levels of other substrates such as fatty acids, and varying concentrations of hormones, such as insulin, catecholamines, cortisol, etc. in various combinations. Furthermore, it is not known if autoregulation of glucose transport occurs in man. Nevertheless, it is

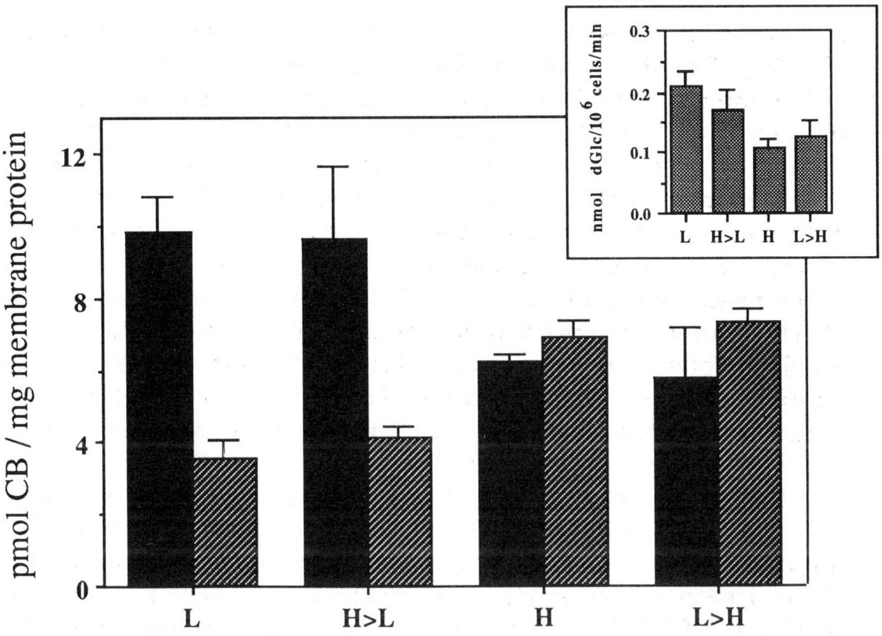

Fig. 7 Regulation of hexose transport rate and the cellular location of glucose transporters by glucose in myocytes. L8 myocytes were incubated overnight in the presence of a high (H) or low (L) concentration of glucose. Thereafter, the medium of the H myocytes was changed to a low glucose one (H>L) and that of the L myocytes was changed to one containing a high glucose concentration (L>H), and incubated for a further 5 h. The inset shows the 2-deoxyglucose uptake rates of these myocytes: overnight incubation at high glucose reduced the transport rate by 50 per cent; switching media for 5 h reversed the effect of overnight incubation almost entirely. The main figure shows the specific cytochalasin B (CB) binding of plasma membrane (dark bars) and microsomal (hatched bars) fractions of the same myocytes. It is clearly seen that the glucose-induced reduction in hexose transport rate is accompanied by shift of glucose transporters (as assessed by CB binding) from the plasma membrane to the intracellular membranes, while glucose withdrawal increases the transporters at the plasma membranes (Adapted from Greco-Perotto *et al.* 1992)

astonishing that the quantitative impact of the *in vitro* down-regulatory effect of glucose on rat muscle glycolysis (~50 per cent reduction at 15–20 mmol/l glucose) fits so well the degree of reduction in glucose utilization found in uncontrolled type II diabetics. It is therefore tempting to speculate that at least part of the post-receptor insulin resistance of type II diabetics is the result of a physiological adaptative mechanism driven by the hyperglycaemic state itself. That this assumption may be correct is shown by the studies of Rossetti *et al.* (1990): insulin resistance in streptozotocin-diabetic rats could be corrected either by insulin administration, which causes hyperinsulinaemia, normoglycaemia, and low fatty acid levels, or by the administration of phlorizin, which by its glycosuric effect induced normoglycaemia in the presence of hypoinsulinaemia and otherwise unchanged substrate levels. A clinical counterpart of these experiments is the demonstration by several groups that the insulin sensitivity of type II diabetics is greatly improved when the hyperglycaemia of patients is regulated by either diet alone, or treatment with various oral antidiabetic agents or insulin (reviewed by Rossetti *et al.* 1990).

The above information strongly suggests that the major part of the insulin resistance in type II diabetes is a secondary event initiated by hyperglycaemia *per se*. Once established, it obviously aggravates the hyperglycaemia, thus activating a vicious cycle.

Two additional mechanisms for insulin resistance in type II diabetes, although speculative at this stage, should be mentioned. Laakso *et al.* (1990) have shown that insulin augments the skeletal muscle blood flow in a dose-dependent manner. This effect seems to be of physiological importance, since muscle glucose uptake is a function of muscle blood flow. In insulin-resistant obese subjects, the effect of insulin on blood flow was reduced to the same extent as the reduction in glucose disposal (Laakso *et al.* 1990). If a similar insensitivity of blood flow to regulation by insulin exists in type II diabetes, it could indeed contribute to the reduction of whole-body glucose utilization. The second mechanism concerns the β-cell peptide, amylin. Studies with muscle and liver preparations have shown that amylin may markedly reduce the sensitivity to insulin (Leighton and Cooper 1990). Admittedly, these effects were obtained with concentrations of amylin unlikely to be present in the circulation; however, it is not known whether type II diabetics deposit abnormal amounts of amylin in peripheral tissues. Sensitive immunoassays for amylin have become available, which will permit the testing of this hypothesis in the immediate future.

11.3.5 Insulin resistance v. insulin deficiency in type II diabetes: clinical implications

Whether primary or secondary events, in the fully developed diabetic state β-cell deficiency and resistance to insulin coexist. Can their respective roles in maintaining (if not causing) the level of hyperglycaemia be evaluated? In general terms, the decrease in the functional capacity of the β-cell is of the order of 80–90 per cent, while the maximal response to insulin is diminished by 40–50 per cent. However, usually different experimental protocols in different groups of patients are used to obtain these figures, which do not permit the quantitation, in the same person, of the impact of each factor. We subjected a large group of obese and lean diabetic and non-diabetic subjects to two similar non-steady-state protocols to determine the dose–response relationships for intravenous glucose-induced insulin release, and intravenous exogenous insulin-mediated glucose clearance (Nesher *et al.* 1987). Individual dose–response curves were mathematically transformed to numeric values, and used for multiple logistic analysis of the data. It was found that only the β-cell responsiveness could distinguish between normal and diabetic subjects, classifying correctly 90 per cent of the patients as diabetics. The insulin sensitivity values could distinguish only obese from non-obese individuals, since the degree of impairment in insulin action was very similar in diabetic and non-diabetic obese subjects. Furthermore, fasting plasma glucose in the total population was found to be significantly correlated conjointly to the slopes of the dose–response curves for β-cell function and insulin sensitivity ($r = 0.48$); however, the β-cell parameter had a threefold greater impact than the peripheral sensitivity parameter in this multiple regression equation (Nesher *et al.* 1987). These results thus support the view expressed in this chapter, that the diabetic state is much more closely related to a failure of the secretion of insulin, than to diminished efficiency of the circulating hormone level.

If it is accepted that insulin resistance is less important than insulin deficiency in maintaining the hyperglycaemia, and that glucose-induced down-regulation of glucose transport is a reversible phenomenon, it should be possible to normalize blood glucose levels in type II diabetics with near-physiological doses of insulin past a transitional adaptive stage. We have recently demonstrated this to be the case (Cerasi *et al.* 1989). A group of newly diagnosed, untreated type II diabetics, and a group of patients with long-standing secondary failure to oral antidiabetic agents, were studied; both groups consisted of mixed obese and lean patients, all were markedly hyperglycaemic (fasting blood glucose 12–18 mM, HbA1c 11–14 per cent). Patients were initiated on continuous subcutaneous insulin infusion while on a weight-maintaining diet. Following a 1–4 day adaptation period, near-normal glycaemic control could be obtained, fasting blood glucose remaining below 6 mM, while the mean of all pre- and postprandial glycaemias was less than 7 mM (Fig. 8). This excellent degree of diabetic control was obtained with total daily insulin doses of 0.43–0.60 U/kg in lean patients, and 0.49–0.87 U/kg in obese patients (Cerasi

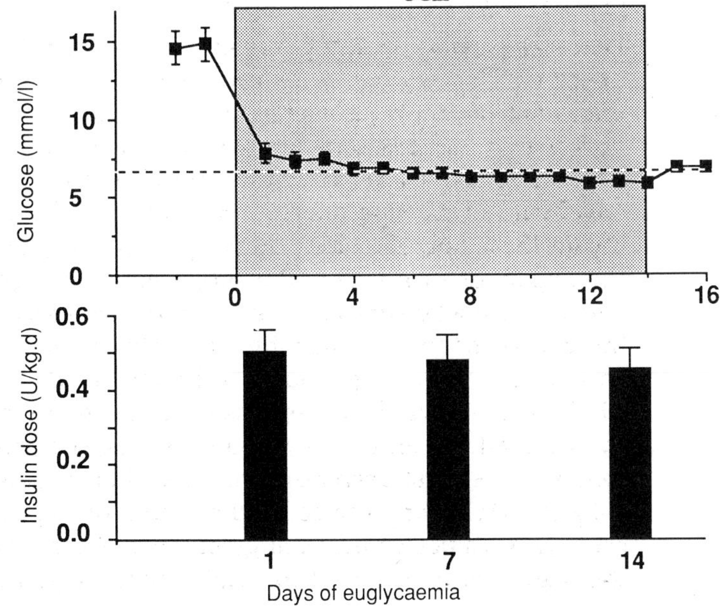

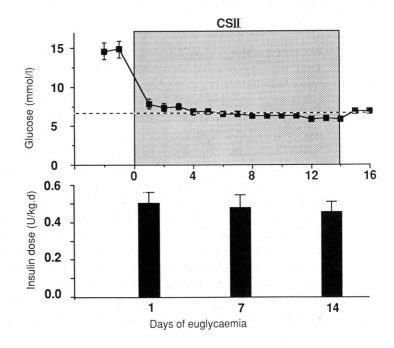

Fig. 8 Maintenance of normoglycaemia in obese NIDDM patients by physiological doses of insulin. Ten obese new-onset type II diabetics were treated with continuous subcutaneous infusion of insulin (CSII). The upper panel shows the mean (pre- and postprandial) glycaemias measured seven times daily; the horizontal line is the upper normal limit. The lower panel presents the daily insulin doses (per kg body weight) used to achieve normoglycaemia. Similar results were obtained in another group of long-standing NIDDM patients with secondary failure to sulphonylureas (not shown) (Data rearranged from Cerasi *et al.* 1989)

et al. 1989). These values, even assuming a certain supplement of endogenous insulin, are fully within the calculated insulin secretion rates of non-diabetic subjects (0.6–1.2 U/kg per 24 h; Kruszynska *et al.* 1987). Thus, if given in a near-physiological manner (basal and multiple preprandial doses), modest amounts of insulin can overcome the insulin resistance of type II diabetics and regulate the blood glucose levels. These results strengthen the hypothesis that most of the insulin resistance of type II diabetes is a secondary event related to the deranged metabolic state; furthermore, they demonstrate that, even in obese diabetics, the resistance is not of such a magnitude that it cannot be readily reversed by physiological doses of insulin.

11.4 Conclusions

The aetiology of type II diabetes is very probably multifactorial, both genetic and environmental factors acting both at the level of the β-cell, and in the periphery at the level of insulin action/glucose utilization. Despite this, a certain hierarchy does exist in the impact of these factors. Much of the evidence suggests that the development of a β-cell secretory defect is a major, genetically controlled factor in type II diabetes. Whether β-cell function is reduced from early on in life or diminishes with age as a consequence of increased demand is not clear since, in man, very long-term prospective studies on the changes in β-cell capacity and glucose tolerance are not available (and unlikely to be performed). Nevertheless, by the time glucose intolerance or mild hyperglycaemia is established, β-cell function is found to be much more impaired than peripheral responsiveness to insulin. It can be suggested, therefore, that the following chain of events leads to type II diabetes. Decreased β-cell function may lead to minimal postprandial hyperglycaemia, demonstrable as normal, but somewhat reduced, glucose tolerance. Either a further impairment of the insulin secretion due to age or other factors, or a gradual reduction in insulin sensitivity due to age, physical inactivity, or obesity, accentuates the effect of decreased insulin output and augments the hyperglycaemic response to meals. At this stage it is likely that down-regulation of glucose transport is activated, reducing glucose clearance and therefore further elevating the blood glucose levels. Thus, a first vicious cycle is established, which in the presence of a β-cell unable to cope with increased demands, either by increasing the secretion rate or enhancing β-cell replication, leads to further down-regulation of glucose utilization and augmented hyperglycaemia. Above a critical level of hyperglycaemia, the β-cell is further impaired due to glucose toxicity, thus initiating the second vicious cycle in the development of type II diabetes. The insulin response is gradually further reduced, thus leading to the acceleration of these two vicious cycles. Major hyperglycaemia then becomes established.

From the above hypothetical, yet plausible, chain of events one clinical conclusion seems inevitable. Initial therapy in type II diabetes should be aimed at the two glucose-dependent vicious cycles, since their neutralization could return the

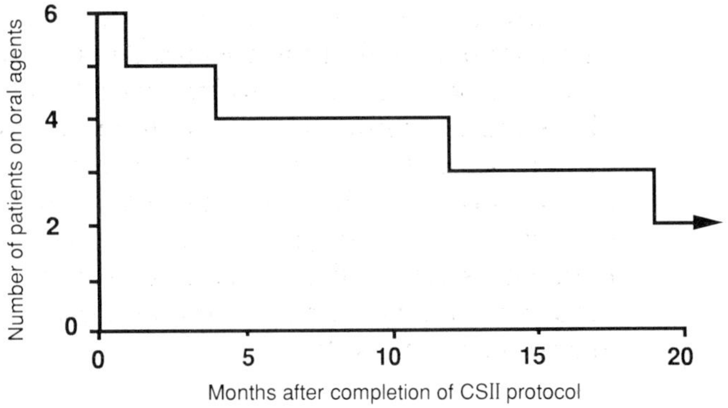

Fig. 9 Pilot study on the long-term effects of transient normoglycaemia in NIDDM. Twelve long-standing diabetics in whom secondary failure to oral agents had been established were treated with subcutaneous infusion of insulin (as in Fig. 8), strict normoglycaemia being maintained for 2 weeks. After stopping insulin treatment, in six patients good glycaemic control by oral agents was obtained for considerable periods of time

metabolic situation to the previous stages of mild diabetes, or even glucose intolerance only. Present therapeutic approaches, based on initial dietary restriction followed after a period of up to several months by oral antidiabetic agents, seem rather unsuited for this purpose. I propose initial, short-term (one to a few weeks) intensified insulin treatment aimed at achieving euglycaemia very rapidly, in order to block down-regulation of glucose transport and improve β-cell function. In a proportion of long-standing diabetics with secondary failure to oral agents, durable positive effects of this approach were obtained (Fig. 9). Its possible effect in newly diagnosed type II diabetics is under investigation.

Acknowledgements

Studies from my laboratory cited here were supported by the Wolfson Foundation, the United States–Israel Binational Science Foundation, the Israel Academy of Sciences and Humanities, and the George Grandis Fund for Medical Research. My sincere thanks are due to Dr Susan Chayen for her help in retrieving the literature, to Dr Benjamin Glaser for his assistance in preparing the illustrations and to Ms Liza Granot for the secretarial work.

References

Andreone, T., Fajans, S., Rotwein, P., Skolnick, M., and Permutt, M. A. (1985). Insulin gene analysis in a family with maturity-onset diabetes of the young. *Diabetes* **34,** 108–14.

Asmal, A. C., Dayal, D., Yialal, I., Learly, W. P., Omar, M. A. K., Pillay, N. L., and Thandoroyen, F. T. (1981). Non-insulin dependent diabetes mellitus with early onset in Blacks and Indians. *S. Afr. Med. J.* **60,** 93–6.

Assan, R., Efendic, S., Luft, R., and Cerasi, E. (1981). Dose-kinetics of pancreatic glucagon responses to arginine and glucose in subjects with normal and impaired pancreatic B cell function. *Diabetologia* **21,** 452–9.

Barnard, R. J., Youngren, J. F., Kartel, D. S., and Martin, D. A. (1990). Effects of streptozotocin-induced diabetes on glucose transport in skeletal muscle. *Endocrinology* **126,** 1921–6.

Barnett, A. H., Eff, C., Leslie, R. D. G., and Pyke, D. A. (1981*a*). Diabetes in identical twins. A study of 200 pairs. *Diabetologia* **20,** 87–93.

Barnett, A. H., Spiliopoulos, A. J., Pyke, D. A., Stubbs, W. A., Burrin, J., and Alberti, K. G. M. M. (1981*b*). Metabolic studies in unaffected co-twins of non-insulin-dependent diabetics. *Br. Med. J.* **282,** 1656–8.

Bell, G. I., Murray, J. C., Nakamura, Y., Kayano, T., Eddy, R. L., Fan, Y. S., Byers, M. G., and Shows, T. B. (1989). Polymorphic human insulin-responsive glucose-transporter gene on chromosome 17p13. *Diabetes* **38,** 1082–75.

Bell, G. I., Kayano, T., Buse, J. B., Furant, C. F., Takeda, J., Lin. D., Fukimoto, H., and Seino, S. (1990). Molecular biology of mammalian glucose transporters. *Diabetes Care* **13,** 198–208.

Bennett, P. H., Rushforth, N. B., Miller, M., and Le Compte, P. M. (1976). Epidemiologic studies of diabetes in the Pima Indians. *Rec. Progr. Hormone Res.* **32,** 333–71.

Berger, J., Biswas, C., Vicario, P. P., Strout, H. V., Saperstein, R., and Pilch, P. F. (1989). Decreased expression of the insulin-responsive glucose transporter in diabetes and fasting. *Nature* **340,** 70–2.

Berntorp, K., Eriksson, K. F., and Lindgarde, F. (1986). The importance of diabetes heredity in lean subjects on insulin secretion, blood lipids and oxygen uptake in the pathogenesis of glucose intolerance. *Diabetes Res. Clin. Pract.* **3,** 231–6.

Blondel, O., Bailbe, D., and Portha, B. (1989). Relation of insulin deficiency to impaired insulin action in NIDDM adult rats given streptozotocin as neonates. *Diabetes* **38,** 610–17.

Bogardus, C., Lillioja, S., Nyomba, B. L., Zurlo, F., Swinburn, B., Esposito del Puente, A., Knowler, W. C., Ravussin, E., Mott, D. M., and Bennett, P. H. (1989). Distribution of *in vivo* insulin action in Pima Indians as mixture of three normal distributions. *Diabetes* **38,** 1423–32.

Byström, C., Larsson, C., Blomberg, C., Sandelin, K., Falkmer, U., Skogseid, B., Öberg, K., Werner, S., and Nordenskjöld, M. (1990). Localization of the MEN1 gene to a small region within chromosome 11q13 by deletion mapping in tumors. *Proc. Natl. Acad. Sci. USA* **87,** 1968–72.

Carroll, R. J., Hammer, R. E., Chan, S. J., Swift, H. H., Rubenstein, A. H., and Steiner, D. F. (1988). A mutant human proinsulin is secreted from islets of Langerhans in increased amounts via an unregulated pathway. *Proc. Natl. Acad. Sci. USA* **85,** 8943–7.

Cerasi, E. (1975). Potentiation of insulin release by glucose in man. III. Normal recognition of glucose as potentiator in subjects with low insulin response and mild diabetes. *Acta Endocrin.* **79,** 511–34.

Cerasi, E. (1985). A la recherche du temps perdu – epilogue to the Minkowski Award Lecture 1974. *Diabetologia* **28,** 547–55.

Cerasi, E. (1988). Insulin secretion in diabetes mellitus. In *The pathology of the endocrine pancreas in diabetes* (ed. P. J. Lefèbvre and D. G. Pipeleers), pp. 191–218. Springer-Verlag, Berlin.

Cerasi, E. and Luft, R. (1967*a*). Insulin response to glucose infusion in diabetic and non-diabetic monozygotic twin pairs. Genetic control of insulin response? *Acta Endocrin.* **55,** 330–45.

Cerasi, E. and Luft, R. (1967*b*). Plasma insulin response to glucose infusion in healthy subjects and in diabetes mellitus. *Acta Endocrin.* **55,** 278–304.

Cerasi, E. and Luft, R. (1970). The occurrence of low insulin response to glucose infusion in children. *Diabetologia* **6,** 85–9.

Cerasi, E. and Luft, R. (1974). Follow-up of non-diabetic subjects with normal and decreased insulin response to glucose infusion—first report. *Horm. Metab. Res.* (Suppl. 5), 113–20.

Cerasi, E., Luft, R., and Efendic, S. (1972). Decreased sensitivity of the pancreatic beta cells to glucose in prediabetic and diabetic subjects. A glucose dose-response study. *Diabetes* **21,** 224–34.

Cerasi, E., Efendic, S., and Luft, R. (1973). Dose–response relationship of plasma insulin and blood glucose levels during oral glucose loads in prediabetic and diabetic subjects. *Lancet* **i,** 794–7.

Cerasi, E., Efendic, S., Thornqvist, C., and Luft, R. (1979). Effect of two sulphonylureas on the dose kinetics of glucose-induced insulin release in normal and diabetic subjects. *Acta Endocrin.* **91,** 282–93.

Cerasi, E., Glaser, B., Del Rio G., Sasson, S., and Della Casa, L. (1989). The clinical significance of insulin resistance in NIDDM: studies with continuous subcutaneous insulin infusion. In *Frontiers of diabetes research: current trends in non-insulin-dependent diabetes mellitus* (ed. K. G. M. M. Alberti and R. Mazze), pp. 309–20. Elsevier, Amsterdam.

Chan, S., Seino, S., Gruppuso, P. A., Gordon, P., and Steiner, D. F. (1987). A mutation in the B chain coding region is associated with impaired proinsulin conversion in a family with hyperproinsulinemia. *Proc. Natl. Acad. Sci. USA* **85,** 2194–7.

Chen, L., Komiya, I., Inman, L., McCorkle, K., Alam, T., and Unger, R. H. (1989). Molecular and cellular responses of islets during perturbations of glucose homeostasis determined by *in situ* hybridization histochemistry. *Proc. Natl. Acad. Sci. USA* **86,** 1367–71.

Chen, L., Alam, T., Johnson, J. H., Hughes, S., Newgard, C. B., and Unger, R. (1990). Regulation of β-cell glucose transporter gene expression. *Proc. Natl. Acad. Sci. USA* **87,** 4088–92.

Chen, M., Bergman, R. N., and Porte, D. (1988). Insulin resistance and β-cell dysfunction in aging: the importance of dietary carbohydrate. *J. Clin. Endocrin. Metab.* **67,** 951–7.

Cocozza, S., Riccardi, G., Monticelli, A., Capaldo, B., Genovese, S., Krogh, V., Celentano, E., Farinaro, E., Varrone, S., and Avvedimento, V. E. (1988). Polymorphism at the 5′ end flanking region of the insulin gene is associated with reduced insulin secretion in healthy individuals. *Eur. J. Clin. Invest.* **18,** 582–6.

Committee on Diabetic Twins, Japan Diabetes Society (1988). Diabetes mellitus in twins: a cooperative study in Japan. *Diabetes Res. Clin. Pract.* **14,** 271–80.

Criffiths, K., McDevitt, D. G., Andrew, M., Baksaas, I., Helgelland, A., Jervell, J., Lunde, P. K. M., Oydvin, K., Agenas, I., Bergman, U., Rosenqvist, U., Sjöqvist, F., Wessling, A., and Wiholm, B. E. (1986). Therapeutic traditions in Northern Ireland, Norway and Sweden: I. Diabetes. *Eur. J. Clin. Pharmacol.* **30,** 513–19.

DeFronzo, R. A. (1988). The triumvirate: β-cell, muscle, liver: a collusion responsible for NIDDM. *Diabetes* **37,** 667–87.

Della Casa, L., Del Rio, G., Glaser, B., and Cerasi, E. (1991). Effect of six month gliclazide treatment on insulin release and sensitivity to endogenous insulin in NIDDM: role of initial CSII-induced normoglycemia. *Am. J. Med.* 90 (suppl. 6A), 375–45S.

Edelman, S. V., Laakso, M., Wallace, P., Brechtel, G., Olefsky, J. M., and Baron, A. D. (1990). Kinetics of insulin-mediated and non-insulin-mediated glucose uptake in humans. *Diabetes* **39,** 955–64.

Efendic, S., Lins, P. E., and Cerasi, E. (1979). Potentiation and inhibition of insulin release in man following priming with glucose and with arginine—effect of somatostatin. *Acta Endocrin.* **90,** 2599–271.

Ekoe, J. M. (1986). Recent trends in prevalence and incidence of diabetes mellitus syndrome in the world. *Diabetes Res. Clin. Pract.* **1,** 249–64.

Elbein, S. C. (1989). Molecular and clinical characterization of an insertional polymorphism of the insulin-receptor gene. *Diabetes* **38,** 737–43.

Elbein, S. C., Corsetti, L., Ullrich, A., and Permut, M. A. (1986). Multiple restriction fragment length polymorphisms at the insulin receptor locus: a highly informative marker for linkage analysis. *Proc. Natl. Acad. Sci. USA* **83,** 5223–7.

Elbein, S. C., Borecki, I., Corsetti, L., Fajans, S. S., Hansen, A. T., Nerup, J., Province, M., and Permutt, M. A. (1987). Linkage analysis of the human insulin receptor gene and maturity-onset diabetes of the young. *Diabetologia* **30,** 641–7.

Elbein, S. C., Corsetti, L., Goldgar, D., Skolnick, M., and Permutt, M. A. (1988*a*). Insulin gene in familial NIDDM. Lack of linkage in Utah Mormon pedigrees. *Diabetes* **37,** 569–76.

Elbein, S. C., Ward, W. K., Beard, J. C., and Permutt, M. A. (1988*b*). Familial NIDDM. Molecular-genetic analysis and assessment of insulin action and pancreatic β-cell function. *Diabetes* **37,** 377–82.

Elton, C. W., Roy, L., Moller, D. E., Pilch, P. F., Pories, W. J., Atkinson, S. M., and Dohm, G. L. (1990). Decreased expression of an insulin-sensitive glucose transporter in muscle from insulin resistant obese and diabetic patients. *Diabetes* **39,** 120A.

Eriksson, J., Franssila-Kallunki, A., Ekstrand, A., Saloranta, C., Widen, E., and Schalin, C., and Groop, L. (1989). Early metabolic defects in persons at increased risk for non-insulin-dependent diabetes mellitus. *New Engl. J. Med.* **321,** 337–43.

Everhart, J., Knowler, W. C., and Bennett, P. H. (1985). Incidence and risk factors for noninsulin-dependent diabetes. In *Diabetes in America* (ed. National Diabetes Data Group), Vol. IV, pp. 1–35. NIH, Bethesda.

Fajans, S. S. (1990). Scope and heterogenous nature of MODY. *Diabetes Care* **13,** 49–64.

Ferrannini, E., Wahren, J., Faber, O. K., Felig, P., Binder, C., and DeFronzo, R. A. (1983). Splanchnic and renal metabolism of insulin in human subjects—a dose-response study. *Am. J. Physiol.* **244,** E517–E527.

Ganda, O. P., Soeldner, J. S., and Gleason, R. E. (1985). Alterations in plasma lipids in the presence of mild glucose intolerance in the offspring of two type II diabetic parents. *Diabetes Care* **8,** 254–60.

Garvey, W. T., Huecksteadt, T. P., and Birnbaum, M. J. (1989). Pretranslational suppression of an insulin-responsive glucose transporter in rats with diabetes mellitus. *Science* **245,** 60–3.

Gerich, J. E., Mitrakou, A., Lelley, D., Mandarino, L., Nurjhan, N., Reilly, J., Jenssen, T., Veneman, T., and Consoli, A. (1990). Contribution of impaired muscle glucose clearance to reduced postabsorptive systemic glucose clearance in NIDDM. *Diabetes* **39,** 211–16.

Glaser, B., Leibovich, G., Nesher, R., Hartling, S., Binder, C., and Cerasi, E. (1988). Improved beta-cell function after intensive insulin treatment in severe non-insulin-dependent diabetes. *Acta Endocrin.* **118,** 365–73.

Greco-Perotto, R., Wertheimer, E., Reiss, N., Jeanrenaud, B., Cerasi, E., and Sasson, S. (1992). Autoregulation of hexose transport in L8 myocyctes by reversible redistribution of glucose transporters. *Biochem. J.*, in press.

Grill, V., Adamson, U., Rundfeldt, M., Andersson, S., and Cerasi, E. (1978). Immediate and time-dependent effects of glucose on insulin release from rat pancreatic tissue. Evidence for different mechanisms of action. *J. Clin. Invest.* **61,** 1034–43.

Grodsky, G. M. (1989). A new phase of insulin secretion. How will it contribute to our understanding of β-cell function? *Diabetes* **38,** 673–8.

Gross, D. J., Halban, P. A., Kahn, R. C., Weir, G. C., and Villa-Komaroff, L. (1989). Partial diversion of a mutant proinsulin (B10 aspartic acid) from the regulated to the constitutive secretory pathway in transfected AtT-20 cells. *Proc. Natl. Acad. Sci. USA* **86,** 4107–111.

Grunnet, J. (1957). Heredity in diabetes mellitus. A proband study. *Opera ex Domo Biologiae Hereditariae Humanae Universitatis Hafniensis* **39,** 1–128.

Gumbiner, B., Polonsky, K. S., Beltz, W. F., Griver, K., Wallace, P., Brechtel, P., and Henry, R. R. (1990). Effects of weight loss and reduced hyperglycemia on the kinetics of insulin secretion in obese non-insulin dependent diabetes mellitus. *J. Clin. Endocrin. Metab.* **70,** 1594–00.

Gutniak, M., Karlander, S. G., and Efendic, S. (1987). Glyburide decreases insulin requirement, increases β-cell response to mixed meal, and does not affect insulin sensitivity: effects of short- and long-term combined treatment in secondary failure to sulfonylurea. *Diabetes Care* **10,** 545–54.

Gutzeit, A., Rabinovitch, A., Karakash, C., Staufacher, W., Renold, A. E., and Cerasi, E. (1974). Evidence for decreased sensitivity to glucose of isolated islets from spiny mice (*Acomys cahirinus*). *Diabetologia* **10,** 661–5.

Gutzeit, A., Renold, A. E., Cerasi, E., and Shafrir, E. (1979). Effect of diet-induced obesity on glucose and insulin tolerance of a rodent with a low insulin response (*Acomys cahirinus*). *Diabetes* **28,** 777–84.

Hansen, B. C., Bodkin, N. L., Schwartz, J., and Jen, K. L. C. (1988). B-cell responses, insulin resistance and the natural history of noninsulin-dependent diabetes in obese rhesus monkeys. In *Frontiers in diabetes research. Lessons from animal diabetes II* (ed. E. Sharir and A. E. Renold), pp. 279–87. John Libbey, London.

Harris, M. I. (1988). Gestational diabetes may represent discovery of preexisting glucose intolerance. *Diabetes Care* **11,** 402–11.

Harris, M. I. (1989). Impaired glucose tolerance in the U.S. population. *Diabetes Care* **12,** 464–74.

Harris, M. I., Hadden, W. C., Knowler, W. C., and Bennett, P. H. (1987). Prevalence of diabetes and impaired glucose tolerance and plasma glucose levels in the U.S. population aged 20–74 yr. *Diabetes* **36,** 523–34.

Henry, R. R., Gumbiner, B., Glynn, T., and Thornburn, A. W. (1990). Metabolic effects of hyperglycemia and hyperinsulinemia on fate of intracellular glucose in NIDDM. *Diabetes* **39,** 149–56.

Hirshman, M. F., Goodyear, L. J., Wardzala, L. J., Horton, E. D., and Horton, E. S. (1990). Identification of an intracellular pool of glucose transporters from basal and insulin-stimulated rat skeletal muscle. *J. Biol. Chem.* **265,** 987–91.

Hitman, G. H., Jowett, N. I., Williams, L. G., Humphries, W. S., Winter, R. M., and Galton, D. J. (1984). Polymorphism in the 5′ flanking region of the insulin gene and non-insulin dependent diabetes. *Clin. Sci.* **66,** 383–8.

Hollenbeck, C. and Reaven, G. M. (1987). Variations in insulin-stimulated glucose uptake in healthy individuals with normal glucose tolerance. *J. Clin. Endocrin. Metab.* **64,** 1169–73.

Hosker, J. P., Burnett, M. A., Davies, E. G., Harris, E. A. and Turner, R. C. (1985). Sulphonylurea therapy doubles B-cell response to glucose in type 2 diabetic patients. *Diabetologa* **28,** 809–14.

Howard, C. F. (1986). Longitudinal studies on the development of diabetes in individual *Macaca nigra*. *Diabetologia* **29,** 301–6.

Iselius, L., Lindsten, J., Morton, N. E., Efendic, S., Cerasi, E., Haegermark, A., and Luft,

R. (1982). Evidence for an autosomal recessive gene regulating the persistence of the insulin response to glucose in man. *Clin. Gen.* **22,** 180–94.

Iselius, L., Lindsten, J., Morton, N. E., Efendic, S., Cerasi, E., Haegermark, A., and Luft, R. (1985). Genetic regulation of the kinetics of glucose-induced insulin release in man. Studies in families with diabetic and non-diabetic probands. *Clin. Gen.* **28,** 8–15.

Iynedjian, P. B., Pilot, P. R., Nouspikel, T., Milburn, J. L., Quaade, J., Hughes, S., Ucla, C., and Newgard, C. B. (1989). Differential expression and regulation of the glucokinase gene in liver and islets of Langerhans. *Proc. Natl. Acad. Sci. USA* **86,** 7838–42.

Johnson, J. H., Newgard, C. B., Milburn, J. L., Lodish, H. F., and Thorens, B. (1990*a*). The high Km glucose transporter of islets of Langerhans is functionally similar to the low affinity transporter of liver and has an identical primary sequence. *J. Biol. Chem.* **265,** 6548–51.

Johnson, J. H., Crider, B. P., McCorkle, K., Alford, M., and Unger, R. H. (1990*b*). Inhibition of glucose transport into rat islet cells by immunoglobulins from patients with new-onset insulin-dependent diabetes mellitus. *New Engl. J. Med.* **322,** 653–9.

Kadowaki, T., Miyake, Y., Hagura, R., Akanuma, Y., Kajinuma, H., Kuzuya, N., Takaku, F., and Kosaka, K. (1984). Risk factors for worsening to diabetes in subjects with impaired glucose tolerance. *Diabetologia* **26,** 44–9.

Kadowaki, T., Bevius, C. L., Cama, A., Ojamaa, K., Marcus-Samuels, B., Kadowaki, H., Beitz, L., McKeon, C., and Taylor, S. I. (1988). Two mutant alleles of the insulin receptor gene in a patient with extreme insulin resistance. *Science* **240,** 787–90.

Kahn, C. B., Soeldner, J. S., Gleason, R. E., Rojas, L., Camerini-Davalos, R. A., and Marble, A. (1969). Clinical and chemical diabetes in the offspring of diabetic couples. *New Engl. J. Med.* **281,** 343–6.

Kahn, S. E., Beard, J. C., Schwartz, M. W., Ward, W. K., Ding, H. L., Bergman, R. N., Taborsky, G. J., and Porte, D. (1989). Increased β-cell secretory capacity as mechanism for islet adaptation to nicotinic acid-induced insulin resistance. *Diabetes* **38,** 562–8.

Kaiser, N., Corcos, A. P., Sarel, I., and Cerasi, E. (1991). Monolayer culture of adult rat pancreatic islets on extracellular matrix: modulation of B-cell function by chronic exposure to high glucose *Endocrinology* **129,** 2067–76.

Kaku, K., Matsutani, A., Muckler, M., and Permutt, M. A. (1990). Polymorphisms of Hep G2/erythrocyte glucose-transporter gene. Linkage relationships and implications for genetic analysis of NIDDM. *Diabetes* **39,** 499–56.

Kanatsuna, T., Baekkeskov, S., Lernmark, Å., and Ludvigsson, J. (1983). Immunoglobulin from insulin-dependent diabetic children inhibits glucose-induced insulin release. *Diabetes* **32,** 520–4.

Karnieli, E., Armoni, M., Cohen, P., Kanter, Y., and Rafaeloff, R. (1987). Reversal of insulin resistance in diabetic rat adipocytes by insulin therapy: restoration of pool of glucose transporters and enhancement of glucose-transport activity. *Diabetes* **36,** 925–31.

Keen, H., Jarrett, R. J., and McCartney, P. (1982). The ten-year follow-up of the Bedford survey (1962–72): glucose tolerance and diabetes. *Diabetologia* **22,** 73–8.

Khan, A., Chandramouli, V., Östenson, C. G., Berggren, P. O., Löw, H., Landau, B. R., and Efendic, S. (1990*a*). Glucose cycling is markedly enhanced in pancreatic islets of obese hyperglycemic mice. *Endocrinology* **126,** 2413–16.

Khan, A., Chandramouli, V., Östenson, C. G., Löw, H., Landau, B. R., and Efendic, S. (1990*b*). Glucose cycling in islets from healthy and diabetic rats. *Diabetes* **39,** 456–9.

King, H., Rao, D. C., Bhatia, K., Koki, G., Collins, A., and Zimmett, P. (1989). Family resemblance for glucose tolerance in a Melanesian population, the Tolai. *Hum. Hered.* **39,** 212–17.

Klinkhamer, M. P., Groen, N. A., van der Zon, G. C. M., Lindhout, D., Sandkuyl, L. A., Krans, H. M. J., Moller, W., and Maassen, J. A. (1989). A leucine-to-proline mutation in the insulin receptor in a family with insulin resistance. *EMBO J.* **8,** 2503–7.

Köbberling, J. and Tillil, H. (1982). Empirical risk figures for first degree relatives of non-insulin dependent diabetics. In *The genetics of diabetes mellitus* (ed. J. Köbberling and R. Tattersal), pp. 201–9. Academic Press, London.

Köbberling, J. and Tillil, H. (1989). Genetics of diabetes mellitus. In *Diabetes mellitus: pathophysiology and therapy,* (ed. W. Creutzfeld and P. Lefèbvre), pp. 27–38. Springer-Verlag, Berlin.

Kosaka, K., Kuzuya, T., Akanuma, Y., and Hagura, R. (1980). Increase in insulin response after treatment of overt maturity-onset diabetes is independent of the mode of treatment. *Diabetologia* **18,** 23–8.

Kruszynska, T. Y., Home, P. D., Hanning, I., and Alberti, K. G. M. M. (1987). Basal and 24-h C-peptide and insulin secretion rate in normal man. *Diabetologia* **30,** 16–21.

Laakso, M., Edelman, S. V., Brechtel, G., and Baron, A. D. (1990). Decreased effect of insulin to stimulate skeletal muscle blood flow in obese man. *J. Clin. Invest.* **85,** 1844–52.

Leighton, B. and Cooper, G. S. (1990). The role of amylin in the insulin resistance of non-insulin-dependent diabetes mellitus. *TIBS* **15,** 295–9.

Leonetti, D. L. and Fukimoto, W. Y. (1989). Type 2 diabetes, impaired glucose tolerance, and hypertension in offspring of migrants and the structure of the population of origin. *Hum. Biol.* **61,** 369–86.

Leslie, R. D., Volkmann, H. P., Poncher, M., Hanning, I., Orskov, H., and Alberti, K. G. M. M. (1986). Metabolic abnormalities in children of non-insulin dependent diabetics. *Br. Med. J.* **293,** 840–2.

Li, S. R., Oelbaum, R. S., Baroni, M. G., Stock, J. and Galton, D. J. (1988). Association of genetic variant of the glucose transporter with non-insulin-dependent diabetes mellitus. *Lancet* **2,** 368–70.

Lilioja, S., Mott, D. M., Zawadzki, J. K., Young, A. A., Abbott, W. G. H., Knowler, W. C., Bennett, P. H., Moll, P., and Bogardus, C. (1987). *In vivo* insulin action is familial characteristic in nondiabetic Pima Indians. *Diabetes* **36,** 1329–35.

Lindsten, J., Cerasi, E., Luft, R., Morton, N., and Ryman, N. (1976). Significance of genetic factors for the plasma insulin response to glucose in healthy subjects. *Clin. Gen.* **10,** 125–34.

McCarthy, S. T., Harris, F., and Turner, R. C. (1977). Glucose control of basal insulin secretion in diabetics. *Diabetologia* **13,** 93–7.

McClain, D. A., Henry, R. R., Ullrich, A., and Olefsky, J. M. (1988). Restriction-fragment length polymorphism in insulin-receptor gene and insulin resistance in NIDDM. *Diabetes* **38,** 1071–5.

Martineau, R., Kohlbacher, M., Shaw, S. N., and Amos, H. (1972). Enhancement of glucose entry into chick fibroblasts by starvation: differential effect of galactose and glucose. *Proc. Natl. Acad. Sci. USA* **69,** 3407–11.

Matschinsky, F. M. (1990). Glucokinase as glucose sensor and metabolic signal generator in pancreatic β-cells and hepatocytes. *Diabetes* **39,** 647–52.

Matthews, D. R., Hosker, J. P., Rudenski, A. S., Naybor, B. A., Treacher, D. F., and Turner, R. C. (1985). Homeostasis model assessment: insulin resistance and β-cell function from fasting plasma glucose and insulin concentrations in man. *Diabetologia* **28,** 412–19.

Mayer, J. (1965). Genetic factors in human obesity. *Ann. NY Acad. Sci. 131,* 412–21.

Medalie, J. H., Herman, J. B., Goldbourt, U., and Papier, C. M. (1978). Variance in incidence of diabetes among 10 000 adult Israeli males and the factors related to their development. In *Advances in metabolic disorders* (ed. R. Levine and R. Luft), pp. 93–101. Academic Press, New York.

Mitsukawa, T., Takemura, J., Asai, J., Nakazato, M., Kangawa, K., Matsuo, H., and Matsukura, S. (1990). Islet amyloid polypeptide. Response to glucose, insulin, and somatostatin analogue administration. *Diabetes* **39,** 639–42.

Mohan, V., Ramachadran, A., Snehalatha, C., Rema, M., Bhreni, G., and Viswanathan, M. (1985). High prevalence of maturity-onset diabetes of young (MODY) among Indians. *Diabetics Care* **8,** 371–4.

Mohan, V., Snehalatha, C., Ramachadran, A., and Viswanathan, M. (1986). Abnormalities in insulin secretion in healthy offspring of Indian patients with maturity-onset diabetes of the young. *Diabetes Care* **9,** 53–6.

Moller, D. E., Yokota, A., and Flier, J. S. (1989). Normal insulin-receptor cDNA sequence in Pima Indians with NIDDM. *Diabetes* **38,** 1496–500.

Nagamatsu, S., Carroll, R. J., Grodsky, G. M., and Steiner, D. F. (1990). Lack of islet amyloid polypeptide regulation of insulin biosynthesis or secretion in normal rat islets. *Diabetes* **39,** 871–4.

Neel, J. V. (1965). Diabetes mellitus. In *The genetics and epidemiology of chronic disease* (ed. J. V. Neel, M. W. Shaw, and W. J. Schull), pp. 105–32. US Public Health Service Publications, Washington.

Nesher, R. and Cerasi, E. (1987). Biphasic insulin release as the expression of combined inhibitory and potentiating effects of glucose. *Endocrinology* **121,** 1017–27.

Nesher, R., Tuch, B., Hage, C., Levy, J., and Cerasi, E. (1984). Time-dependent inhibition of insulin-release: suppression of the arginine effect by hyperglycaemia. *Diabetologia* **26,** 142–5.

Nesher, R., Abramovitch, E., and Cerasi, E. (1985). Correction of diabetic pattern of insulin release from islets of spiny mice (*Acomys cahirinus*) by glucose priming *in vitro. Diabetologia* **28,** 233–6.

Nesher, R., Della Casa, L., Litvin, Y., Sinai, J., Del Rio, G., Pevsner, B., Wax, Y., and Cerasi, E. (1987). Insulin deficiency and insulin resistance in type 2 (non-insulin-dependent) diabetes: quantitative contributions of pancreatic and peripheral responses to glucose homeostasis *Eur. J. Clin. Invest.* **17,** 266–74.

Nesher, R., Praiss, M., and Cerasi, E. (1988). Immediate and time-dependent effects of glucose on insulin release: differential calcium requirements. *Acta Endocrin.* **117,** 409–16.

Nesher, R., Abramovitch, E., and Cerasi, E. (1989). Reduced early and late phase insulin response to glucose in isolated spiny mouse (*Acomys cahirinus*) islets: a defective link between glycolysis and adenylate cyclase. *Diabetologia* **32,** 644–8.

Nishi, M., Sanke, T., Nagamatsu, S., Bell, G. I., and Steiner, D. F. (1990). Islet amyloid polypeptide. A new β-cell secretory product related to islet amyloid deposits. *J. Biol. Chem.* **265,** 4173–6.

Niskanen, L. K., Uusitupa, M. I., Sarlund, H., Siitonen, O., and Pyörälä, K. (1990). Five-year follow-up study on plasma insulin levels in newly diagnosed NIDDM patients and nondiabetic subjects. *Diabetes Care* **13,** 41–8.

Nordlander, S., Östman, J., Cerasi, E., Luft, R., and Ekelund, L. G. (1977). Occurrence of diabetic type of plasma FFA and glycerol responses to physical exercise in prediabetic subjects. *Acta Med. Scand.* **193,** 9–21.

Ober, C., Xiang, K. S., Thisted, R. A., Indovina, K. A., Wason, C. J., and Dooley, S. (1989). Increased risk for gestational diabetes mellitus associated with insulin receptor and insulin-like growth factor II restriction fragment length polymorphism. *Genetic Epidemiol.* **6,** 559–69.

Ohlsén, P., Cerasi, E., and Luft, R. (1971). Glucose tolerance and insulin response to glucose in two large families with diabetic mothers in the first generation. *Horm. Metab. Res.* **3,** 1–5.

Ohlson, L. O., Larsson, B., Björntorp, P., Eriksson, H., Svardsudd, K., Welin, L., Tibblin, G., and Wilhelmsen, L. (1988). Risk factors for type 2 (non-insulin-dependent) diabetes mellitus. Thirteen and one-half years of follow-up of the participants in a study of Swedish men born in 1913. *Diabetologia* **31,** 798–805.

Oka, Y., Asano, T., Shibasaki, Y., Lin, J. L., Tsukuda, K., Akanuma, Y., and Takaku, F. (1990). Increased liver glucose transporter protein and mRNA in streptozotocin-induced diabetic rats. *Diabetes* **39,** 441–6.

O'Rahilly, S. and Turner, R. C. (1988). Early-onset type 2 diabetes versus maturity-onset diabetes of youth: evidence for the existence of two discrete diabetic syndromes. *Diabetic Med.* **5,** 224–34.

O'Rahilly, S., Spivey, R. S., Holman, R. R., Nugent, Z., Clark, A., and Turner, R. C. (1987). Type II diabetes of early onset: a distinct clinical and genetic syndrome? *Br. Med. J.* **294,** 923–8.

O'Rahilly, S., Hosker, J. P., Rudenski, A. S., Matthews, D. R., Burnett, M. A., and Turner, R. C. (1988*a*). The glucose stimulus-response curve of the β-cell in physically trained humans, assessed by hyperglycemic clamps. *Metabolism* **37,** 919–23.

O'Rahilly, S., Turner, R. C., and Matthews, D. R. (1988*b*). Impaired pulsatile secretion of insulin in relatives of patients with non-insulin-dependent diabetes. *New Engl. J. Med.* **318,** 1225–30.

O'Rahilly, S., Wainscoat, J. S., and Turner, R. C. (1988*c*) Type 2 (non-insulin-dependent) diabetes mellitus—New genetics for old nightmares. *Diabetologia* **31,** 407–14.

O'Rahilly, S., Patel, P., Wainscoat, J. S., and Turner, R. C. (1989). Analysis of the Hep G2/ erythrocyte glucose transporter locus in a family with type 2 (non-insulin-dependent) diabetes and obesity. *Diabetologia* **32,** 266–9.

Osei, K. (1990). Increased basal glucose production and utilization in nondiabetic first-degree relatives of patients with NIDDM. *Diabetes* **39,** 597–601.

Owerbach, D. and Nerup, J. (1982). Restriction fragment length polymorphism of the insulin gene in diabetes mellitus. *Diabetes* **31,** 275–7.

Owerbach, D., Thomsen, B., Johansen, K., Lamm, U., and Nerup, J. (1983). DNA insertion sequences near the insulin gene are not associated with maturity-onset diabetes of young people. *Diabetologia* **25,** 18–20.

Panzram, G. and Adolph, W. (1983). Ergebnisse einer Populationsstudie über den nichtinsulin-abhängigen Diabetes mellitus in Kindes- und Jugend-alter. *Schweiz. Med. Wochenschr.* **113,** 779–84.

Pedersen, O., Bak, F., Andersen, P. H., Lund, S., Moller, D. E., Flier, J. S., and Kahn, B. B. (1990). Evidence against altered expression of GLUT 1 or GLUT 4 in skeletal muscle of patients with obesity or NIDDM. *Diabetes* **39,** 865–70.

Permutt, M. A., Rotwein, P., Andreone, T., Ward, W. K., and Porte, D. (1985). Islet beta-cell function and polymorphism in the 5′-flanking region of the human insulin gene. *Diabetes* **34,** 311–14.

Pettit, D. J., Bennett, P. H., Knowler, W. C., Baird, H. R., and Aleck, K. A. (1985). Gestational diabetes mellitus and impaired glucose tolerance during pregnancy. Long-term effects on obesity and glucose tolerance in the offspring. *Diabetes* **34** (Suppl. 2), 119–22.

Pfeiffer, E. F. (ed.) (1983). *Rationale for sulfonylurea therapy.* Excerpta Medica, Amsterdam.

Porte, D. and Kahn, S. E. (1989). Hyperproinsulinemia and amyloid in NIDDM. Clues to etiology of islet β-cell dysfunction? *Diabetes* **38,** 1333–6.

Purrello, F., Vetri, M., Vincic Gatta, C., Buscema, M., and Vigneri, R. (1990). Chronic exposure to high glucose and impairment of K^+ channel function in perfused rat pancreatic islets. *Diabetes* **39,** 397–9.

Radder, J. K. and Terpstra, J. (1975). The incidence of diabetes mellitus in the offspring of diabetic couples. Investigation based on the oral glucose tolerance test. *Diabetologia* **11,** 135–8.

Ravussin, E., Lillioja, S., Knowler, W. C., Christin, L., Freymond, D., Abbott, W. G. H., Boyce, V., Howard, B. V., and Bogardus, C. (1988). Reduced rate of energy expenditure as a risk factor for body-weight gain. *New Engl. J. Med.* **318,** 467–72.

Robertson, R. P. (1989). Type II diabetes, glucose 'non-sense' and islet desensitization. *Diabetes* **38,** 1501–5.

Rosenthal, M., McMahan, C. A., Stern, M. P., Eifler, C. W., Haffner, S. M., Hazirda, H. P., and Franco, L. J. (1985). Evidence of bimodality of two hour plasma glucose concentrations in Mexican Americans: results from the San Antonio heart study. *J. Chronic Dis.* **38,** 5–16.

Rossetti, L. and Giaccari, A. (1990). Relative contribution of glycogen synthesis and glycolysis to insulin-mediated glucose uptake. A dose–response euglycemic clamp study in normal and diabetic rats. *J. Clin. Invest.* **85,** 1785–92.

Rossetti, L., Giaccari, A., and DeFronzo, R. A. (1990). Glucose toxicity. *Diabetes Care* **13,** 610–30.

Rotter, J. I. and Rimoin, D. L. (1981). The genetics of the glucose intolerance disorders. *Am. J. Med.* **70,** 116–26.

Rotwein, P. S., Chirgwin, J., Province, M., Knowler, W. C., Pettit, D. J., and Permutt, M. A. (1983). Polymorphism in the 5′-flanking region of the human insulin gene: a genetic marker for non-insulin dependent diabetes. *New Engl. J. Med.* **308,** 65–71.

Saad, M. F., Kahn, S. E., Nelson, R. G., Pettit, D. J., Knowler, W. C., Schwartz, M. W., Kowalyk, S., Bennett, P. H., and Porte, D. (1990). Disproportionately elevated proinsulin in Pima Indians with noninsulin-dependent diabetes mellitus. *J. Clin. Endocrin. Metab.* **70,** 1247–53.

Samanta, A., Burden, A. C., Jowett, N. I., Galton, D. J., Hosker, J. P., and Turner, R. C. (1987). Insulin secretion to glucose infusion in gestational diabetes subjects with differing DNA polymorphism flanking the insulin gene. *Diabetes Res. Clin. Pract.* **4,** 109–12.

Sanz, N., Karam, J. H., Horita, S., and Bell, G. I. (1986). Prevalence of insulin gene mutations in non-insulin-dependent diabetes mellitus. *New Engl. J. Med.* **314,** 1322.

Sartor, G., Schersten, B., Carlström, S., Melander, A., Norden, A., and Persson, G. (1980). Ten-year follow-up of subjects with impaired glucose tolerance: prevention of diabetes by tolbutamide and diet regulation. *Diabetes* **29,** 41–9.

Sasson, S. and Cerasi, E. (1986). Substrate regulation of the glucose transport system in rat skeletal muscle. Characterization and kinetic analysis in isolated soleus muscle and skeletal muscle cells in culture. *J. Biol. Chem.* **261,** 16827–33.

Sasson, S., Edelson, D., and Cerasi, E. (1987). *In vitro* autoregulation of glucose utilization in rat soleus muscle. *Diabetes* **36,** 1041–6.

Seino, S., Seino, M., and Bell, G. I. (1990*a*). Human insulin-receptor gene. *Diabetes* **39,** 129–33.

Seino, S., Seino, M., and Bell, G. I. (1990*b*). Human insulin-receptor gene. Partial sequence and amplification of exons by polymerase chain reaction. *Diabetes* **39,** 123–8.

Serjeantson, S. W., White, B., Bell, G. I., and Zimmet, P. (1989). The glucose transporter gene and type 2 diabetes in the Pacific. In *Diabetes 1988* (ed. R. G. Larkins, P. Z. Zimmet, and D. J. Chisholm), pp. 329–33. Excerpta Medica, Amsterdam.

Shafrir, E. (1990). Diabetes in animals. In *Diabetes mellitus. Theory and practice* (ed. H. Rifkin and D. Porte), pp. 299–340. Elsevier, New York.

Shapiro, E. T., Tillil, H., Rubenstein, A. H., and Polonsky, K. S. (1988). Peripheral insulin parallels changes in insulin secretion more closely than C-peptide after bolus intravenous glucose administration. *J. Clin. Endocrin. Metab.* **67,** 1094–9.

Shibasaki, Y., Asano, T., Shibasaki, M., Kajio, H., Kanazawa, Y., and Takaku, F. (1990). Insulin biosynthesis and glucose transporter expression in RINr cell line. *Diabetes* **39,** 138A.

Shows, T. B., Eddy, R. L., Byers, M. G., Fukishima, Y., DeHaven, C. R., Murray, J. C., and Bell, G. I. (1987). Polymorphic human glucose transporter gene (GLUT) is on chromosome 1p31.3 p 35. *Diabetes* **36,** 546–9.

Shulman, G. I., Rothman, D. L., Jue, T., Stein, P., DeFronzo, R. A., and Shulman, R. G. (1990). Quantitation of muscle glycogen synthesis in normal subjects and subjects with non-insulin-dependent diabetes by 13C nuclear resonance spectroscopy. *New Engl. J. Med.* **322,** 223–2.

Sicree, R. A., Zimmet, P. A., King, H. O. M., and Coventry, J. S. (1987). Plasma insulin response among Nauruans. Prediction of deterioration in glucose tolerance over 6 yr. *Diabetes* **36,** 179–86.

Sivitz, W. I., DeSautel, S. L., Kayano, T., Bell, G. I., and Pressin, J. E. (1989). Regulation of glucose transporter messenger RNA in insulin-deficient states. *Nature* **340,** 72–4.

Smith, C. P., Tarn, A. C., Thomas, J. M., Overdamp, D., Corakir, A., Savage, M. O., and Gale, E. A. M. (1988). Between and within subject variation of the first phase insulin response to intravenous glucose. *Diabetologia* **31,** 123–5.

Snehalatha, C., Ramachandran, A., Mohan, V., and Viswanathan, M. (1986). Insulin response in obese and nonobese offspring of conjugal Indian diabetic parents with increasing glucose intolerance. *Pancreas* **1,** 139–42.

Soeldner, J. S., Gleason, R. E., Williams, R. F., Garcia, M. J., Beardwood, D. M., and Marble, A. (1968). Diminished serum insulin response to glucose in genetic prediabetic males with normal glucose tolerance. *Diabetes* **17,** 17–26.

Steiner, D. F., Bell, G. I., and Tager, H. S. (1989). Chemistry and biosynthesis of pancreatic protein hormones. In *Endocrinology* (ed. De Groot), pp. 1263–89. Saunders, Philadelphia.

Steiner, D. F., Tager, H. S., Chan, S. J., Nanjo, K., Sanke, T., and Rubenstein, A. H. (1990). Lessons learned from molecular biology of insulin-gene mutations. *Diabetes Care* **13,** 600–9.

Strout, H. V., Vicario, P. P., Biswas, C., Saperstein, R., Brady, E. J., Pilch, P. F., and Berger, J. (1990). Vanadate treatment of streptozotocin diabetic rats restores expression of the insulin-responsive glucose transporter in skeletal muscle. *Endocrinology* **126,** 2728–32.

Stunkard, A. J., Sorensen, T. I. A., Harris, C., Teasdale, T. W., Chakaborty, R., Schull, W. J., and Schulsinger, F. (1986). An adoption study of human obesity. *New Engl. J. Med.* **314,** 193–8.

Tager, H., Given, B., Baldwin, D., Mako, M., Markese, J., Rubenstein, A., Ofefsky, J., Kobayashi, M., Kollerman, O., and Poucher, P. (1979). A structurally abnormal insulin causing human diabetes. *Nature* **281,** 122–5.

Tattersall, R., and Fajans, S. S. (1975). Diabetes and carbohydrate tolerance in 199 offspring of 37 conjugal diabetic parents. *Diabetes* **24,** 452–62.

Temple, R. C., Carrington, C. A., Luzio, S. D., Owens, D. R., Schneider, A. E., Sobey, W. J., and Hales, C. N. (1989). Insulin deficiency in non-insulin-dependent diabetes. *Lancet* **i,** 293–5.

Terazono, K., Watanabe, T., and Yonemura, Y. (1990). A novel gene, *reg*, expressed in regenerating islets. In *Molecular biology of the islets of Langerhans* (ed. H. Ikamoto), pp. 301–13. Cambridge University Press, Cambridge.

Thorens, B., Flier, J. S., Lodish, H. F., and Kahn, B. B. (1990) Differential regulation of two glucose transporters in rat liver by fasting and refeeding and by diabetes and insulin treatment. *Diabetes* **39,** 712–19.

Turner, R. C. and Holman, R. R. (1976). Insulin rather than glucose homeostasis in the pathophysiology of diabetes. *Lancet* **i,** 1272–4.

Vinik, A. I., Cox, N. J., Xiang, K., Fajans, S. S., and Bell, G. I. (1988). Linkage studies of maturity onset diabetes of the young: R.W. pedigree. *Diabetologia* **31,** 778–880.

Viswanathan, M., Mohan, V., Snehalatha, C., and Ramachandran, A. (1985). High prevalence of Type 2 (non-insulin-dependent) diabetes among the offspring of conjugal Type 2 diabetic parents in India. *Diabetologia* **28,** 907–10.

Walker, P. S., Ramlal, T., Sarabia, V., Koivisto, U. M., Bilan, P. J., Pessin, J. E., and Klip, A. (1990). Glucose transport activity in L6 muscle cells is regulated by the coordinate control of subcellular glucose transporter distribution, biosynthesis, and mRNA transcription. *J. Biol. Chem.* **265,** 1516–23.

Ward, G. M., Walters, J. M., Aitken, P. M., Best, J. D., and Alford, F. P. (1990). Effects of prolonged pulsatile hyperinsulinemia in humans. Enhancement of insulin sensitivity. *Diabetes* **39,** 501–7.

Ward, W. K., Johnston, C. L. W., Beard, J. C., Benedetti, T. J., and Porte, D. (1985). Insulin resistance and impaired insulin secretion in subjects with histories of gestational diabetes mellitus. *Diabetes* **34,** 861–9.

Ward, W. K., LaCava, E. C., Paquette, T. L., Beard, J. C., Wallum, B. J., and Porte, D. (1987). Disproportionate elevation of immunoreactive proinsulin in type 2 (non-insulin-dependent) diabetes mellitus and in experimental insulin resistance. *Diabetologia* **30,** 698–702.

Watanabe, T., Yonekura, H., Terazono, K., Yamamoto, H., and Okamoto, H. (1990). Complete nucleotide sequence of human *reg* gene and its expression in normal and tumoral tissues. The reg protein, pancreatic stone protein, and pancreatic thread protein are one and the same product of the gene. *J. Biol. Chem.* **265,** 7432–9.

Wertheimer, E., Ben-Neriah, Y., Sasson, S., and Cerasi, E. (1989). Regulation of glucose transporter mRNA levels by glucose in muscle cells. *Diabetes* **38,** 40A.

Wertheimer, E., Sasson, S., Cerasi, E., and Ben-Neriah, Y. (1991). The ubiquitous glucose transporter GLUT-1 is a GRP-like stress-inducible protein. *Proc. Natl. Acad. Sci. USA* **88,** 2525–9.

Winter, W. E., Maclaren, N. K., Riley, W. J., Clarke, D. W., Kappy, M. S., and Spillar, R. P. (1987). Maturity-onset diabetes of youth in black Americans. *New Engl. J. Med.* **316,** 285–91.

Xiang, K. S., Cox, N. J., Sanz, N., Huang, P., Karam, J. H., and Bell, G. I. (1989). Insulin-receptor and apolipoprotein genes contribute to development of NIDDM in Chinese Americans. *Diabetes* **31,** 17–23.

Zimmet, P. and Whitehouse, S. (1978). Bimodality of fasting and two-hour glucose tolerance distributions in a Micronesian population. *Diabetes* **27,** 793–800.

Zimmet, P., Dowse, G., La Porte, R., Finch, C., and Moy, C. (1989). Epidemiology—its contribution to understanding of the etiology, pathogenesis, and prevention of diabetes mellitus. In *Diabetes mellitus: pathophysiology and therapy* (ed. W. Creutzfeldt and P. Lefèbvre). Springer-Verlag, Berlin.

Glossary

Words in italics are defined elsewhere in the glossary

A-current A transient outward K-current

acetyl-CoA carboxylase The enzyme that catalyses the reaction

acetyl-CoA + $CO_2 \rightarrow$ malonyl-CoA.

This is the first reaction in fatty acid synthesis.

action potential A rapid transient change in the electrical potential across a cell membrane caused by an alteration in ion permeability.

adenyl cyclase The plasma membrane-bound enzyme catalysing the formation of *cyclic AMP* from ATP.

ADP-ribosylation The irreversible covalent addition of ADP-ribose from NAD^+ to a protein. Many bacterial toxins (e.g. cholera toxin, pertussis toxin) modify target proteins in this way.

adrenaline A hormone released from the adrenal medulla. Also known as epinephrine.

aetiology The origin or causes of a disease.

affinity chromatography A technique for protein purification in which the protein is selectively absorbed to a ligand which has been immobilized on a solid support.

affinity labelling A method for labelling a protein with a radioactive or fluorescent ligand.

allele One of the two or more forms of a particular gene that occur within a species.

alloxan A chemical which induces diabetes when injected into animals because of its cytotoxic effect on pancreatic β-cells.

α-helix A frequent form of secondary structure in proteins in which the amino acid residues are arranged in a regular spiral stabilized by hydrogen bonds between NH and CO groups of the protein backbone.

alternative splicing The processing of a single mRNA precursor in different ways leading to the production of more than one protein from a single gene.

aminoacyl tRNA synthetase The enzyme responsible for attaching an amino acid to the appropriate transfer RNA for addition to a growing polypeptide chain during protein synthesis.

amphipathic helix A sequence of amino acids in a helical region of a protein in which one side of the helix contains significantly more hydrophilic residues than the other.

amylin A protein co-secreted with insulin from the β-cell, which is believed to be a hormone. Deposition of a polymerized form (amyloid) of amylin is a characteristic finding in the islets of type II diabetics.

ankylosing spondylitis An autoimmune disease resulting in inflammation and eventual immobility of the vertebrae of the spine.

anomer Two forms of the same compound differing only in the orientation of groups at an assymetric carbon atom. For the ring form of glucose, α- and β-anomers exist which differ in their ability to stimulate insulin secretion (α- > β-).

anti-idiotype *Antibodies* or *lymphocyte* receptors specific for an *idiotype*. Anti-idiotypic antibodies or lymphocyte receptors may bear an 'internal image' of the original *antigen* and thus potentially mimic its action.

antibody A protein produced by B-cells in response to an *antigen* and which binds specifically to that antigen. There are five classes of antibodies: IgA, IgD, IgE, IgG, and IgM. A typical antibody is a Y-shaped molecule with two identical antigen-binding sites.

antigen Any substance capable of stimulating the production of an *antibody*.

antigen-presenting cell (APC) Any cell capable of bearing an *antigen* at its surface in a form that stimulates *lymphocytes*.

antilipolytic effect (of insulin) The ability of insulin to decrease the rate of breakdown of *triacylglycerol* in adipose tissue.

ATP citrate lyase The enzyme catalysing the reaction

citrate → acetyl-CoA + oxaloacetate.

ATP-regulated K-channel A class of K-channels which are inhibited by the binding of intracellular ATP. In β-cells these channels play a central role in glucose-stimulated insulin release. Also known as ATP-sensitive or ATP-dependent K-channel.

autoantibody An *antibody* produced against a self-protein.

autophosphorylation The phenomenon whereby a *protein kinase* phosphorylates an amino acid residue in its own structure.

autosomal dominant A mode of inheritance in which the possession of a single *allele* on a non-sex chromosome determines the *phenotype*.

B-cell A class of *lymphocytes* produced in the bone marrow which secrete soluble antibodies.

basal lamina The extracellular matrix surrounding muscle, fat, and nerve cells and separating *epithelial* cell sheets from the *endothelium* surrounding blood vessels.

BB rat The Bio-Breeding rat used as a model for type I diabetes. This strain develops a *T-cell*-mediated *insulitis* linked to both *MHC* and non-MHC genes.

β-pleated sheet A frequent form of secondary structure in proteins in which the amino acid residues are arranged in a sheet stabilized by hydrogen bonds between NH and CO groups of adjacent strands of the protein backbone.

betagranin A protein found in the matrix of β-cell *secretory granules*.

biguanide A class of drug used in the treatment of type II diabetes which lowers blood glucose via effects on peripheral tissues.

BiP A binding protein found in the lumen of the *endoplasmic reticulum* which helps newly synthesized proteins to fold correctly.

blotting The transfer of DNA (*Southern blotting*), RNA (*Northern blotting*), or protein (*Western blotting*) from a gel to a nitrocellulose filter to facilitate detection of the nucleic acid by a radioactive complementary *oligonucleotide* probe or of protein by an *antibody*.

bombesin A tetradecapeptide, originally isolated from the skin of the frog *Bombina bombina*, which stimulates insulin secretion.

C-peptide The sequence of around 30 amino acid residues that in *proinsulin* links the C-terminal of the insulin B-chain to the N-terminal of the insulin A-chain. The C-peptide is retained within the *secretory granule* after conversion of proinsulin to insulin and is co-secreted with insulin.

CCAAT box A consensus sequence found in eukaryotic *promoters* for mRNA precursors.

calcitonin-gene-related peptide (CGRP) A 37-amino acid peptide found in nerve terminals within the islet which inhibits glucose-stimulated insulin secretion.

calmodulin A protein which serves as a major intracellular receptor for calcium ions.

cAMP response element (CRE) A sequence found upstream of eukaryotic genes whose *transcription* can be influenced by *cyclic AMP*. The CRE serves as a specific binding site for a transcription factor whose activity is altered by phosphorylation by *protein kinase A*.

5′ cap A 7-methylguanylate attached by an unusual 5′-5′ triphosphate linkage to the sugar at the 5′-end of a eukaryotic mRNA. It is important for splicing reactions and for translation.

capillary fenestration A gap in the wall of a capillary through which large molecules may pass.

capillary glomeruli Clusters of small interconnected blood vessels.

carbamylcholine A drug which stimulates muscarinic acetylcholine receptors.

carboxypeptidase A *protease* which hydrolyses the C-terminal amino acid from a peptide chain. Carboxypeptidase H is responsible for removal of the basic amino acids exposed during conversion of *proinsulin* to insulin.

carnitine acyl transferase The enzyme which is responsible for the transport of fatty acids into the mitochondrion in the form of fatty-acyl carnitine.

catecholamine A general name for the group of dihydroxyphenols which includes adrenaline and nor-adrenaline.

CD4, CD8 Cell surface *glycoproteins* in the immunoglobulin gene superfamily. CD4, found on *T-helper cells*, is the HIV receptor. CD8 is found on *cytotoxic* and *suppressor T-cells*. They help their cells to recognize class I and II *MHC* proteins on other cells.

cDNA library The total mRNA of a tissue converted into copy DNA (cDNA) using the enzyme reverse transcriptase and inserted into a vector such as a *plasmid* or bacteriophage to produce a library of vectors each containing a different cDNA insert.

cell-attached patch A *patch-clamp* configuration used for single channel recording in which the cell remains intact and attached to the patch pipette. See Section 4.1.5 for details.

cellular immunity Immunity mediated by *T-cell* mechanisms.

cephalic phase The anticipatory response of the β-cell to a meal, mediated by the *vagus*.

charybdotoxin A toxin which binds to and blocks some types of K-channels.

chloramphenicol acetyl transferase (CAT) A bacterial enzyme which can be expressed in eukaryotic cells transfected with the appropriate cDNA in an expression vector. Widely used as a reporter gene by placing its expression under the control of a *promoter* region for a particular eukaryotic gene.

cholecystokinin (CCK) A 33 amino acid peptide released by the gut (principally the duodenum) which stimulates insulin secretion.

chondroitin sulphate A major component of connective tissues composed of an unbranched chain of sulphated disaccharide units.

chromatin The complex of DNA, RNA, and protein which constitutes eukaryotic chromosomes.

chromogranin A peptide expressed in several secretory cells. In the β-cell it is processed to a smaller peptide *betagranin* which is co-secreted with insulin.

class I/II/III MHC molecules The three major classes of protein coded for by the *MHC*. Class I molecules contain one MHC-encoded protein associated with β2-microglobulin. In humans, the class I MHC proteins are coded for by three distinct but *homologous* loci HLA-A, HLA-B, HLA-C. Class II molecules are dimers of two MHC encoded proteins. In humans there are three class II MHC proteins whose α and β chains are encoded by genes designated DPα, DPβ, DQα, DQβ, DRα and DRβ. Class III molecules are components of the *complement* system.

clathrin A non-glycosylated protein which can form a polyhedral framework around the *vesicles* in which proteins are transported between the *endoplasmic reticulum* and their final cellular destination.

clonal anergy The state in which a particular *clone* of *lymphocytes* is unable to be activated.

clonal deletion The elimination of a particular *clone* of *lymphocytes*.

clone Cells containing identical genetic information.

clonidine A drug which stimulates α_2-adrenergic receptors.

co-translational Occurring during the process of *translation* of mRNA into protein.

codon A sequence of three bases in a nucleic acid which constitutes part of the genetic code.

complement A system of enzymes capable of causing cell lysis in the context of immunity.

concanavalin A A sugar-binding protein (lectin) from jack beans.

concordance The occurrence of the same genetic trait in both members of a pair of twins.

congener A chemical closely related structurally to another and having similar or opposing effects.

conservative substitution Replacement of an amino acid in a protein with a similar residue (e.g. aspartate for glutamate).

constitutive secretion Secretion via a non-regulated pathway in which the secretory product is transferred continuously from the *Golgi* to the plasma membrane in transport *vesicles*.

cortisol The main glucocorticoid secreted from the adrenal cortex. Synonymous with hydrocortisone.

CRI G1 cell A β-cell line derived from a rat insulinoma.

crossing over The exchange of DNA between *homologous* chromosomes.

cyanocinnamate An inhibitor of the transport of pyruvate into mitochondria.

cyclic AMP A universal second messenger formed from ATP by the plasma membrane enzyme *adenyl cyclase*.

cyclic AMP-dependent protein kinase See *protein kinase A*.

cycloheximide An inhibitor of protein synthesis via its effect on peptidyl transferase.

cyclooxygenase An enzyme involved in the synthesis of *prostaglandins* from arachidonic acid.

cyclophosphamide A DNA-alkylating drug used in the treatment of cancer.

cyclosporin A A cyclic peptide of fungal origin used as an immunosuppressive agent.

cytochalasin B A mould metabolite that inhibits glucose transporters and disrupts the cytoskeleton in eukaryotic cells.

cytokines Protein hormones secreted by *T-cells*, *macrophages*, and *mononuclear phagocytes* which mediate both natural and acquired immunity. Their functions include

(1) mediation of natural immunity, e.g. *interleukin*-1, type 1 *interferon*, *tumour necrosis factor*;

(2) regulation of *lymphocyte* activation, growth, and differentiation, e.g. interleukin-2 and -4;

(3) activation of inflammatory cells, e.g. interferon γ, lymphotoxin;

(4) stimulation of growth of bone marrow cells e.g. interleukins-3 and -7, granulocyte-macrophage colony-stimulating factor.

(5) anti-viral activity, e.g. interferon γ

Excessive production or action of cytokines can lead to tissue injury or cell death.

cytoskeleton The internal scaffolding of eukaryote cells, composed of *microtubules*, *microfilaments*, and intermediate filaments.

cytotoxic T-cell *Lymphocytes* which bind to and kill virally infected cells. They bind to their targets through recognition of *antigen* in association with *MHC* class I molecules.

degenerate Since many amino acids are specified by more than one *codon*, the genetic code is said to be degenerate.

dehydrouramil A stable analogue of *alloxan*.

delayed rectifier One of several classes of voltage-activated K-channel in the plasma membrane.

dendritic cell A specialized cell believed to play an accessory role in the induction of immune responses.

depolarization Reduction of the voltage across the plasma membrane due to a change in ionic permeability.

dexamethasone An anti-inflammatory steroid derivative.

diacylglycerol A lipid in which two fatty acid chains are esterified to glycerol. It plays a second messenger role because it is generated from *phosphatidylinositol 4,5-bisphosphate* and activates *protein kinase C*.

diazoxide A drug which inhibits insulin secretion because of its ability to open *ATP-regulated K-channels*.

discordance The occurrence of a genetic trait in only one of a pair of twins.

DNAse footprinting A method for detecting the binding of proteins to specific sequences in DNA.

docking protein A protein in the *endoplasmic reticulum* involved in the process whereby newly synthesized protein is discharged directly from the *ribosome* across the endoplasmic reticulum membrane.

double blind trial The testing of a drug or treatment under conditions where neither the administrator nor the recipient knows whether placebo or test substance is being administered.

down regulation A reduction in the number of a receptor or other cellular protein.

ectodomain A region of protein exposed to the outside of a cell.

endocytosis A mechanism whereby extracellular molecules are taken up into cells via coated *vesicles*.

endopeptidase A *protease* that hydrolyses proteins at internal peptide bonds.

endoplasmic reticulum A system of membrane-enclosed spaces within cells through which proteins destined for cellular locations other than the cytosol pass.

endosome A series of compartments through which endocytosed particles and macromolecules pass on their way to *lysosomes*.

endothelium The layer of cells lining blood and lymph vessels.

enhancer A sequence of bases in DNA to which regulatory proteins bind to modify expression of specific genes. Enhancers may be at a considerable distance from the genes whose expression they influence and their activity is independent of their orientation in the DNA sequence.

enzyme-linked immunoabsorption assay (ELISA) An assay technique utilizing an enzyme joined to an *antibody* molecule (see Section 4.1.1).

epidemiology The study of the various factors which determine the frequency and distribution of human disease.

epidermal growth factor (EGF) A protein stimulating growth of cells.

epithelium The layer of cells covering the internal and external body surfaces.

epitope A region of an *antigen* recognized by an *antibody*.

euglycaemia Normal blood glucose concentration.

eukaryote Cells possessing a nucleus.

exocytosis The mechanism whereby *secretory granules* move to and fuse with the plasma membrane to liberate their contents.

exon A sequence of bases within a gene which codes for protein. In eukaryotes such coding sequences are usually interrupted by non-coding sequences (*introns*).

experimental autoimmune encephalomyelitis A model of demyelinating autoimmune disease, induced by injection of animals with myelin basic protein, in which inflammation of the brain and spinal cord occurs.

fatty acid synthase The enzyme catalysing the synthesis of fatty acids from the precursor malonyl-CoA.

flux-generating step A non-equilibrium reaction at the beginning of a metabolic pathway which is saturated with substrate and which determines the rate of flux through the pathway.

forskolin A mould metabolite that activates *adenyl cyclase*.

fura II A fluorescent compound used to measure the concentration of Ca^{2+} in cells.

furin A human protein suggested to be a *protease* specific for dibasic amino acid sites.

fusion pore A putative pore formed between the plasma membrane and a secretory *vesicle* membrane which precedes *exocytosis*.

G-protein One of a number of trimeric guanine nucleotide-binding proteins that play a key role in the responses of cells to stimuli.

galanin A 29 amino acid neuropeptide which inhibits insulin secretion.

ganglioside A lipid-containing oligosaccharide found in plasma membranes.

gap junction An array of tubular particles ('connexons') joining discrete regions of neighbouring cells. The connexons consist of two opposed hexagonal rings of protein subunits surrounding a central pore through which small molecules and ions, but not macromolecules, can pass.

gastro-intestinal peptide (GIP) A gut hormone of 43 amino acids secreted by the duodenum which stimulates insulin secretion.

genome The entire content of genetic information of a cell.

glibenclamide An antidiabetic *sulphonylurea* drug that stimulates insulin secretion.

gliclazide An antidiabetic *sulphonylurea* drug that stimulates insulin secretion.

glucagon A paracrine hormone released by islet α-cells which stimulates insulin secretion.

glucokinase An enzyme catalysing the reaction

$$\text{glucose} + \text{ATP} \rightarrow \text{glucose 6-phosphate} + \text{ADP}$$

It is distinguished from the enzyme hexokinase which also catalyses this reaction by its lower affinity for glucose and its cellular location – glucokinase occurs only in the liver and the pancreatic β-cell.

gluconeogenesis The biosynthesis of glucose from precursors such as lactate.

glucose 6-phosphatase The enzyme present in liver and pancreatic β-cells catalysing the reaction

$$\text{glucose 6-phosphate} \rightarrow \text{glucose} + P_i.$$

glucose clamp The infusion of glucose at a rate that maintains a fixed plasma glucose concentration.

glucose cycling The interconversion of glucose and glucose 6-phosphate by means of the enzymes glucokinase and glucose 6-phosphatase which constitute a *substrate-cycle*.

glucose intolerance A less than normal ability to lower blood glucose after administration of glucose.

glucose priming The phenomenon whereby prior exposure to glucose enhances the secretory response of the β-cell to subsequent exposure to the sugar.

glucose–fatty acid cycle The concept that carbohydrates and fatty acids have reciprocal effects on each other's metabolism.

glutamic acid decarboxylase (GAD) The enzyme catalysing the conversion of glutamate to the inhibitory neurotransmitter γ-aminobutyric acid (GABA). This enzyme is the 64kD antigen to which antibodies are commonly found in the blood of insulin-dependent diabetics (see Section 10.7).

glycogen synthase The enzyme catalysing the formation of glycogen.

glycoprotein Proteins containing covalently associated carbohydrate.

glycosylation The covalent addition of carbohydrate residues to proteins.

Golgi complex A stack of flattened membranous sacs which plays a key role in the processing and targeting of cellular proteins.

Graves disease An autoimmune disease affecting the thyroid (thyrotoxicosis).

haemochromotosis A disorder of iron metabolism associated with liver cirrhosis and diabetes mellitus.

haplotype The *alleles* possessed by an individual at a particular genetic *locus* on one chromosome.

heat shock protein One of a family of proteins whose synthesis is increased on exposure of cells to elevated temperatures.

heritability coefficient A measure of the extent to which a disease can be transmitted genetically.

heterologous Dissimilar in structure or source.

heterozygous Having different *alleles* for a given gene.

hybridization The binding of a nucleic acid to a complementary sequence in a second nucleic acid.

hippocampus A part of the brain playing a role in learning and memory.

histopathology The study of the morphology of diseased tissues.

HIT T15 cell A β-cell line obtained by SV40 transformation of hamster islet cells.

HLA antigens The gene products of the *MHC* class I and II gene complexes.

homozygous Having identical *alleles* for a given gene.

humoral immunity Immunity mediated by *antibodies* produced by *B-cells*.

hydrogen bond A directional intermolecular association between the hydrogen atom of a group such as O–H or N–H and an acceptor atom such as N or O bearing a lone pair of electrons.

hydropathy analysis Prediction of protein structure by means of an analysis of the hydrophobic nature of stretches of constituent amino acid residues. For a globular protein, hydrophobic residues tend to cluster within the centre of the molecule; for membrane proteins they are expected to lie within the membrane.

hyperglycaemia Higher than normal blood glucose level. A fasting blood glucose level of above 5.5 mM is regarded as hyperglycaemic.

hyperglycaemic clamp The maintenance of hyperglycaemia by infusion of glucose.

hyperinsulinaemia A higher than normal blood insulin level.

hyperproinsulinaemia A higher than normal blood *proinsulin* level.

hypoglycaemia A lower than normal blood glucose level.

hypothalamus A part of the brain which produces factors controlling the secretions of the pituitary gland and which is involved in processes such as appetite control.

idiotype The unique set of antigenic determinants associated with the antigen binding site of an *antibody*.

imaging A technique for producing a visual representation of the concentration and spatial distribution of a substance within a cell or tissue.

immunoblotting The detection of proteins by means of specific *antibodies* after transfer of the proteins from a gel to nitrocellulose or other paper.

immunocytochemistry The use of enzyme-linked antibodies to stain tissue sections to localize proteins of interest.

immunoglobulin Another term for *antibody*.

immunoprecipitation The precipitation of proteins by using specific *antibodies*.

immunoradiometric assay A method of measuring the concentration of a substance which involves the use of a radiolabelled *antibody* to that substance. It is capable of detecting very small quantities of the substance. See Section 4.1.1 for details.

impaired glucose tolerance (IGT) An impaired ability to normalize blood glucose after a glucose load.

incretin Any of a number of gut hormones capable of increasing the insulin secretory response to oral glucose.

indo 1 A fluorescent compound used to measure the concentration of Ca^{2+} in cells.

initiation codon The base sequence AUG which signals the start of the translatable sequence of an mRNA.

initiation factor II (eIF2) One of a number of proteins involved in the initiation of protein synthesis.

inositol trisphosphate (IP_3) A second messenger formed from *phosphatidylinositol bisphosphate* by plasma membrane bound phospholipase C. IP_3 elicits release of Ca^{2+} from intracellular membrane stores.

inside-out patch A configuration in the *patch clamp technique* in which an excised patch of membrane has its intracellular membrane surface exposed to the bath solution. See Section 4.1.5.

insulin-like growth factor-1 A 70 amino acid protein, also known as somatomedin-1, which promotes the growth of bone, muscle, and fat cells.

insulin release, first and second phase The response of the β-cell to constant stimulation by glucose is biphasic. It consists of a rapid increase in secretion rate which within 10 min is reduced again (1st phase), followed by a slowly increasing rate of secretion (2nd phase).

insulitis Lymphocytic infiltration and necrosis of the islets of Langerhans.

interferon γ A *cytokine* secreted by stimulated *T-helper cells* which attracts and activates macrophages.

interleukins *Cytokines* were originally regarded as molecules principally produced by and acting on leukocytes and hence called 'interleukins'. This usage is too restricted and the general term 'cytokines' is now preferred, with the interleukins regarded as members of this family.

intravenous glucose tolerance test (IVGTT) The intravenous administration of a standard amount of glucose to assess the ability of an individual to respond with appropriate insulin secretion.

intron A non-coding sequence of bases in DNA which, in eukaryotes, frequently interrupts the coding sequences (*exons*).

ion channel An integral membrane protein consisting of one or more subunits which form a central pore through which ions can traverse the membrane. Ion channels differ from simple aqueous pores in that they show ion selectivity and are not continuously open. The pore may be opened and closed by changes in membrane voltage or by ligand binding.

islet amyloid polypeptide (IAPP) See *amylin*.

islet cell antibody (ICA) *Antibodies* that bind to islet cell components in frozen sections of pancreas.

3-isobutylmethylxanthine (IBMX) An inhibitor of the *phosphodiesterase* which hydrolyses *cyclic AMP* thus raising intracellular cyclic AMP.

isoelectric point (pI) The pH at which a molecule bearing ionizable groups has no net charge. For a protein, the solubility as a function of pH is usually at a minimum at the pI.

isoproterenol A β-adrenergic agonist.

K-ATP channel See *ATP-regulated K-channel*.

ketoacidosis The occurrence of abnormally high concentrations of acetoacetate and β-hydroxybutyrate ('ketone bodies') in the blood, leading to a decrease in blood pH. The condition is symptomatic of uncontrolled type I diabetes.

2-ketoisocaproate (KIC) The initial product of leucine degradation, formed and metabolized within the mitochondrion. Stimulates insulin secretion.

KEX2 A subtilisin-like *protease* from yeast, specific for dibasic amino acid residues.

K_i The dissociation constant of an enzyme–inhibitor complex.

kinesin A force-generating protein complex involved in the unidirectional transport of vesicles and organelles along microtubules. Like myosin, kinesin contains two globular ATPase-containing heads which provide the motive force.

K_m The concentration of substrate at which the velocity of an enzyme-catalysed reaction is half its maximal value.

leprechaunism A lethal congenital disorder characterized by small size and severe endocrine abnormalities.

linkage The extent to which genes on the same chromosome are inherited together, measured by the frequency of recombination between *loci*.

linkage analysis Genetic analysis of the frequency of various combinations of genes.

linkage disequilibrium The tendency for some genes on a chromosome to remain associated rather than undergoing genetic randomization in the population. The result is that the frequency of particular combinations of *alleles* is greater than the product of each individual gene frequency.

lipolysis The hydrolysis of *triacylglycerol* to glycerol and free fatty acids.

lipoprotein lipase An extracellular

enzyme which hydrolyses dietary *triacylglycerols* contained within chylomicrons to permit adipose and muscle tissues to take up free fatty acids for storage.

locus The specific site of a gene or group of genes on a chromosome.

lymphocyte White blood cells which confer acquired immunity. The two major classes of immune response, *cell-mediated* and *humoral* immunity, are mediated by different classes of lymphocytes (*T-cells* and *B-cells* respectively).

lymphokine *Cytokines* produced by activated *T-cells*. These include both *interleukin-2, -3,* and *-4* and *interferon-γ*.

lysosome A cellular organelle containing a variety of hydrolytic enzymes.

macrophage White blood cells which defend against infection by phagocytosing (ingesting) microorganisms, which play an important role in scavenging old and damaged self cells, and which constitute a major population of *antigen-presenting cells*.

major histocompatibility complex (MHC) A region of highly polymorphic genes whose products are expressed at the surface of various cells. In humans the MHC occupies about 3500 kilobases on the short arm of chromosome 6 (roughly the size of the entire E. coli *genome*). It contains genes for the class I (HLA-A, -B, and -C) and class II (HLA-DP, -DQ, and -DR) MHC molecules as well as components of the *complement* system and the *cytokines tumour necrosis factor* and lymphotoxin. MHC-encoded class I and class II molecules bind fragments of foreign *antigens* to form complexes that are recognized by specific *T-cells*. The total set of MHC alleles on a chromosome is called an MHC *haplotype*. Some haplotypes in which several alleles are in *linkage disequilibrium* are associated with autoimmune diseases.

malic enzyme The enzyme catalysing the reaction:

$$\text{malate} + \text{NADP}^+ \rightarrow \text{pyruvate} + \text{CO}_2 + \text{NADPH}$$

This reaction is an important source of NADPH for fatty-acid synthesis.

mannoheptulose A seven-carbon ketose, found in nature in the avocado pear. It acts as a competitive inhibitor of glucokinase and hence glucose metabolism. In the β-cell this leads to inhibition of insulin secretion.

MAP kinase A protein kinase which phosphorylates microtubule associated proteins. Also known as mitogen-activated protein kinase.

meglitinide A member of the *sulphonylurea* group of drugs which stimulates insulin secretion.

membrane potential The potential difference between the inside and the outside of the cell membrane.

Mendelian inheritance The basic laws of inheritance formulated by Gregor Mendel. These state that two *alleles* of each gene are inherited, one from each parent. These alleles segregate during meiosis and are independently transmitted to the next generation. Likewise genes on different chromosomes sort independently.

methionyl tRNA A transfer RNA molecule bearing methionine. Two forms are found, one of which inserts methionine into internal positions in the protein. The other type acts as the initiation tRNA for protein synthesis.

MHC See *major histocompatiblity complex*.

MHC restriction The requirement that an *antigen-presenting cell* (APC) must express *MHC* molecules recognized as self by a *T-cell* in order for that *T-cell* to respond to the antigen presented by the APC.

microfilament A cytoskeletal structure composed of actin.

microtubule A cytoskeletal structure composed of tubulin.

monoclonal antibody A population of identical *immunoglobulins* produced by a *clonal B-cell* line.

mononuclear phagocyte The precursors of *macrophages*.

monozygotic Derived from the same ovum.

morphometry Quantitative microscopy.

multiple endocrine neoplasia A syndrome with tumours in several endocrine glands.

mutagenesis A change in the sequence of bases in DNA.

NADPH-isocitrate dehydrogenase The enzyme catalysing the reaction:

$$\text{isocitrate} + \text{NADP}^+ \rightarrow \text{oxoglutarate} + \text{NADPH}.$$

natural immunity See *humoral immunity*.

negative cooperativity The phenomenon whereby binding of a ligand to a protein possessing several ligand binding sites becomes progressively less favourable with increasing occupancy of sites.

nephropathy Degeneration of the kidney, leading to renal failure. Occurs as a secondary complication in diabetes mellitus, possibly because of poor glucose control.

neuraminidase An enzyme hydrolysing sialic acid residues of *glycoproteins*.

neuropathy Degeneration of the peripheral nerves. Occurs as a secondary complication in diabetes mellitus, possibly because of poor glucose control.

NK cells Natural killer (NK) cells are *lymphocytes* that kill certain tumour cells and virus-infected cells. They are also the principal mediator of antibody-dependent cell-mediated cytotoxicity.

NOD mouse A species of mouse which expresses I-A but not I-E class II *MHC* molecules and manifests an autoimmune disease closely resembling type I diabetes.

Northern blotting A method for detecting specific RNA molecules by transferring RNA from an agarose gel to a nitrocellulose filter on which it can be hybridized to radiolabelled complementary DNA or RNA.

okadaic acid A cyclic polypeptide isolated from shellfish which acts as a potent inhibitor of protein phosphatases I and 2A.

oligonucleotide A sequence of several nucleotides.

omega loop A structure found in many proteins in which 6 to 16 amino acid residues are arranged in an Ω-shaped loop.

oncogene A gene whose expression results in uncontrolled cell division.

open reading frame A DNA sequence in which no stop *codons* are present.

oral glucose tolerance test The oral administration of a standard amount of glucose to assess the ability of an individual to respond with appropriate insulin secretion.

osteoblast Cells which secrete the bone matrix.

osteoclast Cells which control bone reabsorption and which are also involved in cartilage erosion when cartilage is replaced by bone.

oxygen free radical A highly reactive oxygen anion (superoxide, O_2^-) which arises from the acquisition of a single electron by an oxygen molecule. It is a short-lived intermediate in several biological reactions and may have deleterious effects on tissues.

pancreastatin A pancreatic hormone which inhibits insulin secretion.

pancreatic polypeptide A peptide secreted by islet PP cells whose function is obscure.

paracrine A hormonal effect acting within a tissue. For example, *glucagon*, secreted by pancreatic islet α_2-cells, is said to have a paracrine action on islet β-cells (stimulatory) and D-cells (inhibitory).

patch-clamp technique A method for measuring the *single channel* or *whole-cell current*. See Section 4.1.5 for further details.

penetrance The extent to which an autosomal dominant gene is expressed in individuals possessing the gene. For example, a penetrance of 80 per cent implies 80 per cent of individuals with the gene will express it in some way.

perforated patch A variation of the whole-cell configuration of the *patch clamp* method in which the patch membrane is permeabilized with (for example) nystatin in order to obtain electrical access to the cell interior. This method helps preserve cellular metabolism and second messenger systems. See Section 4.1.5.

peroxidase An enzyme which catalyses the oxidation of an electron donor by hydrogen peroxide. It is widely used conjugated to *antibody* for detection and quantification of *antigens* by *ELISA* and microscopy.

phenothiazine A group of tranquillizers, including trifluoperazine, a *calmodulin* inhibitor.

phenotype The appearance of an organism, which results from the interaction of its genetic make-up and the environment.

phorbol esters A group of tumour-promoting drugs which activate *protein kinase C*.

phosphatidylinositol 4,5-bisphosphate (PIP_2) An intrinsic membrane lipid which is split by *phospholipase C* to produce two cellular second messengers, *inositol trisphosphate* (IP_3) and *diacylglycerol* (DAG).

phosphodiesterase A group of enzymes which hydrolyse phosphodiester linkages. Such bridges link adjacent pentose moieties to form the backbone of DNA and RNA. Additionally, a phosphodiester link occurs in cyclic AMP where its hydrolysis produces AMP.

phosphoenolpyruvate carboxykinase The enzyme in the gluconeogenic pathway catalysing the reaction:

$$\text{oxaloacetate} + \text{GTP} \rightarrow \text{phosphoenolopyruvate} + \text{GDP} + CO_2.$$

phosphofructokinase The enzyme that catalyses the reaction.

$$\text{fructose-6-phosphate} + \text{ATP} \longrightarrow \text{fructose-1,6,-diphosphate} + \text{ADP}.$$

phosphoinositol glycan A carbohydrate-containing lipid suggested to be the precursor of an intracellular mediator of insulin action.

phospholipase C A membrane-bound enzyme which hydrolyses the phosphodiester bond of *phosphatidylinositol 4,5-bisphosphate* to form *inositol 1,4,5-trisphosphate* and *diacylglycerol*.

phosphorylase The enzyme that catalyses the breakdown of glycogen to form glucose 1-phosphate.

phosphorylase kinase An enzyme which phosphorylates and thereby activates glycogen phosphorylase.

pI See *isoelectric point*.

Pima Indian An extensively studied tribe of American Indians living in Arizona which have an unusually high incidence of type II diabetes (up to 30 per cent of adults).

plasmid Small, circular double-stranded DNA molecules found naturally in bacteria and yeast and used extensively as vectors for cloning DNA sequences.

platelet-derived growth factor (PDGF) A protein released from platelets which stimulates proliferation of neuroglia and connective tissue cells.

point mutation A change in a single base pair.

polyclonal antibody A heterogeneous population of *antibodies*.

polymerase chain reaction (PCR) A method of selectively amplifying a specific DNA sequence over a million-fold. Part of the selected sequence must already be known. Single stranded DNA is formed by heating, and the strands annealed to *oligonucleotides* (PCR primers) complementary to the two ends of the selected sequence. Addition of DNA polymerase results in the synthesis of the region of DNA between the two primers. The newly synthesized DNA strands are then separated by heating and this cycle is repeated 20 to 30 times.

polymorphic region A region of a gene in which a large number of alternative forms occur in the population.

polymorphism The occurrence in the population of a large number of alternative forms of the same gene.

polysome A complex of *ribosomes* bound along a single mRNA molecule.

preproinsulin *Proinsulin* plus the *signal peptide sequence*. Preproinsulin is cleaved to proinsulin during transfer of the nascent protein into the *endoplasmic reticulum*.

proband The original person who serves as the basis for a genetic study.

proinsulin A precursor of insulin consisting of insulin and connecting (C) peptide. Proinsulin is cleaved to insulin in the *Golgi* complex and the immature *secretory granules*.

promoter A DNA sequence (signal sequence) at which RNA polymerase begins transcription.

pronase A mixture of relatively non-specific *proteases*.

proopiomelanocortin (POMC) A single chain precursor protein which is proteolytically cleaved to produce several different hormones including ACTH, MSH, and β-endorphin.

prostaglandin A family of hormones consisting of 20 carbon fatty acid derivatives which usually act as local signalling molecules, influencing both the cell in which they are made and its immediate neighbours.

protease A family of enzymes that

catalyse the degradation of a protein by severing the linkages between adjacent amino acids.

protein A A bacterial protein which binds to the Fc portion of *immunoglobulin* G.

protein kinase A family of enzymes which catalyse the transfer of a phosphate group from MgATP onto a target protein.

protein kinase A A cytosolic enzyme activated by *cyclic AMP* which phosphorylates proteins on serine and threonine residues. The inactive holoenzyme consists of two catalytic (C) and two regulatory (R) subunits. Cyclic AMP binds to the R-subunit and releases the catalytic subunit. The enzyme is inhibited by a specific protein kinase A inhibitor (Walsh inhibitor) or by re-binding of the R-subunit.

protein kinase C A family of enzymes phosphorylating proteins on serine and threonine residues and which require phosphatidylserine, *diacylglycerol*, and (usually) Ca^{2+} for activity. *Diacylglycerol* reduces the Ca^{2+} requirement and is usually the physiological activator. Protein kinase C is a cytosolic enzyme which moves to the membrane on activation; it contains a pseudosubstrate site which serves to maintain the enzyme in an inactive state prior to stimulation.

protein phosphatase A group of enzymes which catalyse the removal of a phosphate group from a protein. They include PP1, PP2A, PP2B, PP2C, and protein tyrosine phosphatases.

proteoglycan A proteoglycan consists of a protein core to which many glycosaminoglycans (polysaccharide chains) are covalently linked. Proteoglycans contribute to the extracellular matrix surrounding cells.

proteolysis The degradation of a protein by *proteases*.

pseudo-colour image Refers to the method commonly used to display the spatial distribution of, for example, Ca^{2+} in which each value of Ca^{2+} concentration is assigned a colour. In general, red is used to represent a high concentration and blue a low concentration.

pyruvate dehydrogenase A multi-enzyme complex found in mitochondria which catalyses the conversion of pyruvate to acetyl CoA.

pyruvate kinase The cytosolic enzyme catalysing the reaction:

$$\text{phosphoenolpyruvate} + \text{ADP} \rightarrow \text{pyruvate} + \text{ATP}.$$

quin 2 A fluorescent probe which is used to measure the intracellular Ca^{2+} concentration.

quinine An alkaloid drug, used in the treatment of malaria, which blocks a number of different K-channels with differing potencies. In β-cells, the *ATP-regulated K-channel* is strongly blocked by quinine, and the Ca-activated K-channel is less sensitive.

Rabson–Mendenhall syndrome A syndrome manifesting dental dysplasia, acanthosis nigricans, and precocious puberty which is associated with a decreased number of insulin receptors.

radioimmunoassay A method of measuring the concentration of a substance which involves competition between radiolabelled and unlabelled antigen for binding to an antibody. It is capable of detecting very small quantities of the substance. For details see Section 4.1.1.

reg A gene that has been implicated in islet cell regeneration.

regulated pathway A metabolic pathway which is subject to regulation by factors other than the concentrations of substrates and products.

regulator site hypothesis An hypothesis which postulates that glucose stimulates insulin secretion by binding to a specific glucose receptor in the β-cell.

Reiter's syndrome A syndrome consisting of polyarthritis, non-microbial urethritis, and occular inflammation. The prevalence of the syndrome is increased in individuals possessing the HLA B27 *antigen*.

restriction enzyme An enzyme that cleaves double-stranded DNA at a specific site. Over 100 different restriction enzymes have been reported.

restriction fragment length polymorphism (RFLP) The variation, between individuals, in the sequence of a length of DNA produced by cleaving the full-length DNA with a *restriction enzyme*. Such restriction fragment length polymorphisms allow mutations in a gene within a population to be detected.

retinopathy Deterioration of the blood vessels of the retina, leading to blindness. Occurs as a secondary complication in diabetes mellitus, possibly because of poor glucose control.

reverse haemolytic plaque assay A method of detecting secretion from a single cell by use of *complement*-mediated lysis of adjacent red blood cells bearing *antibodies* to the secretory product. See Section 4.1.2.

ribosome A large multienzyme complex, composed of RNA and protein molecules, which is involved in protein synthesis.

RIN cell A cell line derived from a rat insulinoma which secretes both insulin and somatostatin. A number of variants with different properties are known.

Scatchard plot A graphical method for analysis of ligand binding data.

secretory granule In β-cells this term is used to refer to the mature membrane-bound vesicle which contains the stored insulin–zinc complex.

segregation analysis The analysis of pedigrees to assess the pattern of inheritance of particular genetic susceptibility factors (generally assumed to be genes).

self-tolerance The lack of a response to self-*antigens*.

sib A brother or sister.

signal peptidase A *proteolytic* enzyme that removes the *signal peptide sequence* from a newly synthesized peptide.

signal peptide The peptide sequence that directs *ribosomes* bearing the nascent protein to the rough *endoplasmic reticulum*. The 'pre' region of *pre-proinsulin* serves this function.

signal recognition particle (SRP) Association of the cytosolic signal recognition peptide with the *signal sequence receptor* of a growing protein results in the association of the *ribosome*–protein–SRP complex with the *signal sequence receptor* of the *endoplasmic reticulum*. This protein also halts further protein synthesis until docking has occurred.

signal sequence receptor The receptor on the rough *endoplasmic reticulum* (ER) which recognizes the *signal peptide* sequence of a growing protein on a *ribosomal* complex. Interaction of the *signal peptide* with this receptor results in the formation of a pore in the ER membrane which permits the transfer of the nascent protein into the ER lumen.

single channel recording The method of recording the current (flow of ions) through a single ion channel when it is open.

slow wave A cyclic change in β-cell *membrane potential*, consisting of a slow *depolarization* to a threshold, followed by a rapid depolarization to a plateau on which Ca-dependent action potentials are superimposed and which subsequently repolarizes to the original potential (see Fig. 4.8). In 10 mM glucose there are approximately 4 slow waves per minute.

somatomedins A family of hormones secreted by the liver which regulate the growth of muscle and bone, and which influence the carbohydrate, lipid, calcium, and phosphate metabolism in these tissues.

somatostatin A hormone secreted by the D-cells of the pancreatic islets which inhibits insulin secretion.

Southern blot A method of detecting whether DNA contains a specific nucleotide sequence. The DNA is first cleaved into fragments with *restriction enzymes*, the fragments are size-separated by gel electrophoresis, blotted onto nitrocellulose paper and then screened with a radiolabelled probe complementary to the sequence of interest. *Hybridization* is detected by *autoradiography*.

spermine A polycationic amine.

splicing The cleavage and rejoining of a primary RNA transcript to yield an mRNA molecule that encodes a protein. Splicing occurs at special splice sites which ensures that there is no shift in the reading frame.

stiff man syndrome A rare neurological disorder associated with stiffness of the musculature and muscular spasm. It is associated with the presence of antibodies to *glutamic acid decarboxylase*.

streptozotocin A drug which selectively kills pancreatic β-cells and is widely used to produce an experimental model of diabetes in animals. High concentrations of the drug produce type I diabetes. Lower concentrations applied to neonatal animals, which only kill a proportion of the β-cell population, may produce a condition in later life which resembles type II diabetes.

substrate cycle A biochemical cycle in which the concentrations of the reactants remain unchanged but the concentration of cofactors is altered. For example, the glucose → glucose 6-phosphate → glucose cycle hydrolyses ATP. Substrate cycles increase the sensitivity of the metabolic step concerned and may also be important in *thermogenesis*.

substrate site hypothesis An hypothesis which postulates that glucose stimulates insulin secretion as a consequence of its metabolism, and not by binding to a specific glucose receptor in the β-cell.

subtilisin A *protease* that cleaves adjacent to serine residues.

sulphonylurea A class of drugs which stimulates insulin secretion from pancreatic β-cells by virtue of their ability to block *ATP-regulated K-channels*. Used in the treatment of type II diabetes.

suppressor T-cell A class of *T-cell* that

suppresses the response of both *B-cells* and other T-cells to *antigens*.

synapsin I A protein found on the surface of synaptic vesicles in nerve terminals which is released on phosphorylation by calmodulin-dependent protein kinase II. It is believed that this process facilitates the docking of the synaptic *vesicle* with the active zone region of the plasma membrane.

T-cell The class of lymphocyte responsible for cell-mediated immunity which includes *cytotoxic T-cells, helper T-cells,* and *suppressor T-cells*.

T-helper cell A class of T-lymphocytes that are required to enable *B-cells, cytotoxic T-cells,* and *suppressor T-cells* to respond to *antigens*.

TATA box A consensus sequence principally of A and T residues which is located just upstream of the initiation site for *transcription*. Binding of the TATA factor to the TATA box is required before the *promoter* can be recognized by RNA polymerase.

telomere The tip of a chromosome: essential for correct chromosome replication as it ensures that the end of the DNA double helix is completed.

thermogenesis The process by which the body generates heat.

thiazide A class of diuretic drugs, some of which also inhibit insulin secretion.

tolbutamide A member of the *sulphonylurea* family of drugs, which stimulates insulin secretion.

tolerance The ability of the body to recognize self-proteins and thus not to mount an immune reaction against them. Tolerance is acquired during development by continuous exposure to either self (natural immunological tolerance) or non-self (acquired immunological tolerance) *antigens*.

transcription The process in which a messenger RNA molecule is synthesized from DNA.

transfection The process in which cDNA, in a form in which it can be expressed, is introduced into a cell (for example, by viral infection or electropermeabilization).

transgenic mouse A mouse in which an additional copy of a gene has been introduced into all tissues, by injection of the fertilized ovum with the cDNA for that gene.

translation The process in which a protein is synthesized from messenger RNA.

triacylglycerol The storage form of lipid, consisting of three fatty acid chains esterified to glycerol.

triacylglycerol lipase The group of enzymes which catalyse the reaction *triacylglycerol* → glycerol and fatty acids.

tetraethylammonium chloride (TEA) A drug which blocks most potassium channels. In β-cells, it blocks calcium-activated K-channels with high potency whilst *ATP-regulated K-channels* are unaffected by concentrations below 1 mM.

tryptic (phospho)peptide A (phospho)peptide produced by treatment of a protein with trypsin.

tumour necrosis factor A *cytokine* which is the principal mediator of the host response to gram-negative bacteria.

vagus The tenth cranial nerve, which innervates, amongst other organs, the pancreas and the gastrointestinal tract.

vasoactive intestinal peptide (VIP) A 28 amino acid hormone which is secreted by the gut and which stimulates insulin secretion.

vasopressin A neuropeptide present in islets which stimulates insulin secretion.

vesicle A small membrane-bound organelle, often containing some secretory product.

V_{max} The maximum rate of an enzyme reaction, which is achieved at high substrate concentrations.

voltage-dependent channel An *ion channel* opened or closed by a change in the *membrane potential.*

Western blotting A method of detecting a small quantity of protein in a cell, in which the proteins are first separated by running on a gel and then transferred (by blotting) to filter paper; a specific *antibody* to the protein is then added and the complex detected using a second antibody that is specific to the first and which is radiolabelled or fluorescent.

whole-cell current The total current that flows through all the *ion channels* in the cell membrane that are open (sometimes referred to as the macroscopic current).

Index